Advancements in Personalized Cellular Immunotherapy

AF776075

Advancements in Personalized Cellular Immunotherapy

A Comprehensive Guide to Cancer Treatment

Edited by

Arun Kumar Singh

Assistant Professor, University School of Pharmaceutical Sciences,
Rayat Bahra University,
V.P.O. Sahauran, Kharar, Mohali, Punjab, India

Murali M. Yallapu

Associate Professor, Cancer and Immunology Division,
Director, RCMI Research Core, and Co-Director, CPRIT-ICRC Core,
South Texas Center of Excellence in Cancer Research,
School of Medicine, University of Texas Rio Grande Valley,
TX, United States

Ciniraj Raveendran

Associate Professor, Department of Radiation Oncology,
Government Medical College,
Thiruvananthapuram, Kerala, India

Pranav Kumar Prabhakar

Professor & Head, Department of Biotechnology,
School of Engineering and Technology, Nagaland University,
Meriema, Kohima, Nagaland, India

ELSEVIER

ACADEMIC PRESS
An imprint of Elsevier
elsevier.com/books-and-journals

Academic Press is an imprint of Elsevier

125 London Wall, London EC2Y 5AS, United Kingdom
50 Hampshire Street, 5th Floor, Cambridge, MA 02139, United States

Copyright © 2026 Elsevier Inc. All rights are reserved, including those for text and data mining, AI training, and similar technologies.

Publisher's note: Elsevier takes a neutral position with respect to territorial disputes or jurisdictional claims in its published content, including in maps and institutional affiliations.

Books and Journals published by Elsevier comply with applicable product safety requirements. For any product safety concerns or queries, please contact our authorised representative, Elsevier B.V., at productsafety@elsevier.com.

For accessibility purposes, images in electronic versions of this book are accompanied by alt text descriptions provided by Elsevier. For more information, see https://www.elsevier.com/about/accessibility.

No part of this publication may be reproduced or transmitted in any form or by any means, electronic or mechanical, including photocopying, recording, or any information storage and retrieval system, without permission in writing from the publisher. Details on how to seek permission, further information about the Publisher's permissions policies and our arrangements with organizations such as the Copyright Clearance Center and the Copyright Licensing Agency, can be found at our website: www.elsevier.com/permissions.

This book and the individual contributions contained in it are protected under copyright by the Publisher (other than as may be noted herein).

Notices
Knowledge and best practice in this field are constantly changing. As new research and experience broaden our understanding, changes in research methods, professional practices, or medical treatment may become necessary.

Practitioners and researchers must always rely on their own experience and knowledge in evaluating and using any information, methods, compounds, or experiments described herein. In using such information or methods they should be mindful of their own safety and the safety of others, including parties for whom they have a professional responsibility.

To the fullest extent of the law, neither the Publisher nor the authors, contributors, or editors, assume any liability for any injury and/or damage to persons or property as a matter of products liability, negligence or otherwise, or from any use or operation of any methods, products, instructions, or ideas contained in the material herein.

ISBN: 978-0-443-27731-3

For Information on all Academic Press publications visit our website at https://www.elsevier.com/books-and-journals

Publisher: Megan Ball
Acquisitions Editor: Linda Buschman
Editorial Project Manager: Bhavya Rattan
Production Project Manager: Sathyavani Deivasigamani
Cover Designer: Greg Harris

Working together to grow libraries in developing countries
www.elsevier.com • www.bookaid.org

Typeset by Aptara, New Delhi, India

Contents

Contributors xix

Preface xxv

CHAPTER 1 Introduction to personalized cellular immunotherapy **1**

Reyed Mohamed Reyed, Ashok Kumar Sah, Satyajit Mohanty and Pranav Kumar Prabhakar

1.1 Introduction to personalized immunotherapy 1

1.1.1 Evolution of immunotherapy in modern medicine 2

1.1.2 Rationale for personalization 3

1.2 Overview of the immune system 4

1.2.1 Innate and adaptive immunity 4

1.2.2 Cellular players in immune defense 5

1.2.3 Immune surveillance and tumor evasion 8

1.3 Principles of cellular immunotherapy 15

1.3.1 Adoptive cell transfer 15

1.3.2 Chimeric antigen receptor T-cell therapy 16

1.3.3 T-cell receptors-engineered T cells 17

1.3.4 Dendritic cell-based vaccines 17

1.3.5 Natural killer cell and macrophage-based therapies 18

1.4 Personalization strategies in cellular therapy 19

1.4.1 Role of genomics, transcriptomics, and proteomics 19

1.4.2 Tumor antigen profiling and neoantigen prediction 20

1.4.3 Biomarker-driven patient selection 21

1.4.4 Artificial intelligence and computational tools for therapy design 22

1.5 Clinical translation and manufacturing 22

1.5.1 Workflow of patient-specific cell manufacturing 23

1.5.2 Quality control and regulatory guidelines 24

1.5.3 Safety, toxicity, and monitoring 25

1.6 Challenges and future directions 25

1.6.1 Tumor heterogeneity and immune escape 26

1.6.2 Cost, accessibility, and ethical considerations 27

1.6.3 Emerging technologies: Gene editing, synthetic biology, and next-gen chimeric antigen receptor platforms 27

References 28

CHAPTER 2 Understanding the therapeutic significance of tumor-infiltrating immune cells and their characteristics: A historical overview of cancer immunotherapy 35

Lakshyajeet Nath, Dhrubajyoti Borah, Jyotirmoy Kashyap, Shreyansh Saha, Dipjyoti Sharma, Pollobi Porasar, Jitendra Pandey, Suman Lamchhane, Aadesh Upadhyay and Koushik Nandan Dutta

2.1 Introduction 35
- 2.1.1 Overview of cancer immunotherapy and its success over the past 20 years 35
- 2.1.2 Importance of tumor-infiltrating immune cells in cancer treatment 36
- 2.1.3 Objectives of the chapter: Historical evolution, functions, and therapeutic potential of tumor-infiltrating immune cells 37

2.2 Functional roles of tumor-infiltrating immune cells in tumor immunity 37
- 2.2.1 Immunotherapeutic strategies 46
- 2.2.2 Modulation of tumor-infiltrating immune cells for therapy 47

2.3 Short-term versus long-term effects of reactive oxygen species on therapy outcomes 51

2.4 Potential role of antioxidant systems in improving immunotherapy efficacy 51
- 2.4.1 Antioxidants and immune system modulation 51
- 2.4.2 Antioxidant therapy and combination approaches 52

2.5 Challenges and future directions 52
- 2.5.1 Remaining challenges in understanding and targeting tumor-infiltration immune cells 52
- 2.5.2 Leukocyte migration within tissues 52
- 2.5.3 Extracellular matrix barriers 53
- 2.5.4 Matrix metalloproteinases 53
- 2.5.5 Emerging therapeutic approaches, including personalized immunotherapy and combination strategies 53

2.6 Future research directions to optimize tumor-infiltration immune cells modulation and reactive oxygen species integration 54

2.7 Conclusion 54

References 54

CHAPTER 3 A groundbreaking development in cancer treatment: Chimeric antigen receptor (CAR) T-cell therapy 63

Zahid Ahmad Thoker, Lovleen Marwaha and Ajit Prakash

3.1 Introduction 63
- 3.1.1 Evolution of cancer immunotherapy 63
- 3.1.2 Emergence of adoptive cell therapies 64
- 3.1.3 Rationality and scopes of chimeric antigen receptor T-cell therapy 64

3.2 Immunological basis of chimeric antigen receptor T-cell therapy 64
3.2.1 T-cell biology and antitumor immunity 65
3.2.2 Tumor antigen recognition mechanisms 65
3.2.3 Immune evasion by cancer cells 66
3.3 Design and structural components of chimeric antigen receptors 67
3.3.1 Antigen-binding domains (scFv and alternatives) 67
3.3.2 Hinge and transmembrane regions 67
3.3.3 Domains of the intracellular signaling 67
3.3.4 Evolution of chimeric antigen receptor generations 68
3.4 Manufacturing and engineering of chimeric antigen receptor T-cells 69
3.4.1 T-cell collection and isolation 69
3.4.2 Genetic modification strategies 69
3.4.3 Ex vivo expansion and quality control 70
3.4.4 Regulatory and good manufacturing practice considerations 70
3.5 Clinical applications of chimeric antigen receptor T-cell therapy 70
3.5.1 Chimeric antigen receptor T-cell therapy in hematological malignancies ...71
3.5.2 Current status in solid tumors 72
3.5.3 FDA-approved chimeric antigen receptor T-cell products 73
3.6 Therapeutic efficacy and clinical outcomes 73
3.6.1 Response rates and survival benefits 74
3.6.2 Long-term persistence and memory formation 74
3.6.3 Real-world evidence and postmarketing data 74
3.7 Toxicities and safety challenges 75
3.7.1 Cytokine release syndrome 75
3.7.2 Immune effector cell-associated neurotoxicity syndrome 75
3.7.3 On-target off-tumor effects 76
3.7.4 Strategies for toxicity management 76
3.8 Resistance mechanisms and limitations 76
3.8.1 Antigen escape and tumor heterogeneity 76
3.8.2 Immunosuppressive tumor microenvironment 77
3.8.3 Manufacturing failures and patient-related constraints 77
3.9 Advances and innovations in chimeric antigen receptor T-cell therapy 77
3.9.1 Dual-target and multitarget chimeric antigen receptors 78
3.9.2 Armored chimeric antigen receptor T-cells 78
3.9.3 Universal and allogeneic chimeric antigen receptor T-cell platforms 78
3.9.4 Switchable and logic-gated chimeric antigen receptor systems 79
3.10 Integration with emerging therapeutic strategies 79
3.10.1 Combination with immune checkpoint inhibitors 79
3.10.2 Gene-editing–enhanced chimeric antigen receptor T cells 80
3.10.3 Synergy with personalized and precision oncology 80

3.11 Ethical, economic, and accessibility considerations 81
3.11.1 Cost, manufacturing scalability, and health economics 81
3.11.2 Ethical challenges and patient equity 81
3.11.3 Global accessibility and low-resource settings 82
3.12 Future perspectives and research directions 82
3.12.1 Next-generation chimeric antigen receptor designs 82
3.12.2 Artificial intelligence and systems biology in chimeric antigen receptor development 83
3.12.3 Clinical translation and personalized immunotherapy 83
3.13 Conclusion 83
References 84

CHAPTER 4 Recent progress in CAR-T cell therapy for B-cell cancers with resistant tissues 89
Rajesh Kumar, Seetha Harilal, Mohd Aftab Siddiqui, Mohammad Ramzan and Sumel Ashique

4.1 Introduction to chimeric antigen receptor-T cell therapy 89
4.1.1 Overview 89
4.1.2 Tumor burden of B-cell cancer 90
4.1.3 Role of chimeric antigen receptor-T cell therapy in B-cell cancer 91
4.2 Chimeric antigen receptor-T cell mechanism, structure, and manufacture 92
4.2.1 Chimeric antigen receptor structure 92
4.2.2 Chimeric antigen receptor-T cell manufacture 95
4.3 Recent clinical trials and currently approved chimeric antigen receptor-T cell therapies 95
4.3.1 Recent clinical trials of chimeric antigen receptor-T cell therapies 95
4.3.2 Chimeric antigen receptor-T cell therapies approved for B-cell cancers 95
4.4 Toxicities and side effects 99
4.4.1 Chimeric antigen receptor-T cell-associated toxicity and side effects 99
4.5 Recent research and developments in chimeric antigen receptor-T cell therapy for B-cell cancers 101
4.5.1 Targeting antigens beyond CD19 102
4.5.2 Dual-targeting as well as multiantigen chimeric antigen receptor cells 102
4.5.3 Enhancing chimeric antigen receptor-T cell persistence and overcoming resistance 102
4.5.4 Development of allogeneic (off-the-shelf) chimeric antigen receptor-T cells 103
4.5.5 Reducing toxicities of chimeric antigen receptor-T cell therapy 103
4.5.6 Chimeric antigen receptor-T therapy for early-stage B-cell cancers 103
4.5.7 Expanding chimeric antigen receptor-T cell therapy beyond B-cell malignancies 104

4.6 Limitations, challenges, and future perspectives of chimeric antigen receptor-T cell therapies 104
4.7 Conclusion 105
References 106

CHAPTER 5 Targeting the tumor vasculature using mesenchymal stem cells in cancer therapy: New perspectives 115
Mohammed Sikander, Shabnam Malik, Murali M. Yallapu and Subhash C. Chauhan

5.1 Introduction 115
5.2 Tumor-driven angiogenesis and signaling pathways 116
5.3 Interactions between tumor vasculature and tumor microenvironment 117
5.4 Signaling pathways regulated by mesenchymal stem cells 118
5.5 Mesenchymal stem cells: Ideal candidates for targeting tumors 119
5.6 Significance of mesenchymal stem cell for cancer therapy 121
5.7 Delivery of chemotherapeutic agents using mesenchymal stem cells 121
5.8 Mesenchymal stem cell-derived exosome as a cell-free therapeutic tool 123
5.9 Utilizing oncolytic viruses with mesenchymal stem cells: An innovative approach 123
5.10 Mesenchymal stem cell-mediated suicidal genes: A novel strategy in cancer therapy 124
5.11 Preclinical and clinical evidence supporting mesenchymal and stem cell therapy 124
5.12 Challenges and limitations of mesenchymal stem cell-based therapies 126
5.13 Conclusion 126
References 127

CHAPTER 6 Mesenchymal stromal cell-based delivery of oncolytic viruses: A novel strategy for immunotherapy of metastatic tumors 133
Ubaid Gani, Awadhesh Kumar Verma, Sahar Saleem Bhat and Pranav Kumar Prabhakar

6.1 Introduction 133
6.2 Biological basis of oncolytic virotherapy 135
6.2.1 Mechanisms of tumor selectivity and replication 135
6.2.2 Viral platforms in clinical and preclinical use 136
6.2.3 Oncolytic viruses as immune stimulants: Cytokine release, antigen spreading, and T-cell priming 138
6.2.4 Limitations of naked viral delivery 140
6.3 Mesenchymal stromal cells as therapeutic carriers 141
6.3.1 Biological characteristics and sources of mesenchymal stromal cells 141
6.3.2 Homing capacity to tumor microenvironments 143

6.3.3 Immunomodulatory and regenerative properties 143
6.4 Engineering mesenchymal stromal cell–oncolytic virus systems 145
6.4.1 Virus loading, infection efficiency, and replication within mesenchymal stromal cells 145
6.4.2 Genetic modification of mesenchymal stromal cells for enhanced viral delivery 146
6.4.3 Impact on mesenchymal stromal cell viability, differentiation, and immune behavior 147
6.4.4 Optimization of virus-to-cell ratios and release kinetics 148
6.4.5 Ex vivo versus in vivo viral loading strategies 148
6.5 Mechanisms of antitumor action in mesenchymal stromal cell–oncolytic virus therapy 150
6.5.1 Tumor targeting and viral propagation in metastatic niches 150
6.5.2 Immune activation cascade 151
6.5.3 Disruption of the tumor stroma and vascular networks 151
6.5.4 Modulation of immunosuppressive cells (regulatory T cells, myeloid-derived suppressor cells, and tumor-associated macrophages) 153
6.5.5 Synergistic potential with checkpoint inhibitors and cytokine therapies 154
6.6 Preclinical and clinical studies 155
6.6.1 In vitro evidence: Tumor cell killing and viral replication profiles 155
6.6.2 In vivo animal models: Biodistribution and efficacy studies 156
6.6.3 Clinical translation: Current and completed trials of mesenchymal stromal cell–oncolytic viruses 156
6.6.4 Case studies of successful preclinical-to-clinical transition 158
6.6.5 Comparative outcomes with direct viral delivery 159
6.7 Challenges and limitations 160
6.7.1 Safety concerns 160
6.7.2 Immunogenicity and premature clearance 161
6.7.3 Heterogeneity of mesenchymal stromal cell sources and patient-specific responses 162
6.7.4 Ethical considerations in cellular–viral combinatorial therapies 163
6.8 Innovations and future perspectives 164
6.8.1 Next-generation viral engineering 164
6.8.2 Synthetic biology approaches for programmable mesenchymal stromal cell–virus interactions 164
6.8.3 Combination therapies with chimeric antigen receptor-T, immune checkpoint blockade, or radiotherapy 165
6.8.4 Personalized oncolytic virotherapy using patient-derived mesenchymal stromal cells 166
6.8.5 Prospects for artificial intelligence-assisted modeling and precision delivery systems 167

6.9 Conclusion 167
Glossary 168
References 169

CHAPTER 7 Cancer treatment advancements in the era of immune checkpoint inhibitors via cellular immunotherapies 173
Hariharan Thirumalai Vengateswaran, Mohammad Habeeb, Huay Woon You, Kiran Balasaheb, Girija Balasaheb Bhavar and Govind Sarangdhar Asane

7.1 Introduction 173
7.2 Tumor microenvironment in cancer 174
7.3 Cancer immunotherapy 174
7.3.1 Mechanisms of action in immunotherapy 175
7.4 Immune checkpoint inhibitors 177
7.4.1 Traditional immune checkpoint inhibitors: Programmed cell death protein 1, programmed death-ligand 1, cytotoxic T-lymphocyte-associated protein 4 177
7.4.2 Emerging immune checkpoint inhibitors: TIGIT, LAG-3, TIM-3, and others 178
7.4.3 Immune checkpoint inhibitors resistance action 179
7.5 Immune system modulators 181
7.5.1 Agents enhancing immune responses 181
7.5.2 Cytokine therapies in cancer treatment 181
7.6 Advantages and disadvantages of immune checkpoint inhibitors 182
7.7 Role of artificial intelligence 183
7.8 Conclusion 184
References 184

CHAPTER 8 The future of cancer immunotherapy: Combining gene and cellular approaches 187
Balachander Lokeshwari, Pazhani Saranraj, Karthikeyan Gayathri, Alexander Machado Cardoso and Kesavardhini Kesavardhini

8.1 Introduction 187
8.2 Immunological basis of cancer and tumor immune evasion 188
8.2.1 Cancer-immune system interactions 188
8.2.2 Tumor antigens and immune recognition 189
8.2.3 Mechanisms of immune response 189
8.3 Cancer immunotherapy modalities 190
8.3.1 Immune checkpoint inhibitors 190
8.3.2 Cancer vaccines 192
8.3.3 Adoptive cell therapy 193
8.4 Gene therapy in cancer immunotherapy 195

8.4.1 Gene editing technologies 195
8.4.2 Viral and non-viral gene delivery systems 197
8.4.3 Therapeutic genes in immuno-oncology 198
8.5 Cellular immunotherapy approaches 199
8.5.1 Chimeric antigen receptor-based therapies 199
8.6 Synergistic integration of gene and cellular therapies 200
8.6.1 Gene-edited immune cells 200
8.7 Tumor microenvironment modulation 202
8.7.1 Gene-modified cells targeting tumor microenvironment barriers 202
8.7.2 Hypoxia, fibrosis, and metabolic suppression 202
8.7.3 Immunosuppressive cells 202
8.8 Emerging technologies 203
8.8.1 Synthetic biology and immune cells 203
8.8.2 Nanotechnology-assisted gene and cell delivery 204
8.9 Clinical translation and regulatory challenges 204
8.9.1 Manufacturing scalability and standardization 204
8.9.2 Safety concerns 204
8.9.3 Regulatory frameworks and ethical considerations 205
8.10 Future perspectives 205
8.11 Conclusion 206
References 206

CHAPTER 9 Platelet shielding of cancer cells: Implications for immunotherapy strategies 211
Vivek P. Chavda, Disha Joshi, Arya Vyas and Manish Patel

9.1 Introduction 211
9.2 Platelet biology 212
9.2.1 Structure and function of platelets 212
9.2.2 Platelet activation and signaling pathways 213
9.2.3 Platelets in hemostasis and thrombosis 213
9.3 Platelets as immune cells 216
9.3.1 Role of platelets in innate immunity 216
9.3.2 Platelets in adaptive immune responses 217
9.3.3 Mechanisms of platelet-immune cell interactions 217
9.4 The concept of platelet shielding 218
9.4.1 Definition and mechanisms of platelet shielding 218
9.4.2 Pathological contexts: Inflammation and infection 222
9.5 Implications for immunotherapy 223
9.5.1 Enhancing efficacy of immunotherapeutic agents 224
9.5.2 Role of platelet shielding in tumor microenvironment 225
9.5.3 Case studies: Successful integration of platelet shielding in therapy 227

9.6 Challenges and limitations ... 228
9.6.1 Potential adverse effects of platelet activation ... 228
9.6.2 Variability in patient response to immunotherapy ... 229
9.6.3 Technical challenges in targeting platelets ... 229
9.7 Conclusion ... 230
References ... 231

CHAPTER 10 Cancer cell-platelet interactions: Harnessing therapeutic opportunities in cancer treatment ... 241
Sanya Jain and Gudhanti Siva Naga Koteswara Rao

10.1 Introduction ... 241
10.1.1 Overview of cancer and platelet interactions ... 241
10.1.2 Significance of platelet interactions in cancer progression ... 241
10.2 Receptors involved in platelet–cell interactions ... 243
10.2.1 Adhesion receptors ... 244
10.2.2 Immunoreceptor tyrosine-based activation motif and hemITAM-associated receptors ... 246
10.2.3 C-type lectin receptor families ... 247
10.2.4 P-selectins ... 247
10.2.5 Toll-like receptors ... 247
10.2.6 Nod-like receptors ... 248
10.2.7 G protein-coupled receptors ... 248
10.3 Role of platelet–cancer cell interactions in cancer progression ... 248
10.3.1 Cancer cell survival and growth ... 248
10.3.2 Angiogenesis and tumor vascularization ... 248
10.3.3 Metastatic cascade and platelet support ... 249
10.4 Mechanisms of platelet–cancer cell interactions ... 249
10.4.1 Platelet activation by cancer ... 249
10.4.2 Cancer cell-induced platelet aggregation ... 249
10.4.3 Signaling pathways ... 250
10.4.4 Release of platelet-derived growth factors ... 251
10.4.5 Immune modulation and evasion ... 252
10.5 Types of cancer–cell platelet interactions ... 252
10.6 Signaling pathways involved in cancer cell–platelet interactions ... 253
10.6.1 TGF-β/Smad and NF-κB pathway ... 253
10.6.2 Integrin signaling (particularly $\alpha v\beta 3$) ... 254
10.6.3 Lysophosphatidic acid-lysophosphatidic acid receptor pathway ... 254
10.6.4 CLEC-2-podoplanin pathway ... 254
10.6.5 JAK/STAT pathway ... 254
10.6.6 P38MAPK-MMP-9 pathway ... 255

10.6.7 Protease-activated receptor pathway ... 255
10.6.8 Cathepsin K signaling ... 255
10.6.9 Lysophosphatidic acid signaling ... 255
10.7 Therapeutic targeting of platelet–cancer cell interactions ... 256
10.7.1 Anti-P-selectin antibodies ... 256
10.7.2 Antibodies targeting cancer cell surface markers ... 256
10.7.3 Small molecule inhibitors targeting cancer cell signaling pathways ... 257
10.7.4 Platelet inhibition and immunotherapy ... 257
10.7.5 Emerging therapies ... 257
10.8 Overcoming drug resistance through targeting platelets ... 258
10.9 Challenges, limitations, and future directions ... 258
10.10 Conclusion ... 259
Abbreviations ... 260
References ... 260

CHAPTER 11 Myeloid-derived suppressor cells: A promising target for anticancer immunotherapy ... 271
Vinit Sharma, Sakshi Bansal and Anjali Aggarwal

11.1 Introduction ... 271
11.2 Myeloid-derived suppressor cells inhibit antitumor immunity and promote resistance to therapies ... 272
11.3 Myeloid-derived suppressor cells heterogeneity and challenges in targeting ... 274
11.4 Strategies for targeting myeloid-derived suppressor cells ... 274
11.5 New strategies for targeting myeloid-derived suppressor cells: Beyond depletion ... 275
11.6 Innovative approaches: Engineering CAR-T cells ... 277
11.7 MDSC-modulating vaccines to promote immune activation ... 284
11.8 Clinical trials and future directions ... 285
11.9 Conclusion ... 286
Abbreviations ... 286
References ... 287

CHAPTER 12 Utilizing nanomedicines to modulate myeloid-derived suppressor cells: Enhancing cancer immunotherapy ... 297
Smita Kumbhar and Manish Bhatia

12.1 Introduction ... 297
12.2 Myeloid-derived suppressor cells: Key players in tumor immune evasion ... 298
12.3 Nanomedicine: A revolutionary approach to modulating myeloid-derived suppressor cells ... 300
12.3.1 Overview of nanomedicine and its applications in cancer therapy ... 300

12.3.2 Types of nanomedicines used to target myeloid-derived suppressor cells 301
12.3.3 Mechanisms of myeloid-derived suppressor cells modulation via nanomedicines 301
12.3.4 Translational challenges and future perspectives 303
12.4 Preclinical and clinical advancements 303
12.4.1 Preclinical studies: Key findings in nanomedicine for inhibiting myeloid-derived suppressor cells activity 303
12.4.2 Clinical trials: Emerging evidence of efficacy and safety 305
12.4.3 Case studies: Successful modulation of myeloid-derived suppressor cells by nanomedicines 305
12.4.4 Summary and clinical implications 306
12.5 Combination therapies: Enhancing efficacy 306
12.5.1 Integrating nanomedicines with immune checkpoint inhibitors and CAR-T cell therapy 306
12.5.2 Synergistic effects: Overcoming myeloid-derived suppressor cells-mediated immunosuppression 308
12.5.3 Clinical translation and future directions 308
12.6 Challenges, regulatory pathways, and future perspectives 310
12.6.1 Scientific and translational challenges 310
12.6.2 Regulatory pathways and policy considerations 311
12.6.3 Future directions and strategic roadmap 312
12.6.4 Concluding perspective 313
12.7 Conclusion 313
References 314

CHAPTER 13 Unlocking the potential of macrophages: Advancements in cellular immunotherapy 319
Vivek P. Chavda, Dixa Vaghela, Arya Vyas, Pankti Balar, Tathya J Sheth and Vaishvi Chokshi

13.1 Introduction 319
13.2 Macrophage biology 320
13.2.1 Origin and development of macrophages 320
13.2.2 Macrophage phenotypes 322
13.2.3 Functional roles in the immune system 322
13.3 Mechanisms of action in cellular immunotherapy 324
13.3.1 Phagocytosis and antigen presentation 324
13.3.2 Cytokine production and immune modulation 325
13.3.3 Interaction with other immune cells 326
13.4 Engineering macrophages for therapeutic applications 327
13.4.1 Chimeric antigen receptor macrophages 328

13.5 Tumor microenvironment and macrophage dynamics 332
13.5.1 Role of tumor-associated macrophage 332
13.5.2 Mechanism of tumor progression vs suppression 334
13.5.3 Strategies for reprogramming TAMs for antitumor activity 336
13.6 Clinical application and case studies 336
13.6.1 Current clinical trials involving macrophage therapies 336
13.6.2 Success stories and outcomes 337
13.6.3 Challenges in translation to clinical practice 338
13.7 Challenges and future directions 339
13.7.1 Limitations of current approaches 339
13.7.2 Research gaps and opportunities for innovation 339
13.7.3 Future perspectives on macrophage-based therapies 340
13.8 Conclusion 340
13.8.1 Summary of key insights 340
13.8.2 Implications for future research and clinical practice 340
References 341

CHAPTER 14 Combination therapy: A new strategy to manage cancer 351
Reyed Mohamed Reyed, Ashok Kumar Sah, Satyajit Mohanty and Pranav Kumar Prabhakar

14.1 Introduction 351
14.2 Overview of cancer and its treatment modalities 352
14.2.1 Molecular basis and pathophysiology of cancer 352
14.2.2 Current standard treatments 353
14.2.3 Challenges in existing therapies 354
14.3 Concept of combination therapy 355
14.3.1 Definition and rationale 356
14.3.2 Historical background and evolution 356
14.3.3 Mechanisms of action and therapeutic synergy 357
14.3.4 Advantages over single-agent therapy 358
14.4 Types of combination therapy in cancer treatment 359
14.4.1 Chemotherapy–chemotherapy combinations 359
14.4.2 Chemotherapy–radiotherapy combinations 360
14.4.3 Chemotherapy–immunotherapy combinations 361
14.4.4 Chemotherapy–hormonal therapy combinations 362
14.4.5 Chemotherapy–targeted therapy combinations 363
14.4.6 Multimodal therapy (triple- or multidrug approaches) 363
14.5 Role of natural products in combination therapy 364
14.5.1 Importance of plant-derived compounds in oncology 364
14.5.2 Mechanisms of synergism between phytochemicals and anticancer drugs 365

14.5.3 Examples of effective plant–drug combinations 366
14.5.4 Advantages of phytochemical combinations 367
14.5.5 Challenges in clinical translation 368
14.6 Mechanisms underlying synergistic effects 369
14.6.1 Modulation of apoptosis and cell cycle pathways 369
14.6.2 Inhibition of multidrug resistance transporters 370
14.6.3 Targeting the tumor microenvironment 371
14.6.4 Antiinflammatory and antioxidant effects 372
14.6.5 Epigenetic and gene regulatory mechanisms 372
14.7 Clinical studies and case examples 374
14.7.1 Preclinical studies demonstrating efficacy 374
14.7.2 Ongoing clinical trials 374
14.7.3 Approved combination therapies in clinical practice 375
14.7.4 Limitations and lessons learned 376
14.8 Challenges and future perspectives 377
14.8.1 Pharmacokinetic and pharmacodynamic interactions 377
14.8.2 Toxicity management in combination regimens 377
14.8.3 Personalized medicine approaches 378
14.8.4 Role of nanotechnology and drug delivery systems 378
14.8.5 Regulatory and ethical considerations 379
14.8.6 Future perspectives 379
14.9 Conclusion 380
14.9.1 Summary of current advancements 380
14.9.2 Future directions in combination therapy research 380
14.9.3 Toward safer and more effective cancer management 382
References 383

CHAPTER 15 Future perspective of cellular immunotherapy in cancer treatment 389
Rideb Chakraborty, Abirami Seetharaman, Aniruddha Sen, Darshan S and Ciniraj Raveendran

15.1 Introduction 389
15.1.1 Brief overview of cancer and current treatment approaches 389
15.1.2 Introduction to cellular immunotherapy 389
15.1.3 Cellular immunotherapy: Future of cancer treatment with potential advancements in efficacy, accessibility, and personalization 390
15.2 Current state of cellular immunotherapy 390
15.2.1 Overview of existing cellular immunotherapy approaches 390
15.2.2 Successes and limitations of current treatments 392
15.3 Future advancements in cellular immunotherapy 392
15.3.1 Enhancing efficacy 392

15.3.2 Expanding target range 394
15.3.3 Overcoming resistance mechanisms 395
15.4 Improving accessibility and scalability 395
15.4.1 Developing off-the-shelf cellular therapies 395
15.4.2 Streamlining manufacturing processes 395
15.4.3 Reducing production costs 396
15.5 Personalization and precision medicine 396
15.5.1 Tailoring treatments to individual patients 396
15.5.2 Utilizing genomic and proteomic profiling 396
15.5.3 Implementing artificial intelligence and machine learning 397
15.6 Emerging cellular immunotherapy approaches 397
15.6.1 Natural killer cell therapies 397
15.6.2 Gamma delta T-cell therapies 398
15.6.3 Macrophage-based therapies 398
15.7 Ethics and challenges of ethical consideration 398
15.7.1 Patient safety 399
15.7.2 Equitable access to treatment 400
15.7.3 Regulatory approval and process 400
15.8 Conclusion 401
15.8.1 Recap of key future perspectives 401
15.8.2 Potential impact on cancer treatment landscape 401
15.8.3 Call for continued research and development 402
References 402

Index 407

Contributors

Anjali Aggarwal
Department of Anatomy, Post Graduate Institute of Medical Education and Research, Chandigarh, India

Govind Sarangdhar Asane
Department of Pharmaceutics, SNJB Shriman Sureshdada Jain College of Pharmacy, Neminagar, Chandwad, Nashik, Maharashtra, India

Sumel Ashique
School of Pharmaceutical Sciences, Lovely Professional University, Phagwara, Punjab, India

Pankti Balar
Department of Pharmaceutics and Pharmaceutical Technology, L. M. College of Pharmacy, Ahmedabad, Gujarat, India

Kiran Balasaheb
Department of Pharmaceutical Quality Assurance, School of Pharmacy and Technology Management, SVKM NMIMS Global University, Dhule, Maharashtra, India

Sakshi Bansal
Department of Anatomy, Post Graduate Institute of Medical Education and Research, Chandigarh, India

Sahar Saleem Bhat
Division of Animal Biotechnology, Faculty of Veterinary Sciences and Animal Husbandry, Sher-e-Kashmir University of Agricultural Sciences and Technology of Kashmir, Srinagar, Jammu and Kashmir, India

Manish Bhatia
Department of Pharmaceutical Chemistry, Bharati Vidyapeeth College of Pharmacy, Kolhapur, Maharashtra, India

Girija Balasaheb Bhavar
Department of Pharmaceutical Chemistry, School of Pharmacy and Technology Management, SVKM NMIMS Global University, Dhule, Maharashtra, India

Dhrubajyoti Borah
NETES Institute of Pharmaceutical Science, NGI-Mirza, Kamrup, Assam, India

Alexander Machado Cardoso
Department of Biology, Rio de Janeiro State University (UERJ), Rio de Janeiro, Brazil

Rideb Chakraborty
NSHM Institute of Pharmaceutical Technology, NSHM Knowledge Campus, Durgapur, West Bengal, India

Subhash C. Chauhan
Division of Cancer Immunology and Microbiology, South Texas Center of Excellence in Cancer Research, Medicine and Oncology Integrated Service Unit, School of Medicine, University of Texas Rio Grande Valley (UTRGV), McAllen, TX, United States

Vivek P. Chavda
Department of Pharmaceutics and Pharmaceutical Technology, L. M. College of Pharmacy, Ahmedabad, Gujarat, India

Vaishvi Chokshi
Pharm.D Section, L. M. College of Pharmacy, Ahmedabad, Gujarat, India

Koushik Nandan Dutta
NETES Institute of Pharmaceutical Science, NGI-Mirza, Kamrup, Assam, India

Ubaid Gani
Department of Zoology, School of Bioengineering and Biosciences, Lovely Professional University, Phagwara, Punjab, India; Division of Animal Biotechnology, Faculty of Veterinary Sciences and Animal Husbandry, Sher-e-Kashmir University of Agricultural Sciences and Technology of Kashmir, Srinagar, Jammu and Kashmir, India

Karthikeyan Gayathri
PG and Research Department of Microbiology, Sacred Heart College (Autonomous), Tirupattur, Tamil Nadu, India

Mohammad Habeeb
Department of Pharmaceutics, Crescent School of Pharmacy, B.S. Abdur Rahman Crescent Institute of Science and Technology, Chennai, Tamil Nadu, India

Seetha Harilal
Faculty of Pharmaceutical Sciences, Kerala University of Health Sciences, Thrissur, Kerala, India

Sanya Jain
Shobhaben Pratapbhai Patel School of Pharmacy & Technology Management, SVKM's NMIMS (Deemed to be University), Mumbai, Maharashtra, India

Disha Joshi
Department of Pharmaceutical Technology, L. J. Institute of Pharmacy, Lok Jagruti Kendra (LJ) University, Ahmedabad, Gujarat, India

Jyotirmoy Kashyap
NETES Institute of Pharmaceutical Science, NGI-Mirza, Kamrup, Assam, India

Kesavardhini Kesavardhini
PG and Research Department of Microbiology, Sacred Heart College (Autonomous), Tirupattur, Tamil Nadu, India

Rajesh Kumar
Faculty of Pharmaceutical Sciences, Kerala University of Health Sciences, Thrissur, Kerala, India

Smita Kumbhar
Department of Pharmaceutical Chemistry, Sanjivani College of Pharmaceutical Education and Research (Autonomous), Kopargaon, Maharashtra, India

Suman Lamchhane
Department of Pharmacy, Shree College of Technology, Bharatpur, Chitwan, Nepal

Balachander Lokeshwari
PG and Research Department of Microbiology, Sacred Heart College (Autonomous), Tirupattur, Tamil Nadu, India

Shabnam Malik
Division of Cancer Immunology and Microbiology, South Texas Center of Excellence in Cancer Research, Medicine and Oncology Integrated Service Unit, School of Medicine, University of Texas Rio Grande Valley (UTRGV), McAllen, TX, United States

Lovleen Marwaha
Department of Zoology, School of Bioengineering and Biosciences, Lovely Professional University, Phagwara, Punjab, India

Satyajit Mohanty
Department of Pharmacy, School of Medical and Allied Science, Galgotias University, Greater Noida, Uttar Pradesh, India

Lakshyajeet Nath
NETES Institute of Pharmaceutical Science, NGI-Mirza, Kamrup, Assam, India

Jitendra Pandey
Department of Chemistry, University of Hawaii at Manoa, Honolulu, HI, United States

Manish Patel
Department of Pharmaceutics and Pharmaceutical Technology, L. M. College of Pharmacy, Ahmedabad, Gujarat, India

Pollobi Porasar
NETES Institute of Pharmaceutical Science, NGI-Mirza, Kamrup, Assam, India

Pranav Kumar Prabhakar
Department of Biotechnology, School of Engineering and Technology, Nagaland University, Meriema, Kohima, Nagaland, India

Ajit Prakash
Department of Biochemistry and Biophysics, University of North Carolina, Chapel Hill, NC, United States

Mohhammad Ramzan
School of Pharmaceutical Sciences, Lovely Professional University, Phagwara, Punjab, India

Gudhanti Siva Naga Koteswara Rao
Shobhaben Pratapbhai Patel School of Pharmacy & Technology Management, SVKM's NMIMS (Deemed to be University), Mumbai, Maharashtra, India

Ciniraj Raveendran
Department of Radiation Oncology, Government Medical College, Thiruvananthapuram, Kerala, India

Reyed Mohamed Reyed
Genetic Engineering and Biotechnology Research Institute (GEBRI), City of Scientific Research and Technological Applications (SRTA-City), Alexandria, Egypt

Darshan S
Rajiv Gandhi University of Health Sciences, Bengaluru, Karnataka, India

Ashok Kumar Sah
Department of Medical Laboratory Sciences, College of Applied and Health Sciences, A' Sharqiyah University, Ibra, Oman

Shreyansh Saha
NETES Institute of Pharmaceutical Science, NGI-Mirza, Kamrup, Assam, India

Pazhani Saranraj
PG and Research Department of Microbiology, Sacred Heart College (Autonomous), Tirupattur, Tamil Nadu, India

Abirami Seetharaman
Department of Biotechnology, Indian Institute of Technology Madras, Chennai, Tamil Nadu, India

Aniruddha Sen
Department of Biochemistry, Shri Gorakshnath Medical College Hospital and Research Center, Gorakhpur, Uttar Pradesh, India

Dipjyoti Sharma
NETES Institute of Pharmaceutical Science, NGI-Mirza, Kamrup, Assam, India

Vinit Sharma
Department of Anatomy, Post Graduate Institute of Medical Education and Research, Chandigarh, India

Tathya J Sheth
Pharm.D Section, L. M. College of Pharmacy, Ahmedabad, Gujarat, India

Mohd Aftab Siddiqui
Faculty of Pharmacy, Integral University, Lucknow, Uttar Pradesh, India

Mohammed Sikander
Division of Cancer Immunology and Microbiology, South Texas Center of Excellence in Cancer Research, Medicine and Oncology Integrated Service Unit, School of Medicine, University of Texas Rio Grande Valley (UTRGV), McAllen, TX, United States

Zahid Ahmad Thoker
Department of Zoology, School of Bioengineering and Biosciences, Lovely Professional University, Phagwara, Punjab, India

Aadesh Upadhyay
Project Lead-Oncology, One Clinical Research, Perth, WA, Australia

Dixa Vaghela
Department of Pharmaceutics and Pharmaceutical Technology, L. M. College of Pharmacy, Ahmedabad, Gujarat, India

Hariharan Thirumalai Vengateswaran
Department of Pharmaceutics, Crescent School of Pharmacy, B.S. Abdur Rahman Crescent Institute of Science and Technology, Chennai, Tamil Nadu, India

Awadhesh Kumar Verma
Department of Biotechnology and Bioinformatics, School of Bioengineering and Biosciences, Lovely Professional University, Punjab, Phagwara, India

Arya Vyas
Department of Pharmaceutics and Pharmaceutical Technology, L. M. College of Pharmacy, Ahmedabad, Gujarat, India; PhD Section, Gujarat Technological University, Ahmedabad, Gujarat, India

Murali M. Yallapu
Division of Cancer Immunology and Microbiology, South Texas Center of Excellence in Cancer Research, Medicine and Oncology Integrated Service Unit, School of Medicine, University of Texas Rio Grande Valley (UTRGV), McAllen, TX, United States

Huay Woon You
Faculty of Science and Technology, Universiti Kebangsaan Malaysia, Bangi, Selangor, Malaysia; Pusat PERMATA@Pintar Negara, Universiti Kebangsaan Malaysia, Bangi, Selangor, Malaysia

Preface

Cancer remains one of the most formidable challenges to global health, marked by profound biological heterogeneity, sophisticated mechanisms of immune evasion, and frequent resistance to conventional therapeutic modalities. In recent decades, oncology has undergone a fundamental paradigm shift from nonspecific cytotoxic treatments toward precision-driven, immune-based strategies that harness the body's own defense mechanisms. Within this evolving landscape, personalized cellular immunotherapy has emerged as one of the most transformative and promising approaches, offering durable responses and, in some cases, long-term remission for patients with otherwise refractory malignancies.

The book *Advancements in Personalized Cellular Immunotherapy: A Comprehensive Guide to Cancer Treatment* has been conceived to present a thorough, up-to-date, and integrative overview of this rapidly evolving landscape of cellular immunotherapy. This volume brings together fundamental concepts, historical perspectives, cutting-edge technological advancements, and emerging therapeutic strategies that collectively shape the future of cancer treatment. By bridging basic immunology with translational and clinical insights, the book aims to serve as a comprehensive resource for researchers, clinicians, academicians, and advanced students working in oncology, immunology, biotechnology, and allied disciplines.

The opening chapters establish the conceptual and historical foundation of the field. Beginning with an introduction to personalized cellular immunotherapy, the book traces the historical evolution of cancer immunotherapy, emphasizing the therapeutic significance of tumor-infiltrating immune cells and their functional characteristics. These chapters provide critical context for understanding how early observations of immune surveillance have culminated in sophisticated, patient-specific cellular interventions. A significant portion of the book is devoted to chimeric antigen receptor (CAR) T-cell therapy, one of the most groundbreaking advances in modern oncology. The scope of the volume further explores the scientific rationale behind CAR-T cell engineering, followed by an in-depth discussion on recent progress in CAR-T therapies for B-cell malignancies, particularly in cases involving resistant or immune-protected tissues. These chapters highlight both clinical successes and ongoing challenges, such as toxicity, antigen escape, and durability of response. The book further expands its scope to include mesenchymal stem/stromal cells as innovative therapeutic vehicles. Chapters focusing on targeting tumor vasculature using mesenchymal stem/stromal cells and the delivery of oncolytic viruses through stromal cells provide novel perspectives on exploiting the tumor microenvironment for therapeutic gain, particularly in metastatic disease contexts.

In recognition of the growing importance of immune modulation, the volume comprehensively addresses immune checkpoint inhibitors, their integration with cellular immunotherapies, and their role in overcoming immune exhaustion. Complementing this, future-oriented chapters explore the convergence of gene therapy and cellular approaches, underscoring how genome editing, synthetic biology, and next-generation cell engineering are redefining the boundaries of personalized cancer treatment.

A unique strength of this book lies in its focused attention on cellular interactions within the tumor microenvironment. Dedicated chapters examine platelet-mediated shielding of cancer cells, cancer cell–platelet interactions, and their implications for immune evasion and therapeutic targeting. Additionally, the critical role of myeloid-derived suppressor cells and macrophages in tumor progression is explored, alongside innovative strategies including nanomedicine-based interventions to modulate these cells and enhance antitumor immunity. The later sections of the book emphasize combination therapies, reflecting the growing consensus that effective cancer treatment often requires synergistic integration of multiple therapeutic modalities. The concluding chapters provide a forward-looking perspective on the future of cellular immunotherapy, addressing current limitations, ethical considerations, manufacturing challenges, and the translational roadmap toward more accessible and effective cancer treatments.

This volume represents the collective efforts of leading experts and contributors who have shared their insights, research findings, and clinical experiences. We gratefully acknowledge their dedication and scholarly contributions. We are also grateful to Elsevier for providing an excellent platform to disseminate this work to the global scientific community.

It is our hope that this book will not only deepen understanding of personalized cellular immunotherapy but also inspire innovation, collaboration, and critical thinking in the ongoing quest to develop safer, more effective, and truly personalized cancer therapies.

CHAPTER

Introduction to personalized cellular immunotherapy

1

Reyed Mohamed Reyed[a], Ashok Kumar Sah[b], Satyajit Mohanty[c] and Pranav Kumar Prabhakar[d]

[a]*Genetic Engineering and Biotechnology Research Institute (GEBRI), City of Scientific Research and Technological Applications (SRTA-City), Alexandria, Egypt,* [b]*Department of Medical Laboratory Sciences, College of Applied and Health Sciences, A' Sharqiyah University, Ibra, Oman,* [c]*Department of Pharmacy, School of Medical and Allied Science, Galgotias University, Greater Noida, Uttar Pradesh, India,* [d]*Department of Biotechnology, School of Engineering and Technology, Nagaland University, Meriema, Kohima, Nagaland, India*

1.1 Introduction to personalized immunotherapy

Personalized immunotherapy is a transformative approach in biomedical science that tailors immune-based treatments to the unique genetic, molecular, and immunological characteristics of each patient. It represents the convergence of immunology, genomics, data science, and precision medicine, aiming to optimize therapeutic efficacy while minimizing adverse effects. The central premise of personalized immunotherapy is that each individual possesses a distinct immunological landscape—shaped by their genome, microbiome, disease microenvironment, and prior antigenic exposures—that influences how their immune system responds to disease and treatment (Schreiber et al., 2011; Sharma et al., 2021).

Immunotherapy itself leverages the body's immune mechanisms to recognize and eliminate diseased or malignant cells. Traditional immunotherapeutic strategies—such as cytokine therapy, monoclonal antibodies, or checkpoint inhibitors—have achieved significant clinical success but are often limited by interpatient variability in response and toxicity profiles. Personalized immunotherapy addresses this limitation by integrating patient-specific data to design tailored interventions. Such personalization may involve identifying tumor-specific neoantigens through next-generation sequencing (NGS), profiling the patient's immune repertoire using high-throughput immunogenomics, or engineering autologous immune cells for targeted action (Chen & Mellman, 2017).

At its core, personalized immunotherapy embodies the principles of **precision medicine**, wherein treatment is customized not only based on the disease subtype but also on the patient's unique biological context. This approach acknowledges the dynamic interplay between host and disease, particularly in complex conditions such as cancer, autoimmune disorders, and chronic infections. The goal is to move beyond generalized therapies toward patient-specific regimens that can achieve durable remission with minimal immune-related adverse events (IRAE).

In oncology, for instance, the identification of tumor-specific antigens has facilitated the development of *chimeric antigen receptor (CAR)-T cell therapies* and *personalized cancer vaccines* that target neoantigens unique to each tumor. Similarly, in infectious diseases, immunogenetic profiling

Advancements in Personalized Cellular Immunotherapy: A Comprehensive Guide to Cancer Treatment.
DOI: https://doi.org/10.1016/B978-0-443-27731-3.00011-5

Copyright © 2026 Elsevier Inc. All rights are reserved, including those for text and data mining, AI training, and similar technologies.

can help design individualized vaccine regimens optimized for immune responsiveness. With advancements in single-cell analysis, multiomics integration, and machine learning, the field of personalized immunotherapy is evolving rapidly, promising a new era of precision immune medicine (June & Sadelain, 2018).

1.1.1 Evolution of immunotherapy in modern medicine

The roots of immunotherapy trace back more than a century, beginning with the pioneering observations of William B. Coley in the 1890s. Coley demonstrated that bacterial extracts, known as "Coley's toxins," could induce tumor regression in sarcoma patients by stimulating immune activation (Hoption Cann et al., 2003). Although rudimentary, these early experiments laid the conceptual foundation for modern immuno-oncology: the idea that the immune system could be harnessed to combat malignancy. During the mid-20th century, immunotherapy evolved in tandem with discoveries in cellular and molecular immunology. The identification of **T lymphocytes** as central mediators of adaptive immunity and the elucidation of **major histocompatibility complex (MHC)** restriction provided crucial mechanistic insights. Early therapeutic applications included **interferon-α** and **interleukin-2 (IL-2)** therapies in melanoma and renal carcinoma (Rosenberg et al., 1998). These agents demonstrated that nonspecific immune activation could elicit antitumor effects, though often accompanied by severe toxicity.

A major milestone occurred with the development of **monoclonal antibody (mAb)** technology in the 1970s and 1980s (Köhler & Milstein, 1975). This breakthrough enabled the production of highly specific antibodies against tumor-associated antigens (TAAs), leading to the approval of therapeutic antibodies such as **rituximab** (anti-CD20) and **trastuzumab** (anti-HER2). These biologics transformed oncology by offering targeted mechanisms of immune-mediated cytotoxicity, bridging the gap between classical chemotherapy and immune modulation. The next transformative wave arose from the discovery of **immune checkpoint pathways**—notably cytotoxic T-lymphocyte-associated protein 4 (**CTLA-4**) and programmed cell death protein 1 (**PD-1**)/programmed death-ligand 1 (**PD-L1**)—which act as brakes on T-cell activation. The introduction of checkpoint inhibitors such as **ipilimumab, nivolumab**, and **pembrolizumab** revolutionized cancer treatment by restoring antitumor immunity and achieving durable remissions in malignancies once considered incurable (Pardoll, 2012). However, only a subset of patients derived lasting benefits, highlighting the need for patient-specific predictive biomarkers such as PD-L1 expression, tumor mutational burden (TMB), and microsatellite instability (MSI).

Simultaneously, advances in **cellular engineering** ushered in the era of **adoptive cell therapy (ACT)**. Pioneered through **tumor-infiltrating lymphocyte (TIL)** therapy and later **CAR-T cell technology**, these approaches involve the ex vivo modification or expansion of autologous immune cells before reinfusion (June et al., 2018). CAR-T therapies targeting CD19 and BCMA have demonstrated remarkable success in hematologic malignancies, marking a shift toward living drugs designed at the individual level (Hegde et al., 2016). The emergence of **genomic and proteomic profiling** further accelerated the personalization of immunotherapy. With NGS, researchers can now identify patient-specific mutations and immune signatures that predict therapeutic response. Computational tools integrating *transcriptomics, epigenomics*, and *single-cell RNA sequencing* allow precise mapping of immune cell states within tumors, offering new insights into resistance mechanisms. In recent years, **personalized cancer vaccines, neoantigen-based therapies**, and **microbiome-modulating strategies** have broadened the scope of immunotherapy beyond cancer to encompass autoimmune diseases, allergies, and chronic infections (Sahin & Türeci, 2018). The field continues to evolve with the incorporation of **CRISPR-Cas9 gene editing,synthetic biology**, and **AI-driven drug discovery**, all aimed at designing immune interventions that are as unique as the patients they serve. Thus, the historical trajectory of immunotherapy reflects a progressive shift—from empirical immune stimulation to rational, data-driven personalization—culminating in the modern vision of **personalized cellular immunotherapy**, where each therapeutic construct is tailored to an individual's molecular and immunological profile.

1.1.2 Rationale for personalization

The rationale for personalizing immunotherapy arises from the inherent **heterogeneity of human diseases and immune responses**. Even within a single diagnostic category, such as lung or breast cancer, patients display profound differences in tumor genetics, antigen presentation, immune cell infiltration, and response to treatment. Conventional immunotherapies that apply uniform regimens to all patients often fail to achieve consistent efficacy because they overlook this biological diversity (Blank et al., 2016). Personalization seeks to optimize therapeutic outcomes by matching the right immune intervention to the right patient, at the right time. This involves several interrelated rationales:

1. **Tumor heterogeneity and neoantigen diversity:** Every tumor evolves through unique mutational processes that give rise to distinct **neoantigens**—peptides presented on MHC molecules that can be recognized by T cells. Because no two tumors share an identical mutational landscape, designing therapies that target patient-specific neoantigens can significantly enhance specificity and reduce off-target toxicity. Neoantigen-based vaccines and TCR-engineered cells exemplify this principle, enabling highly selective targeting of malignant cells.
2. **Variability in immune microenvironment:** The **tumor microenvironment (TME)** profoundly influences therapeutic efficacy. Factors such as cytokine profiles, immune checkpoint expression, stromal composition, and metabolic constraints differ between patients. Personalized immunotherapy employs **immunoprofiling**—including flow cytometry, single-cell sequencing, and spatial transcriptomics—to characterize the immune ecosystem and guide treatment selection (Binnewies et al., 2018).
3. **Host genetics and immunogenomics:** Host genetic variation, particularly in **HLA haplotypes**, determines antigen presentation capacity and susceptibility to autoimmune side effects. Incorporating immunogenomic data allows clinicians to predict immunotherapy responsiveness and tailor dosing regimens (Chowell et al., 2018). Pharmacogenomics further refines safety by identifying polymorphisms affecting immune signaling or cytokine metabolism.
4. **Predictive biomarkers and monitoring:** Personalized immunotherapy relies heavily on **biomarker-driven decision-making**. Predictive biomarkers such as PD-L1 expression, TMB, interferon-γ signatures, or circulating immune cell ratios are used to select optimal therapies. Dynamic monitoring using **liquid biopsies** and **immune repertoire sequencing** provides real-time feedback, enabling adaptive therapeutic adjustments (Topalian et al., 2016).
5. **Reducing toxicity and enhancing safety:** irAEs remain a major limitation of checkpoint blockade and ACT. By stratifying patients based on immune tolerance mechanisms and cytokine sensitivity, personalized regimens can minimize systemic inflammation and organ toxicity. Computational modeling of cytokine networks and predictive algorithms aid in preempting adverse reactions.

6. **Integration of AI and systems biology:** Artificial intelligence (AI) and systems immunology enable the integration of massive biological datasets—genomic, transcriptomic, proteomic, and clinical—to generate individualized immune models. Machine learning algorithms can predict response likelihood, optimize antigen selection, and even design synthetic CAR constructs for maximal potency (Lanz et al., 2022).

In essence, the rationale for personalizing immunotherapy is both scientific and clinical: It leverages the unique biological identity of each patient to maximize therapeutic benefit while reducing unnecessary toxicity. As our capacity to decode immune complexity expands, personalization is transitioning from an experimental paradigm to a clinical necessity, laying the foundation for a new era of precision immunomedicine.

1.2 Overview of the immune system

The immune system is a highly coordinated and dynamic network of cells, tissues, and molecular mediators that collectively protect the host against pathogenic insults while maintaining tolerance to self-components. Its sophisticated architecture enables rapid first-line responses to invading organisms and longer-term, highly specific adaptive responses to persistent or recurring threats. This dual functionality underscores the complexity of immune regulation and the challenges inherent in manipulating immune components for therapeutic purposes. In the context of personalized cellular immunotherapy, understanding the biological basis of immune responses is essential for designing interventions that maximize antitumor activity while minimizing toxicity. The following sections provide an in-depth overview of the innate and adaptive branches of the immune system, the principal cellular contributors to immune defense, and the mechanisms governing immune surveillance and tumor immune evasion.

1.2.1 Innate and adaptive immunity

The immune system consists broadly of two interconnected arms: **Innate immunity** and **adaptive immunity**. These divisions are defined by their kinetics, specificity, and memory capabilities. The innate immune system provides rapid but nonspecific protection, whereas the adaptive immune system mounts highly specific responses that are refined through clonal selection and memory formation.

1. **Innate immunity:** Innate immunity constitutes the body's first line of defense, facilitating rapid recognition and elimination of invading pathogens or abnormal cells. Its components include physical barriers such as the skin and mucosa, chemical defenses such as antimicrobial peptides, and cellular defenders, including macrophages, dendritic cells (DCs), neutrophils, natural killer (NK) cells, and innate lymphoid cells. Pattern-recognition receptors (PRRs)—such as toll-like receptors (TLRs), RIG-I-like receptors, and NOD-like receptors—enable innate cells to detect conserved molecular motifs known as pathogen-associated molecular patterns (PAMPs) and damage-associated molecular patterns (DAMPs) (Takeuchi & Akira, 2010). Activation of PRR pathways triggers intracellular signaling cascades that culminate in the secretion of cytokines, chemokines, and interferons, orchestrating an immediate inflammatory response. Although innate immunity lacks antigen specificity, recent evidence highlights "trained immunity," wherein innate cells such as monocytes and NK cells exhibit enhanced responses following prior exposure to certain stimuli

(Netea et al., 2020). This epigenetically encoded memory-like behavior has implications for vaccine design and cancer immunotherapy, where innate reprogramming can enhance the efficacy of immune-based interventions.

2. **Adaptive immunity:** Adaptive immunity is characterized by antigen specificity, immunological memory, and the capacity to generate amplified responses upon reexposure to the same antigen. It is mediated primarily by lymphocytes—T cells and B cells—which undergo somatic gene rearrangements enabling recognition of diverse antigens. Activation of adaptive immunity requires antigen presentation through MHC molecules for T cells and direct binding of soluble antigens to B-cell receptors (BCRs). Cell-mediated immunity, orchestrated by T cells, includes cytotoxic T lymphocytes (CD8+ T cells) that eliminate infected or malignant cells and helper T cells (CD4+ T cells) that regulate immune responses through cytokine signaling. Humoral immunity, mediated by B cells and plasma cells, involves the production of antigen-specific antibodies that neutralize pathogens, facilitate opsonization, and activate complement pathways (Janeway et al., 2022). Central to adaptive immunity is the development of immunological memory. Memory T and B cells persist long after antigen exposure and deliver rapid, robust responses upon repeat encounters. In cancer immunology, exploiting T-cell memory through vaccines or engineered T cells enhances the durability of antitumor immunity.

1.2.1.1 Crosstalk between innate and adaptive immunity

The innate and adaptive branches function synergistically. DCs serve as a bridge by capturing antigens, processing them, and presenting them to naïve T cells within lymph nodes. Cytokines secreted by innate cells influence the differentiation of T-helper subsets—such as Th1, Th2, Th17, and regulatory T cells (Tregs)—thereby shaping the nature of the adaptive response. This integration is foundational to immunotherapy, as therapies such as DC vaccines, ACT, and CAR-T cell therapy exploit these interdependencies to mount targeted antitumor responses.

1.2.2 Cellular players in immune defense

The human immune system functions as an integrated and dynamic network comprising numerous specialized cells that collectively provide protection against pathogens, maintain tissue homeostasis, and detect and eliminate malignant cells. A central component of this system is the diverse repertoire of immune cells, each equipped with distinct molecular machinery, receptors, and functional programs necessary to recognize and respond to antigens. Understanding the biological roles, mechanisms of action, and intercellular communication among these cellular populations is fundamental for the development of personalized cellular immunotherapies. Modern immunotherapeutic strategies, particularly those involving adoptive cell transfer (ACT), engineered immune cells, and DC vaccines, are grounded in the precise manipulation of these immune cells to enhance antitumor efficacy while minimizing systemic toxicity (June et al., 2018; Mellman et al., 2011). This section provides a detailed overview of the major immune cell types participating in immune defense, focusing on their origins, effector functions, and relevance to immunotherapy. These include innate immune cells such as DCs, macrophages, NK cells, and neutrophils, as well as adaptive immune components, including T lymphocytes, B lymphocytes, and regulatory subsets that orchestrate immune homeostasis. In addition, we discuss the interplay between these cells during tumor surveillance and how tumors coopt or evade immune recognition.

1.2.2.1 *Innate immune cellular players*

1. **Dendritic cells:** DCs are the most potent antigen-presenting cells (APCs) of the immune system, bridging innate and adaptive immunity. They arise from both myeloid and lymphoid precursors and are specialized for capturing, processing, and presenting antigens to naïve T cells (Steinman & Banchereau, 2007). DCs express high levels of MHC class I and II molecules as well as costimulatory molecules such as CD80, CD86, and CD40, which are essential for T-cell priming. DCs are often classified into several subsets:
   a. **Conventional DCs (cDC1 and cDC2)**: cDC1 cells are critical in cross-presentation of antigens to $CD8^+$ T cells, a mechanism central to anti-tumor responses (Sánchez-Paulete et al., 2017). cDC2 cells mainly activate $CD4^+$ T-cell responses.
   b. **Plasmacytoid DCs (pDCs)**: Known for robust type I interferon production, particularly during viral infections.
   c. **Monocyte-derived DCs (moDCs)**: Generated during inflammation and frequently used in DC-based vaccines.

In cancer immunotherapy, DCs serve as key elements for priming tumor-specific T cells. Personalized DC vaccines leverage patient-specific tumor neoantigens and have demonstrated promising outcomes in melanoma and glioblastoma (Palucka & Banchereau, 2012). Their importance lies not only in antigen presentation but also in shaping T-cell polarization, an essential consideration in individualized therapy design.

2. **Macrophages:** Macrophages are multifunctional phagocytes derived from circulating monocytes or embryonic precursors. They exist in virtually every tissue and demonstrate remarkable plasticity, adapting to microenvironmental cues. These cells are central to pathogen clearance, tissue remodeling, and orchestrating immune responses through cytokine and chemokine secretion (Gordon & Martinez, 2010). Macrophages are broadly categorized into:
   a. **M1 (classically activated)**: Proinflammatory, microbicidal, and antitumoral.
   b. **M2 (alternatively activated)**: Antiinflammatory, tissue-repairing, and often protumoral within cancer microenvironments.

Tumor-associated macrophages (TAMs), typically resembling the M2 phenotype, can suppress T-cell activity, promote angiogenesis, and facilitate tumor metastasis (Noy & Pollard, 2014). Personalized immunotherapy approaches increasingly focus on reprogramming TAMs into M1-like macrophages or depleting protumoral macrophage subsets. Novel therapies under investigation include macrophage-based CAR (CAR-M) therapies, which engineer macrophages to directly phagocytose cancer cells and modulate the TME (Klichinsky et al., 2020).

3. **Natural killer cells:** NK cells are cytotoxic lymphocytes of the innate immune system capable of eliminating virus-infected and malignant cells without prior sensitization. Their activity is regulated by a balance of activating receptors (e.g., NKG2D, NKp30, and NKp46) and inhibitory receptors such as killer-cell immunoglobulin-like receptors (KIRs) that recognize MHC class I molecules (Vivier et al., 2011). NK cells utilize perforin- and granzyme-mediated cytotoxicity, death receptor ligand interactions (e.g., FasL and TRAIL), and cytokine production (e.g., IFN-γ) to orchestrate antitumor immune responses. Their ability to target MHC-I–low tumor cells, which often evade T-cell surveillance, makes them valuable in cancer immunotherapy. Personalized strategies include:

a. Ex vivo–expanded autologous NK cells
b. Allogeneic NK-cell infusions
c. Cytokine-induced memory-like NK cells
d. CAR-NK cell therapies, which may offer safer profiles than CAR-T cells (Liu et al., 2020)

Given their reduced risk of graft-versus-host disease (GVHD), NK cells are ideal candidates for "off-the-shelf" therapies, an important dimension of customized treatment design.

4. **Neutrophils:** Neutrophils are the most abundant circulating leukocytes and the first responders to infection or tissue injury. Beyond their classical functions of phagocytosis and pathogen killing via reactive oxygen species (ROS) and proteases, neutrophils play complex roles in tumor progression. Tumor-associated neutrophils (TANs) can exhibit both antitumoral (N1) and protumoral (N2) phenotypes depending on cytokine signals (Uribe-Querol & Rosales, 2015). While traditionally overlooked in immunotherapy, emerging evidence implicates neutrophils in modulating responses to checkpoint inhibitors and influencing T-cell infiltration into tumors. Personalized immunotherapy models increasingly consider neutrophil dynamics and their predictive value in patient stratification.

1.2.2.2 *Adaptive immune cellular players*

1. **T lymphocytes:** T cells are central mediators of adaptive immunity, responsible for recognizing antigenic peptides via T-cell receptors (TCRs). Their roles in cancer immunity and the success of T-cell–based immunotherapies highlight their importance. T cells originate in the bone marrow, mature in the thymus, and differentiate into various functional subsets upon antigen encounter.

$CD8^+$ cytotoxic T lymphocytes (CTLs): $CD8^+$ T cells recognize antigens presented by MHC class I molecules and directly kill infected or malignant cells. CTLs employ cytotoxic granules containing perforin and granzymes, and express FasL to induce apoptosis. Their pivotal role in tumor immunity has made them key targets in ACT therapies, cancer vaccines, and checkpoint inhibitor treatments (Smyth et al., 2006).

$CD4^+$ helper T cells (Th cells): $CD4^+$ T cells recognize peptides presented by MHC class II molecules and coordinate immune responses through cytokine secretion. Subsets include:

- **Th1 cells**, promoting cellular immunity via IFN-γ
- **Th2 cells**, supporting humoral responses
- **Th17 cells**, involved in inflammation
- **T follicular helper cells (Tfh)**, facilitating B-cell maturation

Th1 responses are generally associated with antitumor immunity, whereas Th17 cells may exhibit context-dependent protumor or antitumor effects.

Regulatory T cells (Tregs): Tregs ($CD4^+CD25^+FOXP3^+$) maintain immune tolerance and prevent overactivation. However, in cancer, Tregs frequently accumulate within tumors and suppress antitumor T cells, aiding immune evasion (Sakaguchi et al., 2008). Personalized immunotherapies often aim to reduce Treg infiltration or inhibit their suppressive functions for improved clinical outcomes.

Memory T cells: Memory T cells include central memory (Tcm), effector memory (Tem), and tissue-resident memory (Trm) subsets. These cells provide long-term immunity and are foundational for durable cancer remission. Tcm cells, with high proliferative potential, are often preferred in ACT-based immunotherapies (Gattinoni et al., 2011).

2. **B lymphocytes and plasma cells:** B cells participate in immune defense through antibody production, antigen presentation, and regulation of immune responses. Upon activation, B cells undergo affinity maturation in germinal centers and differentiate into plasma cells that secrete high-affinity antibodies (Victora & Nussenzweig, 2022).

In tumors, B cells can exert both antitumor and protumor effects. Tumor-infiltrating B cells (TIBs) in tertiary lymphoid structures are correlated with improved response to immunotherapy (Helmink et al., 2020). Moreover, mAb therapies—such as rituximab or trastuzumab—illustrate the therapeutic potential of B-cell biology. Personalized immunotherapy increasingly incorporates tumor-specific antibody responses and B-cell repertoire profiling for patient stratification.

1.2.2.3 *Interplay among immune cells in antitumor immunity*

Effective immune defense relies not on isolated cell functions but on coordinated interactions among multiple immune populations. For example:

- DCs prime $CD8^+$ T cells through cross-presentation.
- NK cells secrete cytokines that amplify T-cell responses.
- Macrophages modulate inflammation and shape T-cell infiltration.
- B-cell–derived antibodies contribute to antigen presentation and complement activation.

TMEs disrupt these coordinated responses by releasing immunosuppressive cytokines (e.g., IL-10 and TGF-β), recruiting Tregs, polarizing macrophages to M2 phenotypes, and inducing exhaustion in T cells (Joyce & Fearon, 2015). Understanding these interactions is essential for personalized cellular therapies, which aim to restore or enhance antitumor immunity by selectively modifying or replacing dysfunctional immune cell subsets.

1.2.2.4 *Relevance to personalized cellular immunotherapy*

Each immune cell type presents unique therapeutic opportunities:

- **T cells** enable personalized ACT, CAR-T, and TCR-engineered therapies.
- **NK cells** are attractive for allogeneic or universal cellular products.
- **DCs** support patient-specific cancer vaccines using neoantigen profiling.
- **Macrophages** can be engineered to target solid tumors resistant to T-cell–based therapies.
- **B-cell responses** provide valuable biomarkers for predicting immunotherapy efficacy.

Personalized approaches integrate genomic profiling, phenotypic characterization of immune cell subsets, and computational modeling to design therapies tailored to the patient's immunological landscape.

By comprehensively understanding cellular players in immune defense, personalized immunotherapies can be strategically designed to exploit strengths, overcome limitations, and synergistically target tumors using multiple immune cell populations.

1.2.3 Immune surveillance and tumor evasion

The concept of immune surveillance is foundational to understanding how the immune system continuously monitors and eliminates potentially malignant cells. First proposed by Burnet and Thomas in the mid-20th century, immune surveillance posits that transformed cells arise frequently within tissues but are usually detected and removed by innate and adaptive immune mechanisms before they develop into clinically apparent tumors (Dunn et al., 2002). With the progression of immunology and cancer biology, this idea has matured into the broader framework of **cancer immunoediting**, which encompasses three phases: elimination, equilibrium, and escape (Schreiber et al., 2011). These phases collectively describe the dynamic interactions between tumor cells and the immune system over time, shaping tumor immunogenicity and influencing therapeutic responsiveness. Understanding immune surveillance is essential for the development of personalized cellular immunotherapies, as the success of such treatments often depends on the ability to restore or augment natural immune functions that have been subverted by tumors. Conversely, tumor evasion strategies pose major barriers to effective immunotherapy, making it critical to dissect the molecular and cellular mechanisms that enable cancer cells to avoid detection, resist immune-mediated killing, and exploit immunosuppressive pathways within the TME. This section provides an in-depth exploration of both the protective mechanisms of immune surveillance and the sophisticated strategies tumors employ to evade immune destruction.

1.2.3.1 *Immune surveillance: Detection and elimination of malignant cells*

Immune surveillance is executed through cooperative interactions between innate and adaptive immune cells. These cells are equipped to recognize stress signals, abnormal antigens, and cellular transformation markers that indicate malignant potential.

1.2.3.2 *Role of innate immunity in tumor surveillance*

Innate immune components provide the first layer of tumor detection:

Natural killer cells: NK cells play a pivotal role in identifying malignant cells that have downregulated MHC class I molecules, a common strategy tumors use to evade cytotoxic T lymphocytes (CTLs). NK cells detect this "missing-self" phenotype and eliminate such cells through perforin–granzyme pathways and death receptor engagement (Vivier et al., 2011). They also produce cytokines such as IFN-γ, which enhance antigen presentation and promote Th1 polarization.

Macrophages and dendritic cells: Macrophages can recognize stress-induced ligands and apoptotic bodies derived from tumor cells. DCs capture TAAs and present them to T cells, initiating adaptive anti-tumor immunity. During early tumor formation, activated macrophages and DCs produce inflammatory cytokines (e.g., IL-12 and TNF-α) that potentiate T-cell responses (Mellman et al., 2011).

$\gamma\delta$ T cells and innate lymphoid cells (ILCs): $\gamma\delta$ T cells survey tissues for danger signals and phosphoantigens expressed by stressed or transformed cells. They provide rapid cytotoxic responses independent of MHC restriction (Silva-Santos et al., 2019). Similarly, certain ILC subsets modulate antitumor immunity by regulating inflammation and shaping adaptive responses.

Role of adaptive immunity in tumor surveillance: Adaptive immune surveillance provides antigen-specific recognition and long-term memory necessary for sustained tumor control.

Cytotoxic CD8$^+$ T cells in tumor clearance: CD8$^+$ T cells identify tumor-specific or tumor-associated peptides presented on MHC class I molecules. Upon activation, these CTLs proliferate and differentiate into effector cells capable of inducing apoptosis in tumor cells. CTLs are considered the most important effector cells in adaptive tumor immunity (Smyth et al., 2006).

$CD4^+$ T cells and helper functions: Th1 $CD4^+$ cells provide essential cytokines (e.g., IL-2 and IFN-γ) that enhance CTL cytotoxicity and macrophage activation. Tfh cells promote antibody responses, whereas memory $CD4^+$ T cells support long-term surveillance.

B cells and antitumor antibodies: B cells contribute to immune surveillance through antibody production, antigen presentation, and complement activation. TIBs within tertiary lymphoid structures are associated with improved survival in several cancers (Helmink et al., 2020).

1.2.3.3 *Cancer immunoediting: Elimination, equilibrium, and escape*

Schreiber and colleagues refined the concept of immune surveillance into a three-phase model known as **cancer immunoediting**:

1. **Elimination phase:** The immune system detects and destroys early transformed cells using combined innate and adaptive mechanisms. This phase corresponds to classic immune surveillance.
2. **Equilibrium phase:** Some tumor cells survive elimination but remain growth-restricted because of continuous immune pressure. During this phase, tumor variants with reduced immunogenicity may emerge through mutations, antigen loss, or altered signaling (Dunn et al., 2002).
3. **Escape phase:** Tumor cells accumulate adaptations that allow them to proliferate despite immune control, leading to clinically detectable cancer. Immune escape represents a failure of surveillance mechanisms and is central to tumor progression.

1.2.3.4 *Mechanisms of tumor immune evasion*

Tumors employ a diverse array of strategies to escape immune recognition and resist effector mechanisms. These strategies are often coordinated by the TME, which becomes immunosuppressive and hostile to antitumor responses.

1.2.3.4.1 Loss or alteration of tumor antigens

One of the most common mechanisms of immune evasion is the reduction or loss of antigen expression:

- **Antigen loss variants** arise through genetic instability, enabling tumors to avoid recognition by CTLs targeting specific neoantigens (Rosenthal et al., 2019).
- **Downregulation of antigen-processing machinery**, such as TAP1/2 or β2-microglobulin, impairs peptide loading onto MHC class I molecules.

These changes reduce the visibility of tumor cells to T cells and compromise the effectiveness of antigen-specific therapies.

1.2.3.4.2 Impaired MHC class I expression

Tumors frequently downregulate MHC class I molecules to avoid $CD8^+$ T-cell recognition:

- Mutations in *B2M, TAP*, or *HLA* genes lead to partial or complete loss of MHC I expression.
- Some tumors selectively lose HLA alleles most effective at presenting immunodominant antigens.

Although this may allow tumors to escape CTLs, it can increase susceptibility to NK cells. However, tumors often counteract this by upregulating inhibitory ligands for NK receptors (del Campo et al., 2014).

1.2.3.4.3 Expression of immune checkpoint molecules

Tumors exploit immune checkpoints to inhibit T-cell activation. Key molecules include:

1. **PD-L1 (Programmed death ligand-1):** Many tumors upregulate PD-L1, which binds PD-1 on T cells, leading to exhaustion, reduced cytokine production, and impaired cytotoxicity (Ribas & Wolchok, 2018).
2. **CTLA-4 and other checkpoints:** CTLA-4 competes with CD28 for binding to CD80/CD86, limiting T-cell priming. Tumors may also express or induce inhibitory molecules such as TIM-3, LAG-3, TIGIT, and VISTA.

Checkpoint blockade therapies target these pathways, and understanding individual tumor checkpoint profiles is essential for personalized approaches.

1.2.3.4.4 Immunosuppressive tumor microenvironment

The TME includes stromal cells, endothelial cells, fibroblasts, and infiltrating immune cells. Many of these elements support immune evasion:

Regulatory T cells (Tregs): Tregs suppress CTLs through IL-10, TGF-β, and CTLA-4-mediated mechanisms. Tumors actively recruit Tregs via chemokines such as CCL22 (Sakaguchi et al., 2008).
Myeloid-derived suppressor cells (MDSCs): MDSCs inhibit T-cell proliferation using arginase-1, nitric oxide, and ROS. Their expansion is driven by tumor-derived factors (Gabrilovich, 2017).
Tumor-associated macrophages (TAMs): M2-like TAMs support angiogenesis, suppress T-cell responses, and promote metastasis (Noy & Pollard, 2014). The accumulation of these suppressive cells diminishes immunotherapy efficacy and represents a major barrier to successful treatment.

1.2.3.4.5 Secretion of immunosuppressive cytokines and metabolites

Tumors employ highly sophisticated and multilayered immunosuppressive mechanisms to subvert host immunity, and one of their most influential strategies involves the secretion of inhibitory cytokines and metabolites that reprogram the TME. Through these biochemical mediators, malignant cells create a niche that actively prevents the activation, proliferation, and cytolytic functions of multiple immune subsets. Among the most potent molecules involved in this suppressive network is transforming growth factor-beta (TGF-β), a pleiotropic cytokine that exerts wide-ranging effects on immune regulation. Within the TME, TGF-β drives $CD4^+$ T cells toward a Treg phenotype and expands the population of immunosuppressive Treg cells, which subsequently inhibit DCs, cytotoxic T lymphocytes (CTLs), and NK cells (Batlle & Massagué, 2019). Simultaneously, TGF-β acts directly on CTLs and NK cells by limiting their expression of perforin, granzymes, and activating receptors, thereby diminishing their capacity to kill tumor cells. This contributes to the establishment of an immune milieu in which effector responses are selectively silenced, enabling tumor progression.

Interleukin-10 (IL-10) is another influential immunosuppressive cytokine frequently overproduced by tumor and stromal cells. Although IL-10 plays a critical physiological role in preventing chronic inflammation, tumors coopt this pathway to inhibit proinflammatory signaling and suppress antigen presentation. Elevated IL-10 levels downregulate MHC class II expression on DCs and macrophages, impairing their ability to activate naïve T cells and initiate Th1-dominated antitumor immune responses (O'Garra & Vieira, 2007). Through this suppression of APC function, IL-10 weakens the priming of effector T cells and supports immune evasion, particularly in tumors that depend on stealth to avoid recognition. Beyond cytokines, tumors also suppress immunity through metabolic reprogramming of the microenvironment. Indoleamine 2,3-dioxygenase (IDO) is one of the most biologically significant immunosuppressive enzymes expressed by many solid tumors. IDO catabolizes tryptophan into kynurenine and related metabolites, resulting in local tryptophan depletion and accumulation of immunosuppressive catabolites. T cells are highly sensitive to tryptophan deprivation, and IDO-induced nutrient depletion drives T-cell anergy, impaired proliferation, and apoptosis (Munn & Mellor, 2016). This metabolic form of immune regulation prevents the expansion of antitumor effector T cells and fosters an environment conducive to tumor persistence. Moreover, kynurenine itself acts directly on T cells and NK cells to suppress function, creating an additional layer of immunometabolic control. Prostaglandin E2 (PGE2), derived from cyclooxygenase (COX) pathways, also contributes significantly to tumor-mediated immune suppression. PGE2 enhances tumor growth by promoting angiogenesis, inhibiting DC differentiation, and suppressing NK-cell recruitment and cytotoxic activity (Wang & DuBois, 2010). Through these actions, PGE2 alters both innate and adaptive immunity and reshapes the tumor niche into a permissive environment where immune activation is dampened and tumor progression is facilitated. The combined immunosuppressive influence of TGF-β, IL-10, IDO, and PGE2 orchestrates a biochemical environment that blocks multiple axes of antitumor immunity, enabling tumors to evade recognition and destruction. In addition to inhibitory cytokines and metabolites, tumors significantly impair host immunity by inducing T-cell exhaustion, a dysfunctional state arising from persistent antigen stimulation. Unlike transient T-cell activation during acute infections, chronic exposure to tumor antigens leads to progressive loss of effector functions. Exhausted T cells exhibit markedly reduced cytokine production, decreased cytolytic capacity, and poor proliferative ability. Over time, these cells lose the ability to control tumor growth despite being present in the TME (Wherry & Kurachi, 2015).

One of the hallmark features of exhausted T cells is the sustained expression of inhibitory receptors, most notably programmed cell death protein-1 (PD-1), lymphocyte activation gene-3 (LAG-3), and T-cell immunoglobulin and mucin-domain containing-3 (TIM-3). Permanent engagement of these checkpoints by tumor-expressed ligands transmits suppressive signals that weaken TCR signaling, reduce IL-2 synthesis, and limit T-cell survival. For example, PD-1 signaling inhibits phosphatidylinositol-3-kinase (PI3K) pathway activity, leading to decreased glucose uptake and impaired mitochondrial metabolism, both of which are essential for T-cell effector functions (Pauken & Wherry, 2015). As a result, exhausted T cells display a metabolic profile characterized by reduced glycolytic flux and compromised oxidative phosphorylation—constraints directly linked to their functional impairment. Furthermore, the TME exacerbates T-cell exhaustion by exposing lymphocytes to hypoxia, nutrient depletion, and high levels of lactic acid. Tumor cells outcompete T cells for glucose, an essential fuel source for effector T-cell activation. In the absence of adequate metabolic support, T cells fail to sustain cytotoxic functions and gradually adopt an exhausted phenotype. Hypoxia stabilizes hypoxia-inducible factor-1α (HIF-1α), which alters T-cell transcriptional programs toward tolerance rather than activation. Over time, exhausted T cells undergo a form of epigenetic remodeling that becomes increasingly irreversible, making therapeutic reversal challenging even with immune checkpoint blockade therapy (McLane et al., 2019). The persistence of inhibitory signals and adverse metabolic conditions locks exhausted T cells into a nonresponsive state that severely undermines antitumor immunity. This is particularly evident in solid tumors, where the combination of chronic antigen exposure, immunosuppressive cytokines, hypoxia, and metabolic competition creates conditions that force effector T cells into exhaustion. As these dysfunctional T cells accumulate, the tumor gains a significant survival advantage, often becoming resistant to conventional immune-based therapies. Together, the secretion of immunosuppressive cytokines and metabolites and the induction of profound T-cell exhaustion represent two of the most formidable barriers to effective antitumor immunity. These mechanisms operate synergistically, with soluble mediators shaping the TME to prevent immune activation and exhausted T cells further diminishing adaptive immune responses. Understanding the molecular and cellular basis of these suppressive pathways is essential for developing therapeutic interventions that can overcome immune resistance, reinvigorate effector T cells, and restore the capacity of the immune system to eliminate malignant cells.

1.2.3.4.6 Tumor-induced metabolic reprogramming

Tumors profoundly reshape the metabolic architecture of the TME, creating a highly competitive landscape that directly undermines immune function. Because tumor cells exhibit elevated rates of glycolysis—even under normoxic conditions, a hallmark known as the Warburg effect—they consume disproportionate amounts of glucose, thereby depriving CTLs of the metabolic substrates required for their activation, proliferation, and effector functions. T cells depend heavily on aerobic glycolysis during activation, and glucose scarcity severely compromises their capacity to generate ATP and critical metabolic intermediates necessary for cytokine synthesis and cytotoxic granule formation. As a result, nutrient depletion represents a potent form of immunosuppression that stems directly from tumor metabolic dominance. In addition to glucose competition, the accumulation of lactate within the TME further disrupts immune competence. Lactate, the by-product of enhanced glycolysis, acidifies the local microenvironment and directly inhibits the differentiation, maturation, and antigen-presenting ability of DCs, thereby impairing the priming of adaptive immune responses. High lactate levels similarly suppress CTL and NK cell activity, limiting their capacity to produce interferon-gamma and execute tumor cell lysis. The combined effects of nutrient restriction and metabolic by-products thus establish a biochemical milieu profoundly detrimental to effective antitumor immunity. Hypoxia constitutes another central feature of tumor-induced metabolic reprogramming. As proliferating malignant cells outpace their blood supply, oxygen levels decline, stabilizing hypoxia-inducible factor 1-alpha (HIF-1α). HIF-1α activation orchestrates a broad transcriptional program that promotes angiogenesis, glycolysis, and immunosuppression. In immune cells, hypoxia and HIF-1α signaling impair effector T-cell function while promoting the differentiation of Tregs and myeloid-derived suppressor cells (MDSCs), further weakening immune responses. Hypoxia also reduces the susceptibility of tumor cells to cytotoxic killing by altering antigen-processing and presentation pathways. Collectively, metabolic constraints—including glucose competition, lactate accumulation, and hypoxia—act in concert to suppress immune activation and facilitate tumor survival (Semenza et al., 2014).

1.2.3.4.7 Physical barriers and stromal remodeling

In addition to biochemical and metabolic suppression, tumors often create formidable physical and structural barriers that hinder immune cell infiltration. The extracellular matrix (ECM), typically enriched with dense collagen fibers, becomes extensively remodeled as tumors progress. Excessive collagen deposition increases tissue stiffness and generates a physical barrier that restricts the movement of T cells into tumor nests. Even when T cells are present in the peritumoral region, they frequently fail to penetrate deeply into the tumor core because of this rigid and disorganized matrix structure. Stromal components, particularly cancer-associated fibroblasts (CAFs), further contribute to immune exclusion through both structural and signaling mechanisms. CAFs secrete a range of chemokines and cytokines that misdirect effector T cells away from tumor cells, often guiding them toward stromal-rich regions rather than malignant cell clusters. By altering chemokine gradients, CAFs create a form of immunologic decoy that prevents effective immune surveillance. Moreover, CAFs deposit additional ECM proteins and promote crosslinking, further increasing the physical impedance to immune cell migration. Abnormal tumor vasculature compounds these challenges. Tumor blood vessels are often tortuous, leaky, and poorly perfused, resulting in irregular delivery of oxygen, nutrients, and immune cells. This dysfunctional vasculature limits the trafficking of lymphocytes into the TME and prevents efficient recruitment of APCs. Because CAR-T cells and other adoptively transferred immune cells depend on efficient vascular access to infiltrate tumors, physical barriers and vascular abnormalities represent major obstacles to the success of cellular immunotherapies in solid tumors. In fact, physical exclusion is considered one of the primary reasons why many solid tumors demonstrate suboptimal responses to CAR-T-cell therapy, despite the remarkable efficacy observed in hematologic malignancies (Joyce & Fearon, 2015).

1.2.3.5 Clinical implications for personalized cellular immunotherapy

A robust understanding of tumor immune evasion mechanisms is essential for the development of personalized cellular immunotherapies that address patient-specific barriers. Because tumors differ widely in their metabolic profiles, stromal composition, immunosuppressive mediators, and antigenic landscapes, individualized evaluation of each patient's TME is necessary to optimize therapeutic design. Comprehensive genomic, transcriptomic, proteomic, and immunophenotypic profiling enables clinicians to assess the functional status of antigen presentation machinery, checkpoint molecule expression, inflammatory signals, suppressive cell populations, and metabolic constraints. These datasets collectively provide critical predictive indicators of patient response to various immunotherapeutic modalities and facilitate rational selection of the most effective approaches. Personalized immunotherapy therefore involves not only matching the right therapy to the right tumor type but also addressing the specific immune escape pathways active within a patient's tumor. For example, tumors that express high levels of TGF-β may benefit from CAR-T cells engineered to be resistant to TGF-β signaling, enabling them to retain cytotoxic function even in a suppressive environment. Similarly, CAR-NK cells may be particularly effective against tumors with downregulated MHC class I expression, a common strategy used by cancer cells to avoid CTL-mediated killing. In cases where Tregs or MDSCs dominate the TME, therapies aimed at depleting or reprogramming these populations can restore antitumor immunity and enhance the efficacy of ACT.

Macrophage-based interventions also align with personalized strategies. TAMs, especially the M2 phenotype, promote tumor growth and immune evasion. Personalized therapeutic approaches can include reprogramming TAMs toward an M1 proinflammatory phenotype or using engineered macrophages capable of directly killing tumor cells. In addition, metabolic modulators that target glycolytic pathways, lactate dehydrogenase activity, or hypoxia-driven signaling may be integrated with adoptive cell therapies to overcome metabolic suppression within the TME. Checkpoint inhibitors can also be combined synergistically with cellular therapies, particularly in patients whose tumors exhibit high levels of PD-1, LAG-3, or TIM-3 expression on infiltrating T cells. By blocking inhibitory receptor pathways, checkpoint inhibitors can restore effector function and enhance the persistence of transferred immune cells. Therefore personalized immunotherapy represents an evolving paradigm that integrates molecular profiling, cellular engineering, microenvironmental modulation, and combinatorial strategies to overcome the unique immune evasion mechanisms employed by each patient's tumor.

1.2.3.6 Improving cell infiltration and persistence

Gene editing tools (e.g., CRISPR) can be used to enhance T-cell homing, cytotoxicity, and resistance to exhaustion. Immune surveillance plays an indispensable role in preventing tumor initiation and shaping tumor immunogenicity. However, malignant cells evolve numerous mechanisms to evade immune destruction, from altering antigen expression to remodeling the TME and inducing immune suppression. These evasion strategies are major obstacles to effective immunotherapy but also represent actionable targets in personalized treatment design. By integrating high-resolution immune profiling with advanced cell engineering technologies, personalized cellular immunotherapy aims to restore or enhance immune surveillance, overcome tumor escape pathways, and deliver durable clinical responses tailored to each patient's unique tumor biology.

1.3 Principles of cellular immunotherapy

Cellular immunotherapy has emerged as one of the most significant developments in modern cancer treatment, representing a shift from conventional cytotoxic therapies toward approaches that harness and enhance the body's own immune defenses. The fundamental premise of cellular immunotherapy is to modify, expand, or activate immune cells ex vivo and return them to the patient with the capacity to mediate more effective and durable antitumor responses. These strategies capitalize on advances in immunobiology, genetic engineering, synthetic biology, and personalized medicine to create treatments capable of overcoming tumor-induced immune suppression and targeting malignant cells with precision. The major cellular immunotherapy modalities include ACT, CAR T-cell therapy, TCR-engineered T cells, DC-based vaccines, and therapies involving NK cells and macrophages. Each of these approaches rests on distinct immunological principles and exhibits unique therapeutic strengths and limitations, collectively forming the foundation of personalized cellular immunotherapy.

1.3.1 Adoptive cell transfer

ACT is one of the earliest and most extensively explored approaches in cellular immunotherapy. It involves isolating autologous or allogeneic immune cells with antitumor potential, expanding or modifying them ex vivo, and reinfusing them into the patient. The rationale behind ACT is to augment the quantity and functional quality of tumor-reactive lymphocytes that may be insufficient in the patient because of immune exhaustion, low precursor frequency, or tumor-mediated immune suppression. ACT has demonstrated remarkable clinical benefits, particularly in advanced melanoma, and remains a benchmark for personalized immune-based treatments. TILs represent the first and most clinically validated form of ACT. TIL therapy involves isolating lymphocytes that have naturally migrated into the TME, indicating a preexisting recognition of tumor antigens. After surgical resection of a tumor lesion, these lymphocytes are isolated, expanded ex vivo with high-dose IL-2, and reinfused into the patient following a lymphodepleting preparative regimen. The lymphodepletion serves multiple purposes, including reducing Tregs, eliminating cytokine sinks, and creating homeostatic space that supports the engraftment and proliferation of transferred cells. Clinical studies have shown that TIL therapy can induce durable complete responses in patients with metastatic melanoma, with some responses persisting for more than a decade (Rosenberg & Restifo, 2015; Rosenberg et al., 2011). These results highlight the potential of ACT to provide long-term immune-mediated control of malignancy.

Another ACT strategy involves cytokine-induced killer (CIK) cells. CIK cells are derived from peripheral blood mononuclear cells (PBMCs) that are stimulated ex vivo with interferon-γ, anti-CD3 antibodies, and IL-2. These cells acquire properties of both T cells and NK cells and exert cytotoxicity independent of MHC restrictions. Although their antigen specificity is lower compared to TILs, their ease of expansion, safety profile, and broad antitumor activity have made them attractive for clinical exploration, particularly in hematological cancers (Introna et al., 2017). The development of genetic modification techniques has further expanded the scope of ACT. Genetic engineering allows the introduction of synthetic receptors or enhanced signaling pathways into T cells, transforming them into highly potent immune effectors capable of targeting tumors with unprecedented specificity. This technological advancement paved the way for the development of CAR-T and TCR-engineered T-cell therapies, which are discussed in subsequent sections. Collectively, ACT provides a strong therapeutic foundation for personalized immunotherapy, offering diverse approaches to enhance antitumor immunity using the patient's own immune system.

1.3.2 Chimeric antigen receptor T-cell therapy

CAR T-cell therapy represents one of the most transformative advances in oncology, particularly in the treatment of hematologic malignancies. CAR-T cells are genetically engineered T lymphocytes that express synthetic receptors designed to recognize specific TAAs independently of MHC presentation. This MHC-independent recognition endows CAR-T cells with the ability to target tumor cells that evade immune detection through downregulation of antigen presentation pathways. The structure of a CAR consists of an extracellular antigen-recognition domain, usually a single-chain variable fragment derived from monoclonal antibodies, connected through a hinge and transmembrane region to intracellular signaling domains. Early-generation CARs contained only the CD3ζ signaling motif, but subsequent generations incorporated one or more costimulatory domains such as CD28 or 4-1BB to enhance T-cell activation, proliferation, and persistence (June et al., 2018; Brown et al., 2018). These design improvements have been crucial to the clinical success of CAR-T therapy. Upon encountering their target antigen, CAR-T cells undergo rapid activation, proliferation, and cytotoxic engagement of tumor cells. Their effector mechanisms include the release of perforin and granzymes, secretion of proinflammatory cytokines, and recruitment of additional immune cells. The most clinically validated CAR-T therapies are those that target CD19, a molecule widely expressed on B-cell malignancies. Clinical trials involving CD19-specific CAR-T cells have demonstrated unprecedented response rates in refractory or relapsed acute lymphoblastic leukemia (ALL), with complete remission rates reaching 70%–90% (Maude et al., 2018). These successes led to the approval of tisagenlecleucel, axicabtagene ciloleucel, and other CAR-T products for multiple indications, redefining the treatment landscape for B-cell malignancies.

Despite its success, CAR-T therapy is associated with significant challenges. Cytokine release syndrome (CRS) and immune effector cell–associated neurotoxicity syndrome (ICANS) are serious toxicities that can occur due to excessive immune activation (Neelapu et al., 2018). Moreover, antigen escape, wherein tumor cells downregulate or lose the targeted antigen, contributes to relapse. The application of CAR-T therapy in solid tumors remains limited due to heterogeneity of antigen expression, physical barriers imposed by the TME, and the presence of immunosuppressive cellular components. Nonetheless, ongoing research in armored CARs, dual-antigen targeting, gene-edited allogeneic CAR-T cells, and synthetic biology–driven control switches continues to push the boundaries of CAR-T therapy and enhance its personalized nature.

1.3.3 T-cell receptors-engineered T cells

TCR-engineered T-cell therapy represents another major branch of ACT that capitalizes on the natural TCR's ability to recognize intracellular antigens presented by MHC molecules. Unlike CARs, which are restricted to surface antigens, TCRs can target a vast repertoire of intracellular oncogenic proteins, including neoantigens and cancer-testis antigens (CTAs), significantly broadening the therapeutic scope. This makes TCR therapies particularly attractive for cancers driven by intracellular mutations such as KRAS, TP53, or HPV-derived oncoproteins. The process of generating TCR-engineered T cells involves identifying a high-affinity TCR specific to a TAA, transferring the corresponding TCR genes into patient T cells using viral or nonviral vectors, expanding the modified cells, and reinfusing them into the patient following lymphodepletion (Rapoport et al., 2015). These engineered T cells possess enhanced affinity and avidity for their target antigen, allowing them to overcome tolerance mechanisms that often suppress endogenous T-cell responses.

Clinical studies have demonstrated notable successes with TCR-based therapies. TCRs targeting NY-ESO-1, a CTA expressed in synovial sarcoma, melanoma, and multiple myeloma, have produced durable clinical responses (Rapoport et al., 2015). Another groundbreaking demonstration involved the targeting of mutant KRAS, where TCR-engineered T cells induced regression of epithelial tumors bearing KRAS G12D mutations (Tran et al., 2016). These findings underscore the potential of TCR therapies as personalized treatments for cancers with defined antigenic landscapes. However, TCR-based therapies also face distinct challenges. MHC restriction requires that the patient express the specific HLA allele necessary for antigen presentation, limiting the generalizability of each TCR product. Tumors may downregulate MHC molecules as an immune escape mechanism, reducing therapeutic efficacy. Off-target toxicity is another concern when TCRs cross-react with peptides from healthy tissues. Nevertheless, advancements in neoantigen discovery, high-throughput TCR screening, and CRISPR-mediated editing to eliminate endogenous TCR chains continue to enhance the precision and safety of TCR engineering.

1.3.4 Dendritic cell-based vaccines

DCs play a central role in initiating and regulating adaptive immunity, making them powerful tools for therapeutic cancer vaccination. As the most potent APCs, DCs are capable of processing tumor-derived antigens and activating naïve T cells to generate robust cytotoxic and helper responses. DC-based vaccines seek to exploit this biological function by loading ex vivo–generated DCs with tumor antigens and reintroducing them into patients to stimulate antigen-specific immunity. DC vaccines can be prepared using several antigen-loading strategies. These include pulsing DCs with defined tumor-associated peptides, using whole-tumor lysates or apoptotic tumor cells to present a broad antigen repertoire, and transfecting DCs with tumor-derived RNA or DNA to induce endogenous antigen presentation (Palucka & Banchereau, 2013). The maturation status of DCs is critical for vaccine efficacy, as mature DCs express high levels of costimulatory molecules and secrete cytokines essential for T-cell priming.

The first FDA-approved DC vaccine was Sipuleucel-T, which demonstrated a significant improvement in overall survival for patients with metastatic castration-resistant prostate cancer (Kantoff et al., 2010). In this therapy, peripheral blood APCs are pulsed ex vivo with a fusion protein consisting of prostatic acid phosphatase (PAP) and granulocyte–macrophage colony-stimulating factor (GM-CSF), enhancing their ability to activate PAP-specific T cells. Beyond Sipuleucel-T, DC vaccines have shown promise in melanoma, glioblastoma, renal cell carcinoma, and ovarian cancer, particularly when combined with immune checkpoint inhibitors that relieve suppression of T-cell responses. However, challenges persist in translating DC vaccines into universally effective treatments. Variability in DC quality, limited migration to lymph nodes, and the immunosuppressive TME can all hinder therapeutic outcomes. The personalization of DC vaccines through the incorporation of patient-specific neoantigens and the use of optimized maturation stimuli represents an important direction for enhancing their clinical efficacy.

1.3.5 Natural killer cell and macrophage-based therapies

In addition to T-cell–based approaches, cellular immunotherapy also leverages innate immune cells such as NK cells and macrophages, which possess distinct mechanisms for recognizing and eliminating malignant cells. NK-cell and macrophage-based therapies offer unique advantages, particularly in overcoming MHC downregulation and infiltrating solid tumors. NK cells are cytotoxic lymphocytes that can recognize stressed or transformed cells without prior sensitization. Their recognition is governed by a balance of activating and inhibitory receptors, enabling them to detect "missing-self" signals when tumor cells downregulate MHC class I molecules (Vivier et al., 2012). Therapeutically, NK cells can be derived from autologous blood, allogeneic donors, umbilical cord blood, induced pluripotent stem cells, or NK-cell lines. Allogeneic NK cells often exhibit superior antitumor activity because of mismatches in KIR ligands, which diminish inhibitory signaling and enhance cytotoxicity. Advancements in genetic engineering have extended CAR technology to NK cells, producing CAR-NK cells with potent antitumor activity and reduced toxicity compared to CAR-T cells. CAR-NK cells secrete lower levels of proinflammatory cytokines, resulting in a significantly lower incidence of CRS and neurotoxicity. Clinical studies involving CAR-NK cells targeting CD19-positive lymphoid tumors have demonstrated encouraging response rates and excellent safety profiles, marking them as promising candidates for off-the-shelf immunotherapies (Liu et al., 2020).

Macrophage-based therapies have also gained attention because of the central role of macrophages in shaping the TME. TAMs frequently adopt an immunosuppressive phenotype that promotes tumor growth, angiogenesis, and metastasis. Reprogramming these macrophages to a proinflammatory, tumoricidal phenotype represents a novel therapeutic strategy. CAR-macrophages (CAR-M) have been developed to endow macrophages with antigen-specific recognition and enhanced phagocytic activity. Studies have shown that CAR-M cells are capable of infiltrating solid tumors, phagocytosing malignant cells, and stimulating adaptive immune responses by presenting tumor antigens to T cells (Klichinsky et al., 2020). Their ability to modify the TME distinguishes them from CAR-T cells, which often struggle to penetrate dense stromal barriers. The application of macrophage therapies extends beyond CAR engineering. Approaches such as M1 polarization, macrophage reprogramming agents, and macrophage-targeted nanoparticles are being evaluated to overcome the immunosuppressive properties of TAMs. Although still in early development, macrophage-based cellular therapies hold significant potential, particularly for the treatment of solid tumors that are resistant to T-cell–based approaches.

1.4 Personalization strategies in cellular therapy

Personalization lies at the heart of modern cellular immunotherapy, fundamentally differentiating next-generation therapies from traditional one-size-fits-all approaches in oncology and immune-mediated diseases. As cancer biology becomes increasingly understood as a complex interplay of genetic mutations, transcriptomic dysregulation, proteomic alterations, and dynamic microenvironmental factors, the need for individualized therapeutic strategies has become more pressing. Personalized cellular immunotherapy integrates high-throughput omics technologies, computational modeling, biomarker stratification, and precision genetic engineering to tailor therapeutic interventions to each patient's molecular and immunological landscape. The goal is not only to enhance treatment efficacy but also to reduce off-target toxicity and overcome tumor heterogeneity and immune escape—two major challenges in cancer treatment. The rapid evolution of sequencing technologies, improved mass spectrometry, and machine learning tools has created unprecedented opportunities to map the molecular characteristics of tumors and the host immune system with high resolution. These insights inform the design of cellular therapies such as CAR-T cells, TCR-engineered lymphocytes, DC vaccines, NK-cell therapies, and combination immunotherapeutic modalities. Personalized strategies include identifying patient-specific tumor antigens, predicting and validating neoantigens, profiling immune cell phenotypes and functional states, and selecting suitable therapy types based on biomarkers. AI and computational immunology also play significant roles, enabling the complex integration of multidimensional data and accelerating the pipeline from target discovery to therapy manufacturing. Personalization is thus not a single step in cellular immunotherapy, but a multilayered, iterative approach that spans diagnostic assessment, target selection, therapy design, cell engineering, patient stratification, and treatment monitoring. By integrating omics-driven insights, biomarker science, and computational modeling, personalized cellular therapies hold the promise of delivering highly specific, potent, and durable immune responses tailored to individual patient characteristics.

1.4.1 Role of genomics, transcriptomics, and proteomics

Genomic profiling forms the foundational layer of personalization in cellular immunotherapy. Genomics enables comprehensive identification of somatic mutations, copy number variations, chromosomal rearrangements, and mutational signatures that define the malignant phenotype. Such genomic aberrations often generate tumor-specific neoantigens that can serve as targets for engineered immune cells, particularly in TCR-based therapies that recognize intracellular peptides presented by MHC molecules. Advances in whole-exome sequencing (WES) and whole-genome sequencing (WGS) allow precise detection of these mutations with high sensitivity and specificity (Mardis, 2017). These genomic insights guide both the selection of target antigens and the design of TCR constructs, making tumor genotyping an indispensable prerequisite for personalized TCR-engineered cell therapies. Beyond identifying neoantigens, genomic analyses can reveal pathways associated with immune evasion, such as defects in antigen presentation machinery, loss of heterozygosity in HLA genes, or mutations affecting interferon signaling. These insights influence both therapeutic design and patient selection, as tumors with compromised antigen presentation may respond differently to CAR-T or TCR therapies (McGranahan et al., 2017). Genomic biomarkers such as MSI and TMB have also gained significant attention as predictors of response to immunotherapies because they correlate with the likelihood of generating neoantigens capable of stimulating T-cell activation.

Transcriptomics provides a complementary layer of information by revealing gene expression patterns, immune signatures, and functional states of tumor cells and immune cell populations. RNA sequencing (RNA-seq) enables quantification of mRNA transcripts, identification of alternative splicing events, and characterization of TME components, including stromal and immune infiltrates (Stubbington et al., 2017). Single-cell RNA sequencing (scRNA-seq), in particular, has transformed the understanding of cellular heterogeneity by profiling transcriptomic states at the individual cell level. This allows researchers to identify exhausted T-cell populations, characterize tumor-specific TCRs, and understand the immunosuppressive elements within the TME. scRNA-seq can also identify clonally expanded T-cell populations that may serve as candidates for adoptive transfer therapies or for designing personalized neoantigen-targeted vaccines. Proteomics adds another crucial dimension by enabling direct measurement of protein abundance, modifications, signaling pathways, and antigen presentation. Although genomics and transcriptomics provide blueprints of cellular potential, proteomics reflects actual functional activity within the tumor and immune system. Advances in mass spectrometry-based proteomics have facilitated the identification of MHC-presented peptides, posttranslational modifications, and phosphoproteomic signaling networks (Aebersold & Mann, 2016). These insights are essential for validating neoantigens predicted from genomic data, as not all mutations result in protein-level expression or presentation on MHC molecules. Integrated multiomics approaches—combining genomics, transcriptomics, and proteomics—offer a comprehensive strategy for identifying and validating therapeutic targets. These methods enable molecular subtyping of tumors, prediction of therapy response, and design of cell products tailored to the functional and phenotypic attributes of each patient's tumor and immune system. Collectively, omics technologies enable a deeply personalized approach by informing antigen selection, cell engineering, and treatment planning.

1.4.2 Tumor antigen profiling and neoantigen prediction

Central to personalized cellular immunotherapy is the precise characterization of tumor antigens. Antigen profiling distinguishes between shared TAAs and patient-specific neoantigens. TAAs, such as WT1, NY-ESO-1, and MAGE proteins, are overexpressed in a range of cancers and serve as targets for both CAR-T and TCR-based therapies (Matthew, William, & Sandra P, 2016). However, these shared antigens pose the risk of off-tumor toxicity because they may be expressed at low levels in normal tissues. Neoantigens, in contrast, arise from somatic mutations unique to individual tumors and are highly immunogenic because they are not subject to central tolerance.

The process of neoantigen prediction relies on sequencing tumor DNA and RNA to identify nonsynonymous mutations, followed by computational prediction of peptide-MHC binding affinity. Tools such as NetMHC, MuPeXI, and pVACseq evaluate the likelihood that mutant peptides will bind to patient-specific HLA molecules and be presented to T cells (Hundal et al., 2020). These predictions enable rational design of personalized TCR-engineered lymphocytes or DC vaccines. Neoantigen prediction is rapidly improving with the incorporation of deep-learning algorithms trained on massive datasets of known peptide-HLA interactions. It is increasingly recognized that neoantigen burden correlates with immune responsiveness. Tumors with high mutational loads, such as melanoma and smoking-associated lung cancer, tend to generate more neoantigens and exhibit stronger T-cell infiltration (Snyder et al., 2014). This understanding informs patient selection for TCR therapies and neoantigen vaccination strategies. Antigen profiling is not limited to genomic and computational approaches. Immunopeptidomics, a subfield of proteomics, directly analyzes peptides presented on MHC molecules using mass spectrometry. This approach provides empirical evidence of antigen presentation and helps validate neoantigen predictions. MHC ligandomics also aids in identifying tumor-specific antigens that arise from alternative splicing, gene fusions, or posttranslational modifications—antigens that may not be detectable through genomic methods alone (Bassani-Sternberg et al., 2016).

In addition to neoantigens, profiling tumor-associated carbohydrate antigens (TACAs), CTAs, and glycopeptides expands the repertoire of potential targets. These antigens are being explored for CAR-T cell engineering and multivalent vaccine platforms. Antigen escape remains a significant concern, as tumors may downregulate targeted antigens or alter antigen-processing pathways. Personalized antigen profiling allows real-time monitoring of antigen evolution and supports the development of multitarget or bispecific cellular therapies that mitigate immune escape. Thus comprehensive tumor antigen profiling and neoantigen prediction provide the molecular foundation for designing personalized cell therapies capable of recognizing diverse and patient-specific tumor signatures with high precision.

1.4.3 Biomarker-driven patient selection

Effective personalization extends beyond target identification to include careful selection of patients who are most likely to benefit from specific cellular therapies. Biomarkers play a critical role in predicting efficacy, minimizing adverse events, and optimizing therapeutic decision-making. Biomarkers used in cellular immunotherapy include genetic, proteomic, immunological, and clinical parameters. One of the most established biomarkers is antigen expression level. For example, successful CD19-directed CAR-T cell therapy requires robust expression of CD19 on malignant B cells. Heterogeneous or low antigen expression increases the risk of relapse because of antigen-negative tumor escape (Majzner & Mackall, 2018). Similarly, TCR-based therapies require that target peptides be presented in the context of specific HLA alleles. Thus, HLA typing is an essential biomarker for TCR therapy eligibility.

Immune competence and functional status of T cells also influence therapeutic outcomes. Patients with heavily exhausted or senescent T-cell populations may demonstrate reduced CAR-T expansion and persistence. Biomarkers such as PD-1 expression, T-cell differentiation status, and cytokine secretion profiles inform the suitability of patients for ACT or CAR-T therapies (Fraietta et al., 2018). TME markers, including levels of immunosuppressive cytokines, regulatory T-cell infiltration, and stromal density, affect the infiltration and activity of cell therapies, particularly in solid tumors. Biomarkers also help predict adverse events such as CRS and neurotoxicity. Elevated tumor burden, high inflammatory baseline states, and specific cytokine signatures have been associated with increased risk of CAR-T–related toxicities. Identifying these biomarkers enables clinicians to tailor dosing strategies, employ prophylactic interventions, and select candidates for safer NK-cell–based therapies. Genomic biomarkers such as TMB, MSI, and alterations in DNA damage repair (DDR) pathways correlate with responsiveness to TCR-based therapies and neoantigen vaccines. In hematological malignancies, cytogenetic markers and molecular subtypes guide the selection of therapeutic targets and predict response durability. Overall, biomarker-driven patient selection ensures that cellular immunotherapies are matched to those patients most likely to derive benefit, thereby maximizing efficacy and safety. As biomarker science advances, integration of multi-omics biomarkers and computational prediction will further refine personalized therapy selection algorithms.

1.4.4 Artificial intelligence and computational tools for therapy design

AI and computational biology have become indispensable in advancing personalization in cellular immunotherapy. The complexity and volume of genomic, transcriptomic, proteomic, and clinical data require sophisticated computational tools capable of extracting meaningful patterns and generating actionable insights. AI supports nearly every stage of therapy design, from antigen discovery and neoantigen prediction to manufacturing optimization and clinical outcome forecasting. Machine-learning algorithms have dramatically improved the accuracy of neoantigen prediction by modeling peptide-MHC interactions and integrating features from millions of experimentally validated epitopes (Jurtz et al., 2017). Deep neural networks enable the prediction of not only binding affinity but also antigen processing, peptide stability, and T-cell recognition likelihood. These computational predictions accelerate the development of personalized vaccines and TCR-engineered therapies.

AI tools also contribute to TCR repertoire analysis by identifying clonally expanded tumor-reactive T-cell populations from bulk or single-cell sequencing data. These analyses help identify endogenous TCRs suitable for engineering or adoptive transfer. Similarly, computational tools can design optimized CAR structures by modeling scFv affinity, linker length, costimulatory domain combinations, and signaling kinetics. In manufacturing, AI-driven quality control (QC) systems monitor cell expansion, phenotype, and viability in real time, enabling early detection of aberrant growth patterns and improving batch consistency. Reinforcement learning algorithms are being explored to automate cell culture optimization and media formulation. Clinical decision-support tools integrate patient-specific biomarkers, omics data, and prior outcomes to predict therapy response and guide treatment planning. Predictive modeling enables risk stratification for adverse events such as CRS and supports personalized dosing algorithms. Computational modeling also extends to simulating immune-tumor interactions. Agent-based models and digital twins of the immune system can forecast how engineered cells will behave in the patient's TME, thereby aiding in therapy customization and experimental design. In summary, AI and computational tools enhance the precision, scalability, and speed of personalized cellular immunotherapy development. Their integration into clinical pipelines represents a key advancement toward more effective and individualized treatment strategies.

1.5 Clinical translation and manufacturing

The successful deployment of personalized cellular immunotherapies depends not only on scientific innovation but also on the establishment of robust clinical and manufacturing frameworks. Translating cell-based therapies from bench to bedside requires a multidisciplinary ecosystem combining molecular biology, bioprocess engineering, regulatory sciences, and clinical medicine. Personalized cellular immunotherapy, unlike conventional pharmaceutical modalities, involves individualized production workflows, complex ex vivo manipulations, and stringent QC checkpoints. Each therapy represents a unique biologic product derived from a specific patient or donor, thereby creating significant logistical and operational challenges. As such, manufacturing and clinical translation are integral determinants of therapeutic efficacy, safety, accessibility, and scalability (Liu et al., 2020; Pan et al., 2021). The translation pipeline for personalized cellular immunotherapies encompasses several sequential stages: patient assessment and cell procurement, upstream processing and cell activation, genetic modification or antigen loading, expansion and differentiation, product formulation, QC release testing, and clinical administration. These steps must be coordinated with regulatory frameworks to ensure compliance with good manufacturing practice (GMP) guidelines, biosafety requirements, and post-infusion monitoring protocols (Hartmann et al., 2022). In addition, as the landscape of advanced therapy medicinal products (ATMPs) expands, standardized manufacturing platforms, automation technologies, and closed-system bioreactors are being incorporated to overcome the limitations associated with manual processing and variability. This section examines the contemporary frameworks guiding the clinical translation and manufacturing of personalized cellular therapies. It also elaborates on the regulatory and quality aspects that are fundamental to ensuring patient safety and therapeutic consistency across diverse clinical settings.

1.5.1 Workflow of patient-specific cell manufacturing

The manufacturing workflow for patient-specific cellular immunotherapies adheres to a meticulous, stepwise sequence, beginning at the point of patient evaluation and extending through final infusion. Each stage is intricately interconnected, and deviations in early phases often alter final product characteristics. Consequently, standardized procedures and stringent in-process controls are essential to maintain the potency, viability, and purity of the therapeutic cells (Heslop et al., 2021). Manufacturing typically commences with patient assessment and eligibility testing. Physicians evaluate the patient's clinical status, tumor burden, and organ function to confirm suitability for cellular therapy. In many cases, particularly with CAR-T and TCR-engineered T cells, patients undergo lymphodepletion chemotherapy prior to infusion to facilitate engraftment and persistence of the infused cells (Fraietta et al., 2018). This preparatory step is critical to weakening suppressive host immune components and creating a cytokine milieu favorable for the therapeutic cells. Cell procurement is conducted primarily through leukapheresis, a procedure that isolates PBMCs. The quality of the starting material is a major determinant of the final product. For example, heavily pretreated or immunocompromised patients may yield lower T-cell counts or functionally exhausted phenotypes, necessitating additional processing steps or donor-derived alternatives (Rao et al., 2022). Once collected, cells are transported to GMP-certified facilities under controlled temperature and sterility conditions. Upstream processing includes the enrichment of specific cell subsets, such as CD3+ T cells, NK cells, DCs, or macrophages. Magnetic-activated cell sorting (MACS) or flow cytometry-based sorting is commonly used, depending on the therapy type. Following enrichment, cells undergo activation procedures to enhance their proliferation capacity and facilitate genetic modification. T cells, for instance, are activated using anti-CD3/CD28 beads or soluble antibodies, whereas NK cells require cytokines such as IL-2 or IL-15 for optimal priming (Li et al., 2023).

Genetic engineering is a central feature of many personalized therapies. Viral vectors—typically lentiviral or γ-retroviral—remain the dominant platforms because of their high efficiency and stable gene integration. Nonviral approaches using CRISPR-Cas9 or transposon systems such as Sleeping Beauty have also gained traction because of lower cost and reduced risk of insertional mutagenesis (Sadelain et al., 2020). Once modified, cells undergo expansion in bioreactors or culture bags. The expansion phase may last from 7 to 21 days, depending on the therapy. Parameters such as nutrient availability, cytokine concentrations, dissolved oxygen, and waste accumulation must be carefully managed. Automated closed systems such as the CliniMACS Prodigy or G-Rex bioreactors increasingly play a role in minimizing contamination and process variability (Rafiq et al., 2020). After expansion, cells are harvested, washed, and formulated into infusion-ready products. Cryopreservation may be used to facilitate logistic coordination, although recent studies have shown that fresh products sometimes retain superior metabolic and functional properties (Hirsch et al., 2022). Prior to release, each lot undergoes rigorous QC testing. Thus the manufacturing workflow is a complex but well-orchestrated sequence integrating advanced biological processes with engineering and regulatory considerations. Ensuring consistency, preventing contamination, and optimizing cell functionality are central goals throughout the production pipeline.

1.5.2 Quality control and regulatory guidelines

QC is indispensable for the safe and effective clinical use of personalized cellular therapies. The regulatory framework governing these therapies is increasingly stringent, reflecting the high biological complexity and potential risks associated with living cell products. Regulatory agencies such as the U.S. Food and Drug Administration (FDA), European Medicines Agency (EMA), and Indian Central Drugs Standard Control Organization (CDSCO) classify cellular immunotherapies as ATMPs, necessitating adherence to GMP and Good Tissue Practice (GTP) standards (EMA, 2021; FDA, 2020). QC testing encompasses identity, purity, potency, viability, sterility, endotoxin levels, vector copy number (VCN), replication-competent virus (RCV) assays, and functional performance assessments. Cell identity tests verify that the final product contains the intended cell type, typically performed using flow cytometry surface marker profiling. Purity testing ensures the absence of undesired cell populations, such as B cells or myeloid contaminants, which may compromise therapeutic function or introduce unintended immune reactions (Hartmann et al., 2022). Potency assays represent one of the most challenging QC components because of the complex mechanisms underlying cellular function. For CAR-T cells, potency assessments may include measuring cytokine secretion, cytotoxicity against target tumor cells, or specific gene expression signatures associated with effector or memory phenotypes. Regulatory agencies require potency assays to be mechanism-specific, reproducible, and predictive of clinical performance (FDA, 2020).

Sterility and endotoxin testing are mandatory to prevent adverse events, particularly sepsis. Mycoplasma screenings are also required due to the high propensity of contamination in cell cultures. VCN and insertion site analysis ensure genomic stability and reduce the risk of insertional oncogenesis. Replication-competent lentivirus (RCL) or replication-competent retrovirus (RCR) assays are conducted for genetically modified products to rule out unintended viral propagation (Suerth et al., 2020). Regulatory guidelines also emphasize traceability and chain-of-identity (COI) systems. Because personalized therapies are patient-specific, COI systems ensure that cells collected from a patient are accurately tracked through manufacturing and returned to the same patient for infusion. Any breach in COI can result in catastrophic cross-contamination or treatment errors (Pan et al., 2021). Environmental monitoring of GMP facilities must meet strict thresholds for particulate matter, microbial load, and air quality. Operators must be highly trained, and all procedures documented under standardized operating procedures (SOPs). Quality assurance (QA) units oversee compliance, deviation reports, and corrective-action processes. The regulatory landscape continues to evolve as newer therapies—such as gene-edited cells and off-the-shelf allogeneic products—challenge traditional approval pathways. Accelerated approval processes, expedited designation programs, and rolling reviews have been introduced to facilitate patient access while maintaining safety standards (Hartmann et al., 2022).

1.5.3 Safety, toxicity, and monitoring

The clinical administration of personalized cellular immunotherapies carries unique safety concerns distinct from conventional anticancer treatments. Immune effector cell therapies, particularly CAR-T and TCR-engineered T cells, can elicit systemic immune activation, off-target effects, or long-lasting immune alterations. Safety management is therefore an essential component of clinical translation, requiring vigilant monitoring, early detection strategies, and standardized treatment algorithms (Neelapu et al., 2018). CRS is the most prominent and potentially severe toxicity associated with T-cell–based therapies. CRS results from rapid activation and proliferation of infused cells, accompanied by massive cytokine secretion, including IL-6, IFN-γ, and TNF-α. Clinically, CRS manifests as fever, hypotension, hypoxia, and organ dysfunction. Management protocols rely on IL-6 blockade using tocilizumab and corticosteroids for severe cases. Early identification through biomarkers such as C-reactive protein (CRP) and ferritin levels is essential for mitigating risk (Lee et al., 2019). Immune effector cell–associated neurotoxicity syndrome (ICANS) is another major safety concern characterized by confusion, aphasia, seizures, or cerebral edema. Its pathophysiology is linked to endothelial activation and blood–brain barrier disruption. While corticosteroids remain the primary treatment, standardized neurologic assessments such as the ICE score are used for early detection (Gust et al., 2020). On-target, off-tumor toxicity occurs when therapeutic cells recognize antigens expressed on healthy tissues. This is a particular concern for TCR-engineered therapies targeting intracellular antigens presented via MHC molecules. Cases of fatal cardiotoxicity have been reported because of cross-reactivity with unrelated peptides expressed on cardiac tissue, highlighting the need for rigorous preclinical antigen specificity studies (Morgan et al., 2013).

For DC vaccines and NK-cell therapies, safety profiles are generally more favorable. Mild injection-site reactions, fever, and fatigue are the most common adverse events. Macrophage-based therapies, though still emerging, require careful monitoring due to potential proinflammatory activation (Zhang et al., 2022). Long-term persistence of cellular products raises additional safety considerations. Although persistence may contribute to durable remission, it can also pose risks of delayed toxicities, insertional mutagenesis, or sustained immune dysregulation. Therefore, long-term follow-up (LTFU) studies, often extending up to 15 years, are mandated by regulatory agencies for gene-modified cell therapies (FDA, 2020). The complexity of personalized therapies necessitates integrated monitoring frameworks combining clinical assessments, laboratory biomarkers, imaging modalities, and immune profiling. Digital health tools, including remote monitoring systems and machine learning algorithms, are being developed to predict toxicity risk and guide personalized management strategies.

1.6 Challenges and future directions

Personalized cellular immunotherapy has transformed the modern therapeutic landscape, offering unprecedented precision and durable responses across a range of malignancies and immune-mediated disorders. Yet, the translation of these advances into universally accessible, safe, and long-lasting treatments remains constrained by scientific, manufacturing, economic, and ethical challenges. The field must confront the multifactorial obstacles derived from tumor heterogeneity, immune escape mechanisms, scalability limitations, and disparities in global accessibility. Furthermore, integrating emerging technologies such as CRISPR gene editing, synthetic biology, and next-generation chimeric platforms necessitates rigorous safety evaluation and ethical deliberation. This section explores key challenges and outlines future directions critical for accelerating innovation and improving patient outcomes in personalized cellular immunotherapy.

1.6.1 Tumor heterogeneity and immune escape

Tumor heterogeneity represents one of the most formidable impediments to the efficacy of cellular therapies. Tumors are inherently composed of genetically and phenotypically diverse cell populations, which evolve dynamically under selective pressures from the microenvironment, therapeutic intervention, and immune surveillance. This heterogeneity can manifest at genomic, transcriptomic, epigenetic, and metabolic levels, affecting antigen expression, immune susceptibility, and cellular signaling networks (McGranahan & Swanton, 2017). As a consequence, antigen-targeted cellular therapies such as CAR-T, TCR-engineered T cells, and NK-cell therapies often face challenges in achieving sustained tumor eradication. A major immune escape mechanism is antigen loss or modulation. Tumors frequently downregulate, mutate, or shed the targeted antigen following selective pressure from cell therapy, leading to relapse and treatment resistance. For example, CD19-negative relapse after CD19 CAR-T therapy highlights how tumors adapt by altering surface marker expression or modifying splice variants (Fischer et al., 2022). Similarly, tumors can exploit defects in antigen presentation pathways, notably the loss of β2-microglobulin or HLA molecules, rendering TCR therapies ineffective (Rodriguez et al., 2021.). The TME plays a crucial role in impeding cellular immunotherapy. Solid tumors possess dense stroma, hypoxic niches, and dysregulated vasculature, which limit therapeutic cell infiltration and function. In addition, immunosuppressive elements such as Tregs, myeloid-derived suppressor cells (MDSCs), TAMs-associated macrophages, and inhibitory cytokines (TGF-β, IL-10) collectively restrain antitumor immunity (Hinshaw & Shevde, 2019). Many TME components directly inhibit CAR-T or NK-cell activation through checkpoint signaling pathways such as PD-1/PD-L1, TIM-3, LAG-3, and TIGIT.

Addressing tumor heterogeneity requires multipronged strategies. One promising direction involves multispecific and tandem CAR designs capable of recognizing multiple antigens concurrently, thereby mitigating antigen-loss–mediated escape (Niedermann et al., 2022). Similarly, gene-edited T cells engineered to resist TGF-β signaling, checkpoint inhibition, or metabolic suppression exhibit enhanced persistence and cytotoxicity within hostile microenvironments. Combination therapies leveraging cytokines, immune checkpoint inhibitors, or targeted small molecules are also under intense investigation to modulate the TME and enhance cell infiltration. Moreover, exploiting neoantigens—tumor-specific mutations not expressed in healthy tissues—offers highly personalized yet potentially more durable targets, though identifying and validating neoepitopes remains labor-intensive and computationally demanding (Ott et al., 2019). In the future, integrative single-cell sequencing, spatial transcriptomics, and lineage tracking technologies will deepen understanding of tumor evolution in response to cellular therapy. Such approaches may enable real time, adaptive therapeutic redesign, increasing the likelihood of sustained remission despite tumor plasticity.

1.6.2 Cost, accessibility, and ethical considerations

Although personalized cellular therapies have redefined therapeutic possibilities, their cost and limited accessibility remain substantial barriers to widespread adoption. Current CAR-T therapies for hematologic malignancies often cost between USD 350,000 and USD 500,000 per treatment, excluding hospitalization and management of toxicities (Mullard et al., 2021). High manufacturing expenses stem from autologous workflows, labor-intensive protocols, and complex QA processes. In low- and middle-income countries (LMICs), these treatments are often entirely out of reach, contributing to inequitable global health outcomes. One major cost driver is the individualized production process, which requires unique manufacturing runs for each patient. This model suffers from limited scalability, long production times, and vulnerability to manufacturing failures. The need for specialized GMP facilities further restricts access to major academic centers or high-income regions. Transportation logistics, cold-chain requirements, and COI systems add additional layers of expense and complexity (Rozwadowska et al., 2022). Ethical considerations emerge at multiple levels: patient selection, informed consent, long-term monitoring obligations, and equitable access. Enrolling patients in clinical trials requires transparent communication regarding uncertain long-term risks, including potential insertional mutagenesis, delayed toxicities, and immune dysregulation. In pediatric populations, decisions regarding gene-modified cellular therapies introduce additional ethical complexities, particularly when guardians must weigh experimental risks against limited alternatives (Shukla et al., 2021).

From a societal perspective, the sustainability of healthcare systems is a growing concern. The pricing models adopted for ATMPs have raised debates about reimbursement frameworks, value-based pricing, and long-term outcome-linked payment systems. Moreover, as gene editing and synthetic biology enable more profound manipulation of human cells, concerns regarding enhancement, dual-use technologies, and potential misuse intensify. Ethical oversight must continue to evolve in tandem with scientific advancements, ensuring responsible innovation without hindering progress (Pepper et al., 2018). Strategies to improve access include decentralizing manufacturing, reducing production costs, adopting allogeneic or off-the-shelf product platforms, and developing regulatory harmonization across jurisdictions. Government incentives, public–private partnerships, and national cell therapy programs can further foster equitable distribution and reduce regional disparities. Efforts to integrate automated closed systems and scalable bioreactors hold promise for lowering costs and democratizing therapy availability.

1.6.3 Emerging technologies: Gene editing, synthetic biology, and next-gen chimeric antigen receptor platforms

Emerging technologies are poised to shape the next era of personalized cellular immunotherapy, introducing new levels of precision, safety, and functional sophistication. Innovations in gene editing, synthetic biology, and advanced CAR designs offer powerful solutions to overcome many limitations associated with first-generation therapies.

Gene editing: CRISPR-Cas systems have revolutionized genetic manipulation, enabling rapid, precise, and multiplexed gene edits in immune cells. CRISPR editing can eliminate inhibitory receptors, disrupt immune checkpoint pathways, or remove endogenous TCRs to prevent GVHD in allogeneic settings (Eyquem et al., 2020). Universal or "off-the-shelf" CAR-T cells generated through CRISPR-mediated knockout of TCR and HLA molecules remove donor–recipient mismatch barriers, allowing large-scale production and reduced costs. Furthermore, precise knock-in strategies enable integration of transgenes into safe-harbor loci, producing more consistent and predictable expression patterns. Base editing and prime editing techniques, which minimize double-strand breaks, are being explored to reduce genotoxicity and improve genomic stability (Gaudelli et al., 2020). Gene editing also enables the engineering of advanced NK cells, macrophages, and $\gamma\delta$ T cells with enhanced resistance to immunosuppressive signals.

Synthetic biology: Synthetic biology provides a design framework to reprogram immune cells with sophisticated circuitry, allowing conditional responses to environmental signals. Synthetic receptors such as SynNotch systems permit Boolean logic gating, ensuring activation only when two or more antigens are present, thereby improving specificity and safety (Roybal et al., 2016). Synthetic promoters, tunable gene switches, and kill-switch mechanisms add further layers of control. Engineered metabolic pathways allow therapeutic cells to thrive in nutrient-deprived, hypoxic tumor environments. Synthetic cytokine circuits can provide localized, self-regulated immunostimulation while minimizing systemic toxicity. In addition, bioengineered scaffolds and nanoparticles are being developed to guide cell migration, support in vivo expansion, and enhance antitumor efficacy.

Next-generation CAR platforms: New CAR architectures aim to address limitations of earlier designs. Fourth-generation CARs, also known as TRUCKs ("T cells redirected for universal cytokine killing"), incorporate inducible cytokine expression modules to amplify T-cell function within the TME (Chmielewski et al., 2020). Fifth-generation CARs integrate JAK–STAT signaling domains, enabling more potent and physiologic cytokine signaling. Dual-targeted, bispecific, or switchable CARs reduce antigen escape and permit external regulation of CAR activity through small-molecule drugs. Logic-gated CARs enforce AND/NOT activation patterns, enhancing tumor specificity and reducing off-tumor toxicity. Meanwhile, armored CAR-T cells engineered to secrete checkpoint inhibitors, chemokines, or matrix-degrading enzymes show superior infiltration and persistence in solid tumors (Majzner & Mackall, 2019). Parallel advances are being made in CAR-M, which demonstrate strong phagocytic activity and capacity for antigen presentation, making them promising candidates for solid tumor immunotherapy. NK-cell–based CAR therapies offer innate tumor-targeting capabilities with a favorable safety profile and potential for off-the-shelf production.

References

Aebersold, R., & Mann, M. (2016). Mass-spectrometric exploration of proteome structure and function. *Nature, 537*(7620), 347–355. https://doi.org/10.1038/nature19949.

Bassani-Sternberg, M., Bräunlein, E., Klar, R., Engleitner, T., Sinitcyn, P., Audehm, S., Straub, M., Weber, J., Slotta-Huspenina, J., Specht, K., Martignoni, M. E., Werner, A., Hein, R., Busch, D. H., Peschel, C., Rad, R., Cox, J., Mann, M., & Krackhardt, A. M. (2016). Direct identification of clinically relevant neoepitopes presented on native human melanoma tissue by mass spectrometry. *Nature Communications, 7*(1). https://doi.org/10.1038/ncomms13404.

Batlle, E., & Massagué, J. (2019). Transforming growth factor-β signaling in immunity and cancer. *Immunity, 50*(4), 924–940. https://doi.org/10.1016/j.immuni.2019.03.024.

Binnewies, M., Roberts, E. W., Kersten, K., Chan, V., Fearon, D. F., Merad, M., Coussens, L. M., Gabrilovich, D. I., Ostrand-Rosenberg, S., Hedrick, C. C., Vonderheide, R. H., Pittet, M. J., Jain, R. K., Zou, W., Howcroft, T. K., Woodhouse, E. C., Weinberg, R. A., & Krummel, M. F. (2018). Understanding the tumor immune microenvironment (TIME) for effective therapy. *Nature Medicine, 24*(5), 541–550. https://doi.org/10.1038/s41591-018-0014-x.

Blank, C. U., Haanen, J. B., Ribas, A., & Schumacher, T. N. (2016). The "cancer immunogram". *Science, 352*(6286), 658–660. https://doi.org/10.1126/science.aaf2834.

Brown, C. E., Mackall, C. L., & Brenner, M. K. (2018). CAR T cell therapy: The CD19 paradigm. *Clinical Cancer Research, 24*(19), 4735–4743.

Chen, D. S., & Mellman, I. (2017). Elements of cancer immunity and the cancer-immune set point. *Nature, 541*(7637), 321–330. https://doi.org/10.1038/nature21349.

Chmielewski, M., & Abken, H. (2020). TRUCKs: The fourth generation of CARs. *Cancer Immunology, Immunotherapy, 69*(3), 399–406.

Chowell, D., Morris, L. G. T., Grigg, C. M., Weber, J. K., Samstein, R. M., Makarov, V., Kuo, F., Kendall, S. M., Requena, D., Riaz, N., Greenbaum, B., Carroll, J., Garon, E., Hyman, D. M., Zehir, A., Solit, D., Berger, M., Zhou, R., Rizvi, N. A., & Chan, T. A. (2018). Patient HLA class I genotype influences cancer response to checkpoint blockade immunotherapy. *Science, 359*(6375), 582–587. https://doi.org/10.1126/science.aao4572.

del Campo, A. B., Kyte, J. A., Carretero, J., Zinchenko, S., Méndez, R., González-Aseguinolaza, G., & Garrido, F. (2014). Immune escape of cancer cells with aberrant HLA class I expression. *International Journal of Cancer, 134*(5), 1329–1341.

Dunn, G. P., Old, L. J., & Schreiber, R. D. (2002). The three Es of cancer immunoediting. *Annual Review of Immunology, 20*, 751–778.

EMA. (2021). *Guideline on quality, non-clinical and clinical requirements for investigational advanced therapy medicinal products*. European Medicines Agency.

Eyquem, J., Mansilla-Soto, J., Giavridis, T., van der Stegen, S. C., Hamieh, M., Cunanan, K. M., & Sadelain, M. (2020). Targeting a CAR to the TRAC locus with CRISPR/Cas9 enhances tumor rejection. *Nature, 543*(7643), 113–117.

FDA. (2020). *Chemistry, manufacturing, and control (CMC) requirements for human gene therapy investigational new drug applications*. U.S. Food & Drug Administration.

Fischer, J., Paret, C., & Merz, H. (2022). Antigen escape and resistance mechanisms in CAR T-cell therapy. *Frontiers in Oncology, 12*, 887432.

Fraietta, J. A., Lacey, S. F., Orlando, E. J., Pruteanu-Malinici, I., Gohil, M., Lundh, S., & Maude, S. L. (2018). Determi nants of response and resistance to CD19 CAR T cell therapy. *Nature Medicine, 24*(5), 563–571.

Gabrilovich, D. I. (2017). Myeloid-derived suppressor cells. *Cancer Immunology Research, 5*(1), 3–8. https://doi.org/10.1158/2326-6066.CIR-16-0297.

Gattinoni, L., Lugli, E., Ji, Y., Pos, Z., Paulos, C . M., Quigley, M. F., Almeida, J. R., Gostick, E., Yu, Z., Carpenito, C., Wang, E., Douek, D. C., Price, D. A., June, C. H., Marincola, F. M., Roederer, M., & Restifo, N. P. (2011). A human memory T cell subset with stem cell-like properties. *Nature Medicine, 17*(10), 1290–1297. https://doi.org/10.1038/nm.2446.

Gaudelli, N. M., Komor, A. C., & Rees, H. A. (2020). Base editing: Precision chemistry on the genome and transcriptome of living cells. *Nature Reviews Genetics, 21*(8), 465–482.

Gordon, S., & Martinez, F. O. (2010). Alternative activation of macrophages: Mechanism and functions. *Immunity, 32*(5), 593–604. https://doi.org/10.1016/j.immuni.2010.05.007.

Gust, J., Ponce, R., & Turtle, C. J. (2020). Neurotoxicity associated with CD19-targeted CAR-T cell therapies. *Current Opinion in Neurology, 33*(6), 720–728.

Pepper, M. S., Pope, A., Kling, S., Alessandrini, M., Van Staden, W., & Green, R. J. (2018). Ethical considerations in the application of cell and gene therapies in children. South African Medical Journal, 108(12), 1027-1029.

Hartmann, J., Schüßler-Lenz, M., Bondanza, A., & Buchholz, C. J. (2022). Clinical development of CAR T cells—Challenges and opportunities in translating innovative treatment concepts. *EMBO Molecular Medicine, 14*(1), e15389.

Helmink, B. A., Reddy, S. M., Gao, J., Zhang, S., Basar, R., Thakur, R., Yizhak, K., Sade-Feldman, M., Blando, J., Han, G., Gopalakrishnan, V., Xi, Y., Zhao, H., Amaria, R. N., Tawbi, H. A., Cogdill, A. P., Liu, W., LeBleu, V. S.,

Kugeratski, F. G., … Patel, S. (2020). B cells and tertiary lymphoid structures promote immunotherapy response. *Nature, 577*(7791), 549–555. https://doi.org/10.1038/s41586-019-1922-8.

Heslop, H. E., Gottschalk, S., & Brenner, M. K. (2021). Cellular immunotherapy: An overview. *Nature Reviews Immunology, 21*(12), 700–716.

Hinshaw, D. C., & Shevde, L. A. (2019). The tumor microenvironment innately modulates cancer progression. *Cancer Research, 79*(18), 4557–4567. https://doi.org/10.1158/0008-5472.CAN-18-3962.

Hirsch, A. E., Di Roberto, R. B., & Caimi, P. F. (2022). Fresh versus cryopreserved CAR T cells: Does it matter? *Journal of Cellular Therapy, 28*(2), 89–101.

Hoption Cann, S. A, van Netten, J. P, & van Netten, C. (2003). Dr William Coley and tumour regression: A place in history or in the future. *Postgraduate Medical Journal, 79*(938), 672–680. https://doi.org/10.1093/postgradmedj/79.938.672.

Hundal, J., Kiwala, S., McMichael, J., Miller, C. A., Xia, H., Wollam, A. T., Liu, C. J., Zhao, S., Feng, Y. Y., Graubert, A. P., Wollam, A. Z., Neichin, J., Neveau, M., Walker, J., Gillanders, W. E., Mardis, E. R., Griffith, O. L., & Griffith, M. (2020). PVACtools: A computational toolkit to identify and visualize cancer neoantigens. *Cancer Immunology Research, 8*(3), 409–420. https://doi.org/10.1158/2326-6066.CIR-19-0401.

Introna, M., Correnti, F., & O'Hare, R. (2017). Cytokine-induced killer cells as immunotherapy for hematologic malignancies. *Immunotherapy, 9*(4), 333–345.

Joyce, J. A., & Fearon, D. T. (2015). T cell exclusion, immune privilege, and the tumor microenvironment. *Science, 348*(6230), 74–80. https://doi.org/10.1126/science.aaa6204.

June, C. H., O'Connor, R. S., Kawalekar, O. U., Ghassemi, S., & Milone, M. C. (2018). CAR T cell immunotherapy for human cancer. *Science, 359*(6382), 1361–1365. https://doi.org/10.1126/science.aar6711.

June, C. H., & Sadelain, M. (2018). Chimeric antigen receptor therapy. *New England Journal of Medicine, 379*(1), 64–73. https://doi.org/10.1056/NEJMra1706169.

Jurtz, V., Paul, S., Andreatta, M., Marcatili, P., Peters, B., & Nielsen, M. (2017). NetMHCpan-4.0: improved peptide–MHC class I interaction predictions integrating eluted ligand and peptide binding affinity data. *The Journal of Immunology, 199*(9), 3360–3368. https://doi.org/10.4049/jimmunol.1700893.

Kantoff, P. W., Higano, C. S., Shore, N. D., Berger, E. R., Small, E. J., Penson, D. F., Redfern, C. H., Ferrari, A. C., Dreicer, R., Sims, R. B., Xu, Y., Frohlich, M. W., & Schellhammer, P. F. (2010). Sipuleucel-T immunotherapy for castration-resistant prostate cancer. *New England Journal of Medicine, 363*(5), 411–422. https://doi.org/10.1056/NEJMoa1001294.

Klichinsky, M., Ruella, M., Shestova, O., Lu, X. M., Best, A., Zeeman, M., Schmierer, M., Gabrusiewicz, K., Anderson, N. R., Petty, N. E., Cummins, K. D., Shen, F., Shan, X., Veliz, K., Blouch, K., Yashiro-Ohtani, Y., Kenderian, S. S., Kim, M. Y., O'Connor, R. S., … Gill, S. (2020). Human CAR macrophages for cancer immunotherapy. *Nature Biotechnology, 38*(8), 947–953.

Köhler, G., & Milstein, C. (1975). Continuous cultures of fused cells secreting antibody of predefined specificity. *Nature, 256*(5517), 495–497. https://doi.org/10.1038/256495a0.

Lanz, T. V., Klimatcheva, E., & Rao, D. A. (2022). Systems immunology and machine learning: The new frontier for precision immunotherapy. *Nature Reviews Immunology, 22*(8), 465–480. https://doi.org/10.1038/s41577-022-00683-5.

Lee, D. W., Santomasso, B. D., & Locke, F. L. (2019). Systemic inflammatory toxicities following CAR T-cell therapy: Recognition and management. *Blood, 134*(6), 510–518.

Li, Y., Li, S., & Chen, Z. (2023). Enhancing NK cell activation for cancer immunotherapy. *Frontiers in Immunology, 14*, 1122485.

Liu, E., Tong, Y., & Li, J. (2020). Manufacturing challenges in cell therapy. *Cell & Gene Therapy Insights, 6*(5), 731–745.

Majzner, R. G., & Mackall, C. L. (2018). Tumor antigen escape from car t-cell therapy. *Cancer Discovery, 8*(10), 1219–1226. https://doi.org/10.1158/2159-8290.CD-18-0442.

Majzner, R. G., & Mackall, C. L. (2019). Clinical lessons learned from the first leg of the CAR T cell journey. *Nature Medicine, 25*(9), 1341–1355. https://doi.org/10.1038/s41591-019-0564-6.

Mardis, E. R. (2017). DNA sequencing technologies: 2006-2016. *Nature Protocols, 12*(2), 213–218. https://doi.org/10.1038/nprot.2016.182.

Matthew, D., William, D Tap, & Sandra P, D'A. (2016). Current status of engineered T-cell therapy for synovial sarcoma. *Immunotherapy, 8*(9), 1073–1080.

Maude, S. L., Laetsch, T. W., Buechner, J., Rives, S., Boyer, M., Bittencourt, H., Bader, P., Verneris, M. R., Stefanski, H. E., Myers, G. D., Qayed, M., De Moerloose, B., Hiramatsu, H., Schlis, K., Davis, K. L., Martin, P. L., Nemecek, E. R., Yanik, G. A., Peters, C., … Baruchel, A. (2018). Tisagenlecleucel in children and young adults with B-cell lymphoblastic leukemia. *New England Journal of Medicine, 378*(5), 439–448. https://doi.org/10.1056/NEJMoa1709866.

McGranahan, N., Rosenthal, R., Hiley, C. T., Rowan, A. J., Watkins, T. B. K., Wilson, G. A., … Swanton, C.TRACERx Consortium. (2017). Allele-Specific HLA Loss and Immune Escape in Lung Cancer Evolution. *Cell, 171*(6), 1259–1271.e11. https://doi.org/10.1016/j.cell.2017.10.001.

McGranahan, N., & Swanton, C. (2017). Clonal heterogeneity and tumor evolution: Past, present, and the future. *Cell, 168*(4), 613–628. https://doi.org/10.1016/j.cell.2017.01.018.

McLane, L. M., Abdel-Hakeem, M. S., & Wherry, E. J. (2019). CD8 T cell exhaustion during chronic viral infection and cancer. *Annual Review of Immunology, 37*, 457–495. https://doi.org/10.1146/annurev-immunol-041015-055318.

Mellman, I., Coukos, G., & Dranoff, G. (2011). Cancer immunotherapy comes of age. *Nature, 480*(7378), 480–489. https://doi.org/10.1038/nature10673.

Morgan, R. A., Chinnasamy, N., & Rosenberg, S. A. (2013). Genetic therapy for cancer: TCR targeting. *Nature Reviews Cancer, 13*(12), 804–817.

Mullard, A. (2021). FDA approves first CAR-T therapy for adult leukemia. *Nature Reviews Drug Discovery, 20*(5), 337–339.

Munn, D. H., & Mellor, A. L. (2016). IDO in the tumor microenvironment: Inflammation, counter-regulation, and tolerance. *Trends in Immunology, 37*(3), 193–207. https://doi.org/10.1016/j.it.2016.01.002.

Neelapu, S. S., Tummala, S., Kebriaei, P., Wierda, W., Gutierrez, C., Locke, F. L., Komanduri, K. V., Lin, Y., Jain, N., Daver, N., Westin, J., Gulbis, A. M., Loghin, M. E., De Groot, J. F., Adkins, S., Davis, S. E., Rezvani, K., Hwu, P., & Shpall, E. J. (2018). Chimeric antigen receptor T-cell therapy-assessment and management of toxicities. *Nature Reviews Clinical Oncology, 15*(1), 47–62. https://doi.org/10.1038/nrclinonc.2017.148.

Niedermann, G., et al. (2022). Tandem CAR strategies for antigen escape mitigation. *Journal of Immunotherapy, 45*(1), 1–12.

Noy, R., & Pollard, J. W. (2014). Tumor-associated macrophages: From mechanisms to therapy. *Immunity, 41*(1), 49–61. https://doi.org/10.1016/j.immuni.2014.06.010.

O'Garra, A., & Vieira, P. (2007). TH1 cells control themselves by producing interleukin-10. *Nature Reviews Immunology, 7*(6), 425–428. https://doi.org/10.1038/nri2097.

Ott, P. A., & Wu, C. J. (2019). Cancer vaccines: Steering T cells down the right path. *Nature Reviews Clinical Oncology, 16*(6), 347–348.

Palucka, K., & Banchereau, J. (2012). Cancer immunotherapy via dendritic cells. *Nature Reviews Cancer, 12*(4), 265–277. https://doi.org/10.1038/nrc3258.

Palucka, K., & Banchereau, J. (2013). Dendritic-cell-based therapeutic cancer vaccines. *Immunity, 39*(1), 38–48. https://doi.org/10.1016/j.immuni.2013.07.004.

Pan, Y., Zhang, X., & Shen, J. (2021). Logistics and chain of identity in personalized cell manufacturing. *Biotechnology Advances, 53*, 107849.

Pardoll, D. M. (2012). The blockade of immune checkpoints in cancer immunotherapy. *Nature Reviews Cancer, 12*(4), 252–264. https://doi.org/10.1038/nrc3239.

Pauken, K. E., & Wherry, E. J. (2015). Overcoming T cell exhaustion in infection and cancer. *Trends in Immunology, 36*(4), 265–276. https://doi.org/10.1016/j.it.2015.02.008.

Hegde, M., Mukherjee, M., Grada, Z., Pignata, A., Landi, D., Navai, S. A., Wakefield, A., Fousek, K., Bielamowicz, K., Chow, K. K., Brawley, V. S., Byrd, T. T., Krebs, S., Gottschalk, S., Wels, W. S., Baker, M. L., Dotti, G., Mamonkin, M., Brenner, M. K., et al., (2016). Tandem CAR T cells targeting HER2 and IL13Rα2 mitigate tumor antigen escape. The Journal of clinical investigation, 126(8), 3036–3052. https://doi.org/10.1172/JCI83416.

Rafiq, A., Hackett, C. S., & Brentjens, R. J. (2020). Engineering strategies to overcome CAR T-cell therapy resistance. *Nature Reviews Clinical Oncology, 17*(6), 372–385.

Rao, S., Kakarla, S., & Nayak, R. (2022). Patient factors influencing CAR-T manufacturing outcomes. *Cancer Gene Therapy, 29*(2), 87–96.

Rapoport, A. P., Stadtmauer, E. A., Binder-Scholl, G. K., Goloubeva, O., Vogl, D. T., Lacey, S. F., Badros, A. Z., Garfall, A., Weiss, B., Finklestein, J., Kulikovskaya, I., Sinha, S. K., Kronsberg, S., Gupta, M., Bond, S., Melchiori, L., Brewer, J. E., Bennett, A. D., … Gerry, A. B. (2015). NY-ESO-1-specific TCR-engineered T cells mediate sustained antigen-specific antitumor effects in myeloma. *Nature Medicine, 21*(8), 914–921. https://doi.org/10.1038/nm.3910.

Ribas, A., & Wolchok, J. D. (2018). Cancer immunotherapy using checkpoint blockade. *Science, 359*(6382), 1350–1355. https://doi.org/10.1126/science.aar4060.

Rodriguez, G. M., et al. (2021). HLA loss and immune escape in cancer. *Nature Reviews Cancer, 21*(7), 421–432.

Rosenberg, S. A., & Restifo, N. P. (2015). Adoptive cell transfer as personalized immunotherapy for human cancer. *Science, 348*(6230), 62–68. https://doi.org/10.1126/science.aaa4967.

Rosenberg, S. A., Yang, J. C., & Restifo, N. P. (1998). Cancer immunotherapy: Moving beyond current vaccines. *Nature Medicine, 4*(3), 321–327. https://doi.org/10.1038/nm0398-321.

Rosenberg, S. A., Yang, J. C., Sherry, R. M., Kammula, U. S., Hughes, M. S., Phan, G. Q., Citrin, D. E., Restifo, N. P., Robbins, P. F., Wunderlich, J. R., Morton, K. E., Laurencot, C. M., Steinberg, S. M., White, D. E., & Dudley, M. E. (2011). Durable complete responses in heavily pretreated patients with metastatic melanoma using T-cell transfer immunotherapy. *Clinical Cancer Research, 17*(13), 4550–4557.

Rosenthal, R., Cadieux, E. L., Salgado, R., Al-Bakir, M., Moore, D. A., Hiley, C. T., Lund, T., Tanić, M., Reading, J. L., Joshi, K., Henry, J. Y., Ghorani, E., Wilson, G. A., Birkbak, N. J., Jamal-Hanjani, M., Veeriah, S., Szallasi, Z., Loi, S., Hellmann, M. D., … Feber, A. (2019). Neoantigen-directed immune escape in lung cancer evolution. *Nature, 567*(7749), 479–485. https://doi.org/10.1038/s41586-019-1032-7.

Roybal, K. T., Rupp, L. J., Morsut, L., Walker, W. J., McNally, K. A., Park, J. S., & Lim, W. A. (2016). Precision tumor recognition by T cells with combinatorial antigen-sensing circuits. *Cell, 164*(4), 770–779. https://doi.org/10.1016/j.cell.2016.01.011.

Rozwadowska, N. (2022). Logistics for personalized cell-based therapies: Challenges and solutions. *Bioprocessing Journal, 21*(4), 18–32.

Sadelain, M., Rivière, I., & Riddell, S. (2020). The 25-year journey of CAR T-cell therapy. *Nature Reviews Cancer, 20*(2), 89–109.

Sahin, U., & Türeci, Ö (2018). Personalized vaccines for cancer immunotherapy. *Science, 359*(6382), 1355–1360. https://doi.org/10.1126/science.aar7112.

Sakaguchi, S., Yamaguchi, T., Nomura, T., & Ono, M. (2008). Regulatory T cells and immune tolerance. *Cell, 133*(5), 775–787. https://doi.org/10.1016/j.cell.2008.05.009.

Sánchez-Paulete, A. R., Cueto, F. J., Martínez-López, M., Labiano, S., Morales-Kastresana, A., Rodríguez-Ruiz, M. E., & Melero, I. (2017). Cancer immunotherapy with immunomodulatory antibodies requires intact cDC1 function. *Cancer Cell, 31*(5), 626–636.

Schreiber, R. D., Old, L. J., & Smyth, M. J. (2011). Cancer immunoediting: Integrating immunity's roles in cancer suppression and promotion. *Science, 331*(6024), 1565–1570. https://doi.org/10.1126/science.1203486.

Semenza, G. L. (2014). Hypoxia-inducible factor 1 and cancer pathogenesis. *IUBMB Life, 66*(4), 263–274.

Sharma, P., Hu-Lieskovan, S., Wargo, J. A., & Ribas, A. (2021). Primary, adaptive, and acquired resistance to cancer immunotherapy. *Cell, 184*(21), 5309–5327. https://doi.org/10.1016/j.cell.2021.09.016.

Shukla, V., et al. (2021). Ethical considerations in pediatric gene and cell therapy. *Pediatrics, 148*(3), e2021050211.

Silva-Santos, B., Serre, K., & Norell, H. (2019). $\gamma\delta$ T cells in cancer. *Nature Reviews Immunology, 19*(8), 431–447.

Smyth, M. J., Dunn, G. P., & Schreiber, R. D. (2006). Cancer immunosurveillance and immunoediting. *Immunity, 21*(1), 10–22.

Snyder, A., Makarov, V., Merghoub, T., Yuan, J., Zaretsky, J. M., Desrichard, A., Walsh, L. A., Postow, M. A., Wong, P., Ho, T. S., Hollmann, T. J., Bruggeman, C., Kannan, K., Li, Y., Elipenahli, C., Liu, C., Harbison, C. T., Wang, L., Ribas, A., … Wolchok, J. D. (2014). Genetic basis for clinical response to CTLA-4 blockade in melanoma. *New England Journal of Medicine, 371*(23), 2189–2199. https://doi.org/10.1056/NEJMoa1406498.

Steinman, R. M., & Banchereau, J. (2007). Taking dendritic cells into medicine. *Nature, 449*(7161), 419–426. https://doi.org/10.1038/nature06175.

Stubbington, M. J. T., Rozenblatt-Rosen, O., Regev, A., & Teichmann, S. A. (2017). Single-cell transcriptomics to explore the immune system in health and disease. *Science, 358*(6359), 58–63. https://doi.org/10.1126/science.aan6828.

Suerth, J. D., Morgan, M. A., & Kloess, S. (2020). Replication-competent virus detection in genetically modified cell therapies. *Molecular Therapy, 28*(1), 78–90.

Topalian, S. L., Taube, J. M., Anders, R. A., & Pardoll, D. M. (2016). Mechanism-driven biomarkers to guide immune checkpoint blockade in cancer therapy. *Nature Reviews Cancer, 16*(5), 275–287. https://doi.org/10.1038/nrc.2016.36.

Tran, E., Robbins, P. F., Lu, Y. C., Prickett, T. D., Gartner, J. J., Jia, L., Pasetto, A., Zheng, Z., Ray, S., Groh, E. M., Kriley, I. R., & Rosenberg, S. A. (2016). T-cell transfer therapy targeting mutant KRAS in cancer. *New England Journal of Medicine, 375*(23), 2255–2262. https://doi.org/10.1056/NEJMoa1609279.

Uribe-Querol, E., & Rosales, C. (2015). Neutrophils in cancer: Two sides of the same coin. *Journal of Immunology Research, 2015*, 1–21. https://doi.org/10.1155/2015/983698.

Victora, G. D., & Nussenzweig, M. C. (2022). Germinal centers. *Annual Review of Immunology, 40*, 413–442. https://doi.org/10.1146/annurev-immunol-120419-022408.

Vivier, E., Raulet, D. H., Moretta, A., Caligiuri, M. A., Zitvogel, L., Lanier, L. L., & Ugolini, S. (2012). Innate or adaptive immunity? The example of natural killer cells. *Science, 331*(6013), 44–49.

Vivier, E., Raulet, D. H., Moretta, A., Caligiuri, M. A., Zitvogel, L., Lanier, L. L., & Ugolini, S. (2011). Innate or adaptive immunity? The example of natural killer cells. *Science, 331*(6013), 44–49.

Wang, D., & DuBois, R. N. (2010). The role of COX-2 in intestinal inflammation and colorectal cancer. *Oncogene, 29*(6), 781–788. https://doi.org/10.1038/onc.2009.421.

Wherry, E . J., & Kurachi, M. (2015). Molecular and cellular insights into T cell exhaustion. *Nature Reviews Immunology, 15*(8), 486–499. https://doi.org/10.1038/nri3862.

Zhang, Q., DeNardo, D. G., & Coussens, L. M. (2022). CAR-Macrophages: Broadening the horizon of immunotherapy. *Nature Reviews Immunology, 22*(7), 420–435.

CHAPTER

2

Understanding the therapeutic significance of tumor-infiltrating immune cells and their characteristics: A historical overview of cancer immunotherapy

Lakshyajeet Nath[a], Dhrubajyoti Borah[a], Jyotirmoy Kashyap[a], Shreyansh Saha[a], Dipjyoti Sharma[a], Pollobi Porasar[a], Jitendra Pandey[b], Suman Lamchhane[c], Aadesh Upadhyay[d] and Koushik Nandan Dutta[a]

[a]*NETES Institute of Pharmaceutical Science, NGI-Mirza, Kamrup, Assam, India,* [b]*Department of Chemistry, University of Hawaii at Manoa, Honolulu, HI, United States,* [c]*Department of Pharmacy, Shree College of Technology, Bharatpur, Chitwan, Nepal,* [d]*Project Lead-Oncology, One Clinical Research, Perth, WA, Australia*

2.1 Introduction

2.1.1 Overview of cancer immunotherapy and its success over the past 20 years

Cancer immunotherapy has revolutionized oncological treatment over the past two decades by harnessing the body's immune system to combat malignancies. Unlike conventional cancer treatment, such as chemotherapy and radiation, which directly target tumor cells but often lead to systemic toxicity, immunotherapy enhances the body's natural defense mechanisms to selectively eliminate cancer cells. The primary modes of immunotherapy include immune checkpoint inhibitors (ICIs), monoclonal antibodies, adoptive cell transfer (ACT), cancer vaccines, and cytokine therapies (Sharma & Allison, 2015). The initial generation of antibody-mediated immunotherapy, referred to as immune-checkpoint blockade (ICB), operates by inhibiting the receptor and ligand interactions of molecules such as cytotoxic T-lymphocyte-associated protein 4 (CTLA-4) and PD-1 that are responsible for blunting T-cell activation or function. ICB treatments have established strong clinical benefit in a minority of patients, who exhibit long-lasting responses. Regrettably, an unmet clinical need still exists for most patients, who are not responsive to ICB. Retrospective patient population analyses treated with ICB have identified that there are classes of TIME that correlate with those tumors more likely to be responsive to ICB (Binnewies et al., 2018).

The clinical innovation of immunotherapy dates back to the release of the ICI, ipilimumab, a CTLA-4 inhibitor, in 2011 for metastatic melanoma (Hodi et al., 2013). This represented a paradigm shift, as those with late-stage cancers and few therapeutic options were starting to show long-term survival. Then came programmed cell death protein 1 (PD-1) and programmed death-ligand 1 (PD-L1) inhibitors, including pembrolizumab and nivolumab, which showed outstanding effectiveness in a wide range of

Advancements in Personalized Cellular Immunotherapy: A Comprehensive Guide to Cancer Treatment.
DOI: https://doi.org/10.1016/B978-0-443-27731-3.00007-3

Copyright © 2026 Elsevier Inc. All rights are reserved, including those for text and data mining, AI training, and similar technologies.

cancers, including non-small cell lung cancer (NSCLC), renal cell carcinoma, and Hodgkin's lymphoma (Topalin et al. 2012).

The years have seen tremendous progress in the field, with the emergence of chimeric antigen receptor (CAR) T-cell therapies, which have been incredibly successful in hematological malignancies (June et al., 2018). The Nobel Prize in Physiology or Medicine 2018, given to James P. Allison and Tasuku Honjo for their work on immune checkpoint inhibition, further highlighted the world reach of immunotherapy (Ledford et al., 2018). In spite of its successes, immune-related adverse effects, differential patient response rates, and resistance mandate continued research aimed at maximizing therapeutic effectiveness (Brahmer et al., 2013).

2.1.2 Importance of tumor-infiltrating immune cells in cancer treatment

Tumor-infiltrating immune cells (TIICs) are critical in the tumor microenvironment (TME) that determines cancer progression, immune evasion, and response to therapy. Immune cells are composed of T lymphocytes, B lymphocytes, natural killer (NK) cells, dendritic cells (DCs), macrophages, and myeloid-derived suppressor cells (MDSCs), each of which plays a role in either the suppression or promotion of the tumor based on its functional status (Fridman et al., 2012). Cytotoxic CD8+ T cells are central effectors of antitumor immunity, recognizing and destroying malignant cells through the secretion of perforin and granzyme B (Chen & Mellman, 2017). CD4+ T helper cells regulate immune responses, whereas regulatory T cells (Tregs) have the capacity to suppress antitumor immunity, enabling immune escape (Sakaguchi et al., 2010). Tumor-associated macrophages (TAMs) play dual functions: M1-polarized macrophages enhance inflammation and destruction of tumors, whereas M2-polarized macrophages aid in tumor growth and metastasis (Biswas & Mantovani, 2010).

B cells in the TME have been reported to play both pro- and antitumor roles, producing antibodies that facilitate tumor clearance or secreting immunosuppressive cytokines that promote tumor growth (Shalapour & Karin, 2015). NK cells present a natural immune defense mechanism against tumor cells, although their function is normally repressed in the immunosuppressive tumor microenvironment (Vivier et al., 2016). DCs, being professional antigen-presenting cells, play a key role in priming antitumor T-cell responses but can be disabled within the immunosuppressive TME (Wculek et al., 2020).

The novel prognostic predictive models were constructed from the numbers of infiltrating TANs, Tregs, and TAMs. These models could facilitate the evaluation of prognosis and identification of 318 patients with colorectal cancer (CRC) who are more suitable for chemotherapy. These findings reveal the significant role of TIICs in CRC and could have implications for postoperative personalized follow-up, care, action, management, and decision-making regarding individualized chemotherapy for CRC (Ye et al., 2019). The dynamic interaction between these TIICs and cancer cells dictates disease progression and treatment response. Leveraging the therapeutic potential of TIICs through modulation, such as immune checkpoint blockade, adoptive cell therapy, and tumor vaccines, is a promising area in cancer therapeutics (Chen & Mellman, 2017). Furthermore, major advances in the development of humanized mouse models have made xenografts with matching patient tumor and immune compartments possible, thereby enabling studies in which a patient's own adoptively transferred TILs can be used to recapitulate the exhaustion or the introduction of targeted gene reporter loci to visualize intravital tumor immune interactions. Although these models have downsides, being able to implant human tumor tissue with a native mutational landscape into a partially reconstituted human immune repertoire represents

Table 2.1 Historical development of TIIC research.

Era	Key developments	Contribution	Reference
Late 19th–Early 20th century	Early histopathological observations	Identification of immune cell infiltration in tumors	Virhow (1989)
1950s–1960s	Tumor immunosurveillance hypothesis	Proposal that the immune system detects and eliminates tumors	Burnet (1957)
1980s–1990s	Characterization of TIICs	Use of monoclonal antibodies and flow cytometry to identify immune cells	van den Broek et al. (1996)
2000s–Present	Single-cell and spatial transcriptomics approaches	High-resolution mapping of immune cell heterogeneity	Tirosh et al. (2016); Schumacher & Thommen (2022)
Recent advances (2020s)	Role of artificial intelligence in TIIC research	AI-driven analysis of immune infiltration for personalized cancer treatment	Elkhader & Elemento (2022)

AI, *Artificial intelligence;* TIIC, *tumor-infiltrating immune cells.*

major progress. Even if mouse models fail to ever fully recapitulate human disease, it is important to understand the minutiae that make the most difference in dictating therapeutic response versus nonresponse. Distilling a disease into a few critical parameters is challenging, but understanding what cell types can be modulated and when may enable the next biggest improvements in immunotherapy (Binnewies et al., 2018).

2.1.3 Objectives of the chapter: Historical evolution, functions, and therapeutic potential of tumor-infiltrating immune cells

This chapter will attempt to offer a thorough examination of the historical development, functional roles, and therapeutic meaning of TIICs in cancer immunotherapy. Precisely, the goals are:

1. **Historical development of TIIC research**: Following the scientific progress made in elucidating the functions of TIICs in cancer, from the initial histopathological observations to contemporary single-cell sequencing methodologies and spatial transcriptomics (Schumacher & Thommen, 2022) (Table 2.1).

2.2 Functional roles of tumor-infiltrating immune cells in tumor immunity

Protumorigenic and antitumorigenic functions: The tumor microenvironment (TME) is infiltrated by diverse immune cell types that either promote or suppress tumor progression. Understanding the dynamic roles of TIICs is crucial for developing targeted immunotherapies (Galon & Bruni, 2019).

Table 2.2 Functional roles of TIICs in tumor immunity.

Immune cell type	Function	Protumorigenic role	Antitumorigenic role	Reference
Cytotoxic T cells (CTLs) ($CD8^+$)	Recognize and eliminate tumor cells	Suppressed by immunosuppressive signals in TME	Release perforin and granzymes to kill tumor cells	Wherry (2011)
Regulatory T cells (Tregs)	Suppress effector T-cell activity	Maintain immuno-suppressive environment	-	Sakaguchi et al. (2010)
Tumor-associated macrophages (TAMs)	Highly plastic macrophages	M2 phenotype promotes tumor growth and immune evasion	M1 phenotype supports antitumor immunity	Mantovani et al. (2005)
Dendritic cells (DCs)	Bridge innate and adaptive immunity	Dysfunctional DCs fail to prime effective responses	Present antigens to activate T cells	Wculek et al. (2020)
Myeloid-derived suppressor cells (MDSCs)	Inhibit T-cell activation	Suppress antitumor immunity and support tumor growth	-	Bronte et al. (2016)
Natural killer (NK) cells	Mediate cytotoxicity and cytokine secretion	Function impaired in TME	Directly kill tumor cells and enhance immune response	Vivier et al. (2018)
Tumor-associated neutrophils (TANs)	Influence tumor progression	N2 phenotype promotes tumor growth	N1 phenotype enhances antitumor immunity	Fridlender et al. (2010)
B cells	Modulate tumor immunity and antigen presentation	Can promote immune evasion through immunoregulatory cytokines	Produce antibodies and activate T-cell responses	Shalapour & Karin (2015)

TIIC, *tumor-infiltrating immune cells.*

Therapeutic potential
Assessing methods to modulate TIICs for therapeutic effect, such as immune checkpoint blockade, cell-based therapies, and new immunomodulatory drugs. This part will also describe current clinical trials and future directions in targeting TIICs for better patient outcomes (Tables 2.2 and 2.3).

Tumor-infiltrating immune cells

1. **Antitumor immune cells**
   (a) **Cytotoxic T lymphocytes:** Cytotoxic T lymphocytes (CTLs), especially $CD8^+$ cells, engage with and eliminate malignant neoplasms by seizing the MHC class I (MHC-I) protein's tumor-specific antigens (TSA). CTLs employ multiple methods to destroy tumor cells; these include perforin and granzyme action, apoptosis via the Fas/FasL pathway, and the inflammatory

Table 2.3 Therapeutic potential.

Therapeutic approach	Description	References
Immune checkpoint blockade (ICB)	Monoclonal antibodies targeting immune checkpoints (e.g., PD-1/PD-L1 and CTLA-4) to restore T-cell function and enhance antitumor immunity	Topalian et al. (2012); Sharma & Allison (2020)
Adoptive cell therapy (ACT)	CAR-T cell therapy and tumor-infiltrating lymphocyte (TIL) therapy involve reinfusing ex vivo-expanded cytotoxic T cells to enhance antitumor immunity	Rosenberg et al. (2018)
Cancer vaccines and oncolytic viruses	Strategies such as dendritic cell vaccines and engineered oncolytic viruses to stimulate the immune system against tumors	Finn (2008)
Combination therapies	Combining immune-based therapies with chemotherapy, radiation, and targeted drugs to enhance efficacy and overcome resistance	Chen & Mellman (2017)
Targeting the tumor microenvironment (TME)	Modulating the TME using anti-VEGF agents, cytokine therapy, and metabolic reprogramming to enhance TIIC-mediated tumor destruction	Joyce & Fearon (2015)
Epigenetic and metabolic Regulation of TIICs	Investigating the impact of epigenetic modifications and metabolic reprogramming on TIIC function for targeted therapy	Zheng et al. (2021)
Clinical trials and future directions	Ongoing trials exploring novel immunotherapies, including personalized neoantigen vaccines, bispecific antibodies, and engineered immune cells	Wei et al. (2018); June et al. (2018)
Role of microbiota in modulating TIIC function	Gut microbiome influences immunotherapy effectiveness, highlighting microbiota-targeting strategies for improved patient outcomes	

TIIC, *Tumor-infiltrating immune cells;* CTLA-4, *cytotoxic T-lymphocyte-associated protein 4;* PD-1, *programmed cell death protein 1;* VEGF, *vascular endothelial growth factor.*

cytokines IFN-γ and TNF-α. Documented that some types of neoplasm are found with prolonged survival due to the presence of CD8 + T cells, which are sparse in the periphery and are seen in the diagnostic tissues of the tumors because of their prognostic value (Bruni et al., 2020). On the contrary, their CTL evading malignancies suffer with various immune evasion strategies such as PD-L1 enhanced immune checkpoint augmentation, immunosuppressive protein secretion, and downregulation of antigen presentation pathways. The development of techniques such as ICB, adoptive T cell therapy, and cancer vaccines, which overcome these challenges to control tumors for longer periods with CTLs, has greatly improved the efficacy of cancer treatment (Bruni et al., 2020; Schumacher & Schreiber, 2015)

(b) Natural killer cells: A system under infection defense, NK cells possess the ability to destroy cancerous cells through direct Immune Modulation and cytotoxicity. NK cells identify malignant cells via the "missing self" strategy, which activates tumors that have downregulated MHC class I molecules, as well as mediate antibody-dependent cell cytotoxicity (ADCC) through the CD16 receptor, making them crucial effectors in monoclonal antibody therapies (Myers & Miller, 2021). NK cells also release different cytokines, including IFN-γ and TNF-α, thus capturing and enhancing adaptive immunity and the tumoral microenvironment (Chiossone et al., 2018). However, tumors mute NK cell responses via the suppressive dysfunction of their activatory cytokines, immune checkpoints, and metabolic barriers (Myers & Miller, 2021). Restoration of NK cell activity and improvement of their longevity in vivo is achieved by using ICIs, such as anti-NKG2A, anti-PD-1, cytokine-based therapies such as IL-15 super agonists, and CAR NK cells with cytotoxic activity (Shemesh et al., 2022). Consequently, they have begun to stand behind NK cells as the answer for immunotherapy targeted at the cancer, providing fast, wide-ranging, destructive ability to cancer cells and less stringent requirements than T-cell microsurgery (Chiossone et al., 2018; Myers & Miller, 2021; Shemesh et al., 2022)

2. **Protumor and immunosuppressive cells**
   (a) **Regulatory T cells:** Regulatory T cells (Tregs) are heavily involved in immune tolerance and balance as they prevent overactive immune responses within the body (Kim & Lee, 2020; Li et al., 2023;). On the contrary, Tregs subdue essential antitumor immune reactions, enabling them to elude tumor microenvironments (TME) in the scope of cancer (Wang, Li, & Nakahata, 2024). Tregs or regulatory T cells are distinguished by the forkhead box P3 (FOXP3) transcription factor. This factor is very important in determining the various functions and actions of Tregs (Li et al., 2020). With respect to numerous types of carcinomas, it is noted that higher concentrations of Tregs within the tumor tend to correlate with worse outcomes in the disease, as Tregs inhibit CTLs alongside many other immune effector cells that are important for antitumor immunity (Kim & Lee, 2020).

During the TME, there are multiple ways that Tregs help with immune suppression. They elevate the number of obstructing receptors such as CTLA-4, PD-1, LAG-3, and TIGIT, which reinforce the suppression of effector T cells and antigen-presenting cells (Li et al., 2023). Tregs are also responsible for producing immunosuppressive cytokines such as TGF-β, IL-10, and IL-35. This weakens antitumor responses even more (Li et al., 2024; Wang, Li, & Nakahata, 2024). They also alter some metabolic pathways by utilizing IL-2, an important resource for the survival and function of effector T cells, in addition to expressing CD39 and CD73, which generate adenosine, an immunosuppressive compound (Li et al., 2020).

While Tregs are mostly associated with aiding in tumoral immune evasion and growth, some context-specific evidence suggests that they could also have an anti-tumor effect by modulating the inflammation and decreasing the immune responses that aid in tumor progression (Li et al., 2024). The diversity of Tregs from different tumor types and their adaptability in the TME adds another layer of complexity to the treatment strategies targeting them. The main hurdle in targeting Tregs for cancer immunotherapy is the selective depletion or inhibition of Tregs within the tumor without affecting the peripheral immune system, as autoimmune disorders may develop due to the uncontrolled removal of Tregs (Li et al., 2023). Current approaches of treatment involve targeting Treg-specific surface proteins such as GITR and OX40, as well as using modulatory checkpoint inhibitors that alter Treg activity (Li et al., 2020). Further, the understanding of the progression and resistance: the two sides of Tregs for the effective design of immunotherapeutic models is critical. More precision immunotherapies that strengthen and enhance antitumor attack at reduced risk of inadvertent immune damage will be possible through closer study of Treg diversity, their dynamics in the TME, and the molecular mechanisms that control their actions (Li et al., 2024; Wang, Li, & Nakahata, 2024)

1. **Myeloid-derived suppressor cells:** MDSCs enable tumor growth and mediate resistance to immunotherapy while exercising potent immune suppression within the tumor microenvironment (TME) (Krishnamoorthy et al., 2021; Ma et al., 2022). This phenomenon is caused because MDSCs functionally arise from disrupted myelopoiesis. Throughout myelopoiesis, the tumor factors that are secreted externally from the tumor's encapsulation halt the normal developmental progression into DCs, granulocytes, and monocytes, and result in the contempotilation of suppressive MDSCs (Wu et al., 2022; Zhao et al., 2023). MDSCs are widely categorized into two major subsets: polymorphonuclear (PMN)-MDSCs, which resemble neutrophils, and monocytic (M)-MDSCs, which resemble monocytes (Kai Li et al., 2021). MDSCs accomplish their immune suppression by harnessing multiple mechanisms at once. The generation of reactive oxygen species (ROS), nitric oxide (NO), and arginase-1 is formed, which all inhibit T cell proliferation and activity (Ma et al., 2022; Wu et al., 2022). Moreover, MDSCs also deplete essential amino acids such as arginine and cysteine, which impairs T-cell receptor signaling and antigen presentation (Krishnamoorthy et al., 2021). They also increase regulatory T-cell expansion, which further results in decreasing antitumor immunity (Zhao et al., 2023). The inclusion and expansion of MDSCs is directed by tumor-secreted cytokines and growth factors such as granulocyte-macrophage colony-stimulating factor (GM-CSF), interleukin 6 (IL-6), vascular endothelial growth factor (VEGF), and prostaglandin E2 (PGE2), which activate critical signaling pathways JAK/STAT and C/EBP β (Kai Li et al., 2021; Wu et al., 2022). Notably, the functions of MDSCs extend beyond immune suppression. They are also involved in tumor angiogenesis and metastasis by secreting matrix metalloproteinases (MMPs) as well as other potent angiogenic stimulators (Krishnamoorthy et al., 2021; Zhao et al., 2023). The latest approaches toward targeting MDSCs involve their depletion using chemotherapeutic agents and tyrosine kinase inhibitors, the inhibition of enzyme function through phosphodiesterase 5 or arginine inhibitors, and the alteration of differentiation in non-suppressive myeloid cells (Veglia et al., 2021; Wu et al., 2022). The complex interactions of MDSCs with other immune cells reinforce the view of their usefulness in forming the suppressive TME; therefore these findings should be explored in the future for enhancing cancer immunotherapy (Kai Li et al., 2021; Zhao et al., 2023)
2. **Tumor-associated macrophages:** TAMs can either aid in the progression of cancer or hinder it. Being the most prominent type of immune cell within the TME, macrophages have contradictory effects in augmenting cancer as well as mediating therapy-defiance resilience (Mantovani et al., 2018). Signals from the tumor and its microenvironment, overwhelmingly determined by tumor biology, govern these cells' functions. They can indeed be summoned to tumors through inflammation and schooled by chemokines CCL2 and CSF1, which enhance their migration (Shao et al., 2024). Macrophages are ipso facto divided into two categories: Procancer immune evasive or M2 macrophages versus anticancer immune M1 macrophages (Tan et al., 2021). The amount of available energy, oxygen, and prevailing environmental conditions dictate the tumor stage, which gives rise to different directions of macrophage metabolism (DeNardo & Ruffell, 2019). The presence of a high number of TAMs negatively correlates with the survival rate of the patients, indicating that they assist advanced stages of cancer by facilitating angiogenesis and metastasis (Chunxiao Li et al., 2021). Most of the patients with advanced cancer tend to succumb to the disease, and the

stage at which they succumb is often marked by the release or secretion of TAM factors such as VEGF and TGF that support the growth of the tumor (Mantovani et al., 2018). Moreover, these macrophages also regulate immune response by inhibiting the functions of the CD8+ T-lymphocytes and modulating other immune checkpoints, thus forming an environment rich in immunity to attack the tumor fundamentally (Galon & Bruni, 2020). The different types of macrophages, whether they are tissue resident macrophages or derived from monocytes, may have specific functional roles in cancer (DeNardo & Ruffell, 2019). Because of the control that TAMs exert over the TME, they have become important targets in cancer immunotherapy (Tan et al., 2021). Changing how TAMs are managed has been attempted by modifying their functions or recruitment, including turning off macrophage recruitment, depleting existing TAMs, or reprogramming them to an antitumor function (Chunxiao Li et al., 2021). It has been demonstrated that blocking CSF1 or its receptor CSF1R signaling tends to decrease the number of TAMs and improve the effectiveness of immune checkpoint blockade therapy (Shao et al., 2024). Also, better results in overcoming resistance and improving patient outcomes can be achieved by combining other types of treatment, including chemotherapy and radiotherapy, with those directed at TAMs. While there is still a problem of targeting TAMs because of their heterogenic nature, advances in single-cell profiling and immune modulation research continue to shed more light on the role of TAMs in cancer progression and therapeutic resistance (DeNardo & Ruffell, 2019; Mantovani et al., 2018). Their presence relates to unfavorable prognosis and treatment results across multiple tumor types, which renders them an interesting target for immunotherapeutic approaches (Ma et al., 2022).

Mechanisms of action

1. **Role in promoting tumor regression (antitumor effects):** The modification of tumor microenvironment (TME) and the growth or regression of cancers relies mostly on TIICs. The TME is a complex and heterogeneous system containing immune cells, stromal cells, and extracellular matrix elements that synergistically interact with cancer cells. The function and presence of TIICs regulate the equilibrium of pro-tumor and antitumor immune responses. Innate and adaptive immune components, including CTLs, NK cells, DCs, and M1 macrophages, play a vital role in strong tumor regression. They actively contribute to cancer elimination through lysis of invading tumor cells, cytokine generation, and antigen presentation (Peña-Romero & Orenes-Piñero, 2022).

The primary component of tumor regression performed by CD8+ T cells is antigen-specific cytotoxicity (Sharma et al., 2018). These cells capture the tumor-associated antigens on the MHC class I molecules within the respective tumor cells. There is activation and clonal expansion of CD8+ T cells, which makes them more proficient in attacking the malignant cells (Peña-Romero & Orenes-Piñero, 2022). The cytotoxic reaction occurs through two principal pathways; they are the perforin-granzyme pathway and the Fas-Fas ligand (FasL) pathway. Cytotoxic CD8 T lymphocytes (CTLs) kill tumor cells by using granzymes and caspases. The point to be focused that granzyme B is released upon perforin, creating pores in the cell membrane during the perforin-granzyme pathway. Therefore, apoptotic granular granzymes would act on caspases III and II to switch on the process of cell death (apoptosis). In the Fas-FasL pathway, the ligand of the T cell's Fas attaches to the Fas receptors of the tumor cells, which activates caspase-8, thus leading to stimulation of apoptosis (Sharma et al., 2018). Furthermore, CD8+ T cells release interferon-gamma (IFN-γ), which enhances the level of antigen presentation together with recruiting effector immune cells (de Visser & Joyce, 2023).

NK cells serve as one of the very first defenses of our innate immune system, as they effectively target and destroy tumor cells that tend to evade immune response by downregulating MHC class I molecules (Lu et al., 2023). NK cells depend on an equilibration of energizing and inhibiting signals to determine whether to establish cytotoxic responses, unlike CD8+ T cells, which need prior sensitization to locate malignant cells. Instead, inhibitory receptors such as killer immunoglobulin-like receptors (KIRs) block tumor autoimmunity by attaching to MHC class I molecules, while activating receptors such as NKG2D and NKp46 target and attach to stress ligands on tumor cells (Waldman et al., 2020). Granzyme-perforin pathways and crosslinking death receptors are the two main pathways by which apoptosis is integrated in tumors by NK cells. The perforin-granzyme mechanism involves the release of perforin, which generates pores in the membrane enabling granzymes to access and induce caspase-dependent apoptosis in the targeted cell. In the death receptor pathway, NK cells express Fas ligand (FasL) and TRAIL, which bind to their respective tumor cell receptors, and apoptosis is initiated via caspase-8. Correspondingly, NK cells are important in the modulation of the TME by unleashing inflammatory cytokines such as IFN-γ and TNF-α, which increase the recruitment and activation of other immune effector cells such as DCs and CD8+ T cells, thereby heightening the body's antitumor reaction (2014a; Bronte et al., 2016; Gopalakrishnan et al., 2018)

As DCs work as APCs, which connect the innate and adaptive immune responses, they are also somewhat important in the regression of tumors. Tumor antigens are recognized, processed, and mounted by the DCs to naïve T cells in lymph nodes, activating and proliferating tumor-specific $CD8^+$ and $CD4^+$ T cells (de Visser & Joyce, 2023). In addition, DCs produce proinflammatory cytokines such as interleukin-12 (IL-12) that also increase the T cell and NK cell activity, promoting a strong antitumor immune response. Regardless of this, there is always an immune avoidance strategy wherein tumors induce an immunosuppressive phenotype within DCs, stifling their capacity to activate T cells. One approach to conquer this problem is using DC-based vaccines, which target the ability of DCs to initiate antitumor T cell responses (Mun et al., 2022).

Checkpoint molecules such as CTLA-4 and PD-L1 play an important role in regulating immune responses within the TME. PD-L1, which is expressed on tumor cells and immune cells, binds to PD-1 on T cells, leading to T-cell exhaustion and immune evasion (Waldman et al., 2020). Equivalently, CTLA-4 competes with CD28 for binding to B7 ligands on antigen-presenting cells, thereby inhibiting T cell activation and decreasing immune responses. ICIs such as anti-PD-1, anti-PD-L1, and anti-CTLA-4 antibodies have revolutionized cancer therapy by blocking the inhibitory pathways, restoring T-cell function, and enhancing antitumor immunity (Sharma et al., 2018; Waldman et al., 2020). Regardless of their success, challenges such as resistance mechanisms and immune-related adverse events demand further investigation into combination therapies that integrate checkpoint blockade with other immunotherapeutic strategies.

1. **Contribution to immune evasion and tumor progression (immunosuppressive roles):** TIICs have a substantial impact on tumor evasion and growth through the establishment of a suppressive TME that inhibits anti-tumor immunity and augments tumor development. The TME is an integrated system of immune and stromal cells, cytokines, and extracellular matrix constituents that form the tumor ecosystem and alter the tumor's behavior (de Visser & Joyce, 2023). A diverse array of immune cells, such as Tregs, TAMs, MDSCs, and some DCs, are central to the inhibition of immune response and tumor advancement (Mun et al., 2022). Tregs exert their suppressive influence through the production of TGF-β and IL-10 that interfere with the activation and expansion of

cytotoxic T cells and NK cells, which leads to effective antitumor responses (Peña-Romero & Orenes-Piñero, 2022). Furthermore, Tregs secrete large amounts of checkpoint molecules such as CTLA-4 and PD-1, which causes another layer of suppression to effector T cells (Waldman et al., 2020). TAMs, particularly the M2 subtype, are major factors of immunosuppression in the TME. Different from M1 macrophages, which are proinflammatory and tumoricidal, M2 TAMs release immunosuppressive cytokines, including IL-10 and TGF-β, and aid in angiogenesis through VEG (Mantovani et al., 2018). These TAMs also enhance tumor invasion and metastasis by extracellular matrix breakdown through MMPs and EMT alterations (Peña-Romero & Orenes-Piñero, 2022). Moreover, they modify CAFs to maintain the inflammatory niche for the tumor, because of their capacity to secrete CCL2, CCL5, and the colony-stimulating factor 1 (CSF-1) that further immunosuppressive myeloid cells to the tumor area (Mun et al., 2022). MDSCs, another important immunosuppressed cell population in the TME, use several mechanisms to exercise immunosuppression, including the removal of crucial nutrients such as arginine and cysteine, production of ROS and NO, and inhibition of T-cell activation (Gao et al., 2022). In addition, these cells also increase the expression of immune checkpoint molecules and facilitate T-cell exhaustion, thereby reducing antitumor immunity (de Visser & Joyce, 2023). Furthermore, MDSCs release VEGF, HIF-1α, and other proangiogenic factors which contribute to the progression of angiogenesis and metastasis. Importantly, their presence in malignancies is related with deteriorating prognosis and resistance to immunotherapy. DCs, and more specifically the immature or tolerogenic ones, are also part of the mechanisms involved in immunoediting and assist the tumor in evading immune responses through poorly presenting tumor antigens to T cells, hence restricting functional adaptive immune system responses. These nonfunctional DCs are known to express PD-L1 at an increased rate, which decreases T-cell activation that is also a form of immunosuppression (de Visser & Joyce, 2023; Lu et al., 2023)The connection between these immunosuppressive immune cells and tumor cells is complex and poses a considerable obstacle to the immune system's ability to remove malignancies. Future therapeutic methods should focus on these suppressive immune populations, altering the TME to improve antitumor immunity. The use of MDSC-targeting agents, TAM reprogramming, immune checkpoint blockade, and Treg depletion is potentially effective in reversing resistance to immunotherapy and increasing the effectiveness of cancer treatment (de Visser & Joyce, 2023).

The TME reveals the complicated nature of immune dynamics, as TIICs reveal both antitumor and immunosuppressive capabilities. Although CD8+ T cells and NK cells induce strong antitumor responses via cytotoxic actions, their activity is often antagonized by suppressive immune cells such as Tregs, MDSCs, and TAMs (Peña-Romero & Orenes-Piñero, 2022). This complexity highlights the importance of the development of more effective therapies that seek to reposition the dynamics to focus on antitumor immunology. One fascinating option is the use of immune checkpoint medications targeting PD-1, PD-L1, and CTLA-4 to revive exhausted T cells and restore their cytotoxic function (Waldman et al., 2020). Furthermore, the potential to reverse or abolish immunological suppression by targeting TAMs and MDSCs with CSF-1R and arginase inhibitors supports probable therapeutic efficacy (Gao et al., 2022).

The advancement of technologies, particularly single-cell RNA sequencing and spatial transcriptomics, is paving the way towards developing personalized immunotherapies and improving cancer treatment through advanced cancer immunotherapy techniques (Elhanani et al., 2023; Peña-Romero & Orenes-Piñero, 2022) New approaches that are being studied to integrate immune response and tumor focus on including CAR-T cell therapy, DC-based vaccines, and immune-modulating nanoparticles (Lu et al., 2023). Also, increasing focus is being put on using metabolic modification of HIF-1α and lactate metabolism as means of lessening immune system suppression within the tumor microenvironment (TME) to improve antitumor response (de Visser & Joyce, 2023).

1. Dual role of reactive oxygen species on tumor-infiltrating immune cells and tumor microenvironment
   (a) **Modulation of immune activation suppression in the TME:** ROS have a climacteric dual function in managing TIICs in the TME. ROS can either increase the antitumor response by stimulating immune cells or suppress the immune system to promote tumor growth within the TME. Initially ROS are constructed via mitochondrial oxidative metabolism, nicotinamide adenine dinucleotide phosphate (NADPH) oxidases (NOXs), and nitric oxide synthases (NOSs) inside the immune cells (Shah et al., 2025). When present at neutral levels, ROS work as secondary messengers that improve immune activation; however, an excess amount of ROS results in oxidative stress, metabolic depletion, and immune suppression (Aboelella et al., 2021). An elevated level of ROS impairs CTL functions by inducing mitochondrial dysfunction, increasing immune checkpoint molecules such as PD-1 and CTLA-4, and encouraging the expansion of Treg, all of which together hinder antitumor immunity (Jie Wang et al., 2021; Kennel & Greten, 2021). Moreover, ROS affect macrophage polarization, driving TAMs to an M2 phenotype that facilitates tumor growth and angiogenesis, whereas MDSCs employ ROS to block antigen recognition and T cell activation, aiding in the creation of an immunosuppressive TME (Huang et al., 2021). DCs are also influenced by ROS, as oxidative stress hampers their maturation, lowers antigen presentation efficacy, and pushes them toward a tolerogenic phenotype that facilitates immune evasion (Shah et al., 2025). The protumor effects of ROS go far enough than immune suppression, increasing epithelial-to-mesenchymal transition (EMT), tumor invasion, metastasis, and angiogenesis via stabilizing hypoxia-inducible factors (HIFs) and raising the expression of VEGF (Aboelella et al., 2021). Furthermore, ROS-induced genomic instability drives tumor advancement by encouraging DNA damage and mutations, resulting in heightened tumor heterogeneity and resistance to therapy (Jie Wang et al., 2021; Kennel & Greten, 2021).

Conversely, ROS significantly aid in tumor suppression by improving the cytotoxic effects of NK cells and CD8 T cells while inducing immunogenic cell death (ICD) and promoting antigen presentation. Elevated ROS levels can activate apoptotic pathways by enhancing mitochondrial permeability and stimulating tumor suppressor proteins such as p53, leading to cancer cell death through processes such as ferroptosis, a type of programmed cell death caused by lipid peroxidation (Huang et al., 2021).

It has emerged as a promising approach to target ROS to enhance immune responses and to improve immunotherapy outcomes. ROS facilitates major immune processes, which include T-cell activation, antigen presentation, and polarization of macrophages, making them a key target for therapeutic intervention (Ling Wang et al., 2021; Perillo et al., 2020). Different methods to modulate ROS levels include the use of prooxidants to promote cancer cell death while concurrently protecting immune cells from oxidative damage. For example, ROS-inducing treatments, such as radiation and some chemotherapeutic agents, use oxidative stress to increase tumor cell apoptosis. Nevertheless, excessive ROS levels could also damage the immune function (Ling Wang et al., 2021). To counterbalance this problem, antioxidant therapies that involve N-acetylcysteine (NAC) and superoxide dismutase mimetics have been researched to maintain immune cell viability and function while especially increasing ROS levels in cancer cells (Perillo et al., 2020). Also, small molecule inhibitors that target NADPH oxidases (NOXs) have been developed to decrease the ROS-mediated immunosuppression by holding down the suppressive effects of MDSCs and Tregs in the TME (Ling Wang et al., 2021). In addition, combination therapies using ROS modulation along with ICIs have flaunted promise in restoring exhausted T cells and improving antitumor activity (Perillo et al., 2020). Contemporary studies say that fine-tuning ROS levels can optimize the TME for immune cell infiltration and function, which makes ROS targeted treatments an eye-catching avenue for future cancer immunotherapies (Ling Wang et al., 2021). As a conclusion, while ROS present a complex challenge in cancer treatment, controlled manipulation of their level offers a unique therapeutic door to boost immune responses and improve patient outcomes.

2.2.1 Immunotherapeutic strategies

The immune system can either limit tumor development and progression by eliminating cancer cells or obstructing their proliferation, or it can facilitate tumor progression by establishing a conducive milieu for tumor growth and metastasis. During the course of tumor formation, cancer immunosurveillance involves the identification and elimination of abnormal cancer cells by elements and cells of the innate and adaptive immune systems. Tumors have the ability to alter the immune system of the host, advance, and become clinically apparent (Kim et al., 2007).

1. **Development of major therapies**
   - (a) **Immune checkpoint inhibitors:** Despite significant advancements in ACT treatments, a novel category of monoclonal antibodies (mAbs), known as immune checkpoint inhibitors (ICIs), is now being integrated into medical practice and has emerged as a crucial kind of immunotherapy. Immune checkpoints are molecules involved in coinhibitory signaling pathways that preserve immunological tolerance; yet, cancer cells frequently exploit them to escape immunosurveillance (Pardoll, 2016). ICIs are engineered to restore anticancer immune responses by disrupting coinhibitory signaling pathways and to facilitate the immunologically mediated eradication of malignant cells (Sharma & Allison, 2015). The predominant targets for ICIs are CTLA-4, PD-1, and PD-L1.
   - (b) **Adoptive cell transfer**
     There are many advantages associated with adoptive T-cell transfer. To acquire anticancer capabilities, antitumor T cells that have a high level of specificity for tumor antigens can be cultivated in vitro in significant quantities, genetically manipulated, and/or activated outside of the body. On top of that, the host can be altered before the transfer of cells in order to eliminate suppressor cells, which include T-regulatory lymphocytes and MDSC, and to facilitate the in vivo proliferation of transferred lymphocytes (through a process known as "homeostatic expansion"). This is accomplished by removing endogenous lymphocytes that serve as a cytokine sink, competing for identical survival and stimulatory factors, particularly cytokines such as IL7 and IL15. During the ACT procedure, the lymphocytes that are delivered function as "living drugs" that are able to trigger long-lasting shielding (Rosenberg et al., 1988).
   - (c) **Oncolytic virus therapies**
     The oncolytic viruses (OVs) are organisms able to recognize, infect, and lyse distinct cells in the tumor environment, seeking to stabilize and diminish the tumor growth

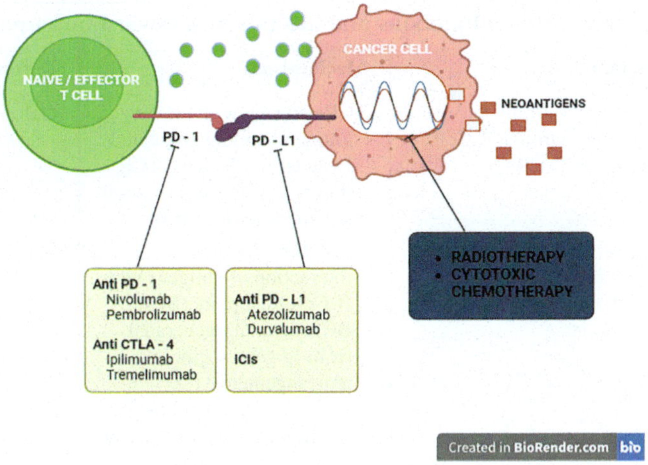

FIGURE 2.1

Immune checkpoint inhibitors in cancer treatment.

(Apolonio et al., 2021). They can provide a natural tropism to the malignant cells or be orientated genetically to find certain targets (Virotherapy, 2014). OVs are currently being explored as potential cancer therapies in clinical investigations. These viruses are being used to treat cancer. Cancer patients are currently being treated with these viruses. There are a number of OVs that have the ability to increase the activation of the immune system against tumor cells, which consequently has an effect on the creation of an anticancer response. This is an additional topic of interest (Desjardins et al., 2016) (Fig. 2.1).

2.2.2 Modulation of tumor-infiltrating immune cells for therapy

Immune infiltrates, as a crucial component of the TME, have been shown to affect tumor proliferation and responses to immunotherapy (Darvin et al., 2018). It has been demonstrated that immune infiltrates, serving as a considerable component of the TME, have the power to influence the genesis of tumors as well as the responses to immunotherapy (Balkwill et al., 2012). In addition, to achieve the goals of elucidating the mechanisms of immunotherapies, developing predictive biomarkers, and discovering novel therapeutic targets, it is essential to have a complete understanding of both innate and adaptive immune cells that are present within the microenvironment of the tumor (Zhang & Zhang, 2020).

1. **Strategies to enhance antitumor immune response**
   T cells, B cells, NK cells, and myeloid cells contribute significantly to efforts aimed at augmenting the antitumor immune response. Multiple roles of different cells are accountable for this. Some of the cell types, along with their functions and mechanisms toward immunotherapy, are outlined in Table 2.4
2. **Targeting immunosuppressive tumor-infiltrating immune cell subsets to overcome immune evasion**

Table 2.4 Cell type along with its functions and mechanisms toward immunotherapy.

Cell type	Function	Mechanisms	References
T cells	Potent tumor-killing capability through T-Cell Receptors (TCRs)	TCRs engage with tumor antigen peptides presented by MHC molecules; TCR diversity generated through genetic rearrangements	Galon et al. (2006); Bassing (2002)
B cells	Humoral immunity, secretion of antibodies to neutralize target antigens	Activation through interaction of antigens with B-Cell Receptors (BCR); BCR diversity generated from Ig gene segment rearrangement; class-switch recombination and somatic hypermutation optimize antibodies	Nutt et al. (2015)
NK cells	Innate immune response with cytotoxic functions, complementing MHC-restricted tumor lysis	Activation by combined action of activating and inhibitory receptors; sensing of stress signals when tumor cells lose MHC class I expression; impaired function in the TME	Sun et al. (2009); Paul & Lal (2017)
Myeloid cells	Heterogeneous cells (granulocytes and mononuclear phagocytes) with critical roles in immunity	Neutrophils and other myeloid cells interact with tumor cells, can escort Circulating Tumor Cell (CTCs), and polarize unconventional T cells. States of activation include N1 (antitumor) and N2 (protumor) phenotypes	Geissmann et al. (2011); Engblom et al. (2016); Coffelt et al. (2016)

There are several different populations and functions that T cells perform within the TME. T cells are the fundamental cell type in anti-tumor immunity. Consequently, in the field of tumor immunotherapy, they are regarded as a primary target cell type due to this reason. According to the established PD-1/PD-L1 inhibition, T cells are the target of the treatment. In spite of this, only a small fraction of people actually reap the benefits of this treatment being administered to them. In circumstances like these, it is extremely necessary to have a grasp of the mechanism of T cell-mediated tumor immunity in order to find new targets or targets of interest (Xie et al., 2021).

3. Antitumor targets associated with CD8+ T cells

CD8+ T cells are instrumental in tumor immunity by eliminating tumor cells and producing cytokines such as IFN-γ and IL-2. Tumors circumvent immunological elimination by suppressing CD8+ T-cell activity. Coinhibitory checkpoint receptors significantly contribute to immune suppression (Anderson et al., 2017; Blackburn et al., 2009; Workman et al., 2002). LAG3, present on exhausted CD8+ T cells, impedes their functionality, whereas its ligand FGL-1, which is elevated in malignancies, may obstruct immunological clearance (Chen et al., 2020). The ablation of FGL-1 enhances CD8+ T cell populations and decelerates tumor advancement, suggesting its involvement in immune evasion and resistance to PD-1 inhibition (Wang et al., 2020).

NRP-1, which is found in T cells and DCs, is responsible for promoting the survival of Treg cells and suppressing the function of CD8+ T cells in cancer patients. Inhibiting NRP-1 can restore CD8+ T-cell functionality and improve the effectiveness of blocking PD-1. B7S1, which is found on myeloid cells, also suppresses CD8+ T cells and may work in conjunction with PD-1 to reduce immune responses. Targeting both B7S1 and PD-1 simultaneously may enhance antitumor immunity (Kikutani & Kumanogoh, 2003).

Additional checkpoint proteins, such as NKG2A, which improves CD8+ T-cell tolerance, and Siglec-15, found in tumor and myeloid cells, also limit CD8+ T-cell activity. There is preliminary evidence that inhibiting NKG2A and Siglec-15 could be effective in suppressing tumor growth. In addition, PSGL-1, which is a T cell adhesion protein, is a contributor to inflammation and has the potential to alter the presence of CD8+ T cells in the TME (Gray et al., 2021; Montfoort et al., 2019; Sreeramkumar et al., 2015)

4. **Tumor-infiltrating CD4+ T cells in the tumor microenvironment**

   CD4+ T cells, which include subsets of Th1, Th2, Th17, and Treg cells, play a crucial role in the immune response to tumors. To enhance the immune system's ability to fight tumors, certain CD4+ T cells facilitate the infiltration and activation of CD8+ T cells. The function of effector T cells is inhibited by Treg cells within the TME, which results in the obstruction of antitumor immunity. One of the most important aspects of cancer immunotherapy is the concentration of attention on tumor-infiltrating regulatory T cells (Borst et al., 2018; Kennedy & Celis, 2008; Pyke et al., 2019). Signaling pathways associated with Treg metabolism present potential strategies for mitigating immunosuppression. Sphingosine-1-phosphate (S1P) suppresses FoxO1 and facilitates Treg development via the PPARγ pathway (Cartier et al., 2020; Pyne & Pyne, 2010). Inhibition of SphK1, which activates S1P, has demonstrated anti-tumor benefits in murine models, positioning the SphK1-S1P pathway as a prospective therapeutic target.

   CD36, a receptor for fatty acids, facilitates the adaptability of Treg cells within the TME and inhibits tumor-infiltrating CD8+ T cells (Pascual et al., 2017; Wang et al., 2020). Inhibiting CD36 with tumor-specific carriers or bispecific antibodies has demonstrated potential in clinical trials, providing a method to diminish Treg cells and augment antitumor immunity (Xiao et al., 2022).

   Activation of TLR8 reverses Treg-mediated regulation of effector T cells inside the tumor microenvironment, which occurs through the inhibition of glucose uptake and glycolysis, highlighting the potential of TLR8 agonists as antitumor agents (Peng et al., 2007). The NF-κB pathway, especially via c-Rel and TNFR2, is essential for Treg survival and functionality (Eiler et al., 2021; Grinberg-Bleyer et al., 2018). Inhibiting TNFR2 and synergizing with GITR activation with PD-1 inhibition has shown significant antitumor efficacy by diminishing Tregs and augmenting CD8+ T cell functionality.

5. **Integration of reactive oxygen species biology in therapy:** ROS include a category of extremely reactive chemical entities that contain oxygen. ROS are typically classified into two categories: Free oxygen radicals, characterized by the presence of one or more unpaired electrons in their outermost electron shell, and nonradical ROS, which lack unpaired electrons. Superoxide anion (O2 •−), hydrogen peroxide (H_2O_2), and hydroxyl radical (OH•) are some examples of ROS that have been the subject of the most extensive research. In the presence of transition metals,

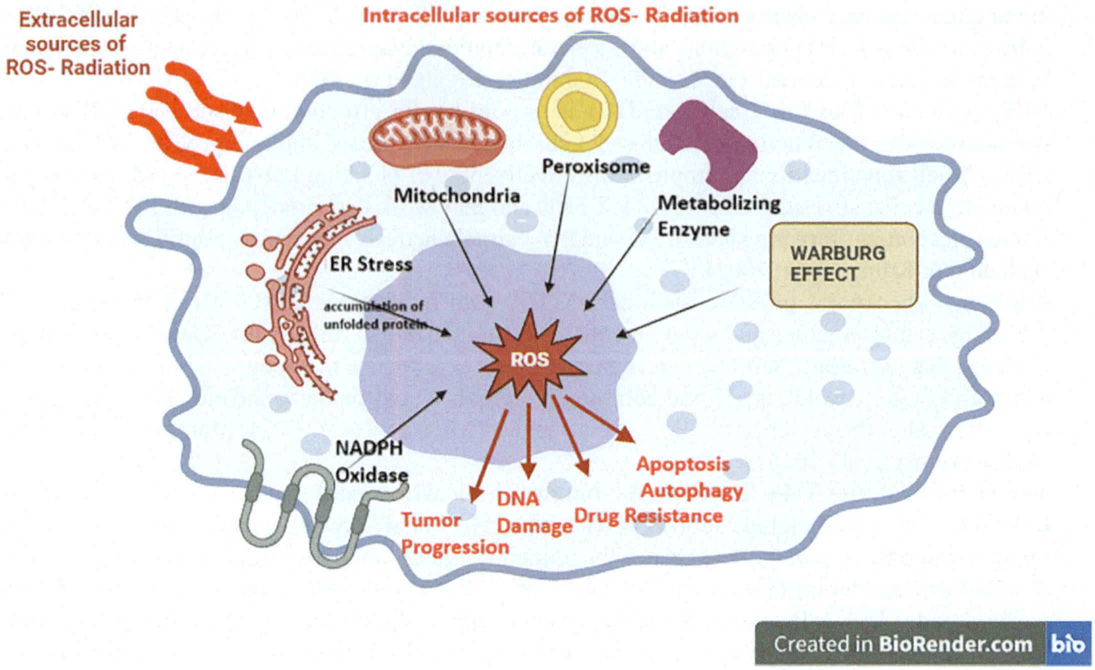

FIGURE 2.2

Reactive oxygen species targeted strategies via extracellular and intracellular sources.

specifically iron, the production of hydrogen peroxide (H_2O_2), which is derived from superoxide (O2 •−) by superoxide dismutase (SOD), can result in the generation of hydroxyl radicals (OH•) through the Fenton (Haber–Weiss) reaction. This transformation of non-radical ROS into radical ROS is a potentially significant phenomenon ($H_2O_2 + Fe^{2+} \rightarrow OH^- + OH\bullet + Fe^{3+}$) (Freinbichler et al., 2011). High levels of reactivity are exhibited by ROS, which are able to efficiently interact with a wide variety of intracellular macromolecules. These macromolecules include nucleic acids, lipids, and proteins. Considering the substantial body of evidence that establishes a connection between the production of ROS and a variety of human issues, such as persistent inflammation, several age-related diseases, and cancer, ROS are typically considered to be harmful consequences of cellular respiration (Dröge, 2002). The research that has been carried out over the course of the last several decades has proven that ROS play an essential role in a wide variety of biological processes. These processes include cellular survival, growth, proliferation, differentiation, and immunological response. It is vital for normal cellular functions to keep a ROS homeostasis that is finely regulated; any change in the equilibrium between oxidation and antioxidation results in oxidative stress, which in turn leads to cellular harm. It is therefore of great relevance to identify the sources of ROS, the signaling events that are triggered by ROS, and the functional implications that these events have across a wide range of cell types (Fig. 2.2).

2.3 Short-term versus long-term effects of reactive oxygen species on therapy outcomes

It is possible that ROS will serve a dual purpose in the outcomes of immunotherapy in the short run. On the one hand, elevated levels of ROS in the TME could potentially cause ICD, which would then result in the release of tumor antigens and the activation of dendritic cells, ultimately leading to an increase in T-cell priming (Zheng et al., 2023). For immunotherapies to be effective, it is necessary to stimulate the immune system. This is especially true for ICIs, which rely on the immunological response of the patient to recognize and remove cancer cells. Nevertheless, in the near term, an excessive synthesis of ROS might also result in negative impacts on the surrounding environment. DNA damage can be caused by ROS, which can lead to increased mutation rates in immune cells as well as neoplastic cells (Zhao et al., 2023). These modifications can result in immune evasion mechanisms, which make it more difficult for the immune system to recognize and eliminate cancer cells. Furthermore, oxidative stress that is caused by ROS has the potential to cause immune cell dysfunction, particularly in T-cells, by impeding their ability to produce cytokines and their capacity to kill cells, thus lowering the overall effectiveness of immunotherapy (Malla et al., 2024).

In the long run, the effects of ROS on the outcomes of immunotherapy become more complex. It is possible for immune suppression to occur as a consequence of persistent oxidative stress in the microenvironment of the tumor. Long-term exposure to ROS might result in the recruitment of immunosuppressive cells, such as Tregs and MDSCs, which hinder the immune system's ability to mount an effective antitumor response (Aboelella et al., 2021). Furthermore, ROS have the ability to increase the expression of immune checkpoint proteins, such as PD-L1, on both tumor cells and immune cells. This, in turn, makes immune evasion strategies easier to implement, which in turn reduces the sustained efficacy of therapies such as anti-PD-1/PD-L1 and anti-CTLA-4 inhibitors (Cui et al., 2024). In addition to this, ROS have been linked to the development of a tumor phenotype that is more aggressive and potentially metastatic. DNA damage caused by chronic ROS may contribute to the heterogeneity of tumors, which consequently leads to the formation of subclonal populations that are able to avoid treatment (Perillo et al., 2020).

2.4 Potential role of antioxidant systems in improving immunotherapy efficacy

Antioxidant systems are absolutely necessary for the function of modifying the TME and immune responses in the context of cancer immunotherapy. The existence of ROS has the potential to enhance the effectiveness of immunotherapies by enabling immunological activation. However, an excessive amount of ROS may undermine the activity of immune cells and lead to immune suppression. Adjusting antioxidant systems to control levels of ROS may thereby improve immunotherapy effects (Aboelella, Brandle, Kim, & Ding, 2021).

2.4.1 Antioxidants and immune system modulation

It is common for the microenvironment surrounding a tumor to be characterized by elevated levels of oxidative stress. This can have a negative impact on the efficacy of immunotherapies, such as ICIs and adoptive T-cell therapy. Through the reduction of oxidative damage, antioxidants may be able to assist in the restoration of immune cell function. In addition to reducing the effects of oxidative stress, antioxidant enzymes, such as SOD and catalase, protect immune cells from malfunctioning (Reilly et al., 2024). Increasing antitumor immunity may be accomplished by antioxidants through the reduction of oxidative stress, which in turn promotes the proliferation and activation of T-cells. To add insult to injury, antioxidants have the potential to reduce the immune-suppressive effects of the TME by lowering the concentration of Tregs and MDSCs, which are typically generated by oxidative stress (Oberholtzer et al., 2024). The findings of this study suggest that targeted antioxidant therapy has the potential to enhance the activity of effector T-cells and reduce the presence of immune-suppressive cells, thus establishing an immunological environment that is more favorable for cancer immunotherapy (Zhou et al., 2023).

2.4.2 Antioxidant therapy and combination approaches

Certain research indicate that the integration of antioxidant treatments with immunotherapy may enhance therapeutic success (Liu et al., 2022). For instance, the integration of antioxidants with ICIs may diminish ROS-mediated immune evasion mechanisms, including the increase of PD-L1 expression on tumor cells (Liu et al., 2022). The use of antioxidants alongside adoptive T-cell therapies may mitigate T-cell depletion induced by oxidative stress, hence enhancing the durability and efficacy of these treatments.

2.5 Challenges and future directions

2.5.1 Remaining challenges in understanding and targeting tumor-infiltration immune cells

The interaction between cancer and immune cells in the TME is thought to be crucial for the control of the development and progression of malignant tumors. Accordingly, tumor-infiltrating lymphocytes (TILs) have been identified in primary tumor tissue, tumor-bearing lymph nodes, and metastases of numerous cancer types. TILs are defined as lymphocytes within and around cancer cells and have been associated with a survival benefit.

To control cancers, infiltrating T cells must be retained there and traffic among and interact with tumor cells. Patients with diffuse immune cell infiltration among tumor cells have a better prognosis than patients whose infiltrates are confined to perivascular spaces. T cell movement is driven by both intrinsic motile capacity and extrinsic environmental organization and cues, including matrix proteins that provide a physical "roadmap" for T cells and chemical guidance cues in the form of chemokines. This makes T-cell motility highly dependent on activation status, antigen density, chemokine gradients, and target tissue type. Tumor cells can overexpress cell–cell adhesion proteins, possibly forming physical barriers to T-cell motility among tumor cells.

2.5.2 Leukocyte migration within tissues

Amoeboid migration enables T cells to move rapidly with undiminished migration velocities, such that they generate fast, low-affinity, and directional crawling. Alternatively, T cells can adapt a mesenchymal-like motility phenotype, characterized by adhesive spreading and nondirectional slow motility. In healthy and tumor tissues, T-cell motility is a dynamic balance between these two phenotypes, described as amoeboid-mesenchymal plasticity, which allows T cells to overcome Extracellular Matrix (ECM) barriers through adaptability and shape change.

2.5.3 Extracellular matrix barriers

The ECM is composed of collagen, fibronectin, elastin, glycosaminoglycans, and proteoglycans, many of which are produced and organized by fibroblasts and macrophages. ECM is a major determinant of tissue architecture, and ECM organization itself determines the direction and efficiency of T-cell movement. Tumors often express higher levels of matrix metalloproteinases (MMPs) than normal epithelial tissue, leading to increased remodelling of ECM fibers.

2.5.4 Matrix metalloproteinases

MMPs are enzymes that degrade collagens and other ECM components and, thus can enable tumor invasion or immune cell infiltration. Activated T cells express MMP RNA, but there is minimal MMP expression at the cell surface. This suggests that increased MMP activity can enhance immune cell migration in the tumor.

2.5.5 Emerging therapeutic approaches, including personalized immunotherapy and combination strategies

The potential role of the immune system to prevent or control cancer has been recognized for over 100 years. However, cancer immunotherapy has become a reality only within the last two decades. Direct targeting of the tumor has largely focused on the development of monoclonal antibodies.

Many cancer immunotherapy agents have recently been shown to induce durable clinical remissions in a variety of cancers. But, only a fraction of the patients appear to respond when these agents are used as monotherapy. Combining immunotherapy with other cancer treatments has emerged as a powerful strategy to enhance the effectiveness of treatments and overcome the limitations of monotherapies. These combination approaches aim to create synergistic effects that improve response rates, reduce resistance, and provide longer-lasting benefits to patients. The following are three key combinations: immunotherapy with chemotherapy, immunotherapy with radiation, and the integration of immunotherapy with targeted therapies.

The combination of immunotherapy with chemotherapy has shown significant promise in improving cancer treatment outcomes. Chemotherapy, which is regarded as the cornerstone of cancer treatment, may impair the immune system, which can hinder the body's ability to generate a strong immune response against cancer. Despite this drawback, recent research has shown that chemotherapy-induced ICD can prime the immune system to be more receptive to immunotherapy.

Radiation has been a standard treatment for localized tumors. Studies have demonstrated that radiation can also have profound immunomodulatory effects. By inducing ICD, radiation therapy releases tumor-associated antigens (TAAs) and damage-associated molecular patterns (DAMPs), which enhance the immune system's ability to detect and target cancer cells.

2.6 Future research directions to optimize tumor-infiltration immune cells modulation and reactive oxygen species integration

One of the major goals of cancer therapy is to induce targeted cytotoxic effects in the tumor tissue without damaging normal tissues. While achieving this using traditional chemotherapies has been challenging, ROS-mediated cancer cell killing has brought new hope. This is because the dose and location of ROS can be controlled by external stimuli, such as laser, ultrasound, and ionizing radiation, making it possible to restrict the killing effect within the tumor tissue.

2.7 Conclusion

TIICs assume a crucial dual role in cancer immunotherapy, acting as both protectors against tumors and promoters of tumor advancement, contingent upon their phenotype and microenvironmental setting. Cytotoxic T cells, NK cells, and DCs facilitate tumor regression by direct cytotoxicity and immunological activation, while suppressive populations such as regulatory T cells, MDSCs, and M2-type TAMs promote immune evasion and therapeutic resistance. Comprehending and altering the functional variety of TIICs is crucial for enhancing immunotherapeutic approaches. Recent technological advancements, including single-cell sequencing and spatial transcriptomics, alongside innovative therapeutic strategies such as checkpoint inhibitors, CAR-T cells, and macrophage-targeting agents, offer potential for personalizing treatments and addressing existing challenges in immunotherapy. Future initiatives should concentrate on modifying the tumor microenvironment to enhance antitumor immunity while reducing systemic immune-related side effects.

However, ROS demonstrate a contradictory role in the TME, functioning as both immune stimulators and inhibitors. Moderate amounts of ROS enhance essential immunological processes, including antigen presentation, T cell activation, and ICD, therefore fostering antitumor responses. In contrast, increased levels of ROS cause oxidative stress, hinder effector immune cells, encourage regulatory immunological subsets (such as Tregs and MDSCs), and facilitate tumor development, metastasis, and resistance to treatment. Addressing ROS with antioxidant treatments, NADPH oxidase inhibitors, and combination methods with ICIs constitutes a potential method to rejuvenate TIICs and modify the TME. Although progress has been made, considerable obstacles persist in enhancing TIIC regulation, surmounting extracellular matrix impediments, and regulating ROS levels to prevent unintended immune dysfunction. Future research should focus on individualized and spatiotemporally regulated ROS modulation techniques, along with current immunotherapies, to improve therapy specificity, effectiveness, and longevity in cancer management.

References

Aboelella, N.S., Brandle, C., Kim, T., & Ding, Z. (2021). Oxidative stress in the tumor microenvironment and its relevance to cancer immunotherapy. Cancers, 13(5), Article 986. https://doi.org/10.3390/cancers13050986

Anderson, A. C., Joller, N, & Kuchroo, V. K. (2017). Lag-3: Functions in immune regulation. *Immunity, 44*(5), 989–1004. https://doi.org/10.1016/j.immuni.2016.05.001.Lag-3.

Apolonio, J. S., Lima, V., Gonçalves, D. S., Luísa, M, Santos, C, Luz, M. S., Santos, M. L., Silva, A. M., & Silva, J. P. (2021). Oncolytic virus therapy in cancer: A current review. *World Journal of Virology, 10*(5), 229–255. https://doi.org/10.5501/wjv.v10.i5.229.

Balkwill, F. R., Capasso, M, & Hagemann, T (2012). The tumor microenvironment at a glance. *Journal of Cell Science, 125*, 5591–5596. https://doi.org/10.1242/jcs.116392.

Bassing, C. H., Swat, W., & Alt, F. W. (2002). The mechanism and regulation of chromosomal V(D)J recombination. *Cell, 109*(Suppl), S45–S55. https://doi.org/10.1016/s0092-8674(02)00675-x.

Binnewies, M, Roberts, E. W., Kersten, K, Chan, V., Douglas, F, Merad, M, Coussens, L. M., Gabrilovich, D. I., Ostrand-Rosenberg, S., Hedrick, C. C., Vonderheide, R. H., Pittet, M. J., Jain, R. K., Zou, W., Howcroft, T. K., Woodhouse, E. C., Weinberg, R. A., & Krummel, M. F. (2018). Understanding the tumor immune microenvironment (TIME) for effective therapy. *Nature Medicine, 24*, 541–550. https://doi.org/10.1038/s41591-018-0014-x.

Biswas, S. K., & Mantovani, A. (2010). Macrophage plasticity and interaction with lymphocyte subsets: Cancer as a paradigm. *Nature Immunology, 11*, 889–896. https://doi.org/10.1038/ni.1937.

Blackburn, S. D, Shin, H, Haining, W. N., Zou, T, & Workman, C. J. (2009). Coregulation of CD8+ T cell exhaustion by multiple inhibitory receptors during chronic viral infection. *Nature Immunology, 10*, 29–37. https://doi.org/10.1038/ni.1679.Coregulation.

Borst, J, Ahrends, T, Bąbała, N, Melief, C. J. M., & Kastenmüller, W (2018). CD4(+) T cell help in cancer immunology and immunotherapy. *Nature Reviews Immunology, 18*, 635–647. https://doi.org/10.1038/s41577-018-0044-0.

Brahmer, J. R., Tykodi, S. S., Chow, L. Q. M., Hwu, W.-J., Topalian, S. L., Hwu, P., Drake, C. G., Camacho, L. H., Kauh, J., Odunsi, K., Pitot, H. C., Hamid, O., Bhatia, S., Martins, R. G., Eaton, K., Chen, S., Salay, T. M., Alaparthy, S., Grosso, J. F., … Wigginton, J. M. (2013). Safety and activity of anti–PD-L1 antibody in patients with advanced cancer. *New England Journal of Medicine, 366*, 2455–2465. https://doi.org/10.1056/NEJMoa1200694.

Bronte, V, Brandau, S, Chen, S, Colombo, M. P., Frey, A. B., Greten, T. F., Mandruzzato, S., Murray, P. J., Ochoa, A., Ostrand-Rosenberg, S., Rodriguez, P. C., Sica, A., Umansky, V., Vonderheide, R. H., & Gabrilovich, D. I. (2016). Recommendations for myeloid-derived suppressor cell nomenclature and characterization standards. *Nature Communications, 7*, 1–10. https://doi.org/10.1038/ncomms12150.

Bruni, D., Angell, H. K., & Galon, J. (2020). The immune contexture and immunoscore in cancer prognosis and therapeutic efficacy. *Nature Review Cancer, 20*, 662–680. https://doi.org/10.1038/s41568-020-0285-7.

Burnet, S. M (1957). *British Medical Journal, 1*(5022), 779–786. doi:10.1136/bmj.1.5022.779.

Cartier, A., & Hla, T. (2020). Sphingosine 1-phosphate: Lipid signaling in pathology and therapy. *Science, 366*, 1–30. https://doi.org/10.1126/science.aar5551.

Chen, D. S., & Mellman, I (2017). Elements of cancer immunity and the cancer–immune set point. *Nature, 541*(7637), 321–330. https://doi.org/10.1038/nature21349.

Chen, Y., Zhu, G., Yin, W., Zheng, L., Zhou, T., & Badri, T. (2020). Fibrinogen-like protein 1 is a major immune inhibitory ligand of LAG-3. *Cell, 176*, 334–347. https://doi.org/10.1016/j.cell.2018.11.010.

Chiossone, L, Dumas, P-Y, Vienne, M, & Vivier, E (2018). Natural killer cells and other innate lymphoid cells in cancer. *Nature Reviews Immunology, 18*, 671–688. https://doi.org/10.1038/s41577-018-0061-z.

Coffelt, S. B., Wellenstein, M. D., & de Visser, K. E. (2016). Neutrophils in cancer: Neutral no more. *Nature Reviews Cancer, 16*, 431–446. https://doi.org/10.1038/nrc.2016.52.

Cui, J.-W., Li, Y., Yang, Y., Yang, H.-K., Dong, J.-M., Xiao, Z.-H., Liu, X., Zhang, Q., Wang, H., Li, J., Zhao, L., Chen, Y., Zhou, W., Sun, T., Huang, X., & Zhang, Y (2024). Tumor immunotherapy resistance: Revealing the mechanism of PD-1 /PD-L1-mediated tumor immune escape. *Biomedicine and Pharmacotherapy, 171*, 116203. https://doi.org/10.1016/j.biopha.2024.116203.

Darvin, P, Toor, S. M., Nair, V. S., & Elkord, E (2018). Immune checkpoint inhibitors: Recent progress and potential biomarkers. *Experimental & Molecular Medicine*, 1–11. https://doi.org/10.1038/s12276-018-0191-1.

DeNardo, D. G., & Ruffell, B. (2019). Macrophages as regulators of tumour immunity and immunotherapy. *Nature Reviews Immunology, 19*, 369–382. https://doi.org/10.1038/s41577-019-0127-6.

Desjardins, A., Vlahovic, G., & Friedman, H. S. (2016). Vaccine therapy, oncolytic viruses, and gliomas. *Oncology (Williston Park), 30*, 211–218.

de Visser, K. E, & Joyce, J. A. (2023). The evolving tumor microenvironment: From cancer initiation to metastatic outgrowth. *Cancer Cell, 41*, 374–403. https://doi.org/10.1016/j.ccell.2023.02.016.

Dröge, W. (2002). Free radicals in the physiological control of cell function. *Physiological Reviews, 82*, 47–95. https://doi.org/10.1152/physrev.00018.2001.

Eiler E., Sun H., Ghosh S., Jin J., & Murali R. c-Rel is a myeloid checkpoint for cancer immunotherapy 2021;1:507–517. https://doi.org/10.1038/s43018-020-0061-3.

Elhanani, O, Ben-uri, R, & Keren, L (2023). Review spatial profiling technologies illuminate the tumor microenvironment. *Cancer Cell, 41*, 404–420. https://doi.org/10.1016/j.ccell.2023.01.010.

Elkhader, J, & Elemento, O (2022). Artificial intelligence in oncology: From bench to clinic. *Seminars in Cancer Biology, 84*, 113–128. https://doi.org/10.1016/j.semcancer.2021.04.013.

Engblom, C, Pfirschke, C, & Pittet, M. J (2016). The role of myeloid cells in cancer therapies. *Nature Reviews Cancer, 16*, 447–462. https://doi.org/10.1038/nrc.2016.54.

Finn, O. J. (2008). Cancer immunology. *The New England Journal of Medicine, 358*, 2704–2715. https://doi.org/10.1056/NEJMra072739.

Freinbichler, W, Colivicchi, M. A., Stefanini, C, Linert, W, & Vares, D (2011). Highly reactive oxygen species: Detection, formation, and possible functions. *Cellular and Molecular Life Sciences, 68*(12), 2067–2079. https://doi.org/10.1007/s00018-011-0682-x.

Fridman, W. H., Pagès, F, & Sautès-fridman, C (2012). The immune contexture in human tumours: Impact on clinical outcome. *Nature Reviews Cancer, 12*. https://doi.org/10.1038/nrc3245.

Galon, J., & Bruni, D. (2019). Approaches to treat immune hot, altered and cold tumours with combination immunotherapies. *Nature Reviews Drug Discovery, 18*, 197–218. https://doi.org/10.1038/s41573-018-0007-y.

Galon, J, & Bruni, D (2020). Tumor immunology and tumor evolution: Intertwined histories. *Immunity, 52*, 55–81. https://doi.org/10.1016/j.immuni.2019.12.018.

Galon, J, Costes, A, Sanchez-Cabo, F, Kirilovsky, A, Mlecnik, B, Lagorce-Pagès, C, Tosolini, M., Camus, M., Berger, A., Wind, P., Zinzindohoué, F., Bruneval, P., Cugnenc, P. H., Trajanoski, Z., & Pagès, F. (2006). Type, density, and location of immune cells within human colorectal tumors predict clinical outcome. *Science, 313*, 1944–1960. https://doi.org/10.1126/science.1129139.

Gao, J, Liu, W-R, Tang, Z, Fan, J, & Shi, Y-H (2022). Myeloid-derived suppressor cells in cancer. *iLIVER, 1*, 81–89. https://doi.org/10.1016/j.iliver.2022.06.002.

Geissmann, F., Manz, M. G., Jung, S., Sieweke, M. H., & Ley, K. (2011). Development of monocytes, macrophages, and dendritic cells. *Science, 327*, 656–661. https://doi.org/10.1126/science.1178331.

Gopalakrishnan, V., Spencer, C. N., Nezi, L., Reuben, A., Andrews, M. C., Karpinets, T. V., Prieto, P. A., Vicente, D., Hoffman, K., Wei, S. C., Cogdill, A. P., Zhao, L., Hudgens, C. W., Hutchinson, D. S., Manzo, T., Petaccia de Macedo, M., Cotechini, T., Kumar, T., Chen, W. S., … Wargo, J. A. (2018). Gut microbiome influences efficacy of cancer immunotherapy. *Science, 359*, 97–103. https://doi.org/10.1126/science.aan4236.

Gray, MA, Stanczak, MA, Mantuano, NR, Xiao, H, Johan, FA, Malaker, SA, Miller, C. L., Weidenbacher, P. A., Tanzo, J. T., Ahn, G., Woods, E. C., Läubli, H., & Bertozzi, C. R. (2021). *Nature Chemical Biology, 16*, 1376–1384. https://doi.org/10.1038/s41589-020-0622-x.

Grinberg-bleyer, Y, Oh, H, Desrichard, A, Bhatt, D. M., Caron, R, Chan, T. A., Schmid, R. M., Klein, U., Hayden, M. S., & Ghosh, S. (2017). NF-κB c-Rel is crucial for the regulatory T cell immune checkpoint in cancer. *Cell, 170*, 1096–1108. https://doi.org/10.1016/j.cell.2017.08.004.

Hodi, F. S., Day, S. J. O., Mcdermott, D. F., Weber, R. W., Sosman, J. A., Haanen, J. B., Gonzalez, R., Robert, C., Schadendorf, D., Hassel, J. C., Akerley, W., van den Eertwegh, A. J. M., Lutzky, J., Lorigan, P., Vaubel, J. M.,

Linette, G. P., Hogg, D., Ottensmeier, C. H., Lebbé, C., & Wolchok, J. D. (2013). *New England Journal of Medicine, 363*, 711–723. https://doi.org/10.1056/NEJMoa1003466.

Huang, R, Chen, H, Liang, J, Li, Y, Yang, J, Luo, C, Liu, H., Wang, J., Xu, J., Zhao, X., Zhang, Y., & Liu, Z. (2021). Dual role of reactive oxygen species and their application in cancer therapy. *Journal of Cancer, 12*(13), 3844–3857. https://doi.org/10.7150/jca.54699.

Joyce, J. A., & Fearon, D. T. (2015). T cell exclusion, immune privilege, and the tumor microenvironment. *Science, 348*, 74–80. https://doi.org/10.1126/science.aaa6204.

June, C. H., O'Connor, R. S., Kawalekar, O. U., Ghassemi, S., & Milone, M. C. (2018). CAR T cell immunotherapy for human cancer. *Science, 359*, 1361–1365. https://doi.org/10.1126/science.aar6711.

Kennedy, R., & Celis, E. (2008). Multiple roles for CD4+ T cells in anti-tumor immune responses. *Immunological Reviews, 222*, 129–144. https://doi.org/10.1111/j.1600-065X.2008.00616.x.

Kennel, K. B., & Greten, F. R. (2021). Immune cell- produced ROS and their impact on tumor growth and metastasis. *Redox Biology, 42*, 101891. https://doi.org/10.1016/j.redox.2021.101891.

Kikutani, H, & Kumanogoh, A (2003). Semaphorins in interactions between T cells and antigen-presenting cells. *Nature Reviews Immunology, 3*, 159–167. https://doi.org/10.1038/nri1003.

Kim, J, Kim, B. S., Lee, S, Lee, S, Kim, J, & Lee, S (2020). Regulatory T cells in tumor microenvironment and approach for anticancer immunotherapy. *Cellular & Molecular Immunology, 20*, 1–17.

Kim R, Emi M, & Tanabe K. Cancer immunoediting from immune surveillance to immune escape Immunology:121(1)1–14. https://doi.org/10.1111/j.1365-2567.2007.02587.x.

Krishnamoorthy, M., Gerhardt, L., & Vareki, S. M (2021). Immunosuppressive effects of myeloid-derived suppressor cells in cancer and immunotherapy. *Frontiers in Immunology*, 1–21.

Ledford, J. R., Lane, J. D., & Severini, K. E (2017). Systematic use of visual analysis for assessing outcomes in single case design studies. *Behavioral Interventions, 19*, 4–17. https://doi.org/10.1017/BrImp.2017.16.

Li, C., Jiang, P., Wei, S., Xu, X., & Wang, J. (2020). Regulatory T cells in tumor microenvironment: New mechanisms, potential therapeutic strategies and future prospects. *Molecular Cancer, 19*, 116. https://doi.org/10.1186/s12943-020-01234-1.

Li, C., Xu, X., Wei, S., Jiang, P., Xue, L., & Wang, J. (2021). Tumor-associated macrophages: Potential therapeutic strategies and future prospects in cancer. *Journal for ImmunoTherapy of Cancer, 9*(4). https://doi.org/10.1136/jitc-2020-001341.

Li, H., Wang, S., Lian, Z., & Deng, S. (2023). Relationship between tumor infiltrating immune cells and tumor metastasis and its prognostic value in cancer. *Frontiers in Oncology, 13*, 1–19.

Li, K., Shi, H., Zhang, B., Ou, X., Ma, Q., Chen, Y., Li, J., Wang, H., Liu, X., Zhao, Y., Zhang, L., Sun, Q., Yang, C., Xu, W., Wu, D., Chen, L., Tang, Y., Li, F., Zhou, J., … Li, X (2021). Myeloid-derived suppressor cells as immunosuppressive regulators and therapeutic targets in cancer. *Signal Transduction and Targeted Therapy, 6*, 362. https://doi.org/10.1038/s41392-021-00670-9.

Li, Y., Zhang, C., Jiang, A., Lin, A., Liu, Z., Cheng, X., Wang, H., Chen, J., Xu, L., Zhao, M., Liu, Y., Zhang, Q., Yang, F., Li, P., Sun, T., Zhou, W., Wu, D., Li, R., … Zhang, H. (2024). Potential anti tumor effects of regulatory T cells in the tumor microenvironment : A review. *Journal of Translational Medicine*, 1–21. https://doi.org/10.1186/s12967-024-05104-y.

Liu, K., Sun, Q., Liu, Q., Li, H., Zhang, W., & Sun, C. (2022). Biomedicine & Pharmacotherapy Focus on immune checkpoint PD-1 /PD-L1 pathway : New advances of polyphenol phytochemicals in tumor immunotherapy. *Biomedicine & Pharmacotherapy, 154*, 113618. https://doi.org/10.1016/j.biopha.2022.113618.

Liu, R., Peng, L., Zhou, L., Huang, Z., Zhou, C., & Huang, C. (2022). Oxidative stress in cancer immunotherapy: Molecular mechanisms and potential applications. *Antioxidants, 11*, 853.

Lu, C., Liu, Y., Ali, N. M., Zhang, B., & Cui, X. (2023). The role of innate immune cells in the tumor microenvironment and research progress in anti-tumor therapy. *Frontiers in Immunology*, 1–14. https://doi.org/10.3389/fimmu.2022.1039260.

Ma, T., Renz, B. W., Ilmer, M., Koch, D., Yang, Y., Werner, J., Zhao, J., Li, X., Chen, H., Liu, Y., Wang, Z., Zhang, Q., Xu, L., Sun, T., Huang, Y., Li, P., Wu, D., Zhou, W., … Li, H (2022). Myeloid-derived suppressor cells in solid tumors. *Frontiers in Immunology*, 1–22.

Malla, R., Kumari, S., Ganji, S. P., Srilatha, M., Nellipudi, H. R., & Nagaraju, G. P. (2024). Reactive oxygen species of tumor microenvironment: Harnessing for immunogenic cell death. *Biochimica et Biophysica Acta (BBA) – Reviews on Cancer. iochim Biophys Acta Rev Cancer, 1879*, 189154. https://doi.org/10.1016/j.bbcan.2024.189154.

Mantovani, A., Marchesi, F., Malesci, A., & Laghi, L. (2018). Tumor-associated macrophages as treatment targets in oncology Tumor-associated macrophages as treatment targets in oncology. *Nature Reviews Clinical Oncology, 14*, 399–416. https://doi.org/10.1038/nrclinonc.2016.217.Tumor-Associated.

Mantovani, A., Sica, A., & Locati, M. (2005). Macrophage polarization comes of age. *Immunity, 23*, 344–346. https://doi.org/10.1016/j.immuni.2005.10.001.

Mun, J., Leem, S., Lee, J. H., & Kim, H. S. (2022). Dual relationship between stromal cells and immune cells in the tumor microenvironment. *Frontiers in Immunology, 13*, 1–15. https://doi.org/10.3389/fimmu.2022.864739.

Myers, J. A., & Miller, J. S. (2021). Exploring the NK cell platform for cancer immunotherapy. *Nature Reviews Clinical Oncology, 18*, 85–100. https://doi.org/10.1038/s41571-020-0426-7.

Nutt, S. L., Hodgkin, P. D., Tarlinton, D. M., & Corcoran, L. M. (2015). The generation of antibody-secreting plasma cells. *Nature Reviews Immunology, 15*, 160–171. https://doi.org/10.1038/nri3795.

Oberholtzer, N., Mills, S., Mehta, S., Chakraborty, P., & Mehrotra, S. (2024). Role of antioxidants in modulating anti-tumor T cell immune response. *Advances in Cancer Research, 162*, 99–124. https://doi.org/10.1016/bs.acr.2024.05.003.

Pardoll, D. M. (2016). The blockade of immune checkpoints in cancer immunotherapy. *Nature Reviews Cancer, 12*, 252–264. https://doi.org/10.1038/nrc3239.

Pascual, G., Avgustinova, A., Mejetta, S., Martín, M., Castellanos, A., Attolini, C. S.-O., Berenguer, A., Prats, N., Toll, A., Hueto, J. A., Vila, L., Palomero, L., Guzmán, M., Riera, A., Guiu, M., Rodriguez, J. L., … Massagué, J. (2017). Targeting metastasis-initiating cells through the fatty acid receptor CD36. *Nature, 541*, 41–45. https://doi.org/10.1038/nature20791.

Paul, S., & Lal, G. (2017). The molecular mechanism of natural killer cells function and its importance in cancer immunotherapy. *Frontiers in Immunology, 8*. https://doi.org/10.3389/fimmu.2017.01124.

Peña-romero, A. C., & Orenes-piñero, E. (2022). Dual effect of immune cells within tumour microenvironment: Pro- and Anti-tumour effects and their triggers. *Frontiers in Immunology, 13*, 1–15.

Peng, G., Wang, H. Y., Peng, W., Kiniwa, Y., Seo, K. H., & Wang, R.-F. (2007). Tumor-infiltrating gammadelta T cells suppress T and dendritic cell function via mechanisms controlled by a unique toll-like receptor signaling pathway. *Immunity, 27*, 334–348. https://doi.org/10.1016/j.immuni.2007.05.020.

Perillo, B., Di Donato, M., Pezone, A., Di Zazzo, E., Giovannelli, P., Galasso, G., Migliaccio, A., Castoria, G., Vici, P., Sacchi, A., Zazzeroni, F., D'Angelo, C., Pescatore, A., Sacco, A., Sica, V., Del Prete, S., … Giordano, A. (2020). ROS in cancer therapy: The bright side of the moon. *Experimental & Molecular Medicine, 52*, 192–203. https://doi.org/10.1038/s12276-020-0384-2.

Pyke, R. M., Thompson, W. K., Salem, R. M., Font-burgada, J., Zanetti, M., & Carter, H. (2019). Evolutionary pressure against MHC class II binding cancer mutations. *Cell, 175*, 416–428. https://doi.org/10.1016/j.cell.2018.08.048.

Pyne, N. J., & Pyne, S. (2010). Sphingosine 1-phosphate and cancer. *Nature Reviews Cancer, 10*, 489–503. https://doi.org/10.1038/nrc2875.

Reilly, A. O., Zhao, W., Wickström, S., Arnér, E. S. J., & Kiessling, R. (2024). Reactive oxygen species: Janus-faced molecules in the era of modern cancer therapy. *Cell*, 1–13. https://doi.org/10.1136/jitc-2024-009409.

Rosenberg, S. A., Packard, B. S., Aebersold, P. M., Solomon, D, Topalian, S. L., Toy, S. T., Simon, P., Lotze, M. T., Yang, J. C., Seipp, C. A., Rosenberg, S. M., White, D. E., & Morton, K. E. (1988). Use of tumor-infiltrating lymphocytes and interleukin-2 in the immunotherapy of patients with metastatic melanoma. A preliminary report. *The New England Journal of Medicine, 319*, 1676–1680. https://doi.org/10.1056/NEJM198812223192527.

Rosenberg, S. A., & Restifo, N. P. (2018). Adoptive cell transfer as personalized immunotherapy for human cancer. *Science, 348*, 62–68. https://doi.org/10.1126/science.aaa4967.

Sakaguchi, S, Miyara, M, Costantino, C. M., & Hafler, D. A. (2010). FOXP3 + regulatory T cells in the human immune system. *Nature Reviews Immunology, 10*, 490–500. https://doi.org/10.1038/nri2785.

Schumacher, T. N., & Schreiber, R. D. (2015). Neoantigens in cancer immunotherapy. *Science, 348*, 69–74. https://doi.org/10.1126/science.aaa4971.

Schumacher, T. N., & Thommen, D. S. (2022). Tertiary lymphoid structures in cancer. *Science, 375*, eabf9419. https://doi.org/10.1126/science.abf9419.

Shah, R., Ibis, B., Kashyap, M., & Boussiotis, V. A. (2025). The role of ROS in tumor infiltrating immune cells and cancer immunotherapy. *Metabolism, 148*, 1–29. https://doi.org/10.1016/j.metabol.2023.155747.

Shalapour, S., & Karin, M. (2015). Immunity, inflammation, and cancer: An eternal fight between good and evil. *Journal of Clinical Investigation, 125*, 3347–3355. https://doi.org/10.1172/JCI80007.

Shao, Y., Han, S., Hou, Z., Yang, C., & Zhao, Y. (2024). Tumor-associated macrophages within the immunological milieu: An emerging focal point for therapeutic intervention. *Heliyon, 10*, e36839. https://doi.org/10.1016/j.heliyon.2024.e36839.

Sharma, P., & Allison, J. P (2015). The future of immune checkpoint therapy. *Science, 348*, 56–61. https://doi.org/10.1126/science.aaa8172.

Sharma, P., & Allison, J. P. (2015). Immune checkpoint targeting in cancer therapy: Toward combination strategies with curative potential. *Cell, 161*, 205–214.

Sharma, P., & Allison, J. P. (2020). Dissecting the mechanisms of immune checkpoint therapy. *Nature Reviews Immunology, 20*, 75–76. https://doi.org/10.1038/s41577-020-0275-8.

Sharma, P., Hu-lieskovan, S., Wargo, J. A., Ribas, A., Cancer, J. C., & Angeles, L. (2018). *Cell, 168*, 707–723. https://doi.org/10.1016/j.cell.2017.01.017.

Shemesh, A., Pickering, H., Roybal, K. T., & Lanier, L. L. (2022). Differential IL-12 signaling induces human natural killer cell activating receptor-mediated ligand-specific expansion. *Journal of Experimental Medicine, 219*, e20212434.

Sreeramkumar, V., Adrover, J. M., Ballesteros, I., Cuartero, M. I., Bilbao, I., Nácher, M., Jordán, A., Martínez de Lizarrondo, S., Muñoz-López, A., García de Frutos, P., Ferrari, E., Pitari, G., Long, E., Hasegawa, H., Kawamoto, S., del Pozo, M. A., … Hidalgo, A. (2015). Neutrophils scan for activated platelets to initiate inflammation. *Science, 346*, 1234–1238. https://doi.org/10.1126/science.1256478.

Sun, J. C., Beilke, J. N., & Lanier, L. L. (2009). Adaptive immune features of natural killer cells. *Nature, 457*, 557–561. https://doi.org/10.1038/nature07665.

Tan Y., Wang M., Zhang Y., Ge S., Zhong F., & Xia G. Tumor-associated macrophages: A potential target for cancer therapy 2021;11:1–17. https://doi.org/10.3389/fonc.2021.693517.

Tgf-, N, Fridlender, Z. G., Sun, J, Kim, S, Kapoor, V., Cheng, G, Ling, L., Worthen, G. S., & Albelda, S. M. (2010). Polarization of tumor-associated neutrophil phenotype by TGF-β: "N1" versus "N2" TAN. *Cancer Cell, 16*, 183–194. https://doi.org/10.1016/j.ccr.2009.06.017.

Tirosh, I., Izar, B., Prakadan, S. M., Wadsworth, M. W., Treacy, D., Trombetta, J. J., Rotem, A., Rodman, C., Lian, C., Murphy, G., Fallahi-Sichani, M., Dutton-Regester, K., Lin, J.-R., Cohen, O., Shah, P., Lu, D., Genshaft, A. S., Hughes, T. K., Ziegler, C. G. K., … Garraway, L. A. (2016). Dissecting the multicellular ecosystem of metastatic melanoma by single-cell RNA-seq. *Science, 352*, 189–196. https://doi.org/10.1126/science.aad0501.

Topalian, S. L., Drake, C. G., & Pardoll, D. M. (2012). Targeting the PD-1/B7-H1(PD-L1) pathway to activate anti-tumor immunity. *Current Opinion Immunology, 24*, 207–212. https://doi.org/10.1016/j.coi.2011.12.009.

van den Broek, M. E., Kägi, D., Ossendorp, F., Toes, R., Vamvakas, S., Lutz, W. K., Melief, C. J. M., & Hengartner, H. (1996). Decreased tumor surveillance in perforin-deficient mice. *Journal of Experimental Medicine, 184*, 1781–1790. https://doi.org/10.1084/jem.184.5.1781.

van Montfoort, N., Borst, L., Korrer, M. J., Sluijter, M, Koen, A, Santegoets, S. J., van Ham, V. J., Ehsan, I., Charoentong, P., André, P., Wagtmann, N., Welters, M. J. P., Kim, Y. J., Piersma, S. J., van der Burg, S. H., … van Hall, T. (2018). NKG2A blockade potentiates CD8+ T cell immunity induced by cancer vaccines. *Cell, 175*, 1744–1755. https://doi.org/10.1016/j.cell.2018.10.028.NKG2A.

Veglia, F., Sanseviero, E., & Gabrilovich, D. I. (2021). Myeloid-derived suppressor cells in the era of increasing myeloid cell diversity. *Nature Reviews Immunology, 21*, 485–498. https://doi.org/10.1038/s41577-020-00490-y.

Virchow, R. (1989). Cellular pathology: As based upon physiological and pathological histology. Lecture XVI—Atheromatous affection of arteries. 1858. *Nutrition Reviews, 47*, 23–25. https://doi.org/10.1111/j.1753-4887.1989.tb02747.x.

Vivier, E., Artis, D., Colonna, M., Diefenbach, A., Di Santo, J. P., Eberl, G., Koyasu, S., Locksley, R. M., McKenzie, A. N. J., Mebius, R. E., Powrie, F., & Spits, H. (2018). Innate Lymphoid Cells: 10 Years On. *Cell, 174*, 1054–1066. https://doi.org/10.1016/j.cell.2018.07.017.

Vivier, E., Ugolini, S., Blaise, D., & Chabannon, C. (2016). Targeting natural killer cells and natural killer T cells in cancer. *Nature Reviews Immunology, 12*, 239–252. https://doi.org/10.1038/nri3174.

Waldman, A. D., Fritz, J. M., & Lenardo, M. J. (2020). A guide to cancer immunotherapy: From T cell basic science to clinical practice. *Nature Reviews Immunology, 20*, 651–668. https://doi.org/10.1038/s41577-020-0306-5.

Wang, H., Franco, F., Tsui, Y., Xie, X., Trefny, M. P., Mohmood, S. R., Zappasodi, R., Mohmood, S. R., Fernández-García, J., Tsai, C. H., Schulze, I., Metelli, A., Guo, X., Meylan, E., Hwu, P., Ho, P. C., & Li, X. (2020). Europe PMC Funders Group CD36-mediated metabolic adaptation supports regulatory T cell survival and function in tumors. *Nature Immunology, 21*, 298–308. https://doi.org/10.1038/s41590-019-0589-5.

Wang, J, Liu, N, Jiang, H, Li, Q, & Xing, D (2021). Reactive oxygen species in anticancer immunity: A double-edged sword. *Frontiers in Bioengineering and Biotechnology, 9*, 1–7. https://doi.org/10.3389/fbioe.2021.784612.

Wang, L., Kuang, Z., Zhang, D., Gao, Y., Ying, M., & Wang, T. (2021). Reactive oxygen species in immune cells: A new antitumor target. *Biomedicine & Pharmacotherapy, 133*, 110978. https://doi.org/10.1016/j.biopha.2020.110978.

Wang, Y., Li, J., & Nakahata, S. (2024). Complex role of regulatory T cells (Tregs) in the tumor microenvironment: Their molecular mechanisms and bidirectional effects on cancer progression. *International Journal of Molecular Sciences, 25*(13). https://doi.org/10.3390/ijms25137346.

Wculek, S. K., Cueto, F. J., Mujal, A. M., Melero, I., Krummel, M. F., & Sancho, D. (2020). Dendritic cells in cancer immunology and immunotherapy. *Nature Reviews Immunology, 20*, 7–24. https://doi.org/10.1038/s41577-019-0210-z.

Wei, S. C., Duffy, C. R., & Allison, J. P. (2018). Fundamental mechanisms of immune checkpoint blockade therapy. *Science, 359*(6382), 1350–1355. https://doi.org/10.1126/science.aar4060.

Wherry, E. J. (2011). T cell exhaustion. *Nature Immunology, 12*, 492–499. https://doi.org/10.1038/ni.2035.

Workman, C. J., Dugger, K. J., & Vignali, D. A. A (2002). Cutting edge: Molecular analysis of the negative regulatory function of lymphocyte activation gene-3. *Journal of Immunology, 169*, 5392–5395. https://doi.org/10.4049/jimmunol.169.10.5392.

Wu, Y., Yi, M., Niu, M., Mei, Q., & Wu, K. (2022). Myeloid-derived suppressor cells: An emerging target for anticancer immunotherapy. *Molecular Cancer, 21*, 184. https://doi.org/10.1186/s12943-022-01657-y.

Xiao, L., Liu, L., Ye, L., Su, P., Bi, E., Yang, M., Li, Z., Liu, H., Wang, Q., Gao, X., Zhou, B., Liu, X., Wang, H., Huang, H., Zhang, Y., Xu, Y., … Wang, S. (2022). CD36-mediated metabolic adaptation supports regulatory T cell survival and function in tumors. *Cell Metabolism, 33*, 1001–1012. https://doi.org/10.1016/j.cmet.2021.02.015.CD36-mediated.

Xie, Y., Xie, F., Zhang, L., Zhou, X., Huang, J., Wang, F., Li, H., Chen, Y., Liu, Z., Sun, Q., Yang, J., Xu, W., Wu, D., Tang, Y., Li, P., Zhao, X., Zhang, Y., Liu, X., Wang, H., … Liu, Y. (2021). Targeted anti-tumor immunotherapy using tumor infiltrating cells. *Advanced Science, 8*, 1–22. https://doi.org/10.1002/advs.202101672.

Ye, L., Zhang, T., Kang, Z., Guo, G., Sun, Y., & Lin, F. (2019). Tumor-Infiltrating immune cells act as a marker for prognosis in colorectal cancer. *Frontiers in Immunology, 10*, 1–12. https://doi.org/10.3389/fimmu.2019.02368.

Zhang, Y., & Zhang, Z. (2020). The history and advances in cancer immunotherapy: Understanding the characteristics of tumor-infiltrating immune cells and their therapeutic implications. *Cellular & Molecular Immunology*. https://doi.org/10.1038/s41423-020-0488-6.

Zhao, Y., Du, J., & Shen, X. (2023). Targeting myeloid-derived suppressor cells in tumor immunotherapy: Current, future and beyond. *Frontiers in Immunology*, 1–17. https://doi.org/10.3389/fimmu.2023.1157537.

Zheng, L, Qin, S, Si, W, Wang, A, Xing, B, Gao, R, Ren, X., Wang, L., Wu, Q., Sun, Y., Wang, L., Yu, X., Zhao, H., Zhang, Y., Li, W., Fan, J., Wang, J., Zhang, Z., Peng, W., … Zhang, L. (2021). Pan-cancer single-cell landscape of tumor-infiltrating T cells. *Science, 374*, abe6474. https://doi.org/10.1126/science.abe6474.

Zheng, Z., Su, J., & Bao, X. (2023). Mechanisms and applications of radiation-induced oxidative stress in regulating cancer immunotherapy. *Frontiers in Immunology, 14*, 1–16. https://doi.org/10.3389/fimmu.2023.1247268.

Zhou, X., An, B., Lin, Y., Ni, Y., Zhao, X., & Liang, X. (2023). Molecular mechanisms of ROS-modulated cancer chemoresistance and therapeutic strategies. *Biomedicine & Pharmacotherapy, 165*, 115036. https://doi.org/10.1016/j.biopha.2023.115036.

CHAPTER

3 A groundbreaking development in cancer treatment: Chimeric antigen receptor (CAR) T-cell therapy

Zahid Ahmad Thoker[a], Lovleen Marwaha[a] and Ajit Prakash[b]

[a]*Department of Zoology, School of Bioengineering and Biosciences, Lovely Professional University, Phagwara, Punjab, India,* [b]*Department of Biochemistry and Biophysics, University of North Carolina, Chapel Hill, NC, United States*

3.1 Introduction

The concept of cancer immunotherapy has transformed contemporary oncology through the paradigm shift of nonspecific cytotoxic modalities of cancer therapy to nonspecific approaches that utilize the inherent ability of the immune system to identify and destroy malignant cells. However, as opposed to traditional treatment methods, such as chemotherapy and radiotherapy, which work mostly by direct tumor cell killing, yet in many cases damage normal tissues, immunotherapeutic modalities are designed to reestablish or enhance immune surveillance in a more specific and long-lasting fashion. In the last 30 years, tumor immunology, molecular biology, and genetic engineering have come together to allow the creation of individualized immune-based treatment that can cause sustained clinical responses, including in patients with advanced or treatment-refractory disease (Couzin-Frankel, 2013). In this changing world, chimeric antigen receptor (CAR) T-cell therapy has become one of the most disruptive innovations with unseen levels of efficacy in some hematological malignancies and has redefined the expectations of cancer therapy outcomes.

3.1.1 Evolution of cancer immunotherapy

The theoretical basis of cancer immunotherapy dates back to initial observations of the immune system having the ability to identify tumor-associated antigens (TAAs) and have antitumor effects. The first immunotherapeutic activities were directed at nonspecific immune stimulation, such as cytokine therapies, including interleukin-2 (IL-2) and interferon-α, which showed little efficacy and high toxicity (Rosenberg, 2014). Later discoveries in the tumor antigen and regulation of the immune system resulted in the production of cancer vaccines and monoclonal antibodies, which have shown better specificity but lower clinical efficacy in most cases.

One of the breakthroughs was made in the discovery of immune checkpoint pathways, which control T-cell activation and tolerance. Clinical success of immune checkpoint inhibitors (ICIs) against cytotoxic T-lymphocyte-associated antigen 4 (CTLA-4) and programmed cell death protein 1 (PD-1) confirmed the notion that the long-term success in antitumor immunity could be promoted through the manipulation of endogenous immune response (Pardoll et al., 2012). Regardless of these improvements,

Advancements in Personalized Cellular Immunotherapy: A Comprehensive Guide to Cancer Treatment.
DOI: https://doi.org/10.1016/B978-0-443-27731-3.00001-2

Copyright © 2026 Elsevier Inc. All rights are reserved, including those for text and data mining, AI training, and similar technologies.

the checkpoint blockade treatment is still poor in a considerable percentage of patients, especially those with immunologically cold tumors or severe immune deficiency. It is these limitations that made it clear that simpler and more manageable immunotherapeutic approaches were required, which led to the development of adoptive cellular therapies, which directly encode immune effector cells to improve their ability to recognize and kill tumors.

3.1.2 Emergence of adoptive cell therapies

Adoptive cell therapy (ACT) is a breakthrough in the treatment of cancer, as it allows ex vivo manipulation and proliferation of immune cells before they can be returned to the patient. Initial ACT methods used tumor-infiltrating lymphocytes (TILs), which had shown excellent clinical results in metastatic melanoma, but were limited by complicated manufacturing and opposed tumor type application (Rosenberg & Restifo, 2015). Subsequently, T-cell specificity was redirected by the development of genetic engineering technology, which incorporated tumor-reactive receptors, such as transgenic T-cell receptors (TCRs).

TCR-engineered T-cells were superior to other immunotherapy methods in terms of antigen specificity, but the major histocompatibility complex (MHC) presentation limited their generalizability and made them vulnerable to tumor immune evasion pathways. One of the ways CAR T-cell therapy overcame these limitations was by incorporating antibody-derived antigen recognition domains into T-cell signaling machinery, enabling it to target tumors that are not dependent on the MHC. This novelty greatly expanded the therapeutic opportunities of ACT and created a framework of highly personalized and flexible treatment of cancer (June et al., 2018).

3.1.3 Rationality and scopes of chimeric antigen receptor T-cell therapy

The basis of CAR T-cell therapy is that it integrates the specificity of monoclonal antibodies with the cytotoxic capability of T lymphocytes, as well as their proliferative ability. CAR T-cell therapy allows specific antitumor activity and long-lasting engagement with their targets by genetically altering patient-derived or donor-derived T-cells to express a TAA synthetic receptor. In B-cell malignancies, clinical trials against CD19 have shown impressive response rates in patients with relapsed or refractory disease, many of whom had already tried all standard treatment methods (Maude et al., 2018).

In addition to being clinically successful, CAR T-cell therapy has been a scalable and modular platform that persistently advances with novel innovations in receptor design, gene-editing methods, and manufacturing technologies. Nonetheless, issues about toxicity, antigen evasion, complexities of manufacturing, and cost have been a major impediment to extensive adoption. This chapter will critically analyse the scientific principles, clinical successes, and translational issues of CAR T-cell therapy, put it in the context of the overall field of personalized cellular immunotherapy, and identify future directions of research and clinical practice.

3.2 Immunological basis of chimeric antigen receptor T-cell therapy

The therapeutic efficacy of CAR T-cell therapy has its primary foundations in the fundamental principles of cellular immunology, specifically the activation of T cells, antigen recognition, and immune regulation in the tumor microenvironment. T lymphocytes are core adaptive immunity effectors, which have antigen-specific cytotoxic capabilities, clonal expansion, and immunological memory. Nevertheless, endogenous T-cell responses are not adequate in cancer because of tolerance to immunity, antigen heterogeneity, and active immunosuppression. CAR T-cell therapy aims to address these shortcomings by reengineering T-cells with artificial receptors that can allow strong, antigen-specific stimulation, regardless of the traditional limitations of the immune system. To comprehend the challenges and issues that arise in the course of this approach application, it is crucial to understand the immunological underpinnings of the approach and its high efficiency.

3.2.1 T-cell biology and antitumor immunity

The T-cells contribute significantly to antitumor immunity by their capacity to identify antigens in the form of peptides on MHC and carrying out cytotoxic functions. When antigens are engaged, naïve T cells become activated by the concerted action of signaling through the TCR complex and costimulatory molecules, resulting in clonal expansion and the division into effector and memory subsets. Tumor cell elimination by cytotoxic CD8+ T cells mainly involves the discharge of perforin and granzymes and death receptor-mediated signaling, including Fas–Fas ligand interactions.

Although this has an inherent antitumor efficacy, successful T-cell-mediated immunity against cancer is often impaired. Tumors usually develop out of self-tissues and produce an antigen that is loosely immunogenic, leading to peripheral tolerance and T-cell anergy. In addition, long-term exposures to antigens in the tumor microenvironment may cause T-cell exhaustion, which is defined as a loss of effector function, persistent expression of inhibitory receptors, and altered transcriptional programs (Wherry & Kurachi, 2015). Such biological limits restrict the ability of endogenous T cells to generate long-term antitumor responses, and hence, there is a solid justification for adoptive therapies that improve T-cell responses by genetic engineering.

3.2.2 Tumor antigen recognition mechanisms

The cancer immunotherapy requires the optimal identification of TAAs or tumor-specific antigens (TSAs) by immune effector cells. The traditional T-cell recognition is induced through the TCR-peptide-MHC complex on the tumor cell or antigen-presenting cells' surface. These restrictions of MHC restriction necessitate a number of constraints, such as variability of human leukocyte antigen (HLA) expression and regular downregulation of antigen presentation machinery by tumors as a pathway of immune evasion (Garrido et al., 2016).

CAR T-cell therapy circumvents these limitations by utilizing antibody-derived single-chain variable fragments (scFvs), which directly recognize cell surface antigens regardless of MHC presentation. This MHC-nonspecific recognition allows for individualizing patients more widely and increasing accessibility of the target. When antigen-bound, intracellular domains of signaling in the CAR construct trigger T-cell-based cascades of signaling similar to physiological TCR signaling with a resultant release of cytokine, proliferation, and cytotoxicity. The addition of costimulatory domains, including CD28 or 4-1BB, also alters T-cell metabolism, persistence, and memory formation, further highlighting the immunological complexity of CAR design (Kawalekar et al., 2016).

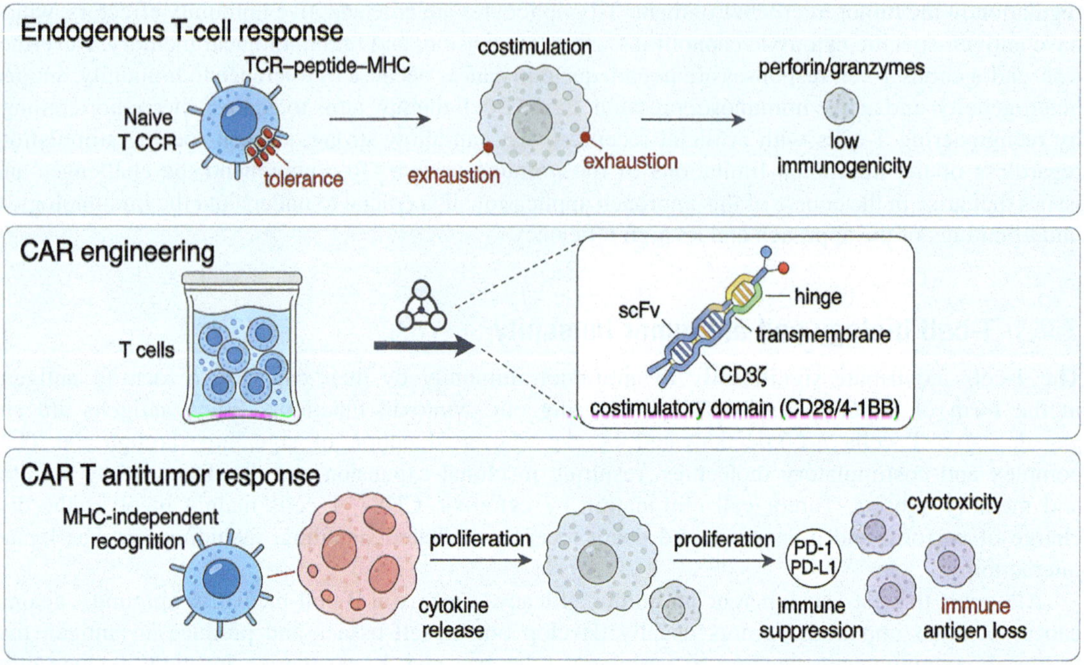

FIGURE 3.1

Schematic overview of endogenous T-cell antitumor responses, CAR T-cell engineering, and enhanced CAR-mediated tumor recognition. *CAR*, Chimeric antigen receptor.

3.2.3 Immune evasion by cancer cells

The development of cancer has a strong association with the capacity of tumor cells to evade immune surveillance by several intrinsic and extrinsic mechanisms. Tumors can silence expression of antigens, distort antigen-processing machinery, or provide an immunosuppressive microenvironment with regulatory T cells, myeloid-derived suppressor cells, and immunosuppressive cytokines such as transforming growth factor 2 and IL-10 (Hanahan & Weinberg, 2011). Moreover, involvement of immune checkpoint pathways such as PD-1/PD-L1 signaling also leads to functional impairment of TILs.

Although CAR T-cells are designed to withstand selected evasion mechanisms, they are prone to the loss of antigens and immunosuppressive cues of the tumor microenvironment. Escape of antigens has become one of the primary relapse conditions after a CAR T-cell therapy, especially in hematologic malignancies, and anti-single antigen approaches such as CD19 (Majzner et al., 2018). These limitations demonstrate the necessity to employ immunologically informed approaches, such as multi-antigen targeting and microenvironment modulation, to maintain long-term remedial actions. Fig. 3.1 summarizes the immunological principles of the CAR T-cell therapy.

3.3 Design and structural components of chimeric antigen receptors

The CAR T-cell therapy has an inherent relationship between the molecular architecture of the CAR construct and its clinical performance. CARs are artificial binding receptors that have been engineered to divert T-cell specificity to tumor-related antigens with a modular design, which incorporates antigen recognition, signal transduction, and cellular activation in a single protein chimeric construct. The CAR has a structural domain, the antigen-binding domain, which then has hinge, transmembrane domain and the intracellular signaling domains, each of which contributes uniquely but interdependently to the determination of antigen sensitivity, antigen signaling strength, antigen persistence and antigen safety. Optimization of these elements has been used rationally to achieve the successive generations of CARs, and is still a focus of enhancing therapeutic efficacy and decreasing toxicity.

3.3.1 Antigen-binding domains (scFv and alternatives)

The antigen-binding domain of a CAR is most often based on the single-chain variable fragment (scFv) of a monoclonal antibody, which provides specificity to a target cell surface antigen and is recognizable by antigens without the need to be presented by MHC. This is an inherent benefit of CAR T-cell therapy compared to other forms of TCR-based targeting because of this MHC-independent targeting (Sadelain et al., 2017).

Although scFv-based CARs are widely used, tonic signaling, aggregation, and immunogenicity have become an issue, especially when using murine-derived antibody sequences. To overcome these drawbacks, other antigen-recognition methods have been considered, such as the fully human scFvs as well as nanobody-based CARs, ligand-based CARs, and receptor-derived binding domains. These include innovations focused on increasing the target specificity, decreasing the immunogenicity, and raising the stability of CARs with retention of potent antitumor activity (June et al., 2018).

3.3.2 Hinge and transmembrane regions

The hinge (or spacer) binds the extracellular antigen-binding domain in CAR and transmembrane regions, which are needed to maintain CAR flexibility, accessibility to epitopes, and stability of the receptors. The hinge region also determines the gap between the T cell and target cell at immune synapse formation and thus affects optimal antigen interaction and minimized signaling responses. Derived domains of immunoglobulin G (IgG), CD8α, or CD28 are the most widespread, and each of them carries a different functional activity (Hudecek et al., 2015).

The transmembrane domain attaches the CAR to the T-cell membrane and aids in the receptor expression, dimerization, and signal transmission. CD3ζ-derived transmembrane regions, CD28-derived transmembrane regions, and CD8α-derived transmembrane regions have been found to selectively influence CAR stability and signaling behaviour. The selection of hinge or transmembrane elements may cause aberrant signaling or decreased persistence due to improper choice, and this fact underlines the significance of careful structural engineering of CARs (Alabanza et al., 2021).

3.3.3 Domains of the intracellular signaling

Activation, proliferation, and effector functions of T cells following antigen engagement are initiated by the intracellular signaling domains of CARs. The early versions of CARs utilized the CD3ζ signaling

Table 3.1 Evolution of CAR T-cell generations and their functional attributes.

CAR generation	Key structural features	Functional advantages	Limitations
First generation	scFv + CD3ζ signaling domain	Proof-of-concept; antigen-specific activation	Poor persistence; limited efficacy
Second generation	CD3ζ + one costimulatory domain (CD28 or 4-1BB)	Enhanced proliferation and persistence; clinical efficacy	Toxicity risk; antigen escape
Third generation	CD3ζ + two costimulatory domains	Improved signaling strength	Limited clinical superiority
Fourth generation (TRUCKs)	Cytokine-secreting CARs (e.g., IL-12, IL-15)	Tumor microenvironment modulation	Risk of cytokine-related toxicity
Fifth generation	Cytokine receptor motifs (e.g., IL-2Rβ, STAT activation)	Sustained activation; physiologic signaling	Early clinical-stage development

CAR, *chimeric antigen receptor;* IL, *interleukin;* scFv, *single-chain variable fragments;* TRUCKS, *T-cells redirected to kill cytokines.*

domain alone, which only gave out activation signals but could not sustain T-cell growth and proliferation. The costimulatory domains, including CD28 or 4-1BB (CD137), were incorporated afterward, resulting in second-generation CARs with significantly higher clinical efficacy (Sadelain et al., 2017).

The costimulatory domains do regulate various signaling pathways and metabolic programs in CAR T-cells. CD28-based CARs normally result in a rapid activation and effector differentiation, whereas 4-1BB-based CARs, in turn, increase mitochondrial biogenesis and long-term survival. The selection and combination of signaling domains, in turn, have an essential impact on the therapeutic longevity, the toxicity rates, and the total clinical results (Kawalekar et al., 2016).

3.3.4 Evolution of chimeric antigen receptor generations

CAR technology has undergone various generations, with each generation having more functional components to improve therapeutic performance. Early CARs, which comprised CD3ζ signaling only, showed low clinical efficacy. Second-generation CARs incorporated one domain of costimulatory and were the foundation of the approved therapies to date. Third-generation CARs also joined several costimulatory domains, but the clinical advantage is yet to be examined.

The fourth-generation CARs are also known as armored CARs or T-cells redirected to kill cytokines (TRUCKs) because they have been programmed to release immunomodulatory cytokines upon activation, altering the tumor microenvironment and thereby facilitating the antitumor immunity. CARs of the fifth generation also include cytokine receptor signaling motifs to induce JAK-STAT pathways to facilitate more physiologic and long-term T-cell activation (Rafiq et al., 2020). This gradual development indicates more advanced knowledge in T-cell biology and leads to innovation in the field of CAR T-cell therapy. Table 3.1 also provides a summary of the sequential development of a CAR T-cell design between the first-generation design and the newly emerging fifth-generation design. It is comparatively discussed that small changes in receptor structure, especially the addition of costimulatory and cytokine-signaling domains, have been translated into enhanced T-cell survival, antitumor efficacy, and clinical relevance, but have also brought new safety and regulation systems.

3.4 Manufacturing and engineering of chimeric antigen receptor T-cells

The manufacturing process of CAR T-cell therapy is of paramount importance to its clinical success because it relies on its strength, reproducibility, and adherence to the regulatory requirements. As a contrast to standard pharmaceuticals, CAR T-cells are biologically active, patient-specific medical goods and demand complex multi-step production pipelines encompassing the fields of cell biology, genetic engineering, and quality control. The variability of manufacturing can have direct effects on therapeutic efficacy, safety, and accessibility, and process optimization is a key concern in the sector. The development of engineering policies and regulatory harmonization has had a gradual enhancement of consistency of the products without loss of biological potency.

3.4.1 T-cell collection and isolation

The production of CAR T-cells usually starts with the harvesting of peripheral blood mononuclear cells (PBMCs) of patients using leukapheresis. This should be handled very carefully, especially in the highly pretreated cancer patients who might display lymphopenia or loss of T-cell fitness. After harvesting, immunomagnetic selection of T cells stained with CD3+ populations is used to enrich the starting material to obtain adequate material to continue with the engineering (Hartmann et al., 2017).

Phenotype and functional quality of T cells that have been harvested play an important role in determining CAR T-cell expansion, persistence, and antitumor activity. There is growing evidence that less-differentiated T-cell subsets, including central memory and stem cell-like memory T-cells, have an enhanced proliferative ability as well as long-term persistence compared to terminally differentiated effector cells. Therefore, production techniques are progressively designed to maintain positive T-cell phenotypes when in isolation and initial activation procedures (Gattinoni et al., 2017).

3.4.2 Genetic modification strategies

The characteristic procedure in CAR T-cell engineering is genetic modification, through which the T-cell specificity against tumor antigens can be redirected by stable expression of synthetic receptors. This process has to be efficient in transferring the gene, and reduced genotoxicity needs to be attained, besides maintaining cellular viability.

3.4.2.1 Viral vector–based gene transfer

Viral vectors, especially gamma-retroviral and lentiviral vectors, have been the most popular system on which CAR genes are delivered because of their high transduction and sustained genomic integration. The benefit of lentiviral vectors is the ability to transduce both dividing and nondividing cells and, hence, the popularity of CAR T-cell products in clinical use (Milone & O'Doherty, 2018).

Although successful, viral vectors have challenges in terms of manufacturing cost, scaling, and risk of insertional mutagenesis. These risks have been significantly reduced by rigorous design of vectors and ensuring safety testing, and there are no oncogenesis cases of any vector-induced oncogenesis, up to the approved CAR T-cell therapies. However, the dependence on viral systems still encourages the search for alternative technologies of gene delivery.

3.4.2.2 Nonviral gene editing approaches

Alternatives to viral vectors have become a promising technology in nonviral gene editing strategies, which have a better safety profile and lower cost of production. Transposon-based systems, such as Sleeping Beauty and piggyBac, make it possible to integrate the gene without viral elements and are shown to be clinically viable in clinical trials at an early stage (Bishop et al., 2021). Moreover, CRISPR–Cas9-based genome editing enables the accurate insertion of CAR constructs in specific loci in the genome, which aids in better control of expression and minimizes the occurrence of insertional mutagenesis. In addition to CAR insertion technology, the editing enables multiplex alterations, such as blocking inhibitory receptors, deleting endogenous TCRs to use in allogenic, and improving resistance to immunosuppressive tumor microenvironment. These functionalities make nonviral solutions emerging as important facilitators of next-generation CAR T-cell solutions.

3.4.3 Ex vivo expansion and quality control

After genetic modification, CAR T-cells are expanded by ex vivo proliferation under regulated culture conditions to obtain clinically pertinent cell dosage. Expansion protocols make use of cytokine supplements, typically IL-2 and IL-7 or IL-15, to attain growth with maintenance of the functional potential. There is a fine optimization of culture duration and conditions to balance between cell yield and phenotypic integrity (June et al., 2018).

Before infusion of patients, extensive quality control testing is required, which involves checks of identity, purity, potency, sterility, and transgene expression. Functional assays of cytotoxicity and cytokine generation are also being included in increasing numbers to have a more accurate performance prediction in vivo. Such quality features are needed to pass the regulatory approval as well as to reduce variability in clinical outcomes.

3.4.4 Regulatory and good manufacturing practice considerations

CAR T-cell products are classified as advanced therapy medicinal products (ATMPs) or biologics, under which there is a high-level regulation by governments, including the U.S. Food and Drug Administration and the European Medicines Agency. Manufacturing needs to adhere to the good manufacturing practice (GMP) principles, which include the design of the facilities, staff training, documentation, and batch traceability (Hartmann et al., 2017). The entire production pipeline is shown in Fig. 3.2, including major steps in the process, genetic modification, and ongoing GMP controls that must be in place to manufacture clinical-grade CAR T-cell products.

The regulatory frameworks are changing due to the special demands of personalized cellular therapies, such as the decentralized models of manufacturing and automated models of production. Regulatory controls and manufacturing standards will need to be harmonized to achieve the expansion of CAR T-cell therapy and increase options of care across the globe.

3.5 Clinical applications of chimeric antigen receptor T-cell therapy

The best clinical effect of CAR T-cells has been observed with hematologic malignancies, wherein target antigens are expressed more uniformly and accessible than with solid tumors. The impressive

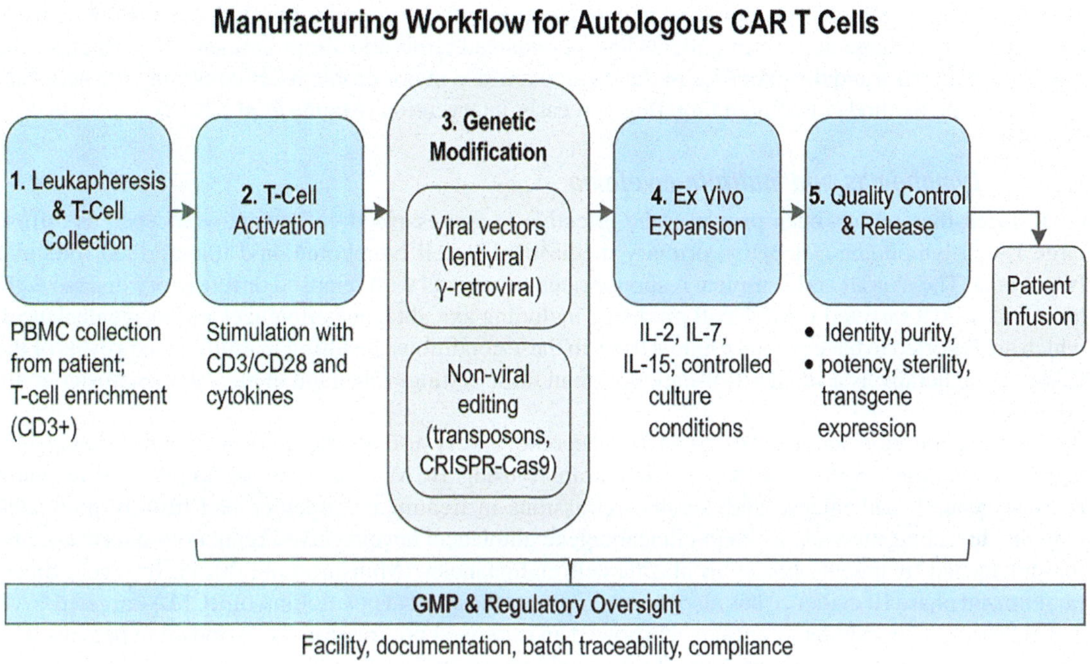

FIGURE 3.2

Manufacturing workflow of autologous CAR T-cells from leukapheresis to final clinical infusion. *CAR*, Chimeric antigen receptor.

response seen in relapsed or refractory disease patients has resulted in several regulatory approvals and has put the CAR T-cell therapy as a standard-of-care therapy in selected indications. The continued research is seeking to apply these achievements to solid tumors and to improve patient selection, timing of therapeutic intervention, and combination therapy.

3.5.1 Chimeric antigen receptor T-cell therapy in hematological malignancies

The initial malignancies to be treated using CAR T-cell therapy were hematological cancers because they have well-defined lineage-specific antigens and because the malignant cells in question are relatively accessible in the circulation and in the bone marrow. Most of these include CD19-directed CAR T-cells, which have demonstrated unparalleled response rates even in schooled patients.

3.5.1.1 Acute lymphoblastic leukemia

The most promising clinical response of CAR T-cell therapy so far is B-cell acute lymphoblastic leukemia (B-ALL). Initial clinical trials showed a 100% remission rate in children and young adult patients with a relapsed or refractory disease, most of whom had gone through several failed therapies (Maude et al., 2014). The sustained response resulted in the original approval of a CAR T-cell product, tisagenlecleucel, for pediatric and young adult B-ALL by the U.S. Food and Drug Administration (FDA).

CAR T-cell therapy has the potential to cure patients with long-term remission, and follow-up trials have shown that a large number of patients have a sustained remission after treatment. Nevertheless, the issue of antigen loss and the toxicities of the therapy are also major problems, highlighting the necessity to enhance the methods of antigen targeting and early intervention (Maude et al., 2018).

3.5.1.2 Lymphomas and multiple myeloma

CAR T-cell therapy has been proven to be effective in aggressive B-cell lymphomas such as diffuse large B-cell lymphoma (DLBCL), primary mediastinal B-cell lymphoma, and transformed follicular lymphoma. The overall and complete response rates in patients with relapsed or refractory disease were high with CD19-targeted CAR T-cell products, including axicabtagene ciloleucel and tisagenlecleucel, which are reported to have pivotal clinical trials in these conditions, resulting in multiple approvals of the CAR T-cell therapy as a standard-of-care option in these settings (Neelapu et al., 2017; Schuster et al., 2019).

B-cell maturation antigen (BCMA) has become a very useful and lineage-specific therapeutic target in multiple myeloma. CAR T-cell therapies using BCMA have shown impressive response rates in general, and intense and durable remissions in treatment-refractory and heavily pretreated patients. Idecabtagene vicleucel and ciltacabtagene autoleucel have received regulatory approvals, and this broadened treatment choices in plasma cell malignancies (Munshi et al., 2021). In recent times, randomized phase III evidence has also again supported the clinical positioning of BCMA-targeted CAR T-cell therapy. Ciltacabtagene autoleucel showed an improved progression-free survival in patients who are refractory to lenalidomide-based multiple myeloma treatments, highlighting the disruptive potential of CAR T-cell therapy even in earlier-stage use during treatment (San-Miguel et al., 2023). Taken together, the results demonstrate the dynamic status of CAR T-cell therapy in lymphoid malignancies, in addition to the importance of the necessity to maximize the lifespan of response and control of recurrence in long-term survivors.

3.5.2 Current status in solid tumors

Although CAR T-cell therapy has been successful in hematological malignancies with the aim of transformation, it has been much harder to translate into solid tumors. Heterogeneous expression of antigens, physical obstacles restricting the infiltration of T-cells, and highly immunosuppressive tumor microenvironments define solid tumors. Antigens, including HER2, GD2, mesothelin, and EGFR, have been the target of clinical trials so far that have shown limited clinical effectiveness and, in certain instances, a high amount of on-target off-tumor toxicity, which emphasizes the challenge in finding truly tumor-specific targets in solid malignancies (Newick et al., 2016).

To eliminate these obstacles, various innovative approaches are now actively being studied. They are local or intratumoral CAR T-cell delivery, genetic engineering of chemokine receptors to promote tumor homing, armored CAR constructs to secrete cytokines, and combinatoric or multiantigen targeting strategies to overcome antigen heterogeneity. Recent systematic reviews suggest that, despite almost 200 CAR T-cell clinical studies in solid tumors undergoing active clinical trials all over the world, objective response rates have been relatively low, which indicates the need to continue optimizing such approaches (Qian et al., 2024). There is also some emerging evidence to suggest that armored versions of CAR T-cells have the potential to promote the local inflammation and persistence of host immune

Table 3.2 Approved CAR T-cell therapies: Indications, targets, and key toxicities.

Product name	Target antigen	Indication	Year approved	Key toxicities
Tisagenlecleucel	CD19	B-ALL, DLBCL	2017	CRS, ICANS
Axicabtagene ciloleucel	CD19	DLBCL	2017	CRS, neurotoxicity
Brexucabtagene autoleucel	CD19	Mantle cell lymphoma	2020	CRS, ICANS
Idecabtagene vicleucel	BCMA	Multiple myeloma	2021	CRS, cytopenias
Ciltacabtagene autoleucel	BCMA	Multiple myeloma	2022	CRS, delayed neurotoxicity

B-ALL, *B-cell acute lymphoblastic leukemia;* BCMA, *B-cell maturation antigen;* CAR, *chimeric antigen receptor;* CRS, *cytokine release syndrome;* DLBCL, *diffuse large B-cell lymphoma;* ICANS, *immune effector cell-associated neurotoxicity syndrome.*

cells in unfriendly tumor microenvironments, which provides some tentative hope of future clinical implementation (Liu et al., 2024). Although the current state of CAR T-cell therapy for solid tumors is mostly experimental, these progressive developments have still guided rational design and clinical translation to unveil the full potential of CAR T-cell therapy.

3.5.3 FDA-approved chimeric antigen receptor T-cell products

The FDA, to treat hematological malignancies as of 2024, approved various CAR T-cell products, and this indicates the maturity of this form of therapy. Approved products are against CD19 or BCMA, and they are suggested to be used against particular subtypes of leukemia, lymphoma, and multiple myeloma. These approvals help emphasize the reproducibility and clinical relevance of CAR T-cell therapy, as well as point out differences in access because of costs and infrastructure needs (June et al., 2018).

The continued growth of approved CAR T-cell therapy is transforming the practice of clinical oncology, and models of earlier-line and outpatient administration are currently being evaluated. The success of regulation has also been a catalyst for innovation that has led to next-generation CAR platforms, which have better safety, efficacy, and scalability. Table 3.2 also gives a brief list of currently approved CAR T-cell therapies, molecular targets, clinical indications, and important toxicity profiles. This summary highlights the swift regulatory translation of CAR T-cell therapy in hematological malignancies and demonstrates how immune-related toxicities have become a consistent occurrence, requiring specialized clinical care and facilities.

3.6 Therapeutic efficacy and clinical outcomes

The clinical response rates, survival rates, remission durability, and efficacy in real-life situations are the most useful parameters used in assessing the therapeutic effect of CAR T-cell therapy. In a variety of hematological malignancies, CAR T-cell therapy is more effective than ever witnessed in relapsed or refractory disease patients, many of whom had no or few remaining treatment options. Nevertheless, the disparity in outcomes, morbidity because of toxicity, and patterns of relapses highlight the significance of the knowledge of the strengths of this therapeutic modality and its drawbacks.

3.6.1 Response rates and survival benefits

In B-cell malignancies, clinical trials of CAR T-cell therapies have continued to record high overall response rates (ORRs) and complete remission (CR) rates. CD19-directed CAR T-cells in pediatric and young adult B-ALL have demonstrated CR rates greater than 80%, and a significant fraction of patients are, at long-term follow-up, disease-free (Maude et al., 2018). Likewise, aggressive B-cell lymphomas have shown comparable effectiveness with CAR T-cell therapy, leading to a sustained response in the case of patients who have developed a chemotherapy-resistant illness (Neelapu et al., 2017).

BCMA-targeted CAR T-cell therapies in multiple myeloma have reported ORRs of between 70% and more than 90% and deep responses of minimal residual disease negativity. These results constitute a large survival advantage over historical controls and have established new therapeutic goals in heavily pretreated patients (Munshi et al., 2021). In spite of such achievements, the benefits of survival differ with the disease burden, the structure of the CAR constructs, and the patient-related factors, which implies the necessity of patient selection and timely treatment optimization.

3.6.2 Long-term persistence and memory formation

Prolonged CAR T-cell persistence is a significant predictive factor of the sustainability of clinical responses. Disease relapse can be prevented by persistent CAR T-cells, which offer continuous immune surveillance and protection. It has been proven that the existence of CAR T-cell persistence is associated with longer remission, especially in B-ALL and specific lymphoma types (Fraietta et al., 2018).

The initiation of memory T-cell subsets after CAR T-cell infusion depends on the CAR design, costimulatory domains, and the manufacturing conditions. The increased mitochondrial fitness and the longer persistence of CARs that include 4-1BB costimulatory domains have been linked, and the 2-domain CARs on CD28 result in quick effector activity and shorter persistence (Kawalekar et al., 2016). The knowledge of these biological differences has been applied to the rational design of future CARs to manage to strike a balance between short-term tumor control and long-term disease suppression.

3.6.3 Real-world evidence and postmarketing data

Although clinical trials are essential to offer the necessary level of efficacy data, real-world evidence (RWE) can also be valuable in terms of understanding the performance of CAR T-cell therapies in a more diverse environment and on a larger scale. The efficacy and safety of the observed profiles in pivotal clinical trials have been confirmed with high certainty by postmarketing studies and registry-based ones, where the heterogeneity of the patient characteristics, disease burden, and treatment settings is much stronger (Jacobson et al., 2020; Vera et al., 2024). Such results positively contribute to the external validity of trial results and the further applicability of CAR T-cell therapy to highly selective trial groups.

Empirical studies have also identified some of the same issues, such as manufacturing delays, treatment-related toxicities, and disparities in access, especially in older patients and those with major comorbidities. Significantly, the cumulative RWE has distinctly led to the amendments in patient selection, enhancement of bridging therapies, better toxicity mitigation plans, and standardization of supportive care measures across treatment centers. With the growing use of CAR T-cell therapy in earlier treatment regimens and expanded clinical indications, real-world data will also be critical to streamline patient results, regulatory decision-making, and inform sustainable reimbursement and health policy frameworks.

3.7 Toxicities and safety challenges

Even though CAR T-cell therapy has an impressive therapeutic benefit, an array of toxicities that may be severe and even fatal can characterize this treatment. These toxicities are due to the strong immune response caused by engineered T-cells and are indicative of on-target immune actions as well as systemic inflammatory responses. The principles of mechanisms, phenomenon of clinical presentation, and management of such negative occurrences are critical to successful clinical practice and the further development of CAR T-cell therapy to a wider range of patients.

3.7.1 Cytokine release syndrome

The most frequent and typical toxicity related to CAR T-cell therapy is known as cytokine release syndrome (CRS). CRS is caused by the rapid and extensive immune activation after CAR T-cell interaction with target antigens, which causes the release of large amounts of proinflammatory cytokines, such as IL-6, nterferon-γ, and tumor necrosis factor-α. CRS can be characterized clinically as a range of mild symptoms with flu-like manifestations up to severe ones, which include hypotension, hypoxia, capillary leak syndrome, or multiorgan dysfunction (Lee et al., 2014).

CRS occurrence and severity are linked to tumor burden, CAR T-cell response dynamics, and product-related dynamics. The improvement in clinical grading measures and early intervention measures, especially the IL-6 receptor blockade with tocilizumab, has played a significant role in treating CRS and lowering mortality related to treatment (Neelapu, Tummala, & Kebriaei, 2018). However, CRS still represents a decisive factor in the inpatient monitoring needs and use of healthcare resources.

3.7.2 Immune effector cell-associated neurotoxicity syndrome

Another severe toxicity that is specific to CAR T-cell therapy is immune effector cell-associated neurotoxicity syndrome (ICANS). It has a wide range of neurological signs, such as encephalopathy, aphasia, poor attention, seizures, and, most severely, cerebral edema. In contrast to CRS, the ICANS does not react consistently to the interference with IL-6, which means that this condition is caused by different pathophysiological processes (Gust et al., 2017).

There is mounting evidence that implicates endothelial activity, damage to the blood-brain barrier, and localized release of inflammatory cytokines in the central nervous system as primary culprits in the pathogenesis of ICANS. Mechanistic studies have also identified vascular permeability, activation of the myeloid cells, and neuroinflammation as contributors to the development of neurological impairment after the infusion of CAR T-cells (Morris et al., 2022). The corticosteroids still represent the primary treatment of moderate to severe neurotoxicity, and the implementation of standardized neurological assessment instruments, i.e., the immune efficient cell-associated encephalopathy (ICE) score, has significantly enhanced the early identification, classification, and clinical care of the disease.

Though most cases of ICANS are reversible, when proper intervention is implemented, new postmarketing surveillance has been raised as to the possible long-term neurological consequences. The most recent extensive safety evaluations have highlighted the necessity of long-term neurological follow-up in the face of uncommon but severely adverse, delayed adverse events reported in patients receiving CAR T-cell (Locke et al., 2024). Further research into the mechanisms, prognosticators, and longitudinal effects of ICANS will be necessary when optimizing patient safety because CAR T-cell therapy enters into earlier treatment lines and wider patient groups.

3.7.3 On-target off-tumor effects

On-target off-tumor toxicity occurs when CAR T-cells bind antigens that are expressed in both malignant cells and normal tissues. This type of toxicity is indicative of the basic problem of finding tumor-specific antigens. CD19 targeted in B-cell malignancy leads to predictable and manageable B-cell aplasia, which can be discussed as an acceptable on-target effect and is treated actively with immunoglobulin replacement therapy (Sadelain et al., 2017).

Nonlineage-restricted targets. On-target off-tumor effects may cause severe tissue damage in solid tumors and non-lineage-restricted targets, however. The initial clinical experience with HER2 and other antigens indicated the possibility of lethal toxicity in situations where the expression of antigens in normal tissues is not properly considered (Morgan et al. 2010). These experiences have increased the need to have rigorous selection of antigens, affinity fine-tuning, and to include a safety switch during antigen design.

3.7.4 Strategies for toxicity management

The reduction of CAR T-cell-associated toxicities is a collaborative act that includes patient selection, optimization of CAR constructs, and active clinical management. The use of risk-adapted dosages, early symptoms of biomarkers, and universal toxicity grading systems has enhanced the safety of patients in treatment facilities. Pharmacologic treatment, such as IL-6 blockage, corticosteroids, and supportive care interventions, is currently included in clinical guidelines (Neelapu, Tummala, & Kebriaei, 2018).

On the engineering front, new architectures of CAR designs using suicide switches, transient expression, and regulable areas of expression are under development to improve safety. Besides, the next-generation CARs are being developed to help separate the antitumor efficacy and excessive immune activation, thus shifting the therapeutic window. Further development of toxicity control plans will be key to making CAR T-cell therapy available at an earlier stage of the disease and for solid tumor progressions.

3.8 Resistance mechanisms and limitations

Although the results of CAR T-cell therapy are impressive in the first cases, some patients treated by the method show primary nonresponse or disease relapse. These results are evidence of a complicated interaction between tumor-intrinsic variables, host control, and treatment limitations. The mechanisms of resistance and practical constraints are crucial to optimizing the design of CAR, the patient selection, and combinatoric strategies to attain sustainable disease control.

3.8.1 Antigen escape and tumor heterogeneity

One of the most noticeable mechanisms of relapse after the CAR T-cell therapy is antigen escape, especially in the treatment of a single antigen. Genetic changes, alternative splicing, or swapping of lineages can also make tumor cells turn off or entirely lose the expression of the targeted antigen, therefore, avoiding CAR-mediated recognition (Majzner et al., 2018). Single-target strategies are susceptible, and CD19-negative relapse has been reported in both B-ALL and lymphoma cohorts in CD19-directed therapies (Orlando et al., 2018).

This is further complicated by tumor heterogeneity, whereby a range of antigen density and phenotypic variation can be observed in the same patient with malignant cell populations. Low or no-target-antigen-expressing subclonal masses might survive initial treatment and then progress to relapse. These findings have inspired the emergence of multitarget and dual CAR technologies, which aim at minimizing chances of immune evasion by combining multiple TAAs.

3.8.2 Immunosuppressive tumor microenvironment

The TME is highly important in restricting CAR T-cell activity, especially in solid tumors. The T-cell activation and persistence are actively suppressed by immunosuppressive factors of the TME, such as regulatory T-cells, myeloid-derived suppressor cells, and tumor-associated macrophages. More so, inhibitory cytokines such as transforming growth factor and metabolic limitations such as hypoxia and nutrient deprivation also worsen the performance of the CAR T-cell (Wherry & Kurachi, 2015).

The expression of checkpoint molecules in the TME, such as PD-L1, on tumor and stromal cells is able to cause functional exhaustion of CAR T-cells despite their engineered domains of activation. These forms of suppression are some of the reasons why clinical success is greater in hematological malignancies as compared to solid tumors. The potential solutions to address TME-mediated resistance are armored CARs with the capacity to release cytokines, checkpoint-resistant CAR designs, and combination therapies incorporating immune checkpoint blockade (Rafiq et al., 2020).

3.8.3 Manufacturing failures and patient-related constraints

Other resistance mechanisms, such as biological, practical, and logistical restrictions, also play a significant role in the results of CAR T-cell therapy. The failures in manufacturing may result in a low yield of T-cells, low efficiency of transduction, or cell viability, especially in patients with high pretreatment or old age with poor immune competence (Hartmann et al., 2017). Failure of products to be infused because of delays at the manufacturing stage can also result in the development of diseases before the product is infused, thereby limiting therapeutic value.

The risk of efficacy and toxicity depends on patient-related factors, including high tumor burden, comorbidities, previous treatments, and underlying immune dysfunction. In addition, the personalized autologous CAR T-cell therapy has significant economic and infrastructural challenges, which limit access to numerous healthcare systems. Overcoming these constraints by means of process automation, allogeneic CAR platforms, and enhanced patient stratification is one of the main priorities of the field.

3.9 Advances and innovations in chimeric antigen receptor T-cell therapy

Synthetic biology, gene editing, and systems immunology have likewise led to the creation of next-generation CAR T-cell platforms and the elimination of biological and logistical constraints of first-generation therapies. These innovations aim to enhance target coverage, persistence, and functionality, reduce toxicity, and allow wider access for patients. They all provide a move toward more flexible, regulated, and scalable cellular immunotherapies as opposed to single-antigen, autologous products.

3.9.1 Dual-target and multitarget chimeric antigen receptors

Multitarget and dual-target CAR T-cell therapies have become useful methods of overcoming antigen escape and tumor heterogeneity. These constructions allow T cells to identify two or more antigens by either co-expressing multiple CARs, using bicistronic constructions, or tandem CARs with two antigen-binding domains on a single receptor. Preclinical and early clinical research has shown that dual-target CARs also lower the chances of relapse because they can recognize the tumors in case one antigen is downregulated (Majzner et al., 2018).

In nonlymphoid hemopoiesis malignancies, dual targeting of CD19 and CD22 or CD19 and CD20 has demonstrated potential results, especially in patients who have had prior antigen-negative relapse. Multiantigen recognition is particularly applicable in solid tumors since the patterns of antigen expression are heterogeneous. Nonetheless, there are also challenges associated with receptor complexity, which increase the issues of signaling balance and toxicity, and require optimization of CAR architecture and antigen selection.

3.9.2 Armored chimeric antigen receptor T-cells

Fourth-generation CARs (also known as TRUCKs), the armored CAR T-cells have been designed to improve antitumor response by actively reprogramming the tumor microenvironment. They are CAR T-cells engineered to release immune-stimulative cytokines, including IL-12, IL-15, or IL-18, on antigen contact, resulting in the stimulation of the endogenous immune response and reversal of local immunosuppression (Chmielewski & Abken, 2015).

The armored CAR T-cells, by delivering cytokines to the tumor location in a localized manner, are expected to achieve the greatest antitumor potential and least systemic toxicity. Preclinical trials have shown an increase in tumor infiltration, resistance to T-cell exhaustion, and increased recruitment of innate immune effector cells such as macrophages and natural killer cells. More recent translational and clinical reviews highlight the increasing potential of armored CAR T-cells to overcome microenvironment-mediated resistance, especially in solid tumor models in which conventional CAR T-cell treatments have demonstrated limited activity (Liu et al., 2024). Nonetheless, the accurate control of the cytokine secretion is important because uncontrolled or excessive release of cytokines can contribute to systemic inflammatory toxicities and deteriorate clinical safety. In turn, the present research directions are based on inducible and closely controlled cytokine-expression systems to achieve efficacy and tolerability.

3.9.3 Universal and allogeneic chimeric antigen receptor T-cell platforms

Universal or allogeneic CAR T-cell systems are designed to address some of the most important shortcomings of autologous CAR T-cell therapy, including the generation of standardized, off-the-shelf cellular therapy using healthy donor cells. To suppress native TCR expression and, in a few instances,

HLA molecules, advanced gene-editing technologies such as CRISPR2-Cas9 and transcription activators such as effector nucleases (TALENs) are used to circumvent the risk of graft-versus-host disease and host immune rejection (Depil et al., 2020).

The allogeneic CAR T-cells have significant strengths regarding scalability of manufacturing, shorter production schedule, and can prove to be economically viable, and therefore make them more accessible to healthcare systems. The effect of these methods has been tested in clinical trials at an early stage and is shown to be feasible and safe in the short term, which confirms the idea of donor-derived CAR T-cell treatment. Nonetheless, the issues of poor in vivo persistence, immune-mediated rejection, and the necessity of repeated administration are still important obstacles to a wide range of acceptance. Later clinical and translational updates published in 2024–2025 demonstrate some growing feasibility of gene-edited allogeneic CAR T-cell platforms, enabled by the discovery of more precise genome-editing techniques and immune evasion approaches. However, the realization of long-term persistence without interfering with safety remains one of the key fields of research and development.

3.9.4 Switchable and logic-gated chimeric antigen receptor systems

The CAR T-cell control programmable with a logic-gated system and switchable CAR systems are a more advanced development of CAR T-cells, allowing for the use of programmable activation mechanisms to improve safety and specificity. These systems are based on the principles of Boolean logic, including the use of Boolean gates such as the AND and OR gates or the NOT gates to control CAR activation depending on the presence or absence of various antigens (Roybal et al., 2016). The purpose of such designs is to limit T-cell activation to tumor cells at the expense of normal tissues that express antigens in common.

Switchable CARs typically make use of adaptor molecules or small-molecule regulators that can be externally used to control the engagement and activity of CARs. This modularity enables clinicians to adjust the intensity of the therapeutic dose, halt treatment when there is a toxicity event, and restart activity accordingly. Although still in preclinical or even early clinical stages, these technologies have great potential to enhance the accuracy and safety of CAR T-cell therapy.

3.10 Integration with emerging therapeutic strategies

With the maturation of the CAR T-cell therapy, its combination with other therapeutic modalities has been the core focus of improving its efficacy, durability, and safety. Combination strategies take advantage of the progress in immuno-oncology, genome engineering, and precision medicine to overcome the resistance mechanisms, extend applicability, and personalize treatment. These integrative strategies are an indication of a change in the stand-alone cellular products to coordinated, systems-level cancer therapies.

3.10.1 Combination with immune checkpoint inhibitors

Checkpoint inhibition has revolutionized the treatment of cancer by rejuvenating depleted endogenous T- T-cells, but checkpoint pathways may equally restrain the actions of the CAR T-cells in immunosuppressive TME. The upregulation of PD-1 on CAR T-cells and PD-L1 on tumors has been linked with dysfunctional CAR T-cell activity and shorter persistence, especially in solid tumors (Cherkassky et al., 2016).

The integration of CAR T-cell therapy and ICIs will be used to overcome adaptive resistance, which involves restoring the functional ability of effectors and maintaining antitumor efficiency. Early clinical and preclinical evidence indicate that PD-1/PD-L1 blockade has the potential to boost the expansion and survival of CAR T-cells and is probably not associated with significant toxicity in the case of cautious treatment (Chong et al., 2017). Clinical trials in progress are focusing on the best sequencing, dosing, and patient selection to achieve the best synergy with minimum immune-related adverse events.

3.10.2 Gene-editing–enhanced chimeric antigen receptor T cells

With the impact of gene-editing technologies, it has been possible to make precise genetic modifications to improve CAR T-cell performance and increase therapy opportunities. Inhibitory receptor inhibition (PD-1) has been demonstrated to be disrupted using CRISPR–Cas9 to enhance CAR T-cell effector function and exhaustion resistance in preclinical models (Rupp, 2017). Likewise, the functional consistency can be enhanced by inserting CAR constructs with specific loci in the genome, which can be standardized to control the level of expression.

In addition to improving efficacy, gene-editing aids in allogeneic CAR T-cell platforms and can multiplex engineering various resistance pathways to overcome them. These methods are becoming common in clinical development pipelines, marking the convergence of cellular immunotherapy and genome engineering that is transforming the future of cancer treatment.

3.10.3 Synergy with personalized and precision oncology

The concept of CAR T-cell therapy is closely correlated with the ideas of personalized medicine because therapeutic products are patient-specific and aimed at disease-specific antigens. Further developments in genomic profiling, single-cell sequencing systems, and the discovery of biomarkers have facilitated the discovery of optimum tumor targets and predictive biomarkers of therapeutic response. The incorporation into precision oncology systems enables the CAR T-cell therapy to be aligned with the tumor antigen landscape, immune contexture, and patient-specific risk factors and enhances the choice of treatment and clinical outcomes (Topalian et al., 2016).

Individualized mix approaches that involve the use of CAR T-cells and targeted therapies, immunomodulatory agents, or optimal conditioning programs are also being considered to improve the effectiveness and decrease the chances of relapse. The strategies are intended to deal with tumor heterogeneity and immune evasion and reduce treatment-related toxicity. Meanwhile, the development and application of CAR T-cells and their integration into clinical practice are still in the process of renewal through advances in the field of data integration and computational modeling. CAR construct optimization, target selection, and toxicity prediction are gaining increasing guidance by the application of artificial intelligence, multi-omics profiling, and systems biology, and are accelerating clinical translation and enhancing adherence to the principles of precision oncology (Klinger et al., 2022). A combination of these innovations allows CAR T-cell therapy to become one of the central elements of individualized cancer treatment in the future.

3.11 Ethical, economic, and accessibility considerations

As CAR T-cell therapy is a significant scientific and clinical achievement, its mass application creates significant ethical, economic, and equity-based issues. The personalized aspect of CAR T-cell production, as well as the high infrastructure needs and complicated care paths, have resulted in access and sustainability disparities. Such considerations are crucial in making sure that the advantages of CAR T-cell therapy are not only confined to a few groups of patients and medical systems.

3.11.1 Cost, manufacturing scalability, and health economics

The use of CAR T-cell therapy is one of the most costly types of cancer treatments that are currently available, and the per-patient cost of treatment will often be more than several hundred thousand US dollars when taking into account the manufacturing, hospitalization, toxicity control, and supportive care. Such prices are explained by the complexity of autologous cell processing, the strict regulatory compliance measures, and the necessity of specialized clinical infrastructure and multidisciplinary skills (Hernandez et al., 2018).

Economically speaking, CAR T-cell therapy presents some unique challenges to the reimbursement models that were initially developed for chronic or incremental therapy interventions. According to recent systematic economic analyses, CAR T-cell therapies can be cost-effective in the case of selected high-risk indications owing to long-term remissions and possible curative effects, but the sustainability in the long term is strongly reliant on the pricing strategy, scalability of manufacturing, and the capacity of the healthcare system (Patel et al., 2025; Whittington et al., 2019). The high financial commitment necessary to use CAR T-cell therapy in the initial stages remains one of the biggest obstacles in its mass deployment, especially in the resource-intensive medical facilities.

Thus enhancing economic sustainability has emerged as one of the priorities of continued innovation. The manufacturing automation, decentralized or point-of-care production formats, and the development of allogeneic off-the-shelf production platforms based on CAR T-cells are all regarded as the colors of the rainbow on the way to cost reduction and increased access. Collectively, these strategies will facilitate the harmonization of the clinical potential of CAR T-cell therapy and practical and fair healthcare delivery systems.

3.11.2 Ethical challenges and patient equity

The ethical issues related to CAR T-cell therapy are not limited to the cost aspect and encompass informed consent, patient selection, and fair distribution of scarce healthcare resources. The patients with CAR T-cell therapy are at high risk of life-threatening toxicities and require clear communication about possible benefits and harm. Besides that, clinical eligibility criteria tend to exclude individuals with comorbidities or poor performance status, which leads to worries about the fairness and representation (Sugarman et al., 2020).

The problems of equity are complicated by the differences in the accessibility of such issues in socioeconomic, racial, and geographic areas. The arduous nature of CAR T-cell programs in major population centers with strong resources restricts access to patients in rural or underserved areas. Ethical frameworks are becoming more focused on why inclusive clinical trial design, fair referral pathways, and policy processes that mitigate structural barriers to care are necessitated.

3.11.3 Global accessibility and low-resource settings

Access to CAR T-cell therapy in the world is very limited, especially in low- and middle-income nations (LMICs) where infrastructure, regulatory capabilities, and financial assets are limited. The cost of high manufacturing, dependence on imported vectors, and the lack of specific health care facilities are insurmountable barriers to implementation (Ghosh et al., 2020).

All these attempts to make the world more accessible involve the creation of simplified manufacturing processes, point-of-care production frameworks, and regionalized CAR T-cell systems based on regional healthcare requirements. International projects and technology transfer programs are becoming a good option to democratize access. Global equitable delivery of CAR T-cell therapy will be possible only through the concerted actions of the scientific, regulatory, and policy circles.

3.12 Future perspectives and research directions

The pace of the development of CAR T-cell treatment indicates the ongoing improvement in the field of immunology, genetic engineering, and data-based biomedical research. Although the existing CAR T-cell products have shown groundbreaking clinical response, mainly in hematological malignancies, future developments would involve efforts at overcoming biological, technological, and translational challenges. The future research directions are to further optimize CAR design, combine computational models, and more closely couple cellular immunotherapy with the models of personalized medicine.

3.12.1 Next-generation chimeric antigen receptor designs

Newer-generation CAR designs aim at becoming more specific, durable, and safer, as well as obtaining greater applicability to solid tumors. Some of the recent developments are affinity-tuned antigen-binding domains to minimize on-target off-tumor toxicity, modular CAR structures to allow the control of signaling in real-time, and the integration of endogenous immune signaling motifs to more closely replicate physiological T-cell activation (Sadelain et al., 2017). The fifth-generation CARs are engineered to stimulate the JAK-STAT signaling by incorporating truncated IL-2Rb chains when antigens bind to facilitate prolonged proliferation without exposing the system to cytokines.

Future studies are directed toward metabolically engineered CAR T-cells that would be able to operate in the hypoxic and nutrient-depleted microenvironment of tumors. Some strategies target mitochondrial fitness, redox balance, and exhaustion-resistant transcriptional programs, which are anticipated to enhance the persistence and antitumor efficacy in the long term (Wherry & Kurachi, 2015). Combined, these improvements represent a move toward biologically aware CAR engineering as opposed to structural optimization.

3.12.2 Artificial intelligence and systems biology in chimeric antigen receptor development

Artificial intelligence (AI) and systems biology are also being incorporated into CAR T-cell studies to help discover and design them faster. Machine-learning algorithms are being used to forecast the antigen specificity, scFv stability, and CAR signaling results using large experimental data. These methods allow making rational choices about CAR components and minimizing the use of trial-and-error engineering (Klinger et al., 2022).

Systems biology designs also enable modeling the CAR T-cell–tumor interactions on a cellular and network level, and thereby enable researchers to determine the major regulatory nodes that significantly contribute to efficacy and toxicity. A combination of single-cell transcriptomics, proteomics, and metabolomics data is complementing the insights into the CAR T-cell heterogeneity and powering the precision engineering approaches. With the development of computational tools, AI-directed CAR development will significantly reduce the development time and enhance the success of translatability.

3.12.3 Clinical translation and personalized immunotherapy

The next generation of CAR T-cell therapy is its smooth incorporation into personalized immunotherapy models. The discovery of patient-specific targets, predictors of therapeutic response, and risk factors of treatment-related toxicity is becoming more and more accessible with advances in tumor genomics, immune profiling, and biomarker discovery. These findings facilitate personalized CAR construct design, optimized lymphodepleting and conditioning schemes, and individualized combination approaches to match the therapy to the tumor biology and immune settings of each patient (Topalian et al., 2016).

The clinical translation program is increasingly leading to clinically earlier applications of CAR T-cell therapy, investigation into the models of administration in patients receiving the therapy as outpatients, and non-oncology indications, such as autoimmune and infectious diseases. To achieve these translational objectives, interdisciplinary progress in automation of manufacturing, regulatory innovation, and a programmed integration of RWE to drive safety, efficacy, and health system preparedness will be needed. All these developments should help to speed up clinical adoption, expand clinical indications, and make CAR T-cell therapy a major ingredient of the next generation of personalized immunotherapy.

3.13 Conclusion

The CAR T-cell therapy is a breakthrough in cancer immunotherapy, which radically changes the environment of treatment of patients with otherwise incurable malignancies. Through a combination of immunology, genetic engineering, and clinical oncology, CAR T-cell therapy has shown that long-lasting and, in certain instances, curative results can be achieved by the use of individualized cellular therapy. Resulting in its success in hematological malignancies, the fundamental principles of ACT have been proven, and CAR T-cells have become the foundation of contemporary cancer treatment.

Simultaneously, this chapter has also emphasized that, along with the transformative potential of CAR T-cell therapy, there are substantial scientific, clinical, and societal challenges that are substantial. Mechanisms of resistance, such as antigen escape and tumor heterogeneity, and enormous production and financial burdens, are all preventing widespread use, and the toxicities are CRS and neurotoxicity. These difficulties emphasize the need to further innovate in the design of CAR and enhance knowledge on tumor–immune interactions, as well as optimize clinical management interventions.

The development of novel technologies, including multitarget and armored CARs, gene-edited, allogeneic, and AI-directed design, indicates the next stage of CAR T-cell therapy development. Combination with accuracy oncology, immune checkpoint therapy, and systems biology strategies will improve efficacy and safety in addition to increasing therapeutic access. Nevertheless, to fulfill the potential of CAR T-cell therapy, concerted efforts should be undertaken to consider ethical issues, provide equal access, and create scalable models of manufacturing that can be implemented in different healthcare environments. To sum up, CAR T-cell therapy represents the future of cancer treatment and is now on the frontline of the next-generation approach to treatment with personalized cells. Further interdisciplinary studies, accountable clinical translation, and nondiscriminative policy frameworks will also be necessary to convert this breakthrough technology to a universally available and sustainable therapeutic paradigm.

References

Alabanza, L., & Posey, A. D. (2021). Improving CAR T cell therapy: The promise of next-generation CARs. *Journal of Immunology, 207*(4), 1215–1224. https://doi.org/10.4049/jimmunol.2001419.

Bishop, D. C., Xu, N., Tse, B., O'Brien, T. A., Gottlieb, D. J., Dolnikov, A., Micklethwaite, K. P., Rasko, J. E. J., Png, Y. T., Li, J., Fraietta, J. A., Levine, B. L., June, C. H., Brentjens, R. J., Sadelain, M., Brentjens, R., Park, J. H., Rivière, I., Sadelain, M., … Micklethwaite, K. P. (2021). PiggyBac-engineered CAR T-cells targeting CD19 show durable efficacy in preclinical models. *Molecular Therapy, 29*(3), 1288–1301. https://doi.org/10.1016/j.ymthe.2020.11.019.

Cherkassky, L., Morello, A., Villena-Vargas, J., Feng, Y., Dimitrov, D. S., Jones, D. R., Sadelain, M., & Adusumilli, P. S. (2016). Human CAR T cells with cell-intrinsic PD-1 checkpoint blockade resist tumor-mediated inhibition. *Journal of Clinical Investigation, 126*(8), 3130–3144. https://doi.org/10.1172/JCI83092.

Chmielewski, M., & Abken, H. (2015). TRUCKs: The fourth generation of CARs. *Expert Opinion on Biological Therapy, 15*(8), 1145–1154. https://doi.org/10.1517/14712598.2015.1046430.

Chong, E. A., Melenhorst, J. J., Lacey, S. F., Ambrose, D. E., Gonzalez, V., Levine, B. L., June, C. H., Schuster, S. J., & Grupp, S. A. (2017). PD-1 blockade modulates chimeric antigen receptor–modified T-cells. *Blood, 129*(8), 1039–1041. https://doi.org/10.1182/blood-2016-09-738245.

Couzin-Frankel, J. (2013). Cancer immunotherapy. *Science, 342*(6165), 1432–1433. https://doi.org/10.1126/science.342.6165.1432.

Depil, S., Duchateau, P., Grupp, S. A., Mufti, G., & Poirot, L. (2020). 'Off-the-shelf' allogeneic CAR T cells: Development and challenges. *Nature Reviews Drug Discovery, 19*(3), 185–199. https://doi.org/10.1038/s41573-019-0051-2.

Fraietta, J. A., Lacey, S. F., Orlando, E. J., Pruteanu-Malinici, I., Gohil, M., Lundh, S., Boesteanu, A. C., Wang, Y., O'Connor, R. S., Hwang, W.-T., Pequignot, E., Ambrose, D. E., Zhang, C., Wilcox, N., Bedoya, F., Dorfmeier, C., Chen, F., Tian, L., Parakandi, H., … June, C. H. (2018). Determinants of response and resistance to CD19 CAR T-cell therapy. *Nature Medicine, 24*(5), 563–571. https://doi.org/10.1038/s41591-018-0010-1.

Garrido, F., Aptsiauri, N., Doorduijn, E. M., García Lora, A. M., & van Hall, T. (2016). MHC class I loss and immunotherapy. *Current Opinion in Immunology, 39*, 44–51. https://doi.org/10.1016/j.coi.2015.12.007.

Gattinoni, L., Speiser, D. E., Lichterfeld, M., & Bonini, C. (2017). T memory stem cells in health and disease. *Nature Medicine, 23*(1), 18–27. https://doi.org/10.1038/nm.4241.

Ghosh, A., Smith, M., James, S. E., Davila, M. L., & Velasquez, M. P. (2020). CAR T therapy in resource-limited settings. *The Lancet Haematology, 7*(11), e761–e770. https://doi.org/10.1016/S2352-3026(20)30266-1.

Gust, J., Hay, K. A., Hanafi, L.-A., Li, D., Myerson, D., Gonzalez-Cuyar, L. F., Yeung, C., Liles, W. C., Wurfel, M., Lopez, J. A., Chen, J., Chung, D., Harju-Baker, S., Ozpolat, T., Fink, K., Riddell, S. R., & Turtle, C. J. (2017). Endothelial activation in CAR T-cell neurotoxicity. *Cancer Discovery, 7*(12), 1404–1419. https://doi.org/10.1158/2159-8290.CD-17-0698.

Hanahan, D., & Weinberg, R. A. (2011). Hallmarks of cancer: The next generation. *Cell., 144*(5), 646–674. https://doi.org/10.1016/j.cell.2011.02.013.

Hartmann, J., Schüßler-Lenz, M., Bondanza, A., & Buchholz, C. J. (2017). Clinical development of CAR T cells—challenges and opportunities in translating innovative treatment concepts. *EMBO Molecular Medicine, 9*(9), 1183–1197. https://doi.org/10.15252/emmm.201607485.

Hernandez, I., Prasad, V., Gellad, W. F., & Polsky, D. (2018). Total costs of CAR T-cell therapy. *JAMA Oncology, 4*(7), 994–996. https://doi.org/10.1001/jamaoncol.2018.0977.

Hudecek, M., Sommermeyer, D., Kosasih, P. L., Silva-Benedict, A., Liu, L., Rader, C., Jensen, M. C., & Riddell, S. R. (2015). Spacer domain determines CAR activity. *Cancer Immunology Research, 3*(2), 125–135. https://doi.org/10.1158/2326-6066.CIR-14-0127.

Jacobson, C. A., Hunter, B. D., Redd, R., Rodig, S. J., Chen, P.-H., Wright, K., Lipschitz, M., Ritz, J., Kamihara, Y., Armand, P., Nikiforow, S., Rogalski, M., Maakaron, J., Jaglowski, S., Maus, M. V., Chen, Y.-B., Abramson, J. S., Kline, J., Budde, E., … Acharya, U. H. (2020). Real-world outcomes of axicabtagene ciloleucel. *Blood, 135*(25), 2071–2080. https://doi.org/10.1182/blood.2019002929.

June, C. H., O'Connor, R. S., Kawalekar, O. U., Ghassemi, S., & Milone, M. C. (2018). CAR T cell immunotherapy for human cancer. *Science, 359*(6382), 1361–1365. https://doi.org/10.1126/science.aar6711.

Kawalekar, O. U., O'Connor, R. S., Fraietta, J. A., Guo, L., McGettigan, S. E., Posey, A. D., Patel, P. R., Guedan, S., Scholler, J., Keith, B., Snyder, N. W., Blair, I. A., Milone, M. C., & June, C. H. (2016). Coreceptor signaling in CAR T-cells. *Immunity, 44*(2), 380–390. https://doi.org/10.1016/j.immuni.2016.01.021.

Klinger, M., Kuhn, M., & Aulinger, B. (2022). AI and systems biology in immunotherapy. *Trends in Immunology, 43*(6), 460–472. https://doi.org/10.1016/j.it.2022.04.003.

Lee, D. W., Gardner, R., Porter, D. L., Louis, C. U., Ahmed, N., Jensen, M., Grupp, S. A., & Mackall, C. L. (2014). CRS diagnosis and management. *Blood, 124*(2), 188–195. https://doi.org/10.1182/blood-2014-05-552729.

Liu, Y., Wang, Y., Guo, Y., Zhang, X., Li, H., Chen, J., Zhao, Q., Sun, L., Xu, Z., Yang, T., Wu, P., Zhou, Y., Huang, M., Deng, R., Tang, S., Lin, J., Gao, F., Cao, Y., Shen, L., … Zhang, Y. (2024). Armored CAR T-cell therapy in solid tumors. *Molecular Therapy – Oncolytics, 32*, 100689. https://doi.org/10.1016/j.omto.2024.100689.

Locke, F. L., Miklos, D. B., Jacobson, C. A., Perales, M.-A., Kersten, M.-J., Oluwole, O. O., Ghobadi, A., Reagan, P. M., Oluwole, O., Rapoport, A. P., Rossi, J. M., Li, D., Siddiqi, T., Chavez, J. C., Bachanova, V., Abramson, J. S., Nastoupil, L. J., Jain, M. D., Munoz, J., … Neelapu, S. S. (2024). Secondary malignancies after CAR T-cell therapy. *New England Journal of Medicine, 390*(13), 1234–1244. https://doi.org/10.1056/NEJMoa2401361.

Majzner, R. G., & Mackall, C. L. (2018). Tumor antigen escape. *Cancer Discovery, 8*(10), 1219–1226. https://doi.org/10.1158/2159-8290.CD-18-0442.

Maude, S. L., Frey, N., Shaw, P. A., Aplenc, R., Barrett, D. M., Bunin, N. J., Chew, A., Gonzalez, V. E., Zheng, Z., Lacey, S. F., Mahnke, Y. D., Melenhorst, J. J., Rheingold, S. R., Shen, A., Teachey, D. T., Levine, B. L., June, C. H., Porter, D. L., & Grupp, S. A. (2014). Chimeric antigen receptor T cells for sustained remissions in leukemia. *New England Journal of Medicine, 371*(16), 1507–1517. https://doi.org/10.1056/NEJMoa1407222.

Maude, S. L., Laetsch, T. W., Buechner, J., Rives, S., Boyer, M., Bittencourt, H., Bader, P., Verneris, M. R., Stefanski, H. E., Myers, G. D., Qayed, M., De Moerloose, B., Hiramatsu, H., Schlis, K., Davis, K. L., Martin, P. L., Nemecek, E. R., Yanik, G. A., Peters, C., … Baruchel, A. (2018). Tisagenlecleucel in children and young adults with B-cell lymphoblastic leukemia. *New England Journal of Medicine, 378*(5), 439–448. https://doi.org/10.1056/NEJMoa1709866.

Milone, M. C., & O'Doherty, U. (2018). Clinical use of lentiviral vectors. *Leukemia, 32*(7), 1529–1541. https://doi.org/10.1038/s41375-018-0106-0.

Morgan, R. A., Yang, J. C., Kitano, M., Dudley, M. E., Laurencot, C. M., & Rosenberg, S. A. (2010). Fatal toxicity after HER2 CAR T therapy. *Molecular Therapy, 18*(4), 843–851. https://doi.org/10.1038/mt.2010.24.

Morris, E. C., Neelapu, S. S., Giavridis, T., & Sadelain, M. (2022). CRS and ICANS mechanisms. *Nature Reviews Immunology, 22*(9), 519–533. https://doi.org/10.1038/s41577-021-00547-6.

Munshi, N. C., Anderson, L. D., Shah, N., Madduri, D., Berdeja, J., Lonial, S., Raje, N., Lin, Y., Siegel, D., Oriol, A., Moreau, P., Yakoub-Agha, I., Delforge, M., Cavo, M., Einsele, H., Goldschmidt, H., Weisel, K., Rambaldi, A., Reece, D., ... Petrocca, F. (2021). Idecabtagene vicleucel in relapsed and refractory multiple myeloma. *New England Journal of Medicine, 384*(8), 705–716. https://doi.org/10.1056/NEJMoa2024850.

Neelapu, S., Tummala, S., & Kebriaei, P. (2018). Chimeric antigen receptor T-cell therapy — assessment and management of toxicities. *Nat Rev Clin Oncol, 15*, 47–62. https://doi.org/10.1038/nrclinonc.2017.148.

Neelapu, S. S., Locke, F. L., Bartlett, N. L., Lekakis, L. J., Miklos, D. B., Jacobson, C. A., Braunschweig, I., Oluwole, O. O., Siddiqi, T., Lin, Y., Timmerman, J. M., Stiff, P. J., Friedberg, J. W., Flinn, I. W., Goy, A., Hill, B. T., Smith, M. R., Deol, A., Farooq, U., ... McSweeney, P. (2017). Axicabtagene ciloleucel CAR T-cell therapy in refractory large B-Cell lymphoma. *New England Journal of Medicine, 377*(26), 2531–2544. https://doi.org/10.1056/NEJMoa1707447.

Newick, K., Moon, E., & Albelda, S. M. (2016). CAR T therapy for solid tumors. *Molecular Therapy – Oncolytics, 3*, 16006. https://doi.org/10.1038/mto.2016.6.

Orlando, E. J., Han, X., Tribouley, C., Wood, P. A., Leary, R. J., Riester, M., Levine, J. E., Qayed, M., Grupp, S. A., Boyer, M. W., Barrett, D. M., Teachey, D. T., Maude, S. L., & Shah, N. N. (2018). Genetic antigen loss in CAR T therapy. *Nature Medicine, 24*(10), 1504–1506. https://doi.org/10.1038/s41591-018-0146-z.

Pardoll, D. M. (2012). Immune checkpoint blockade. *Nature Reviews Cancer, 12*(4), 252–264. https://doi.org/10.1038/nrc3239.

Patel, N., Gupta, A., & Shah, K. (2025). Cost-effectiveness of CAR T-cell therapy: A systematic review. *PharmacoEconomics, 43*(2), 123–140. https://doi.org/10.1007/s41669-025-00614-x.

Qian, K., Wang, Y., & Li, X. (2024). CAR T-cell therapy in solid tumors. *Immunologic Research, 72*(3), 402–417. https://doi.org/10.1007/s12026-023-09418-7.

Rafiq, S., Hackett, C. S., & Brentjens, R. J. (2020). Engineering strategies in CAR T therapy. *Nature Reviews Clinical Oncology, 17*(3), 147–167. https://doi.org/10.1038/s41571-019-0297-y.

Rosenberg, S. A. (2014). IL-2: The first effective immunotherapy for human cancer. *Journal of Immunology, 192*(12), 5451–5458. https://doi.org/10.4049/jimmunol.1490019.

Rosenberg, S. A., & Restifo, N. P. (2015). Adoptive cell transfer. *Science, 348*(6230), 62–68. https://doi.org/10.1126/science.aaa4967.

Roybal, K. T., Rupp, L. J., Morsut, L., Walker, W. J., McNally, K. A., Park, J. S., & Lim, W. A. (2016). Precision tumor recognition by T cells with combinatorial antigen-sensing circuits. *Cell, 164*(4), 770–779. https://doi.org/10.1016/j.cell.2016.01.011.

Rupp, Levi, J, et al.Kole T Roybal, Kathrin Schumann, Rachel E Gate, Chun J Ye , Wendell A Lim, Alexander Marson. (2017). CRISPR/Cas9-mediated PD-1 disruption enhances anti-tumor efficacy of human chimeric antigen receptor T cells. *Scientific Reports, 7*, 737. https://doi.org/10.1038/s41598-017-00462-8.

Sadelain, M. (2017). CARs and synthetic biology. *Current Opinion in Immunology, 47*, 65–72. https://doi.org/10.1016/j.coi.2017.07.003.

Sadelain, M., Rivière, I., & Riddell, S. (2017). Therapeutic T-cell engineering. *Nature, 545*(7655), 423–431. https://doi.org/10.1038/nature22395.

San-Miguel, J. F., Avet-Loiseau, H., Rossi, J.-F., Beksac, M., Dimopoulos, M. A., Spicka, I., Hungria, V., Hajek, R., Weisel, K., Vorhees, P. M., Raje, N., Usmani, S. Z., Quach, H., Goldschmidt, H., Mateos, M.-V., Einsele, H., Cavo, M., Moreau, P., Krishnan, A., & Richardson, P. G. (2023). Cilta-cel vs standard care in multiple myeloma. *New England Journal of Medicine, 388*(14), 1319–1331. https://doi.org/10.1056/NEJMoa2303379.

Schuster, S. J., Bishop, M. R., Tam, C. S., Waller, E. K., Borchmann, P., McGuirk, J. P., Jäger, U., Jaglowski, S., Andreadis, C., Westin, J. R., Fleury, I., Bachanova, V., Foley, S. R., Ho, P. J., Mielke, S., Magenau, J. M., Holte, H., Pantano, S., Pacaud, L. B., … Awasthi, R. (2019). Tisagenlecleucel in adult relapsed or refractory diffuse large B-cell lymphoma. *New England Journal of Medicine, 380*(1), 45–56. https://doi.org/10.1056/NEJMoa1804980.

Sugarman, J., & Sulmasy, D. P. (2020). Ethics of CAR T therapy. *JAMA, 323*(11), 1035–1036. https://doi.org/10.1001/jama.2020.0441.

Topalian, S. L., Taube, J. M., Anders, R. A., & Pard oll, D. M. (2016). Mechanism-driven biomarkers to guide immune checkpoint blockade in cancer therapy. *Nature Reviews Cancer, 16*(5), 275–287. https://doi.org/10.1038/nrc.2016.36.

Vera, D. G., Shah, N. N., Fry, T. J., & Mackall, C. L. (2024). Real-world outcomes of approved CAR T-cell therapies. *Blood Advances, 8*(3), 567–579. https://doi.org/10.1182/bloodadvances.2023001234.

Wherry, E. J., & Kurachi, M. (2015). Molecular and cellular insights into T cell exhaustion. *Nature Reviews Immunology, 15*(8), 486–499. https://doi.org/10.1038/nri3862.

Whittington, M. D., McQueen, R. B., Ollendorf, D. A., Kumar, V. M., Chapman, R. H., Tice, J. A., Pearson, S. D., Campbell, J. D., Ramsey, S. D., Bansal, A., Carlson, J. J., Veenstra, D. L., Sullivan, S. D., Chambers, J. D., Neumann, P. J., Cohen, J. T., Garrison, L. P., Basu, A., Phillips, K. A., … Shankaran, V. (2019). Cost-effectiveness of CAR T therapy. *Journal of Clinical Oncology, 37*(24), 2105–2119. https://doi.org/10.1200/JCO.18.02079.

CHAPTER

4

Recent progress in CAR-T cell therapy for B-cell cancers with resistant tissues

Rajesh Kumar[a], Seetha Harilal[a], Mohd Aftab Siddiqui[b], Mohammad Ramzan[c] and Sumel Ashique[c]

[a]*Faculty of Pharmaceutical Sciences, Kerala University of Health Sciences, Thrissur, Kerala, India,* [b]*Faculty of Pharmacy, Integral University, Lucknow, Uttar Pradesh, India,* [c]*School of Pharmaceutical Sciences, Lovely Professional University, Phagwara, Punjab, India*

4.1 Introduction to chimeric antigen receptor-T cell therapy

4.1.1 Overview

T lymphocytes (T cells) are the white blood cells that play an important function in the adaptive immune response. The T cells can be genetically modified and engineered to produce chimeric antigen receptor (CAR)-T cells, which can recognize cancer cells and destroy them without harming other healthy cells (Feins et al., 2019). The procedure starts with leukapheresis to extract the T cells, which are seen in the blood of the patients. The cells then undergo treatment ex vivo to make them express CARs. In this process, the gene bearing the code for the CAR is introduced into the genome of the T cell through the use of viruses such as lentiviruses/ retroviruses. The T cells that are modified and have the CAR are cultured to expand the amount as they are to be transplanted back into the patient. During this expansion, the CAR-T cells are multiplied, increasing their quantity, thus providing sufficient amounts to directly attack and destroy the cancer cells (Nair & Westin, 2020). All these processes from cell collection to reinfusion are conducted under GMP to ensure that the final CAR-T cell quality is not compromised. The CAR itself is a chimeric protein that has three domains: An extracellular antigen-binding domain, a transmembrane domain, and a cytoplasmic signaling domain. The antigen-binding domain is typically derived from a monoclonal antibody that can interact with a specific tumor-associated antigen (TAA), resulting in CAR-T cells to selectively target those cells expressing that specific antigen, minimizing the damage to normal cells. The transmembrane domain attaches the CAR to the membrane of the T cell. The intracellular signaling domain participates in the activation of T cells as well as mediates the initiation of the signaling cascade leading to the activation of T cells, further cloning, and cytotoxicity. Signaling domains such as CD28, CD137 (4-1BB), and OX40 also play a role in modulating the T cell as well as its survival rate based on the design of CAR. CAR signaling domain selection is one of the most important factors in CAR design since it can considerably influence the treatment's efficacy and safety (Xie et al., 2023). Although CAR-T cell therapy has been effective, it causes side effects, including cytokine release syndrome (CRS), neurotoxicity (CAR-T-cell-related encephalopathy syndrome), as well as B-cell aplasia, which compromises the body's immune system because of low B-cell count, and the condition necessitates immediate management (Ahmad et al., 2020). Moreover, the potency and

Advancements in Personalized Cellular Immunotherapy: A Comprehensive Guide to Cancer Treatment.
DOI: https://doi.org/10.1016/B978-0-443-27731-3.00008-5

Copyright © 2026 Elsevier Inc. All rights are reserved, including those for text and data mining, AI training, and similar technologies.

persistence of CAR-T cells have improved significantly in the past years. First-generation CAR-T cells had a low level of activation and proliferation capability, but this was enhanced in the second generation as well as the third-generation CAR-T cells, which consist of a costimulatory signaling domain. The recent advancement is known as fourth-generation CAR-T, or "CART 2.0," which includes adjustments to CRS, which enhances the efficacy of the antitumor response, but at the same time reduces the risk of relapse, making it an ideal therapy (Ramesh et al., 2023). CAR-T cell therapy showed its efficacy in treating various cancers, such as the B-cell cancers. For example, in acute lymphoblastic leukemia (ALL), CAR-T has been more effective among patients with relapse or refractory diseases, especially among pediatric and young-adult groups. It has also been used to treat diffuse large B-cell lymphoma (DLBCL), which can be considered the commonest subtype of non-Hodgkin lymphoma (NHL). Further, the CAR-T cell therapy can be considered useful for patients who are resistant to the standard first-line therapy. The therapy is also utilized in the treatment of mantle cell lymphoma (MCL). In the case of multiple myeloma (MM), the B-cell maturation antigen (BCMA) targeted therapy has emerged as promising since it offers a ray of hope to the relapsed or refractory patients (Sadelain et al., 2015).

4.1.2 Tumor burden of B-cell cancer

Tumor load is defined as the total volume or mass of the tumor involving a particular patient. It has an important place in the diagnosis, staging, and treatment of cancers, especially for B-cell neoplasms. These cancers are derived from B lymphocytes, a specialized type of white blood cell charged with antibody manufacture. The B-cell cancers can include NHL, MM, and chronic lymphocytic leukemia (CLL), and in these diseases, the immune cells are not able to function optimally. Evaluating the tumor mass provides an understanding of the stages and the aggressiveness of the disease, as well as estimate the outcomes of the therapy (Zhao et al., 2015). B-cell cancers are a group of hematological tumors that differ in some aspects from other cancers and from each other. Hodgkin's disease is related to nodules known as Hodgkin and NHL, particularly destructive subtypes such as DLBCL, which are characterized by rapidly growing cancers and aggressive progression. The CLL is a clinical-pathologic disorder characterized by the build-up of mature B cells but low activity; common signs include high leukocyte count and adenopathy. The MM can have an increased secretion of monoclonal immunoglobulins that can have repercussions such as bone marrowing and anemia. In these diseases, the degree of massiveness of tumor invasion determines the kinds of therapy applied as well as the prognosis of the disease (Davila & Sadelain, 2016). Tumor burden is a good prognostic indicator in B-cell cancers. High tumor load can be suggestive of late-stage cancer, which is normally related to poor prognosis and aggressive therapy. On the other hand, low tumor volume could indicate low-stage disease, which does not require such intensive treatments. Tumor assessment and measurement are also used to assess the treatment outcomes. Reduction in tumor load is often associated with a good therapeutic outcome, and increased tumor mass is associated with treatment failure and requires a change in management strategy (Yuan & Su, 2014). Several methods are used to assess tumor burden in clinical practice. Physical examinations can identify enlarged lymph nodes or visible tumors, offering initial insights into disease extent. Imaging techniques, such as CT scans, MRIs, and PET scans, provide detailed visualization of tumor size, location, and metastasis. Laboratory tests are very relevant where the level of specific protein or blood cell count is indicative of tumor mass in diseases such as CLL and MM. Control tissue biopsies are also commonly done to assess the degree of infiltration by cancer cells. The Ann Arbor classifications for lymphomas and the Rai classifications for CLL divide the disease according to progression and allow for quantifying the degree of the tumor. Technological advancements, such as circulating tumor DNA (ctDNA), enable methods for analyzing the tumor dynamics with minimal invasion and the identification of minimal residual disease for specificity in monitoring (Shapiro et al., 2002).

The CAR-T cell therapy efficacy as well as safety in the case of B-cell cancers depend to a significant extent on the tumor burden. Several issues arise where there is high tumor burden; this comprises of high risk of developing serious CRS, as well as neurotoxicity, low CAR-T cell function due to the immunosuppressive nature of the tumor, and a high likelihood of treatment failure and relapse. On the other hand, a low tumor burden increases the chances of therapy due to the reduced CRS severity, space for CAR-T cell infiltration, better treatment response, and limiting the risks of neurotoxicities. Since high tumor load is associated with poor outcome, strategies to reduce tumor burden prior to the administration of CAR-T cells, by utilizing standard therapies such as surgery and chemotherapy, are usually applied. Assessment of tumor load and its implications is critical to appropriate patient selection, therapy individualization, overall survival, as well as tumor-free survival rates among patients undergoing CAR-T therapy (Magee & Snook, 2014).

4.1.3 Role of chimeric antigen receptor-T cell therapy in B-cell cancer

CAR-T cell therapy has improved the relapsed/refractory B-cell cancer, leading to a better outcome, particularly in ALL and DLBCL. The CAR-T cell therapy involves the modification of T cells, which can detect and interact with specific antigens on malignant B cells, and has shown remarkably high success rates with high response amounts and effectiveness. The use of this breakthrough therapy has made it possible for patients with very few treatment options, especially failure of conventional treatments to have an opportunity of a cure (Almåsbak et al., 2016).

Current clinical practice has demonstrated good efficacy against B-cell cancers through CAR-T cell therapies which target the antigen CD19. This particular protein is widely expressed on the surface of B cells, including the malignant ones. FDA-approved products such as tisagenlecleucel (Kymriah) for B-ALL as well as axicabtagene ciloleucel (Yescarta) for DLBCL have shown remarkable results in clinical trials, offering new options for the management of refractory/relapsed cancers. In the case of B-ALL, response rates of 70%–90% have been reported, with many patients clearing most residual disease or minimal residual disease (MRD) negativity, a greatly relevant measure of the long-term remission. Similarly, for patients suffering from DLBCL, response rates of 50%–80% and complete response rates of around 40%–60% has also been observed; yet again, a few patients recurrently existed and sustained long, healthy lives. Even with this level of success, there are hurdles such as antigen loss in which tumor cells no longer upregulate the expression of CD19 and therefore allow for relapse. This has caused investigation of other targets, including BCMA, as alternatives for treating MM and similar other B-cell malignancies (Zhang et al., 2016). Despite the promising outcomes of CAR-T therapy, a number of limitations have to be addressed. Relapse is one of the key difficulties, which is often caused by antigen escape, in which tumor cells lose or change the target antigen, so they can no longer be attacked by CAR-T cells. In addition, the immunosuppressive tumor microenvironment and CAR-T cell senescence also promote resistance and relapse. Another issue is the toxicity of CAR-T therapy, which can result in CRS, which has severe side effects. CRS has been defined as a cascade of fever, hypotension, and multiorgan failure or dysfunction. Neurotoxicity could be present as confusion, seizures, and encephalopathy. These adverse effects necessitate close monitoring of the patient and supportive care measures. In addition to this, the most notable problem that remains is the cost-effectiveness of this therapy. It is expensive because of the high complexity of the manufacturing procedures. But the advances in developing allogeneic CAR-T cells aim to broaden their application and availability by lowering their costs (Gauthier & Yakoub-Agha, 2017). The other is the applicability of CAR-T therapy in cancers such as solid tumors, where the tumor microenvironment and the tumor morphology, structure, and genetic characteristics shield the CAR-T cells from accessing as well as effectively treating malignant cells. The CAR-T cell therapy can be more effective in blood cancers, in comparison with solid tumors. Nevertheless, the further identification of novel targets, application of combination therapy, and development of novel approaches for CAR T cells is underway to overcome such emerging problems and to expand the applicability of CAR T cells beyond the treatment of blood cancers. In total, CAR-T cell therapy is a revolutionary form of therapy for B-cell cancers, providing hope to patients who do not have many therapeutic options left and encouraging additional investigation to work through the challenges that it presents for improving its accessibility to a greater number of patients (Springuel et al., 2019).

4.2 Chimeric antigen receptor-T cell mechanism, structure, and manufacture

The CAR-T cell therapy targets the surface antigens present in the cancer cells. The CAR interacts with the target antigen resulting in immune synapse formation leading to contact-dependent cytotoxicity (Fig. 4.1). The therapy can result in robust as well as long-lasting anticancer activity, and the recognition as well as clearance of cancer cells depends on the CAR, rather than the normal receptor on the T cells, or the human leukocyte antigen (HLA), thereby overcoming the immune escape produced by down regulated HLA expression in tumor cells (Jiang et al., 2023; Wei et al., 2019). The CAR-T cells also undergo differentiation, resulting in the formation of memory T cells that can produce long-term anticancer activity. Therefore, they are being increasingly utilized in the therapy of hematological malignancies (Bulliard et al., 2023; López-Cantillo et al., 2022). In the case of B-cell acute lymphocytic leukemia (B-ALL), the anti-CD19 CAR-T cell therapy can lead to a complete remission (CR) rate, which can be higher than 90% (Elsallab et al., 2023; Maude et al., 2014). Excellent anticancer activity is also seen in the case of B-cell lymphoma (Cappell & Kochenderfer, 2023; Denlinger et al., 2022; Schuster et al., 2017).

4.2.1 Chimeric antigen receptor structure

The CARs can be considered as synthetic receptors type I created with modular components/ modular design and they have four regions, such as the extracellular domain which can interact and bind to the cancer cell antigen which is targeted, the hinge/ spacer, together known as the ectodomain; the hydrophobic transmembrane, as well as the intracellular signaling domain, known as the endodomain (Bachmann, 2019; Guedan et al., 2019; Sterner & Sterner, 2021; Zhu et al., 2022). The extracellular domain can be seen outside the membrane of the T cell, and the intracellular domain, consisting of signaling components, is seen inside the cell. The part outside the cell membrane is made of fragments

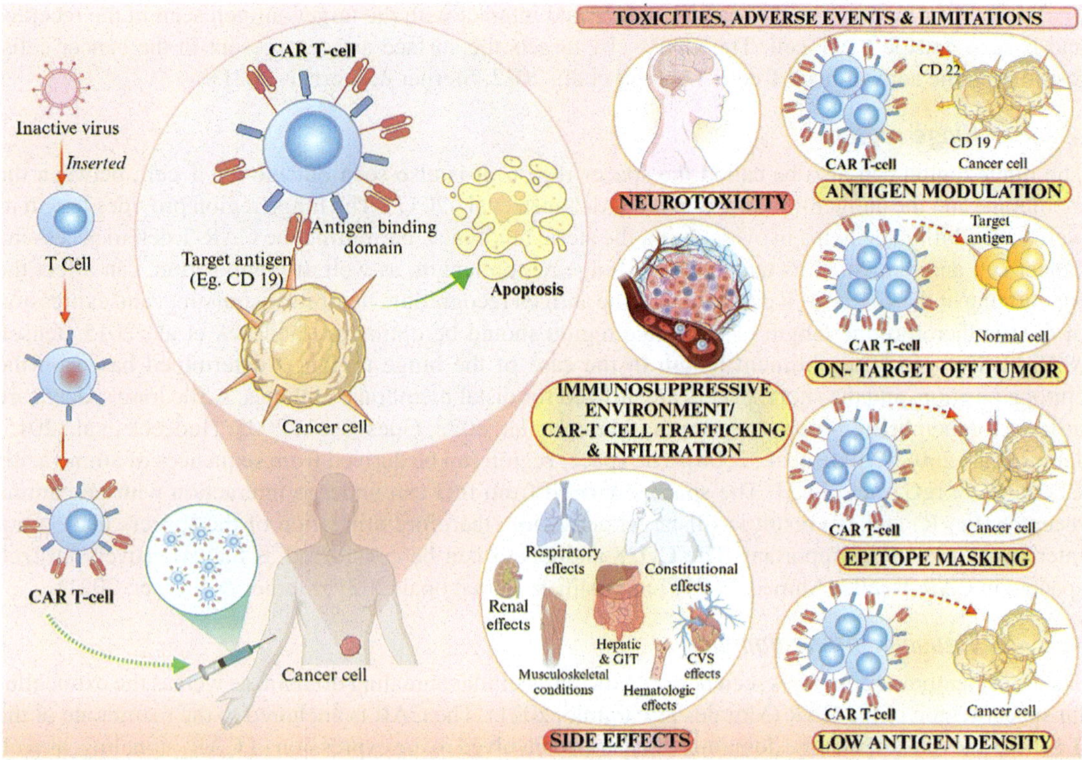

FIGURE 4.1

Mechanism, toxicities, and limitations in CAR-T cell therapy (Created in BioRender.com). *CAR-T*, chimeric antigen receptor-T.

of laboratory-made monoclonal antibodies, and such fragments influence the efficacy of the receptor in recognition and binding to the cancer cell-specific antigens. The part inside the cell membrane consists of signaling as well as costimulatory domains, which produce transmission of signals inside the cell during the interaction of the receptor with the antigen. All the domains utilized can affect the overall function of the CAR-T cells (Guedan et al., 2019; Jayaraman et al., 2020; Sterner & Sterner, 2021; Zhang et al., 2017).

4.2.1.1 ***Target antigen-binding domain***

The target antigen-binding domain is responsible for the specificity of the target antigen. The antigen-binding domain is linked as a single-chain variable fragment (scFv), having variable heavy (V_H) as well as light (V_L) monoclonal antibody chains. The linker peptides, such as serine-glycine, as well as glutamate–lysine can link both the chains (Ahmad et al., 2022; Bachmann, 2019; Sterner & Sterner, 2021).

The antigen-binding domain can recognize and interact with the target antigen seen in the receptor and is seen outside the T cell. The CAR scFv targets the surface antigen present in the cancer cells, resulting in the activation of T cells (Ahmad et al., 2022; Sterner & Sterner, 2021).

4.2.1.2 *Hinge region*

The hinge region can also be called the spacer region. It is also seen outside the T cell, between the membrane and the antigen-binding domain (Alabanza et al., 2017). The hinge region provides length as well as flexibility and helps in overcoming the steric hindrance, permitting the CAR-T cell to access the cancer cell antigen, which is targeted. The hinge region length, as well as composition, can affect the functioning of the CARs, as it can influence the antigen recognition, flexibility, signaling, and expression of CARs, therefore the length of the hinge region should be optimum (Hudecek et al., 2015; Jensen & Riddell, 2015). The optimum length in the case of the hinge region is determined based on the epitope position, and the short spacers are utilized for distal membrane epitopes, while long spacers are utilized for membrane-proximal epitopes (Ahmad et al., 2022; Guest et al., 2005; Hudecek et al., 2015; James et al., 2008; Wilkie et al., 2008). The spacer region can be derived from sequences of amino acids CD28, CD8, IgG1, and IgG4. The spacers derived from IgG can undergo interaction with Fc gamma receptor (FcγR) and can result in off-target activation, therefore utilization of the spacers that do not interact with FcγR is important. The CD28 and CD8α can be considered as the extensively utilized spacers in CAR T-cells (Ahmed, 2021; Lankipalli & Ramagopal, 2021; Sterner & Sterner, 2021).

4.2.1.3 *Transmembrane domain*

The transmembrane domain is seen between the intracellular signaling domain, as well as the extracellular spacer region of the CAR (Morales & Paramio, 2021). The CAR is anchored to the membrane of the T cell via the transmembrane domain, which is also involved in the expression of CAR, signaling, as well as its stability (Bridgeman et al., 2010; Guedan et al., 2019; Zhang et al., 2012). The transmembrane domain can be derived from the CD4, CD8, CD8α, CD3-zeta, and CD28. Even though considered inert, the transmembrane domain is crucially involved in the functions of CAR-T cells. The CD28 transmembrane domain can result in a stable as well as highly expressed CAR, and the commonly used are CD28 and the CD8α transmembrane domains (Julamanee et al., 2021; Muller et al., 2021; Sterner & Sterner, 2021)

4.2.1.4 *Intracellular signaling domain*

The intracellular signaling domain is the endodomain region of the CAR (Lanitis et al., 2020). The CAR-T cells belonging to the first generation utilized FcRγ/ CD3ζ signaling domains; the second generation utilized a costimulatory domain; the third generation utilized two costimulatory domains to overcome the drawback of incomplete activation when one costimulatory domain is utilized. The mainly used 4-1BB (CD137), as well as CD28 costimulatory domains, are approved by the FDA and show good patient response. The CAR T-cells with 4-1BB and CD28 showed strong cytokine generation in the case of lymphoma as well as lung metastasis, resulting in better antitumor activity (Jayaraman et al., 2020; Larson & Maus, 2021; Wu et al., 2020; Whilding & Maher, 2015; Yu et al., 2017). Various other costimulatory domains include CD40, MYD88, CD27, CD134, as well as the inducible T cell costimulator (Mata et al., 2017; Song & Powell, 2012; Sterner & Sterner, 2021). The cytoplasmic domain of CD-3 Zeta can be utilized as the intracellular signaling domain [52]. The activation of CAR T-cells needs costimulatory molecules, including CD-28, CD-27, CD-137, as well as CD-134

(Loskog et al., 2006; Weinkove et al., 2019). When the antigen-binding domain of the CAR-T cells interacts with the target antigen, the receptors can cluster, leading to activation and signaling inside the cell (Korell et al., 2022). The CD3 zeta is crucial for T cell activation, resulting in the anticancer action. The 4-1BB can enhance the T-cell expansion and persistence. When the anti-CD19 CAR-T cell binds with the targeted tumor cells expressing CD19, the CAR-T cell CD3-zeta, as well as CD28 costimulatory domains, produce activation of the downstream signaling pathway leading to T-cell activation, its proliferation, effector functions, as well as the release of chemokines and cytokines. All these events can result in the killing of the cancer cells expressing CD19 (Chang & Chen, 2017; Korell et al., 2022; Philipson et al., 2020).

4.2.2 Chimeric antigen receptor-T cell manufacture

The CAR-T cell manufacture includes steps such as the isolation of lymphocytes, the selection of T cells, the transfer of genes, T-cell expansion, final formulation, and quality control. A complete blood count test is performed, and the whole blood is collected through a leukoreduction filter in which the leukocytes, particularly the lymphocytes, are trapped (Cholette et al., 2012). The next step is centrifugation, such as the Percoll density gradients centrifugation, to separate the lymphocytes from other leukocytes, such as the monocytes. The lymphocytes are then cryopreserved at about a temperature of −80°C (Abou-el-Enein et al., 2021; Ahmed, 2021; Brosseron et al., 2015). The gene can be transferred into the T cell utilizing a viral vector such as adenovirus, adeno-associated virus, lentivirus, or retrovirus. Gene transfer via non-viral techniques such as liposomes, naked DNA, and transposons is also being studied, but viral vectors are mainly utilized because of their efficiency (Ahmad et al., 2022; Atsavapranee et al., 2021; Dogbey et al., 2023). Then, T cell expansion is being carried out utilizing culture systems including wave bags, T flask, and bioreactor systems such as the ambr 250 automate bioreactor. The CAR-T cells then undergo quality control, including toxicity tests, FACS analysis, chromium release assays, and endotoxin detection tests, as they must adhere to the specific regulations laid down by the FDA and should be safe, before administering to the patient and transportation to the clinic (Ahmad et al., 2022; Li et al., 2019; Reddy et al., 2020; Wang & Rivière, 2016).

4.3 Recent clinical trials and currently approved chimeric antigen receptor-T cell therapies

4.3.1 Recent clinical trials of chimeric antigen receptor-T cell therapies

CAR-T cell therapies are being explored in several clinical trials to assess their potential to treat B-cell cancers. Many of the therapies have reached Phase 2 studies and are expected to reach the clinic in the future. Table 4.1 highlights various recent clinical trials of CAR-T cell therapy to treat B-cell cancers.

4.3.2 Chimeric antigen receptor-T cell therapies approved for B-cell cancers

The current FDA-approved CAR-T cell therapies for B-cell cancers include.

Table 4.1 Clinical trials of CAR-T cell therapy to treat B-cell cancers.

Disease	Clinical trials details	Type of study	References
CNS-associated refractory/relapsed CD19-B cell malignancies (Hematological malignancy)	Clinical trial ID: **NCT04287309** Cytokine release syndrome exploration primarily followed by efficacy, completion, and recruitment status is unknown	Observational	Huang, H. (2020)
Refractory/relapsed CD19-B cell lymphoma/leukemia	Clinical trial ID: **NCT04271410** Primarily, adverse events as well as response rates were noted concerning the treatment. Secondarily, the rate, quantity, cellular kinetics, period of response, progression-free survival, as well as overall survival of anti-CD19 CAR-T cell therapy. In the Phase 2 study	Interventional	Chongqing Precision Biotech Co., Ltd (2023a)
Refractory/relapsed CD19-B cell malignancies	Clinical trial ID: **NCT06532630** Primarily, safety as well as tolerability In the early stage of Phase 1	Interventional	Shanghai First Song Biotechnology Co., Ltd (2024)
Refractory/relapsed CD19-B cell tumors	Clinical trial ID: **NCT06375161** Primarily, safety as well as tolerability In the early stage of Phase 1	Interventional	Liang, A. (2024a)
Refractory/relapsed CD19-B cell malignancies (Lymphoma/leukemia)	Clinical trial ID: **NCT02537977** Primarily, adverse events occurrence followed by response rates, progression-free as well as overall survival, and remission duration were noted	Interventional	Liang, A. (2018)
CD19-associated B-cell malignancies	Clinical trial ID: **NCT03952923** Primarily, adverse events occurrence and remission rate were noted, followed by Relapse-free as well as overall survival. In the Phase 1 study	Interventional	Hebei Yanda Ludaopei Hospital (2022)
Refractory/relapsed CD19-B cell malignancies (Lymphoma/leukemia)	Clinical trial ID: **NCT04271800** Primarily, adverse events and response rates were noted concerning the treatment. Secondarily, Response duration, progression-free as well as overall survival, cytokine levels in serum, cellular kinetics, determining the rate and quantity of CD19 CAR-T cells in peripheral blood/ bone marrow. In the Phase 2 study	Interventional	Chongqing Precision Biotech Co., Ltd (2023b)

(*continued on next page*)

Refractory/relapsed CD19-B cell malignancies (Acute lymphoblastic leukemia/ lymphoma)	Clinical trial ID: **NCT03559439** Primarily assess adverse events as well as toxicities, severities, and secondarily, response rate as well as overall survival. In the Phase 1 study.	Interventional	Ligen, L. (2020)
Refractory/relapsed B Cell (Hematological malignancies) due to loss of expression of CD19	Clinical trial ID: **NCT02794961** Primarily assess remission rate. In the Phase 2 study.	Interventional	Zheng, K. L. X. J. N. (2016)
Esophagogastric junction adenocarcinoma/B-cell leukemia/lymphoma	Clinical trial ID: **NCT03302403** Assess tolerability of treatment and CAR-T cell cytokinetics	Interventional	YU, K. (2020)
Refractory/relapsed B-cell (Lymphoma/leukemia)	Clinical trial ID: **NCT04648475** Primarily assess adverse events and response rate of treatment of CD19 as well as CD22 and secondarily, the rate and quantity of CD19 and CD22 CAR-T cells in peripheral blood and bone marrow. In the Phase 2 study.	Interventional	Chongqing Precision Biotech Co., Ltd, (2023c)
Anti-CD22 therapy associated B Cell Malignancies (Lymphoma/leukemia)	Clinical trial ID: **NCT03262298** Primarily assess adverse events as well as severity of toxicities and secondarily, the response rate. In the Phase 2 study	Interventional	Affiliated Hospital to Academy of Military Medical Sciences (2021)
CD19-associated B-cell malignancies (Acute lymphoblastic leukemia/lymphoma)	Clinical trial ID: **NCT02965092** Primarily, adverse events occurrence followed by anti-CD19 CAR-T cells quantity and rate, remission rates, as well as relapse-free, event-free, and overall survival were noted. In the Phase 2 study.	Interventional	Wuhan Sian Medical Technology Co., Ltd (2019)

(*continued on next page*)

Table 4.1 Clinical trials of CAR-T cell therapy to treat B-cell cancers—cont'd

Disease	Clinical trials details	Type of study	References
B-cell acute lymphoblastic leukemia	Clinical trial ID: **NCT06343090** Primarily assess event-free survival of sequential infusion of both CD19 CAR as well as CD22 CAR-T and also hematopoietic stem cell transplantation group-associated CD19 CAR T-cell infusion. In Phase 2 study.	Interventional	Pan (2024)
Diffuse large B-cell lymphoma	Clinical trial ID: **NCT02737085** Primarily assess adverse events. In the Phase 2 study.	Interventional	Chen (2016)
Refractory/diffuse large B-cell lymphoma	Clinical trial ID: **NCT05715606** Primarily, determination of Maximal Tolerable Dose followed by Pharmacodynamics and Pharmacokinetics evaluation. In the early Phase 1 study	Interventional	Anhui Provincial Hospital (2023)
B-cell malignancies	Clinical trial ID: **NCT05432882** Evaluation of safety of bi-4SCARCD19/22 T cells. In the Phase 2 study	Interventional	Shenzhen Geno-Immune Medical Institute (2022a)
Refractory/relapsed B-cell malignancies	Clinical trial ID: **NCT06519344** Evaluate dose-limiting toxicity. In the early Phase 1 study.	Interventional	Liang, A. (2024b)
B-cell malignancies (Diffuse large B-cell lymphoma)	Clinical trial ID: **NCT05436496** Evaluation of adverse events as well as safety of bi-4SCARCD19/70 T cells. In the Phase 2 study	Interventional	Shenzhen Geno-Immune Medical Institute (2022b)
B-cell lymphoma/leukemia	Clinical trial ID: **NCT04649983** Primarily, adverse events occurrence followed by CD19 as well as CD22 CAR-T response rate. Secondarily, CD19 as well as CD22 cellular kinetics, quantity, response duration, progression-free as well as overall survival, and serum cytokine levels were noted. In the Phase 2 study	Interventional	Chongqing Precision Biotech Co., Ltd (2023d)

CAR-T; *Chimeric antigen receptor.*

4.3.2.1 *Axicabtagene ciloleucel (YESCARTA)*

Axicabtagene ciloleucel (YESCARTA) is developed by Kite Pharma Inc. and is a genetically modified CD19-targeted adoptive T cell transfer therapy. It is utilized for the therapy of patients having large B-cell lymphoma which is refractory to the first-line therapy, or relapsed disease within one year of the first-line therapy, refractory/relapsed cancer including high-grade B-cell lymphoma, primary mediastinal large B-cell lymphoma, DLBCL not otherwise specified, as well as the DLBCL developed and transformed from follicular lymphoma, after more than two rounds of systemic treatment (Research, 2024a).

4.3.2.2 *Lisocabtagene maraleucel (BREYANZI)*

Lisocabtagene maraleucel (BREYANZI) is developed by Juno Therapeutics, Inc., and Bristol-Myers Squibb, and is a genetically modified CD19-targeted adoptive T-cell transfer therapy. It is utilized for the therapy of B-cell cancers, including high-grade B-cell lymphoma, large B-cell lymphoma (LBCL), primary mediastinal large B-cell lymphoma, DLBCL, as well as grade 3B follicular lymphoma. The therapy is approved for patients having refractory/relapsed cancer following more than two rounds of systemic treatment, refractory cancer to the first-line therapy, relapsed disease within one year of the first-line therapy, and patients with relapsed cancer not having the eligibility to undergo hematopoietic stem cell transplantation (HSCT) (Research, 2024b).

4.3.2.3 *Tisagenlecleucel (KYMRIAH)*

Tisagenlecleucel (KYMRIAH) is manufactured by Novartis Pharmaceuticals Corporation and is a genetically modified CD19-targeted adoptive T-cell transfer therapy. It is utilized for the therapy of patients having refractory/relapsed B-cell cancers such as large B-cell lymphoma, high-grade B-cell lymphoma, DLBCL not otherwise specified, as well as DLBCL developed and transformed from follicular lymphoma, following more than two rounds of systemic treatment. The therapy is also approved for patients less than age 25 years having refractory/relapsed B-ALL (Research, 2024c).

4.4 Toxicities and side effects

4.4.1 Chimeric antigen receptor-T cell-associated toxicity and side effects

Even though CAR-T cell therapy has considerably improved the management of B-cell cancers, there are concerns regarding its toxicity, including the chances of mortality. The anti-CD19 CAR T-cell therapy has been studied for the chances of toxicity, as it is the first to receive FDA approval. The therapy has produced severe as well as fatal events in patients during the clinical trials, even though they were able to produce exceptional response rates (Cushman-Vokoun et al., 2020; Ma et al., 2019; Park et al., 2016). The toxicity and side effects, including immune effector cell–associated neurotoxicity syndrome (ICANS), CRS, hemophagocytic lymphohistiocytosis, as well as macrophage activation syndrome, can also be associated with the CAR design, its target, and the disease. The toxicity can result in an elevation in the serum ferritin levels, hyper-inflammatory syndrome, pulmonary edema, and hemophagocytosis, affecting various organs such as the spleen, liver, and kidneys. The ICANS can result in high concentrations of cerebrospinal fluid cytokines, as well as the disruption of the blood-brain barrier (BBB). Severe toxicity with CRS can result in hypotension, circulatory collapse, cardiac dysfunction, as well as multiorgan system failure, including renal failure and respiratory failure, resulting

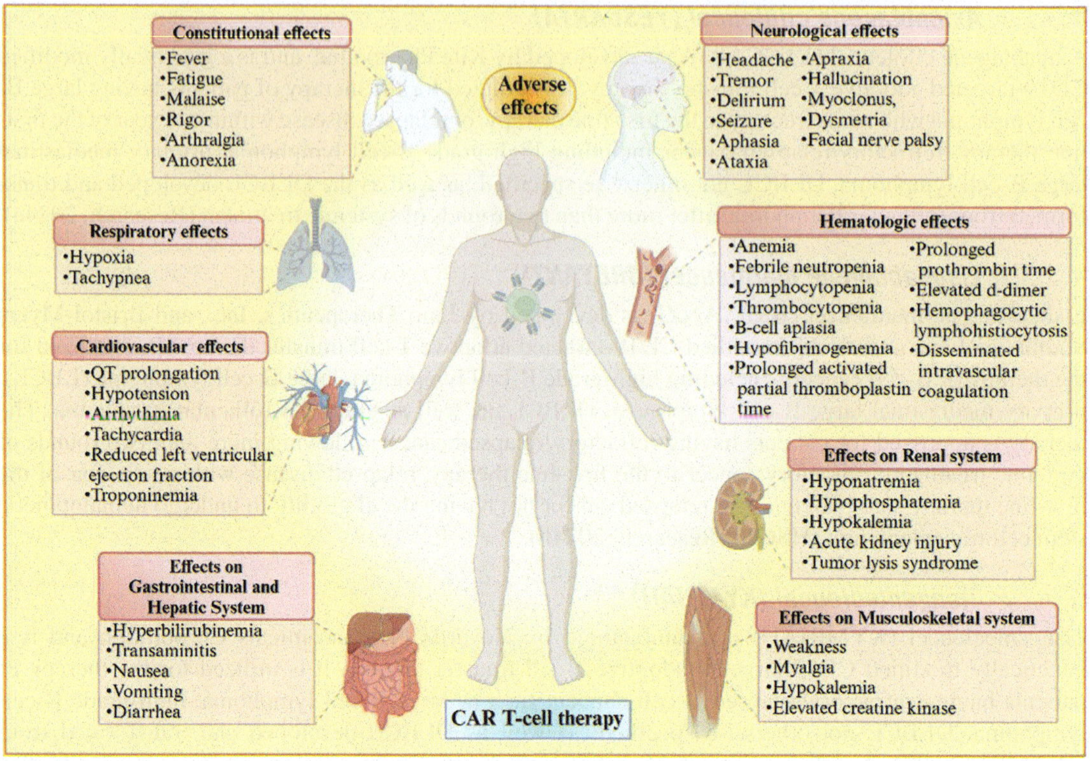

FIGURE 4.2

CAR T-cell therapy adverse effects (Created in BioRender.com). *CAR-T,* chimeric antigen receptor-T.

in mortality (Ahmad et al., 2022; Brudno & Kochenderfer, 2024; Ferreri & Bhutani, 2024; Hines et al., 2023; Walton et al., 2024) (Fig. 4.2). About 37%–93% of the lymphoma patients experienced CRS after CAR- T cell therapy. Among the patients suffering from B-ALL who received CAR T-cell therapy (tisagenlecleucel), about 46% went through grade 3 or more of CRS, and it was also experienced by about 13%–18% of DLBCL patients who received CAR T-cell therapy (tisagenlecleucel, axicabtagene ciloleucel) (Sterner & Sterner, 2021).

4.4.1.1 *Cytokine release syndrome*

The CRS is considered a commonly seen, as well as serious side effect of CAR-T cell therapy, and the consequences depend on the severity of the syndrome. The T cells, which is crucially involved in the immune system functions, produce cytokines facilitating the immune response, therefore in the therapy, the CAR- T cells, which are infused into the blood, can produce a large concentration of cytokines, resulting in the CRS, which can be considered as a serious side effect, resulting in high fever, as well as hypotension, which can be severe and result in mortality. The CRS can produce hypoxia, as well as multiorgan damage affecting renal, respiratory, gastrointestinal, hematological, cardiovascular, and the nervous system. The CRS can happen within a couple of weeks following

CAR T-cell therapy. Its management approaches depend on the grade, monitoring the blood pressure, temperature, oxygen saturation, as well as potential toxicity to various organ systems (Ahmad et al., 2022; Ferreri & Bhutani, 2024; Walton et al., 2024). The CRS can be regarded as an on-target effect of the therapy. Patients with extensive cancer can be prone to CRS during CAR-T cell therapy. There are standard treatment options, such as the utilization of steroids in the management of mild CRS. The supportive care includes intravenous fluid administration to manage dehydration, as well as hypotension, providing oxygen to manage hypoxia, administration of corticosteroids, and IL-6 antagonists. Since the CAR-T cell therapy is widely available, and with more experience, more options are emerging to manage the cases of CRS, such as the drug tocilizumab, which was utilized in the therapy of inflammatory conditions such as juvenile arthritis. It acts by IL-6 blockade, which is a cytokine produced in high levels by the macrophages as well as the T cells (National Cancer Institute, 2013; Walton et al., 2024).

4.4.1.2 Neurological side effects

The CAR T-cell therapy can also produce neurological side effects, such as ICANS, and its mechanism is yet to be clearly understood. The neurological side effects can also include activities similar to seizures, confusion, attention deficits, focal neurological deficits, word-finding difficulties, and speech impairment. Currently, ICANS can be managed with the help of steroids such as dexamethasone, which can reach the central nervous system (CNS) by penetrating the BBB to produce its effects (Gust et al., 2020; Morris et al., 2022; Roth et al., 2021; Sterner & Sterner, 2022). The CAR T-cell–related encephalopathy syndrome (CRES) shows similar characteristics to toxic encephalopathy, and the signs and symptoms include language disturbance, impaired attention, and handwriting, agitation, disorientation, tremor, confusion, somnolence, and aphasia. The CRES can occur within a week following CAR T-cell therapy, and its pathogenesis is not yet completely elucidated. The CRES management is based on assessing the CAR T-cell therapy-associated toxicity, 10-point neurological assessment, motor weakness/seizures, and intracranial pressure. The management includes supportive therapy, IL-6 antagonists, as well as corticosteroids (Castaneda-Puglianini & Chavez, 2021; Gust et al., 2020). Several studies are being carried out for the management and prevention of CRS, as well as ICANS. The approach being utilized currently in the management of CRS and ICANS includes the prophylactic administration of tocilizumab, as well as steroids in small doses. Another strategy to prevent CRS, as well as ICANS, is the modification of CARs. A clinical trial in lymphoma patients reported that a remodeled CAR T-cell therapy targeting CD19 resulted in less neurological side effects when compared to the original form of CAR, in CAR T-cell therapy (Castaneda-Puglianini & Chavez, 2021; Gust et al., 2020).

4.5 Recent research and developments in chimeric antigen receptor-T cell therapy for B-cell cancers

For refractory and relapsed B-cell malignancies, CAR-T cell therapy has dramatically changed the treatment landscape. The further determination of efficacy, accessibility, and safety of the therapy and continued advances in targeting mechanisms, prevention of resistance, allogeneic options, and reduced toxicities promise to be fruitful ways forward. CAR-T technology is evolving, and it may be utilized as an earlier treatment option for various cancers, in the earlier stages of B-cell cancers, and may be solid tumors (Wang et al., 2018).

4.5.1 Targeting antigens beyond CD19

Initial CAR-T therapies were designed to target CD19, an antigen predominantly present on B cells. Some patients were cured using these treatments, but recurrence occurred because of CD19-negative escape variants, cancer cells that have lost their expression of CD19 and hence cannot be targeted by CAR-T cells. Novel targets are currently being explored to overcome such resistance and offer alternative options for resistant patients (Geyer & Brentjens, 2016). Targeting CD19 and CD20 has emerged as a promising target on B cells and has been successful when used in combination. It is recognized that joint targeting of CD19 and CD20 significantly improves the efficiency of therapy in refractory/relapsed B-cell lymphomas, with research reports showing that it is possible to decrease relapse rates (Jacoby et al., 2019).

CD22: This is another target, and in particular, it comes for patients who have relapsed with ALL after CD19 targeted therapy. There has been shown to improve outcomes in some who had already demonstrated resistance to CAR-T based on targeting CD19 (Hutchinson, 2014).

4.5.2 Dual-targeting as well as multiantigen chimeric antigen receptor cells

To combat relapse more effectively, dual-targeting CAR-T cells have been designed that can recognize multiple antigens simultaneously. This approach aims to address antigen escape, a frequent cause of relapse in single-antigen therapies. A very recent clinical trial has demonstrated that CAR-T cells engineered to target CD19 as well as CD20 showed improved durability in patients with DLBCL, thus leading to superior overall survival (Schneider et al., 2016). Multiantigen CAR-T cells target multiple antigens, which reduces the chances of immune evasion. According to studies, this strategy could ensure strong immunity, particularly where antigens are heterogeneously expressed in the cancer (Smith et al., 2013).

4.5.3 Enhancing chimeric antigen receptor-T cell persistence and overcoming resistance

Relapse can be considered a major issue in CAR-T cell therapy because of the loss of activity of CAR-T cells over time. Researchers are testing ways of improving CAR-T cell persistence as well as functionality, thereby ensuring prolonged antitumor effects.

4.5.3.1 Armored chimeric antigen receptor-T cells

To operate in the immunosuppressive tumor microenvironment, CAR-T cells are developed to produce cytokines, including IL-12. IL-12 improves T cell persistence and improves the capability of the immune system to recognize as well as eliminate tumor cells (Chmielewski et al., 2014).

4.5.3.2 Fourth-generation chimeric antigen receptor-T cells (TRUCKs)

These are “proinflammatory cytokine-initiated, antigen-unrestricted T cells” developed to set off the release of proinflammatory cytokines upon recognizing tumor antigens. This mechanism creates an unfavorable environment for the cancer cells—a process that has already shown much promise in making the therapy more effective and increasing the chances of better outcomes for patients (Gao & Li, 2019).

4.5.3.3 *Combination with checkpoint inhibitors*

Immune checkpoint inhibitors, including PD-1/PD-L1 targeted therapy, can be used in improving CAR-T cell activity. This enables checkpoint inhibitors to overcome T cell exhaustion, prolong the functional lifespan of CAR-T cells, and sustain an immune response against cancer (Cherkassky et al., 2016).

4.5.4 Development of allogeneic (off-the-shelf) chimeric antigen receptor-T cells

The difficulty with traditional CAR-T therapies is that they are autologous, which means they are developed from the T cells collected from cancer patients, a lengthy and expensive process. However, allogeneic or "off-the-shelf" CAR-T cell products derived from healthy donors can produce potential benefits:

The premanufactured availability of these CAR-T cells for immediate administration could especially make them appealing to patients who have more aggressive cancers, where a shorter waiting period may be beneficial (Doyle, 2015). Allogeneic CAR-T cells, if effective in early phase clinical trials, are scalable as well as cost-effective and have the advantage of using commonly available umbilical cord blood or even peripheral blood as sources of donor material. This also allows the possibility of treating patients who may not be suitable candidates for autologous CAR-T cell therapy because of a lack of sufficient T cells or other health-related factors (Qasim, 2019).

4.5.5 Reducing toxicities of chimeric antigen receptor-T cell therapy

The CAR-T cell therapy has been proven to be successful to date, and the adverse effects of the treatment, including CRS as well as ICANS, remain significant risks for the patients. There has been an intense focus on the reduction of these side effects: Cytokine modulation: For severe CRS, standard therapy involves pharmacologic interventions: IL-6 receptor blockers such as tocilizumab and anakinra. This helps prevent the potentially lethal side effect of the cytokine storm that can occur in CAR-T cell therapy (Ishii et al., 2016). Next-generation CAR-T cells: To maintain the balance between antitumor efficacy and bias toward proinflammatory cytokine secretion, there is research on CAR-T cells that can release fewer proinflammatory cytokines. The next-generation CAR-T cells might provide a better balance, retaining their capability to detect as well as kill the cancer cells without producing too many side effects (Balakrishnan, 2018).

4.5.6 Chimeric antigen receptor-T therapy for early-stage B-cell cancers

CAR-T cell therapy is a breakthrough in the treatment of B-cell cancers, especially for patients suffering from refractory and relapsed diseases. This traditionally has been considered only at the most advanced stage when first and second lines of treatment have been exhausted. However, recent clinical trials are studying its applicability to earlier stages of the disease (Furukawa, 2019). Early-stage administration of CAR-T therapy may have multiple potential benefits. This may include more profound remissions, reduced risk of relapse, and survival benefit. This approach will also perhaps make subsequent therapies less toxic in the long run, thereby enhancing quality of life for the patients (Adkins, 2019). Thus, in further research, these studies may define whether CAR-T cell therapy can be added safely and successfully to the first-line therapy for the management of B-cell cancers. The outcome may reshape the treatment landscape, providing new hope to the patients at earlier stages of the disease.

4.5.7 Expanding chimeric antigen receptor-T cell therapy beyond B-cell malignancies

While showing miraculous success in the case of B-cell cancers, CAR-T cell therapy has gained much interest in its application in the development of therapies for other types of cancers:

4.5.7.1 *Multiple myeloma*

Encouraging results have come from clinical trials involving CAR-T cells, which target BCMA, which is upregulated in MM. With these studies reporting high response rates to CAR-T cell therapy, it also has a potentially good outcome in this cancer type (Carpenter et al., 2013).

4.5.7.2 *Solid tumors*

On the other hand, extension of the therapy to solid tumors is another challenge because of the immunosuppressive environment present in such tumors. However, barriers affecting the efficacy of CAR-T cells are being reduced owing to innovation in CAR-T cell development, including resistance of T cells to immune suppression. Early signals of efficacy from the first trials in solid tumors are already underway, and much more research is expected (Di & Li, 2016).

4.6 Limitations, challenges, and future perspectives of chimeric antigen receptor-T cell therapies

An excellent anticancer activity requires the targeted antigen to have adequate coverage on the cancer cells and targets, including CD19, CD20, and BCMA, which show excellent results in the case of B-cell cancers. An ideal target should also have high specificity along with high coverage to ensure safety and efficacy. In the case of CD19, from the development of B cells, its expression till the B cells get differentiated into the plasma cells can be observed. Therefore the CD19 is considered to have a high coverage. But in the case of specificity, the anti-CD19 CAR-T cell therapy can also result in the clearance of normal B cells, leading to long-term B-cell aplasia, which needs to be managed. Similar issues can also be seen in the case of BCMA, as well as CD20. These targets are not expressed exclusively on cancer cells; therefore, the targeted CAR-T cell therapy can result in off-target effects, sometimes severe, even though they have a high coverage (Wei et al., 2019; Xu et al., 2018). The antigen escape can be considered a big challenge of CAR-T cell therapy, as the cancer develops resistance toward the CAR-T cell therapy targeting a single antigen. High response rates are possible during the initial phase of the treatment, but later, a large number of patients show cancer with complete/partial loss or downregulation of the expressed antigen, such as CD19, which is targeted. After anti-CD19 CAR-T cell therapy, about 30%–70% of the recurrent cancer patients show loss or downregulation of the antigen. The effect can also be seen in BCMA-targeted CAR-T cell therapy. Therefore to address the limitation, research focuses on targeting multiple antigens. Such therapy utilizes dual CAR constructs, which consist of two scFvs to target multiple antigens, to achieve a durable response. Several such studies are explored in clinical trials, and the outcome seems promising, such as in the case of a clinical trial exploring CD19/CD22 targeted therapy for ALL as well as DLBCL (Han et al., 2019; Lin et al., 2024; Sterner & Sterner, 2021; Xu et al., 2018). The current scenario is rapidly evolving as the area is receiving huge attention and a lot of studies, with more than a hundred clinical trials focusing on improving the efficacy of CAR-T cell therapy. The currently approved CAR T-cell therapies target majorly two antigens, such as CD19 and BCMA. The current research also focuses on developing CAR T-cell therapies that can target various antigens involved in blood cancer, and treatment which can target more than one antigen at the same time (Daei Sorkhabi et al., 2023). Another development in research is the change in the source of T cells, such as from normal healthy donors in place of cancer patients. Such CAR-T cell therapies, called off-the-shelf or allogeneic CAR-T cell therapies, can be made available immediately for treatment instead of being manufactured specifically for each cancer patient. Another area of research is the introduction of genetic material into the collected T cells for the generation of CARs. Newer gene editing strategies, including clustered regularly interspaced short palindromic repeats (CRISPR), as well as transcription activator-like effector nucleases (TALEN), are currently being explored for generating the CARs (Lanitis et al., 2020; Martínez Bedoya et al., 2021; National Cancer Institute, 2013; Song et al., 2024). The research also explores the utilization of another type of cell of the immune system, the natural killer (NK) cells to generate CAR NK cell therapies for cancer treatment, as in the case of a study that utilizes modified NK cells for expressing anti-CD19 CAR to overcome the limitations such as the complexity of manufacture, and high toxicity of CAR-T cells (Liu et al., 2020). The research also explores the utilization of nanotechnology, as well as mRNA-based strategies for producing CAR T-cells in vivo.

4.7 Conclusion

The CAR-T cell therapies, mainly the CD19-targeted CAR-T cell therapy, have made considerable improvements in the management of B-cell cancers, particularly lymphomas such as the NHL, including the relapsed/refractory NHL, indolent as well as aggressive NHL, leading to high and durable response rates in comparison with the standard therapy. The current CAR-T cell therapies, which are FDA-approved, act by targeting the antigens such as CD19/ BCMA present in the B cells. CAR-T cell therapy is currently widely available and can be considered as a standard therapeutic option in the case of patients having aggressive lymphomas. Considering the research directions along with the approval of current CAR-T cell therapies, their utilization in the earlier lines of therapy is also being studied, and the treatment of patients post anti-CD19 CAR-T cell therapy can become frequent, particularly in B-cell cancers. In such cases, understanding and addressing the mechanisms of failure of CAR-T cell therapy, including the resistance, is essential to improve the survival of patients suffering from B-cell cancers. The CAR-T cell therapy can be considered a successful therapeutic strategy because the off-target effects are mostly tolerable. But still, CAR-T cell therapy shows several adverse events, including severe toxicities, which require careful management. CAR-T cell therapy is being utilized in the case of patients after multiple lines of treatment. Such an approach may be modified, as the clinical trials reporting the CAR-T cell therapy to produce better outcomes in comparison with the current standard therapy for the treatment of NHL, relapsed after first-line chemotherapy. In such patients, CAR-T cell therapy can potentially replace the currently standard chemotherapy as the second-line therapeutic option. New advancements are happening at a fast pace in the area of CAR-T cell therapies, for example, the developments to address the current limitations of CAR-T cell therapies, including toxicities, and resistance, such as utilization of nanotechnology, as well as mRNA-based strategies, multiple antigen targeting during the treatment, the source and the delivery of genetic material into the T-cells, utilization of another type of immune cells such as the NK cells, as well as incorporating gene editing strategies. Hundreds of studies, including preclinical and clinical trials, have been carried out to translate the developments from the laboratory to the clinic, which makes the future of CAR-T cell therapies, particularly in the management of B-cell cancers, promising.

References

Abou-el-Enein, M., Elsallab, M., Feldman, S. A, Fesnak, A. D, Heslop, H. E, Marks, P., Cooper, L. J. N., & Lee, D. W. (2021). Scalable manufacturing of CAR T cells for cancer immunotherapy. *Blood Cancer Discovery, 2*, 408–422. https://doi.org/10.1158/2643-3230.BCD-21-0084.

Adkins, S. (2019). CAR T-cell therapy: Adverse events and management. *Journal of the Advanced Practitioner in Oncology, 10*(4), 408–415. https://doi.org/10.6004/JADPRO.2019.10.4.11.

Affiliated Hospital to Academy of Military Medical Sciences. (2021). *Anti-CD22 chimeric antigen receptor (CAR)-modified T cell therapy targeting CD22 in treating patients with B cell malignancies*. Clinicaltrials.gov.

Ahmad, A., Uddin, S., & Steinhoff, M. (2020). CAR-T cell therapies: An overview of clinical studies supporting their approved use against acute lymphoblastic leukemia and large B-cell lymphomas. *International Journal of Molecular Sciences*. https://doi.org/10.3390/IJMS21113906.

Ahmad, U., Khan, Z., Ualiyeva, D., Amissah, O. B, Noor, Z., Khan, A., Akbar, A., & Haider, S. (2022). Chimeric antigen receptor T cell structure, its manufacturing, and related toxicities; A comprehensive review. *Advances in Cancer Biology - Metastasis, 4*, 100035. https://doi.org/10.1016/j.adcanc.2022.100035.

Ahmed, M. M. E. (2021). Car-T cell therapy: Current advances and future research possibilities. *Journal of Scientific Research in Medical and Biological Sciences, 2*, 86–116. https://doi.org/10.47631/jsrmbs.v2i2.234.

Alabanza, L., Pegues, M., Geldres, C., Shi, V., Wiltzius, J. J. W., Sievers, S. A, Yang, S., Kochenderfer, J. N., & Lee, D. W. (2017). Function of novel anti-CD19 chimeric antigen receptors with human variable regions is affected by hinge and transmembrane domains. *Molecular Therapy, 25*, 2452–2465. https://doi.org/10.1016/j.ymthe.2017.07.013.

Almåsbak, H., Aarvak, T., & Vemuri, M. C (2016). CAR T cell therapy: A game changer in cancer treatment. *Clinical & Developmental Immunology*. https://doi.org/10.1155/2016/5474602.

Anhui Provincial Hospital. (2023). *Safety and efficacy of metabolically armed CD19 CAR-T cells (Meta10-19) in the treatment of r/r DLBCL clinical research*. Clinicaltrials.gov.

Atsavapranee, E. S, Billingsley, M. M, & Mitchell, M. J (2021). Delivery technologies for T cell gene editing: Applications in cancer immunotherapy. *EBioMedicine, 67*, 103354. https://doi.org/10.1016/j.ebiom.2021.103354.

Bachmann, M. (2019). The UniCAR system: A modular CAR T cell approach to improve the safety of CAR T cells. *Immunology Letters, 211*, 13–22. https://doi.org/10.1016/j.imlet.2019.05.003.

Balakrishnan, A. (2018). CAR-T cells: Next generation cancer therapeutics. *Journal of the Indian Institute of Science*. https://doi.org/10.1007/S41745-018-0062-8.

Bridgeman, J. S., Hawkins, R. E., Bagley, S., Blaylock, M., Holland, M., & Gilham, D. E. (2010). The optimal antigen response of chimeric antigen receptors harboring the CD3ζ transmembrane domain is dependent upon incorporation of the receptor into the endogenous TCR/CD3 complex. *Journal of Immunology, 184*, 6938–6949. https://doi.org/10.4049/jimmunol.0901766.

Brosseron, F., Marcus, K., & May, C. (2015). Isolating peripheral lymphocytes by density gradient centrifugation and magnetic cell sorting. *Methods in Molecular Biology (Clifton NJ), 1295*, 33–42. https://doi.org/10.1007/978-1-4939-2550-6_3.

Brudno, J. N., & Kochenderfer, J. N. (2024). Current understanding and management of CAR T cell-associated toxicities. *Nature Reviews Clinical Oncology, 21*, 501–521. https://doi.org/10.1038/s41571-024-00903-0.

Bulliard, Y., Andersson, B. S., Baysal, M. A., Damiano, J., & Tsimberidou, A. M. (2023). Reprogramming T cell differentiation and exhaustion in CAR-T cell therapy. *Journal of Hematology & Oncology, 16*, 108. https://doi.org/10.1186/s13045-023-01504-7.

Cappell, K. M., & Kochenderfer, J. N. (2023). Long-term outcomes following CAR T cell therapy: what we know so far. *Nature Reviews Clinical Oncology, 20*, 359–371. https://doi.org/10.1038/s41571-023-00754-1.

Carpenter, R. O, Evbuomwan, M. O, Pittaluga, S., Rose, J. J, Raffeld, M., Yang, S., Gress, R. E., Hakim, F. T., & Kochenderfer, J. N. (2013). B-cell maturation antigen is a promising target for adoptive T-cell therapy of multiple myeloma. *Clinical Cancer Research*. https://doi.org/10.1158/1078-0432.CCR-12-2422.

Castaneda-Puglianini, O., & Chavez, J. C (2021). Assessing and management of neurotoxicity after CAR-T therapy in diffuse large B-cell lymphoma. *Journal of Blood Medicine, 12*, 775–783. https://doi.org/10.2147/JBM.S281247.

Chang, Z. L, & Chen, Y. Y. (2017). CARs: Synthetic immunoreceptors for cancer therapy and beyond. *Trends in Molecular Medicine, 23*, 430–450. https://doi.org/10.1016/j.molmed.2017.03.002.

Chen, J. (2016). *The sequential therapy of CD19-targeted and CD20-targeted CAR-T cell therapy for diffuse large B cell lymphoma (DLBCL)*. clinicaltrials.gov.

Cherkassky, L., Morello, A., Villena-Vargas, J., Feng, Y., Dimitrov, D. S, Jones, D. R, Sadelain, M., Adusumilli, P. S., & Brentjens, R. J. (2016). Human CAR T cells with cell-intrinsic PD-1 checkpoint blockade resist tumor-mediated inhibition. *Journal of Clinical Investigation*. https://doi.org/10.1172/JCI83092.

Chmielewski, M., Hombach, A. A, & Abken, H. (2014). Of CARs and TRUCKs: Chimeric antigen receptor (CAR) T cells engineered with an inducible cytokine to modulate the tumor stroma. *Immunological Reviews*. https://doi.org/10.1111/IMR.12125.

Cholette, J. M., Henrichs, K. F., Alfieris, G. M., Powers, K. S., Phipps, R., Spinelli, S. L., Harris, K. A., Waugh, M., & Blumberg, N. (2012). Washing red blood cells and platelets transfused in cardiac surgery reduces postoperative inflammation and number of transfusions: Results of a prospective, randomized, controlled clinical trial*. *Pediatric Critical Care Medicine, 13*, 290. https://doi.org/10.1097/PCC.0b013e31822f173c.

Chongqing Precision Biotech Co., Ltd. (2023a). *CD19+ targeted CAR-T cell therapy for relapsed/refractory CD19+ B cell leukemia and lymphoma*. clinicaltrials.gov.

Chongqing Precision Biotech Co., Ltd. (2023b). *Safety and efficacy of CD19-targeted CAR-T therapy for relapsed/refractory CD19+ B cell leukemia and lymphoma*. clinicaltrials.gov.

Chongqing Precision Biotech Co., Ltd. (2023c). *CD19 and CD22 targeted CAR-T cell therapy for relapsed/refractory B cell leukemia and lymphoma*. clinicaltrials.gov.

Chongqing Precision Biotech Co., Ltd. (2023d). *Safety and efficacy of CD19 and CD22 targeted CAR-T therapy for relapsed/refractory B cell leukemia and lymphoma*. clinicaltrials.gov.

Cushman-Vokoun, A. M, Voelkerding, K. V, Fung, M. K, Nowak, J. A, Thorson, J. A, Duncan, H. L, & Konnick, E. Q. (2020). A primer on chimeric antigen receptor T-cell therapy: What does it mean for pathologists? *Archives of Pathology & Laboratory Medicine, 145*, 704–716. https://doi.org/10.5858/arpa.2019-0632-CP.

Daei Sorkhabi, A., Mohamed Khosroshahi, L., Sarkesh, A., Mardi, A., Aghebati-Maleki, A., Aghebati-Maleki, L., Khosroshahi, A. Y., & Baradaran, B. (2023). The current landscape of CAR T-cell therapy for solid tumors: Mechanisms, research progress, challenges, and counterstrategies. *Frontiers in Immunology, 14*, 1113882. https://doi.org/10.3389/fimmu.2023.1113882.

Davila, M. L, & Sadelain, M. (2016). Biology and clinical application of CAR T cells for B cell malignancies. *International Journal of Hematology*. https://doi.org/10.1007/S12185-016-2039-6.

Denlinger, N., Bond, D., & Jaglowski, S. (2022). CAR T-cell therapy for B-cell lymphoma. *Current Problems in Cancer, 46*, 100826. https://doi.org/10.1016/j.currproblcancer.2021.100826.

Di, S., & Li, Z. (2016). Treatment of solid tumors with chimeric antigen receptor-engineered T cells: Current status and future prospects. *Science China-Life Sciences*. https://doi.org/10.1007/S11427-016-5025-6.

Dogbey, D. M, Torres, V. E. S., Fajemisin, E., Mpondo, L., Ngwenya, T., Akinrinmade, O. A, Ogunwobi, O. O., & Zhang, J. (2023). Technological advances in the use of viral and non-viral vectors for delivering genetic and non-genetic cargos for cancer therapy. *Drug Delivery and Translational Research, 13*, 2719–2738. https://doi.org/10.1007/s13346-023-01362-3.

Doyle, C. (2015). CAR-T Cells: The transplants of the future. *American Health & Drug Benefits*.

Elsallab, M., Ellithi, M., Hempel, S., Abdel-Azim, H., & Abou-el-Enein, M. (2023). Long-term response to autologous anti-CD19 chimeric antigen receptor T cells in relapsed or refractory B cell acute lymphoblastic leukemia: a systematic review and meta-analysis. *Cancer Gene Therapy, 30*, 845–854. https://doi.org/10.1038/s41417-023-00593-3.

Feins, S., Kong, W., Williams, E. F., Milone, M. C., & Fraietta, J. A. (2019). An introduction to chimeric antigen receptor (CAR) T-cell immunotherapy for human cancer. *American Journal of Hematology*. https://doi.org/10.1002/AJH.25418.

Ferreri, C. J., & Bhutani, M. (2024). Mechanisms and management of CAR T toxicity. *Frontiers in Oncology, 14*, 1396490. https://doi.org/10.3389/fonc.2024.1396490.

Furukawa, F. (2019). CAR-T therapy is a breakthrough for intractable autoimmune disease? *Trends in Immunotherapy, 3*(1). https://doi.org/10.24294/TI.V3.I1.1125.

Gao, J., & Li, F. (2019). Fourth-generation CAR-T cell as well as construction method and application thereof. https://patents.google.com/patent/CN109504660B/en.

Gauthier, J., & Yakoub-Agha, I. (2017). Chimeric antigen-receptor T-cell therapy for hematological malignancies and solid tumors: Clinical data to date, current limitations and perspectives. *Current Research in Translational Medicine*. https://doi.org/10.1016/J.RETRAM.2017.08.003.

Geyer, M. B, & Brentjens, R. J (2016). Review: Current clinical applications of chimeric antigen receptor (CAR) modified T cells. *Cytotherapy*. https://doi.org/10.1016/J.JCYT.2016.07.003.

Guedan, S., Calderon, H., Posey, A. D, & Maus, M. V (2019). Engineering and Design of Chimeric Antigen Receptors. *Molecular Therapy - Methods & Clinical Development, 12*, 145–156. https://doi.org/10.1016/j.omtm.2018.12.009.

Guest, R. D, Hawkins, R. E, Kirillova, N., Cheadle, E. J, Arnold, J., O'Neill, A., Irlam, J., Chester, K. A., Kemshead, J. T., Shaw, D. M., Embleton, M. J., & Stern, P. L. (2005). The role of extracellular spacer regions in the optimal design of chimeric immune receptors: Evaluation of four different scFvs and antigens. *Journal of Immunotherapy, 28*, 203. https://doi.org/10.1097/01.cji.0000161397.96582.59.

Gust, J., Ponce, R., Liles, W. C., Garden, G. A., & Turtle, C. J (2020). Cytokines in CAR T cell–associated neurotoxicity. *Frontiers in Immunology, 11*, 577027. https://doi.org/10.3389/fimmu.2020.577027.

Han, X., Wang, Y., Wei, J., & Han, W. (2019). Multi-antigen-targeted chimeric antigen receptor T cells for cancer therapy. *Journal of Hematology & Oncology, 12*, 128. https://doi.org/10.1186/s13045-019-0813-7.

Hebei Yanda Ludaopei Hospital. (2022). *CD19-CAR-T for patients with B cell malignancies*. ClinicalTrials.gov.

Hines, M. R., Knight, T. E., McNerney, K. O., Leick, M. B., Jain, T., Ahmed, S., Shah, N. N., Hill, J. A., Turtle, C. J., & Maloney, D. G. (2023). Immune Effector Cell-Associated Hemophagocytic Lymphohistiocytosis-Like Syndrome. *Transplant Cell Ther, 29*, 438.e1–438.e16. https://doi.org/10.1016/j.jtct.2023.03.006.

Huang, H. (2020). *CAR-T cellular therapy for B cell malignancies involved in central nervous system*. ClinicalTrials.gov.

Hudecek, M., Sommermeyer, D., Kosasih, P. L., Silva-Benedict, A., Liu, L., Rader, C., Jensen, M. C., Riddell, S. R., & Maloney, D. G. (2015). The nonsignaling extracellular spacer domain of chimeric antigen receptors is decisive for in vivo antitumor activity. *Cancer Immunology Research, 3*, 125–135. https://doi.org/10.1158/2326-6066.CIR-14-0127.

Hutchinson, L. (2014). Immunotherapy: CAR-modified T cells targeting CD19—Curing the incurable. *Nature Reviews Clinical Oncology*. https://doi.org/10.1038/NRCLINONC.2014.187.

Ishii, K., Shalabi, H., Yates, B., Delbrook, C., Mackall, C. L, Fry, T. J., Shah, N. N., & Lee, D. W. (2016). Tocilizumab-refractory cytokine release syndrome (CRS) triggered by chimeric antigen receptor (CAR)-transduced T cells may have distinct cytokine profiles compared to typical CRS. *Blood*. https://doi.org/10.1182/BLOOD.V128.22.3358.3358.

Jacoby, E., Shahani, S. A., & Shah, N. N. (2019). Updates on CAR T-cell therapy in B-cell malignancies. *Immunological Reviews*. https://doi.org/10.1111/IMR.12774.

James, S. E., Greenberg, P. D., Jensen, M. C., Lin, Y., Wang, J., Till, B. G., Raubitschek, A. A., Forman, S. J., & Press, O. W. (2008). Antigen sensitivity of CD22-specific chimeric TCR is modulated by target epitope distance from the cell membrane. *Journal of Immunology, 180*, 7028–7038. https://doi.org/10.4049/jimmunol.180.10.7028.

Jayaraman, J., Mellody, M. P., Hou, A. J., Desai, R. P., Fung, A. W., Pham, A. H. T., Wang, X., & Brentjens, R. J. (2020). CAR-T design: Elements and their synergistic function. *EBioMedicine, 58*, 102931. https://doi.org/10.1016/j.ebiom.2020.102931.

Jensen, M. C., & Riddell, S. R. (2015). Designing chimeric antigen receptors to effectively and safely target tumors. *Current Opinion in Immunology, 33*, 9–15. https://doi.org/10.1016/j.coi.2015.01.002.

Jiang, N., Yu, Y., Wu, D., Wang, S., Fang, Y., Miao, H., Zhang, Y., & Zhang, L. (2023). HLA and tumour immunology: immune escape, immunotherapy and immune-related adverse events. *Journal of Cancer Research and Clinical Oncology, 149*, 737–747. https://doi.org/10.1007/s00432-022-04493-1.

Julamanee, J., Terakura, S., Umemura, K., Adachi, Y., Miyao, K., Okuno, S., Goto, T., Watanabe, K., Hanajiri, R., Yoshida, K., Nishida, T., Murata, M., & Nakazawa, Y. (2021). Composite CD79A/CD40 co-stimulatory endodomain enhances CD19CAR-T cell proliferation and survival. *Molecular Therapy, 29*, 2677–2690. https://doi.org/10.1016/j.ymthe.2021.04.038.

Korell, F., Berger, T. R., & Maus, M. V. (2022). Understanding CAR T cell-tumor interactions: Paving the way for successful clinical outcomes. *Med, 3*, 538–564. https://doi.org/10.1016/j.medj.2022.05.001.

Lanitis, E., Coukos, G., & Irving, M. (2020). All systems go: converging synthetic biology and combinatorial treatment for CAR-T cell therapy. *Current Opinion in Biotechnology, 65*, 75–87. https://doi.org/10.1016/j.copbio.2020.01.009.

Lankipalli, S., & Ramagopal, U. A. (2021). A structural perspective on the design of decoy immune modulators. *Pharmacological Research, 170*, 105735. https://doi.org/10.1016/j.phrs.2021.105735.

Larson, R. C., & Maus, M. V. (2021). Recent advances and discoveries on the mechanisms and functions of CAR T cells. *Nature Reviews Cancer, 21*, 145–161. https://doi.org/10.1038/s41568-020-00323-z.

Li, Y., Huo, Y., Yu, L., & Wang, J. (2019). Quality control and nonclinical research on CAR-T cell products: General principles and key issues. *Engineering, 5*, 122–131. https://doi.org/10.1016/j.eng.2018.12.003.

Liang, A. (2018). *CD19-directed chimeric antigen receptor T cells therapy in relapsed/refractory B cell malignancy*. ClinicalTrials.gov.

Liang, A. (2024a). *An exploratory clinical study evaluating the safety and efficacy of infusion of anti-CD19-CAR-T cells in relapsed/refractory B-cell tumor patients*. ClinicalTrials.gov.

Liang, A. (2024b). *An exploratory clinical study evaluating the safety and efficacy of intravenous anti-CD20/CD30-CAR-T cell infusion in relapsed/refractory lymphoma patients*. ClinicalTrials.gov.

Ligen, L. (2020). *CD19-targeting CAR T cell therapy in the treatment of relapsed or refractory CD19 positive B-cell malignancies*. ClinicalTrials.gov.

Lin, H., Yang, X., Ye, S., Huang, L., & Mu, W. (2024). Antigen escape in CAR-T cell therapy: Mechanisms and overcoming strategies. *Biomedicine & Pharmacotherapy, 178*, 117252. https://doi.org/10.1016/j.biopha.2024.117252.

Liu, E., Marin, D., Banerjee, P., Macapinlac, H. A, Thompson, P., Basar, R., Nassif Kerbauy, L., Overman, B., Thall, P., Kaplan, M., Nandivada, V., Kaur, I., Nunez Cortes, A., Cao, K., Daher, M., Hosing, C., Cohen, E. N., Kebriaei, P., Mehta, R., … Rezvani, K. (2020). Use of CAR-transduced natural killer cells in CD19-positive lymphoid tumors. *New England Journal of Medicine, 382*, 545–553. https://doi.org/10.1056/NEJMoa1910607.

López-Cantillo, G., Urueña, C., Camacho, B. A., & Ramírez-Segura, C. (2022). CAR-T cell performance: How to improve their persistence? *Frontiers in Immunology, 13*, 878209. https://doi.org/10.3389/fimmu.2022.878209.

Loskog, A., Giandomenico, V., Rossig, C., Pule, M., Dotti, G., & Brenner, M. K. (2006). Addition of the CD28 signaling domain to chimeric T-cell receptors enhances chimeric T-cell resistance to T regulatory cells. *Leukemia, 20*, 1819–1828. https://doi.org/10.1038/sj.leu.2404366.

Ma, S., Li, X., Wang, X., Cheng, L., Li, Z., Zhang, C., Ye, Z., Qian, Q., & Yang, Y. (2019). Current progress in CAR-T cell therapy for solid tumors. *International Journal of Biological Sciences, 15*, 2548–2560. https://doi.org/10.7150/ijbs.34213.

Magee, M. S., & Snook, A. E. (2014). Challenges to chimeric antigen receptor (CAR)-T cell therapy for cancer. *Discovery Medicine*.

Martínez Bedoya, D., Dutoit, V., & Migliorini, D. (2021). Allogeneic CAR T cells: An alternative to overcome challenges of CAR T cell therapy in glioblastoma. *Frontiers in Immunology, 12*, 640082. https://doi.org/10.3389/fimmu.2021.640082.

Mata, M., Gerken, C., Nguyen, P., Krenciute, G., Spencer, D. M., & Gottschalk, S. (2017). Inducible activation of MyD88 and CD40 in CAR T cells results in controllable and potent antitumor activity in preclinical solid tumor models. *Cancer Discovery, 7*, 1306–1319. https://doi.org/10.1158/2159-8290.CD-17-0263.

Maude, S. L., Frey, N., Shaw, P. A., Aplenc, R., Barrett, D. M., Bunin, N. J., Chew, A., Gonzalez, V. E., Zheng, Z., Lacey, S. F., Mahnke, Y. D., Melenhorst, J. J., Rheingold, S. R., Shen, A., Teachey, D. T., Levine, B. L., June, C. H., Porter, D. L., & Grupp, S. A. (2014). Chimeric antigen receptor T cells for sustained remissions in leukemia. *New England Journal of Medicine, 371*, 1507–1517. https://doi.org/10.1056/NEJMoa1407222.

Morales, L., & Paramio, J. M. (2021). Cell therapies in bladder cancer management. *International Journal of Molecular Sciences, 22*, 2818. https://doi.org/10.3390/ijms22062818.

Morris, E. C., Neelapu, S. S., Giavridis, T., & Sadelain, M. (2022). Cytokine release syndrome and associated neurotoxicity in cancer immunotherapy. *Nature Reviews Immunology, 22*, 85–96. https://doi.org/10.1038/s41577-021-00547-6.

Muller, Y. D., Nguyen, D. P., Ferreira, L. M. R., Ho, P., Raffin, C., Valencia, R. V. B., Arnon, T. I., Sykes, M., & Seifert, A. M. (2021). The CD28-transmembrane domain mediates chimeric antigen receptor heterodimerization with CD28. *Frontiers in Immunology, 12*. https://doi.org/10.3389/fimmu.2021.639818.

Nair, R., & Westin, J. R. (2020). CAR T-cells:. *Humana*, 131–141. https://doi.org/10.1007/978-3-030-41008-7_10.

National Cancer Institute 2013CAR T cells: Engineering immune cells to treat cancer. 2013. https://www.cancer.gov/about-cancer/treatment/research/car-t-cells.

Pan, J. (2024). *Pragmatic clinical trial of CD19 and CD22 CAR T-cell sequential therapy versus single CD19 CAR T-cell bridging to transplantation for patients with refractory or relapsed B-cell acute lymphoblastic leukemia.* ClinicalTrials.gov.

Park, J. H., Geyer, M. B., & Brentjens, R. J. (2016). CD19-targeted CAR T-cell therapeutics for hematologic malignancies: interpreting clinical outcomes to date. *Blood, 127*, 3312–3320. https://doi.org/10.1182/blood-2016-02-629063.

Philipson, B. I., O'Connor, R. S., May, M. J., June, C. H., Albelda, S. M., & Milone, M. C. (2020). 4-1BB costimulation promotes CAR T cell survival through noncanonical NF-κB signaling. *Science Signaling, 13*, eaay8248. https://doi.org/10.1126/scisignal.aay8248.

Qasim, W. (2019). Allogeneic CAR T cell therapies for leukemia. *American Journal of Hematology*. https://doi.org/10.1002/AJH.25399.

Ramesh, P., Hui, H., Brownrigg, L. M., Fuller, K., & Erber, W. N. (2023). Chimeric antigen receptor T-cells: Properties, production, and quality control. *International Journal of Laboratory Hematology*. https://doi.org/10.1111/ijlh.14121.

Reddy, O. L., Stroncek, D. F., & Panch, S. R. (2020). Improving CAR T cell therapy by optimizing critical quality attributes. *Seminars in Hematology, 57*, 33–38. https://doi.org/10.1053/j.seminhematol.2020.07.005.

Research. (2024a). *C for BE and. YESCARTA*. FDA https://www.fda.gov/vaccines-blood-biologics/cellular-gene-therapy-products/yescarta.

Research. (2024b). *C for BE and. BREYANZI (lisocabtagene maraleucel)*. FDA https://www.fda.gov/vaccines-blood-biologics/cellular-gene-therapy-products/breyanzi.

Research. (2024c). *C for BE and. KYMRIAH*. FDA https://www.fda.gov/vaccines-blood-biologics/cellular-gene-therapy-products/kymriah.

Roth, P., Winklhofer, S., Müller, A. M. S., Dummer, R., Mair, M. J., Gramatzki, D., Schucht, P., & Weller, M. (2021). Neurological complications of cancer immunotherapy. *Cancer Treatment Reviews, 97*, 102189. https://doi.org/10.1016/j.ctrv.2021.102189.

Sadelain, M. (2015). From T-cell engineering to CAR therapy: Progress and prospects. *Blood*. https://doi.org/10.1182/BLOOD.V126.23.SCI-23.SCI-23.

Schneider, D., Xiong, Y., Wu, D., Dropulic, B., & Orentas, R. J (2016). Abstract 2297: Tandem anti-CD20 and -CD19 scFV-based chimeric antigen receptors (CARs) mitigate tumor escape in hematologic malignancies. *Cancer Research, 76*(14 Suppl), 2297. https://doi.org/10.1158/1538-7445.AM2016-2297.

Schuster, S. J, Svoboda, J., Chong, E. A, Nasta, S. D, Mato, A. R, Anak, Ö, Brogdon, J. L., Pruteanu-Malinici, I., Bhoj, V., Landsburg, D. J., Bashir, Q., Flowers, C. R., Merryman, R. W., Ahmed, N., Jäger, U., Levine, B. L., June, C. H., & Porter, D. L. (2017). Chimeric Antigen Receptor T Cells in Refractory B-Cell Lymphomas. *New England Journal of Medicine, 377*, 2545–2554. https://doi.org/10.1056/NEJMoa1708566.

Shanghai First Song Biotechnology Co., Ltd. (2024). *An exploratory clinical study evaluating the safety and efficacy of anti-CD19-CAR-T cell therapy in subjects with relapsed/refractory B cell malignancies*. ClinicalTrials.gov https://clinicaltrials.gov/study/NCT06519344.

Shapiro, M. B, Yun, M., Junkins-Hopkins, J. M, Vittorio, C. C, Schulman, N., Saidman, B. H, & Duvic, M. (2002). Assessment of tumor burden and treatment response by 18F-fluorodeoxyglucose injection and positron emission tomography in patients with cutaneous T- and B-cell lymphomas. *Journal of the American Academy of Dermatology*. https://doi.org/10.1067/MJD.2002.124076.

Shenzhen Geno-Immune Medical Institute. (2022a). *CD19/22 bi-specific CAR-T cells targeting B cell malignancies*. ClinicalTrials.gov.

Shenzhen Geno-Immune Medical Institute. (2022b). *CD19/70 bi-specific CAR-T cells targeting B cell malignancies*. ClinicalTrials.gov.

Smith J., Scharenberg A. M., Mannioui C., & Eyquem J. Multi-chain chimeric antigen receptor and uses thereof, 2013. https://patents.google.com/patent/US20140134142A1/n

Song, D.-G., & Powell, D. J (2012). Pro-survival signaling via CD27 costimulation drives effective CAR T-cell therapy. *OncoImmunology, 1*, 547–549. https://doi.org/10.4161/onci.19458.

Song, P., Zhang, Q., Xu, Z., Shi, Y., Jing, R., & Luo, D. (2024). CRISPR/Cas-based CAR-T cells: Production and application. *Biomarker Research, 12*, 54. https://doi.org/10.1186/s40364-024-00602-z.

Springuel, L., Lonez, C., Alexandre, B., Van Cutsem, E., Machiels, J.-P., Van den Eynde, M., & Romero, P. (2019). Chimeric antigen receptor-T cells for targeting solid tumors: Current challenges and existing strategies. *Biodrugs*. https://doi.org/10.1007/S40259-019-00368-Z.

Sterner, R. C., & Sterner, R. M. (2021). CAR-T cell therapy: Current limitations and potential strategies. *Blood Cancer Journal, 11*, 1–11. https://doi.org/10.1038/s41408-021-00459-7.

Sterner, R. C., & Sterner, R. M. (2022). Immune effector cell associated neurotoxicity syndrome in chimeric antigen receptor-T cell therapy. *Frontiers in Immunology, 13*. https://doi.org/10.3389/fimmu.2022.879608.

Walton, Z. E., Frigault, M. J., & Maus, M. V. (2024). Current and emerging pharmacotherapies for cytokine release syndrome, neurotoxicity, and hemophagocytic lymphohistiocytosis-like syndrome due to CAR T cell therapy. *Expert Opinion on Pharmacotherapy, 25*, 263–279. https://doi.org/10.1080/14656566.2024.2340738.

Wang, J., Hu, Y., & Huang, H. (2018). Current development of chimeric antigen receptor T-cell therapy. *Stem Cell Investigation*. https://doi.org/10.21037/SCI.2018.11.05.

Wang, X., & Rivière, I. (2016). Clinical manufacturing of CAR T cells: Foundation of a promising therapy. *Molecular Therapy Oncolytics, 3*, 16015. https://doi.org/10.1038/mto.2016.15.

Wei, J., Han, X., Bo, J., & Han, W. (2019). Target selection for CAR-T therapy. *Journal of Hematology & Oncology, 12*, 62. https://doi.org/10.1186/s13045-019-0758-x.

Weinkove, R., George, P., Dasyam, N., & McLellan, A. D (2019). Selecting costimulatory domains for chimeric antigen receptors: Functional and clinical considerations. *Clinical & Translational Immunology, 8*, e1049. https://doi.org/10.1002/cti2.1049.

Whilding, L. M, & Maher, J. (2015). CAR T-cell immunotherapy: The path from the by-road to the freeway? *Molecular Oncology, 9*, 1994–2018. https://doi.org/10.1016/j.molonc.2015.10.012.

Wilkie, S., Picco, G., Foster, J., Davies, D. M, Julien, S., & Cooper, L. (2008). Retargeting of human T cells to tumor-associated MUC1: The evolution of a chimeric antigen receptor. *Journal of Immunology, 180*, 4901–4909. https://doi.org/10.4049/jimmunol.180.7.4901.

Wu, L., Wei, Q., Brzostek, J., & Gascoigne, N. R. J. (2020). Signaling from T cell receptors (TCRs) and chimeric antigen receptors (CARs) on T cells. *Cellular & Molecular Immunology, 17*, 600–612. https://doi.org/10.1038/s41423-020-0470-3.

Wuhan Sian Medical Technology Co., Ltd. (2019). *Treatment of relapsed or refractory B-cell malignancies by CD19 chimeric antigen receptor (CAR)-modified T cells*. ClinicalTrials.gov.

Xie, Y., Li, X., Boucetta, H., Wu, J., Wang, B., Yang, P., Zhang, L., & Chen, H. (2023). CAR-T cells for cancer immunotherapy. *Chinese Chemical Letters*. https://doi.org/10.1016/j.cclet.2023.108202.

Xu, J., Wang, Q., Xu, H., Gu, C., Jiang, L., Wang, J., Yang, P., & Li, Y. (2018). Anti-BCMA CAR-T cells for treatment of plasma cell dyscrasia: Case report on POEMS syndrome and multiple myeloma. *Journal of Hematology & Oncology, 11*, 128. https://doi.org/10.1186/s13045-018-0672-7.

Yu, K. (2020). *Clinical Study of Redirected Autologous T Cells With a Chimeric Antigen Receptor in Patients With Malignant Tumors*. clinicaltrials.gov.

Yu, S., Li, A., Liu, Q., Li, T., Yuan, X., Han, X., et al. (2017). Chimeric antigen receptor T cells: a novel therapy for solid tumors. *Journal of Hematology & Oncology, 10*, 78. https://doi.org/10.1186/s13045-017-0444-9.

Yuan, S.-Z., & Su, H. (2014). Clinical translational research of chimeric antigen receptor-T (CAR-T) cells for the treatment of relapsed and refractory B-cell lymphoma/leukemia. *Journal of Experimental Hematology*. https://doi.org/10.7534/J.ISSN.1009-2137.2014.04.048.

Zhang, C., Liu, J., Zhong, J. F, & Zhang, X. (2017). Engineering CAR-T cells. *Biomarker Research, 5*, 22. https://doi.org/10.1186/s40364-017-0102-y.

Zhang, T., Wu, M.-R., & Sentman, C. L (2012). An NKp30-based chimeric antigen receptor promotes T cell effector functions and antitumor efficacy in vivo. *Journal of Immunology, 189*, 2290–2299. https://doi.org/10.4049/jimmunol.1103495.

Zhang, Y., Zhang, W., Dai, H., Wang, Y., Shi, F., Wang, C., Liu, H., Yang, X., Hu, Y., Xu, J., Zhang, M., & Huang, X. (2016). An analytical biomarker for treatment of patients with recurrent B-ALL after remission induced by infusion of anti-CD19 chimeric antigen receptor T (CAR-T) cells. *Science China-Life Sciences*. https://doi.org/10.1007/S11427-016-5035-4.

Zhao, Z., van der Stegen, S. JC., Condomines, M., Perna, F., Gunset, G., Plotkin, J., Milone, M. C., & Sadelain, M. (2015). 207. Optimal activation of both CD28 and 4-1BB in CAR-targeted T cells results in enhanced tumor eradication. *Molecular Therapy*. https://doi.org/10.1016/S1525-0016(16)33812-6.

Zheng, K. L. X. J. N. (2016). *Application of humanized anti-CD22 antibody in CAR-T therapy of B cell hematological malignancies*. ClinicalTrials.gov.

Zhu, I., Liu, R., Garcia, J. M, Hyrenius-Wittsten, A., Piraner, D. I, Alavi, J., Ham, C. M., Robinson, R. P., Miller, J. P., Lim, J. S., Wu, C., Chen, X., Nguyen, T., Patel, S., Kim, H., Gonzalez, L., Wang, Y., Davis, B., Lee, S., … Lim, W. A. (2022). Modular design of synthetic receptors for programmed gene regulation in cell therapies. *Cell, 185*, 1431–1443.e16. https://doi.org/10.1016/j.cell.2022.03.023.

CHAPTER 5

Targeting the tumor vasculature using mesenchymal stem cells in cancer therapy: New perspectives

Mohammed Sikander, Shabnam Malik, Murali M. Yallapu and Subhash C. Chauhan

Division of Cancer Immunology and Microbiology, South Texas Center of Excellence in Cancer Research, Medicine and Oncology Integrated Service Unit, School of Medicine, University of Texas Rio Grande Valley (UTRGV), McAllen, TX, United States

5.1 Introduction

Cancer remains the leading cause of disease-related mortality worldwide, with the global cancer burden projected to rise to 28.4 million cases by 2040, an estimated 47% increase from 2020 (Ke & Shen, 2017; Sung et al., 2021). The selection of an appropriate cancer therapy is a complex process influenced by several factors, including tumor characteristics (benign or malignant), disease stage at diagnosis, and the patient's overall ability to tolerate treatment (Wolinsky et al., 2012). Current treatment strategies include surgery, chemotherapy, immunotherapy, and radiation therapy, which may be administered either as standalone treatments or in combination (Bellon et al., 2005). However, despite their clinical efficacy, these modalities are often accompanied by significant limitations, including severe adverse effects, off-target toxicity, and the increasing prevalence of drug resistance, which can compromise treatment success (Ghadjar et al., 2008; Holohan et al., 2013). One of the critical challenges in cancer therapy is the development of tumor-induced angiogenesis, the formation of new blood vessels that sustain tumor growth and metastasis (Fig. 5.1). Targeting angiogenesis has emerged as a promising strategy to normalize tumor-associated vasculature, thereby not only inhibiting tumor progression but also enhancing the efficacy of conventional treatments. Integrating anti-angiogenic strategies with existing therapeutic modalities holds great potential for improving treatment outcomes in cancer management (Carmeliet & Jain, 2011; Heath & Bicknell, 2009). In light of limitations associated with traditional cancer treatments, cell-based therapies have gained considerable interest as alternative strategies (Zhuang et al., 2021). Among these, MSCs-based approaches are particularly promising because of their ability to influence disease progression. MSCs naturally migrate toward tumor sites, a property known as tumor tropism, which allows for precise targeting. This makes them well-suited for delivering therapeutic agents, both chemical and genetic, directly to tumor tissues, thereby reducing the systemic side effects typically observed with conventional treatments. In addition to their targeting capabilities, MSCs actively contribute to modifying the tumor microenvironment (TME). They release paracrine factors that regulate immune activity by suppressing T-cell proliferation, altering natural killer (NK) cell function, and affecting dendritic cell (DC) maturation. These actions collectively transform the TME into a less favorable environment for tumor growth. This chapter explores recent advancements

Advancements in Personalized Cellular Immunotherapy: A Comprehensive Guide to Cancer Treatment.
DOI: https://doi.org/10.1016/B978-0-443-27731-3.00002-4

Copyright © 2026 Elsevier Inc. All rights are reserved, including those for text and data mining, AI training, and similar technologies.

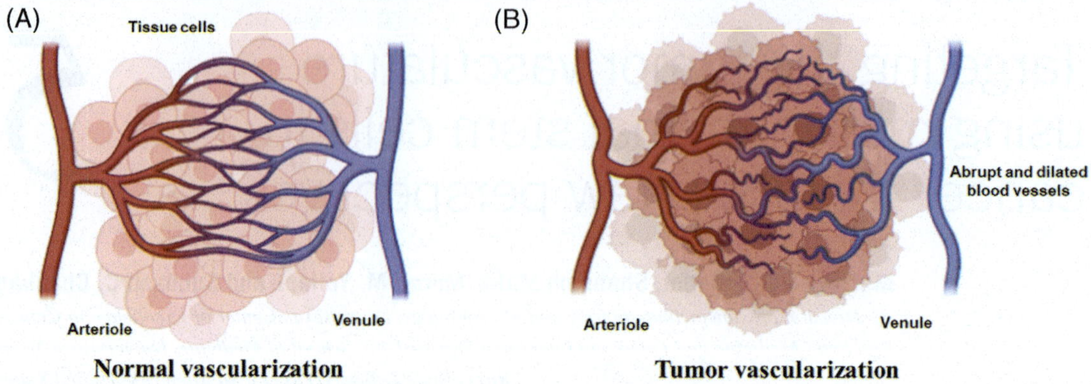

FIGURE 5.1 Comparison of normal vascularization and tumor vascularization.

This figure illustrates the structural differences between normal blood vessel formation and tumor-associated vascularization. (A) Normal vascularization: Depicted are healthy blood vessels characterized by organized structure, uniform diameters, and regular branching patterns, facilitating efficient blood flow and nutrient delivery to surrounding tissues. (B) Tumor vascularization: Shown is the chaotic architecture of tumor blood vessels, which are often irregular, dilated, and poorly organized. These abnormal vessels are characterized by increased permeability and a lack of a functional basement membrane, contributing to inadequate oxygen and nutrient supply to the tumor and facilitating metastasis.

in MSC-based therapies for cancer, with a focus on their role in inhibiting tumor angiogenesis and improving overall treatment outcomes.

5.2 Tumor-driven angiogenesis and signaling pathways

The microarchitecture of normal vascular networks is hierarchically organized, featuring mature vessels that are evenly distributed to ensure adequate perfusion of oxygen and nutrients to all cells. In contrast, the complex tumor vasculature presents a disorganized labyrinth of vessels that lacks the conventional hierarchy of blood vessels, making the identification of arterioles, capillaries, and venules challenging (Konerding et al., 2001). The blood vessels within tumors exhibit inconsistent diameters and irregular shapes, often showing abnormal bulges and blind ends as well as arteriolar-venous shunts and plasma channels devoid of red blood cells (Dewhirst et al., 1996; Konerding et al., 2001; McDonald & Choyke, 2003). Accompanying the aberrant blood vessels, the lymphatic vessels in tumors are typically dilated, leaky, and discontinuous, resulting in fluid-engorged vessels (Leu et al., 2000; Padera et al., 2002).

Tumor progression heavily relies on its capacity to stimulate the formation of new blood vessels, a process referred to as tumor-induced angiogenesis. This vascular development supplies the growing tumor with essential oxygen and nutrients, while also facilitating the removal of metabolic waste, thereby sustaining tumor survival and expansion. Disrupting the angiogenic process has become a key therapeutic objective, as impairing blood vessel formation can starve tumors or critical resources, ultimately leading to cell death through nutrient deficiency and accumulation of toxic by-products. The onset of angiogenesis is primarily triggered by the release of proangiogenic factors from tumor cells, which promote the activation and migration of endothelial cells (ECs) from existing blood vessels. These ECs organized into new capillary networks, converting a previously vascular tumor into a densely vascularized tissue mass (Folkman, 1990).

Angiogenesis is a pivotal process in tumor development and metastasis, as it ensures a sustained supply of oxygen and nutrients necessary to support rapid tumor growth and maintain the microenvironment. This process is orchestrated through a complex network of signaling pathways involving various pro-angiogenic growth factors. Key mediators include acidic and basic fibroblast growth factors (aFGF and bFGF), angiogenin, placental growth factor, epidermal growth factor (EGF), platelet-derived growth factors (PDGF), transforming growth factor-α (TGF-α), and tumor necrosis factor (TNF) and vascular endothelial growth factors (VEGF). These molecules regulate the behavior of EC, promoting their proliferation, migration, and differentiation to form new vascular structures. The resulting vascular network not only nourishes the primary tumor but also provides conduits for metastatic dissemination. Given its central role in cancer progression, angiogenesis has become a strategic therapeutic target. Inhibiting angiogenesis signaling pathways can impair tumor vascularization, restrict nutrient availability, and potentially enhance the effectiveness of conventional therapies. These abnormal features of tumor vasculature contribute to a hostile and irregular microenvironment that often reduces the effectiveness of conventional anti-cancer therapies (Vaupel et al., 1987).

Microregional hypoxia, frequently observed within tumors, is a major factor contributing to resistance against radiotherapy (Mohindra & Rauth, 1976) and chemotherapy (Koch et al., 2003). Moreover, it is well established that tumors are unable to grow beyond a volume of approximately 2 mm^3 without initiating neovascularization to supply essential oxygen and nutrients (Folkman, 1971). This insight has driven the clinical advancement of numerous angiogenesis-inhibiting agents (AIAs), many of which are designed to specifically target VEGF and its receptor (VEGFR) (Ferrara & Kerbel, 2005; Hicklin and Ellis, 2005; Ma & Hidalgo, 2007). Targeting angiogenesis-driven vascular sprouting has significantly transformed the landscape of anticancer drug development over the past decade (Carmeliet & Jain, 2000; Randal, 2000).

5.3 Interactions between tumor vasculature and tumor microenvironment

The TME plays a pivotal role in cancer progression and therapeutic response through mechanisms such as angiogenesis, immune evasion, and the development of treatment resistance. Unlike healthy tissues, which maintain a finely tuned balance between pro- and antiangiogenic factors, tumors exhibit a pathological shift that favors vascular growth. This dysregulation typically involves elevated levels of proangiogenic signals, such as VEGF, alongside a suppression of angiogenesis inhibitors, such as thrombospondin-1 (Jain et al., 2007). In normal tissue, blood vessels are structurally uniform, with low permeability and an intact basement membrane, ensuring efficient oxygen delivery, waste clearance, and drug transport. In contrast, tumor vasculature is disorganized, excessively, and structurally, often resulting in efficient blood flow and limited drug excess. These vascular irregularities create a hypoxic and immune suppressive microenvironment that promotes tumor survival and impairs the efficacy of standard therapies (Potente et al., 2011). Gaining a deeper understanding of these vascular pathologies is essential for developing therapeutic approaches aimed at normalizing tumor blood vessels, thereby improving drug delivery and enhancing overall treatment effectiveness.

The TME is a highly intricate and dynamic system that extends well beyond the vasculature, incorporating components such as the extracellular matrix (ECM), immune cells, fibroblasts, and a wide array of signaling molecules. These elements collectively drive tumor progression, therapeutic resistance, and immune escape. The disorganized and leaky nature of tumor vasculature not only ensures a persistent supply of oxygen and nutrients but also fosters an immunosuppressive milieu that impairs effective surveillance. Consequently, antiangiogenic therapies should not focus solely on normalizing the vasculature but must also account for the complex interaction among ECs, ECM remodeling, and immune cell behavior. A more holistic therapeutic strategy, one that combines vasculature normalization with immune modulation and regulation of ECM, holds promise for improving treatment responses and clinical outcomes in cancer patients.

5.4 Signaling pathways regulated by mesenchymal stem cells

MSCs have emerged as significant modulators in cancer biology, owing not only to their regenerative potential but also to their intricate interactions within the TME. These interactions occur through a network of signaling pathways that can either support tumor development or serve as targets for therapeutic interventions. Among the most prominent pathways modulated by MSCs are Wnt/β-catenin, Hedgehog, Notch, and TGF-β. These pathways plays a distinct role in regulating tumor cell proliferation, differentiation, immune evasion, and resistance to therapy, highlighting the dualistic nature of the MSCs in cancer progression and treatment (Singh et al., 2025; Zhang et al., 2021).

The Wnt/β-catenin signaling pathways play a pivotal role in regulating cell proliferation, differentiation, and migration within the TME. Its activation in cancer stem cells enhances self-renewal and survival, contributing to tumor initiation and progression. MSCs further amplify this effect by secreting Wnt ligands that stimulate β-catenin signaling in neighboring tumor cells. This crosstalk not only supports tumor cell viability but also promotes invasive behavior through processes such as epithelial–mesenchymal transition, a key driver of metastasis. The interaction between MSCs and the Wnt/β-catenin pathway presents a dual challenge; while it contributes to tumor aggressiveness, it also identifies a potential therapeutic target for disrupting oncogenic signaling and limiting disease progression (Chen et al., 2019; Koni et al., 2020; Wang et al., 2015).

In addition to Wnt signaling, the Hedgehog pathway plays a crucial role in mediating the interplay between MSCs and tumor cells. Dysregulated Hedgehog signaling has been linked to the development and progression of various cancers, where it supports tumor growth, survival, and recurrence. MSCs can secrete Hedgehog ligands that activate this pathway in adjacent cancer cells, thereby fostering a microenvironment conducive to tumor expansion. This signaling axis is particularly important in maintaining cancer stem cell populations, enhancing their self-renewal and tumor-initiating potential. As a result, MSCs-mediated activation of Hedgehog signaling contributes not only to tumor progression but also to therapeutic resistance, underscoring its significance as a potential target for anticancer interventions (Du et al., 2023; Qi et al., 2017).

The notch signaling pathway plays an important role in the communication between MSC and tumor cells. It is particularly noted for its involvement in regulating cell differentiation and preserving stem-like properties in various cell types (Melzer, Yang, & Hass, 2016). MSCs can influence Notch signaling in cancer cells by producing specific ligands, thereby fostering interactions that may enhance cell proliferation and survival (Shi et al., 2024).

The TGF-β signaling pathway is known for its complex role in shaping an immune suppressive TME. MSCs contribute to this process by secreting TGF-β, which dampens immune responses through the inhibition of cytotoxic T-cell activity and the induction of regulatory T-cell activity (Tregs), ultimately aiding tumor cells in evading immune surveillance (Seo et al., 2021). TGF-β also plays a key role in modulating stromal cell behavior, helping to create a microenvironment that supports tumor growth and metastasis. The immune suppressive properties of MSCs, largely driven by TGF-β along with other soluble mediators, are central to tumor progression and pose a considerable obstacle to the success of many cancer therapies (de Araújo Farias et al., 2018). The interplay among these signaling pathways highlights the multifaceted role of MSCs in the TME (Yu et al., 2025). Depending on factors such as local cellular context, the availability of signaling ligands, and the specific tumor subtype, MSCs can exert either tumor-promoting or tumor-suppressive effects. This dynamic balance underscores the complexity of MSC function in cancer biology and the need for context-specific therapeutic strategies (Kangari et al., 2024; Zhou et al., 2023). Therefore, a nuanced understanding of these signaling pathways is crucial for designing targeted pathways that harness the biological properties of MSCs to improve cancer treatment efficacy. Targeting these signaling pathways holds substantial promise for optimizing MSC-based therapies. For instance, combining MSCs with pharmacological inhibitors that block Wnt, Hedgehog, or TGF-β signaling may potentiate the efficacy of conventional cancer treatment, including chemotherapy and immunotherapy (Deng et al., 2023). However, the clinical translation of MSCs is limited by considerable heterogeneity within the MSC population, influenced by factors such as donor age, tissue origin, and culture conditions. To fully realize their therapeutic potential, it is essential to establish standardized protocols and perform thorough characterization to ensure consistency, efficacy, and safety across applications (Zhou et al., 2023). Moreover, the secretome of MSCs, comprising a diverse range of cytokines, chemokines, and growth factors, plays a pivotal role in regulating signaling pathways involved in tumor progression and tumor modulation. In addition to soluble factors, MSCs release extracellular vesicles such as exosomes, which are enriched with microRNAs and other bioactive molecules. These vesicles facilitate complex communications between MSCs, tumor cells, and stromal components, influencing various aspects of tumor behavior. Depending on the context, this MSC-mediated signaling network can either support or suppress tumor growth, highlighting the dual nature of its role in TME (Patel et al., 2025).

5.5 Mesenchymal stem cells: Ideal candidates for targeting tumors

A major challenge in oncology is achieving effective and targeted delivery of chemotherapeutic agents, as conventional approaches often encounter physiological barriers and complex pathology. However, recent advancements in cell-based active targeting have opened new avenues for enhancing the precision and efficacy of cancer therapies (Choi et al., 2007; Cheng et al., 2010; Chambers & Mitragotri, 2007; Cheng et al., 2013; Schnarr et al., 2013; Roger et al., 2010; Huang et al., 2015). Among the diverse cellular vehicles under investigation, macrophages, red blood cells, neural stem cells, and MSCs have demonstrated significant potential due to their inherent ability to traverse biological systems and selectively migrate to malignant tissues (Ahmed & Lesniak, 2011; Müller et al., 2006; Nakamizo et al., 2005; Sadhukha et al., 2014; Tang et al., 2010). These adult, multipotent progenitor

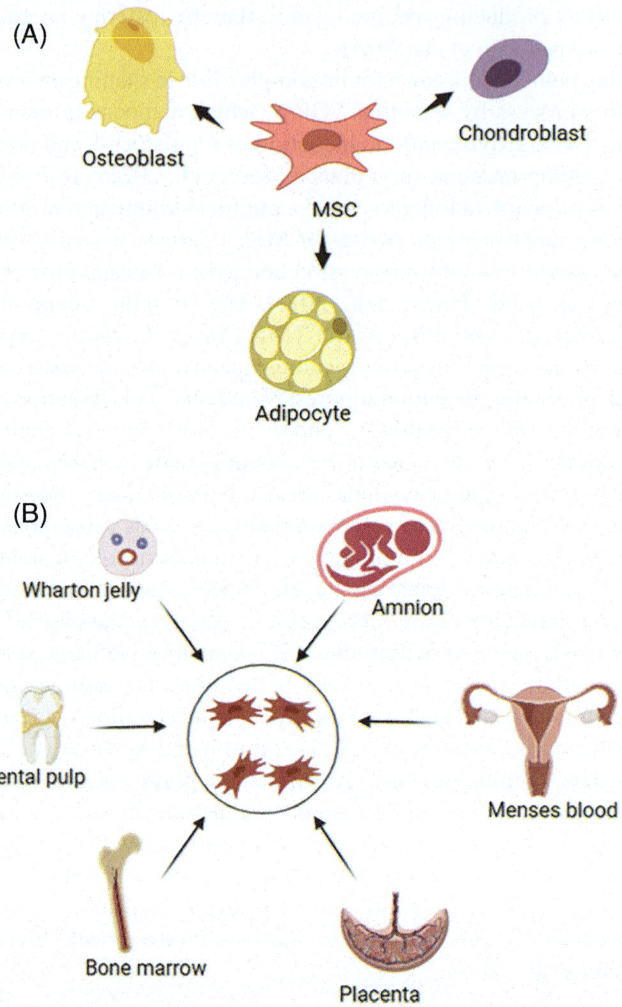

FIGURE 5.2 Differentiation and sources of MSCs.

(A) MSCs can be differentiated into various cell types and (B) sources of mesenchymal stem cells. *MSCs*, Mesenchymal stem cells.

cells are renowned for their remarkable self-renewal capabilities and their ability to differentiate into multiple cell types (osteoblasts, adipocytes, and chondroblasts) (Cheng et al., 2019).

MSCs can be isolated from a wide range of tissues such as dental pulp, adipose tissue, placenta, umbilical cord, Wharton jelly, menses blood, endometrium, and bone marrow (Friedenstein et al., 1987) (Fig. 5.2). These stem cells exhibit remarkable self-renewal and differentiation capabilities, making them a vital component of cell therapy and regenerative medicine. Unlike induced pluripotent stem cells (iPS) and pluripotent embryonic stem (ES) cells, which are associated with risks such as teratoma formation and ethical considerations, MSCs circumvent these challenges, further solidifying their potential for therapeutic applications in oncology (Wei et al., 2013). This diversity of sources not only facilitates the accessibility of MSCs but also allows for their convenient extraction and use in clinical settings. MSCs offer several advantages that solidify their role in therapeutic applications: they are easily accessible, capable of extensive in vitro culture expansions, and inherently multipotent, allowing for differentiation into various cell types (Aghebati-Maleki et al., 2019). These properties, combined with their potential to modulate immune responses and promote tissue regeneration, underscore their significance in contemporary cancer therapies and regenerative approaches. Overall, MSCs present a versatile and ethically favorable option for advancing cell-based treatments in oncology.

5.6 Significance of mesenchymal stem cell for cancer therapy

Resistance to cancer treatment often develops due to the prolonged administration and increased concentrations of anticancer agents, which can inadvertently enhance the toxicity in noncancerous healthy proliferating cells (Xu et al., 2018). MSCs have been considered an effective vehicle for the targeted delivery of chemotherapeutic agents. Their intrinsic tumor-tropic properties and ability to deeply infiltrate the TME make them ideal for chemotherapy drugs, suicide genes, oncolytic viruses (OVs), cytokines, and growth factors (Fig. 5.3) (Almeida-Porada et al., 2020). By leveraging MSCs as delivery systems, it is possible to enhance treatment precision while minimizing off-target effects, ultimately improving therapeutic efficacy and reducing systemic toxicity.

5.7 Delivery of chemotherapeutic agents using mesenchymal stem cells

Tumor vasculature and angiogenesis are pivotal factors in cancer progression, driven by an imbalance between antiangiogenic and proangiogenic factors alongside growth factors expressed in the TME (Folkman et al., 1989; Samant & Shevde, 2011). One innovative approach to limit tumor growth and metastasis involves utilizing MSCs as vehicles to deliver antiangiogenic and chemotherapeutic agents. The efficacy of MSCs in cancer therapy has been validated through various models. Accumulating studies have shown the increasing significance of MSCs in transporting established chemotherapeutics, such as sorafenib, paclitaxel (PTX), doxorubicin (DOX), and gemcitabine, aiming to enhance therapeutic efficacy (Kangari et al., 2024; Petrella et al., 2017) (Fig. 5.3). Notably, endostatin, a prominent endogenous angiogenesis inhibitor, has been extensively studied for various malignancies (Zheng et al., 2012). A significant study employed adenoviral transduction to engineer human placenta-derived MSCs for endostatin delivery. These MSCs were injected into nude mice, effectively homing to tumors, diminishing tumor size without producing notable systemic toxicity. This approach not only promoted apoptosis in cancer cells but also disrupted neovascularization, thereby reducing tumor proliferation and deceleration (Zheng et al., 2012). Moreover, PTX-loaded bone marrow-derived MSCs (BM-MSCs) and dental pulp-derived MSCs (DP-MSCs) have shown effectiveness in retarding glioblastoma and breast cancer growth by hindering cell proliferation and inducing apoptosis (Pessina et al., 2011; Salehi et al., 2018). In addition, the use of conditioned media from adipose tissue-derived MSCs (ASCs)

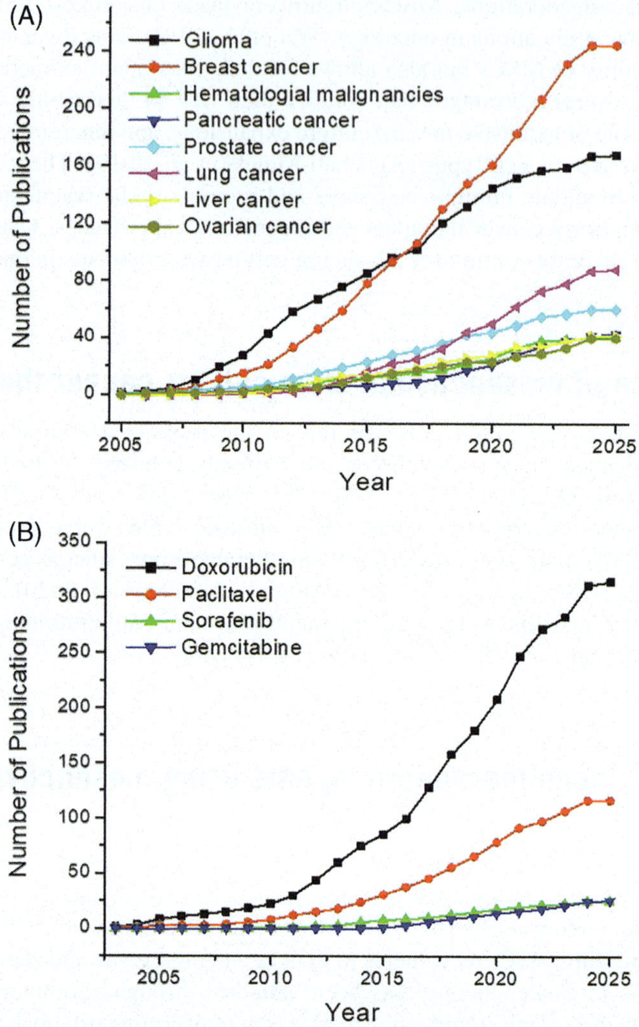

FIGURE 5.3 The role and utilization of mesenchymal stem cells in cancer applications.

(A) The graph depicts the research trends regarding the utilization of mesenchymal stem cells (MSCs) across various cancer types, including glioma, breast cancer, hematological cancer, pancreatic cancer, prostate cancer, lung cancer, liver cancer, and ovarian cancer. Data were derived from PubMed, showcasing the number of relevant publications from 2005 to 2024. The *x*-axis represents the different cancer types, whereas the *y*-axis indicates the total number of studies published for each cancer category. Each cancer type is represented by a distinct color or marker, facilitating comparison of research interest across different malignancies. (B) The graph illustrates the annual publication trends of research studies involving mesenchymal stem cells (MSCs) in relation to four different chemotherapeutic agents: Doxorubicin (DOX), Paclitaxel (PTX), Sorafenib (SRB), and Gemcitabine (GEM). Data were sourced from PubMed, highlighting the number of articles published each year from 2000 to 2024. The *x*-axis represents the years of publication, whereas the *y*-axis indicates the total number of research articles for each drug. Distinct colors or markers represent each chemotherapeutic agent, allowing for easy comparison of research focus over time.

loaded with PTX resulted in a significant reduction in ovarian cancer cell proliferation compared to free PTX, effectively overcoming drug resistance in these cells (Borghese et al., 2020). Furthermore, doxorubicin-loaded BM-MSCs demonstrated substantial cytotoxic effects in xenograft mouse models of thyroid and breast cancer, reinforcing the therapeutic potential of MSCs (Kalimuthu et al., 2018). Lastly, gingival papilla-derived MSCs have been affirmed as a reliable delivery system for gemcitabine, effectively inhibiting oral squamous cell proliferation by releasing the drug in an active form and appropriate quantities (Coccè et al., 2017). In summary, MSCs exhibit great promise as effective delivery systems for antiangiogenic and chemotherapeutic agents, potentially improving cancer treatment outcomes by targeting TMEs and enhancing the efficacy of therapeutic agents.

5.8 Mesenchymal stem cell-derived exosome as a cell-free therapeutic tool

Recent studies emphasize the promise of MSC-derived exosomes, typically ranging from 30–100 nm in diameter, as an innovative cell-free tool for cancer therapy (Szewc et al., 2022). One notable function of MSC-derived exosomes is their ability to inhibit tumor angiogenesis by regulating VEGF expression. Alcayaga-Miranda et al. (2016) demonstrated that MSC-derived exosomes could impede angiogenesis in prostate cancer (PC3) cells by reducing VEGF secretion. Importantly, MSC-derived exosomes are rich in various therapeutic cargoes, including proteins, lipids, nucleic acids (both DNA and RNA), and metabolites, and are reported to express common surface biomarkers such as CD81 and CD9, including MSC-specific markers such as CD29, CD44, CD73, and CD90 (Lin et al., 2022; Oveili et al., 2023; Sun et al., 2021;). Furthermore, these exosomes possess numerous advantageous properties, such as low immunogenicity, biocompatibility, and prolonged circulation time (Lyu et al., 2021; Weng et al., 2021). Accumulating studies have shown that MSC-derived exosomes can modulate the oncogenic signaling pathways and induce apoptosis in cancer cell lines (Rezaeian et al., 2022). MSC-derived exosomes have been explored as vehicles for chemotherapeutic drugs. Studies have shown that exosomes loaded with doxorubicin exhibited significantly higher uptake and cytotoxicity in the osteosarcoma cell line MG63 compared to free DOX (Wei et al., 2019). Notably, Pascucci et al. (2014) were pioneers demonstrating the role of exosomes in encapsulating and releasing paclitaxel for enhancing its therapeutic efficacy. Exosomes containing polo-like kinase 1 (PLK-1) siRNA induced apoptosis and necrosis in bladder cancer cells (Greco et al., 2016). Moreover, MSC-derived exosomes carrying miR-379 were reported to inhibit the growth of T47D breast cancer cell lines, highlighting their potential for targeted therapy in breast cancer (O'brien et al., 2018). The advancement of MSC-derived exosome therapy holds substantial promise for more effective and targeted therapeutic interventions in oncology.

5.9 Utilizing oncolytic viruses with mesenchymal stem cells: An innovative approach

Several studies have highlighted the potential of MSCs as effective carriers for OVs, offering protection against host immune neutralization, promoting targeted delivery to tumor sites, and improving viral infection efficiency and tumor cell eradication (Choi et al., 2022). OVs represent a novel and promising therapeutic strategy for cancer treatment, selectively inducing apoptosis in infected cancer cells

(Bartlett et al., 2013). Preclinical investigations have demonstrated that mesenchymal stem cell (MSC)-mediated delivery of OVs can significantly enhance therapeutic outcomes across a range of solid tumor models. Notable success has been observed in the treatment of hepatocellular carcinoma, glioblastoma, glioma, colorectal cancer, prostate cancer, and breast cancer metastases to the lung, highlighting the broad applicability of this strategy (Belete, 2021; Chastkofsky et al., 2021; Duebgen et al., 2014; Düzgüneş, 2019; Guo et al., 2019; Yoon et al., 2019; Zhang et al., 2022).

5.10 Mesenchymal stem cell-mediated suicidal genes: A novel strategy in cancer therapy

An emerging strategy in cancer treatment is suicide gene therapy, which offers the potential to selectively eliminate tumor cells while sparing healthy tissue (Belete, 2021). Also referred to as gene-directed enzyme prodrug therapy, this approach involves introducing a gene into cancer cells that encodes an enzyme capable of converting a harmless prodrug into a toxic compound. Once the prodrug is administered, it is locally activated within the tumor environment, generating cytotoxic metabolites that selectively destroy malignant cells (Düzgüneş, 2019). Preclinical investigations have demonstrated that MSCs engineered to express the fusion enzyme cytosine deaminase:uracil phosphoribosyl transferase (CD:UPRT) can effectively trigger complete tumor regression in models of prostate cancer. This therapeutic effect is achieved through the enzyme's ability to convert the non-toxic prodrug 5-fluorocytosine into 5-fluorouracil, a potent chemotherapeutic agent that inhibits tumor cell proliferation (Altaner, 2008; Durinikova et al., 2018). This innovative strategy illustrates the potential of MSCs in enhancing the efficacy of cancer treatments and overcoming the challenges posed by drug resistance.

5.11 Preclinical and clinical evidence supporting mesenchymal and stem cell therapy

MSCs have drawn increasing attention in the field of oncology because of their distinct biological characteristics, including their capacity for multilineage differentiation and immune modulation (Dührsen et al., 2019). These properties position MSCs as valuable candidates for targeted cancer therapies (Lan et al., 2021). Notably, their ability to migrate toward tumor sites, often referred to as tumor-homing, is a key advantage (Lan et al., 2021). This targeted migration is primarily guided by chemokine signaling, especially the interaction between CXCR4 receptors on MSCs and the chemokine CXCL12, which is frequently elevated in TME. Such mechanisms allow MSCs to deliver therapeutic agents directly to malignant tissues, potentially minimizing off-target toxicity (Lan et al., 2021; Chaturvedi et al., 2013). In addition to their migratory functions, MSCs play a significant role in shaping the immune landscape within tumors. Research indicates that MSCs can inhibit T lymphocyte proliferation, increase the presence of regulatory T cells, and suppress NK cell activity (Selmani et al., 2008). These immunoregulatory effects can either hinder or support therapeutic goals, depending upon the clinical context. For instance, experimental models have demonstrated that MSCs cocultured with tumor cells can decrease CD8+ T-cells activation, suggesting that MSCs may contribute to immune evasion mechanisms exploited by tumors (Selmani et al., 2008). Beyond their immune-related functions,

MSCs have also been used as carriers for therapeutic agents. Genetically engineered MSCs that express proapoptotic molecules, such as TRAIL, have shown tumor-specific cytotoxicity while sparing healthy tissue (Cheng et al., 2019). Moreover, MSCs secrete a wide range of soluble factors—cytokines such as IL-6 and IL-10, growth factors such as HGF, and extracellular vesicles—that can alter the TME (Cheng et al., 2019). These secreted components may support tumor progression by enhancing survival, angiogenesis, and metastasis.

Early clinical trials have started translating preclinical MSC research into human applications. Several Phase I and II studies have evaluated the safety and feasibility of MSC administration in oncology. In some trials, MSCs have been administered alongside chemotherapy for patients with advanced-stage malignancies (Cheng et al., 2019). These studies have reported improved patient tolerance to chemotherapeutic regimens, with reductions in adverse effects such as myelosuppression and gastrointestinal toxicity, suggesting a protective role for MSCs during intensive treatments (Cheng et al., 2019).

A notable application of MSCs has been their use in patients undergoing hematopoietic stem cell transplantation (HSCT), where they demonstrated benefits in enhancing graft acceptance and lowering the incidence of acute graft-versus-host diseases (GVHD) (Kangari et al., 2024; Lan et al., 2021). Bone marrow-derived MSCs, in particular, have been associated with reduced GVHD-related complications, likely because of their immunomodulatory functions that suppress excessive inflammatory responses and promote tissue repair. In addition to transplant settings, MSC-based therapies are being explored in the treatment of difficult-to-manage cancers, including glioblastoma and pancreatic cancer. Clinical reports have documented slowed disease progression in glioblastoma patients receiving MSC-infused therapies, while those with pancreatic cancer have shown improvements in symptom management and overall quality of life (Lan et al., 2021). The therapeutic outcomes of MSC-based treatments are highly dependent on how and when the cells are administered. Timing in relation to other therapies, such as chemotherapy, can significantly influence MSC behavior, particularly their immunological interactions (Cheng et al., 2019). This highlights the need for careful optimization of dosing schedules and delivery routes. Moreover, combining MSCs with other treatment modalities, such as checkpoint inhibitors or molecularly targeted ligands, may enhance immune system activation against tumors by limiting treatment induced exhaustion (Cheng et al., 2019). Ongoing research is also investigating genetic modifications of MSCs to further their therapeutic value. By engineering MSCs to produce specific proteins or chemokines, these cells can be converted into targeted delivery systems capable of concentrating therapeutic payloads within tumors, thereby increasing efficacy and reducing systemic side effects (Cheng et al., 2019).

The effects of MSCs in the context of cancer therapy can vary based on several factors, including the number of MSCs administered, their source or origin, their differentiation status, and the specific type of tumor being targeted. Hence, the role of MSC in cancer therapies must be carefully considered. Comprehensive investigations are crucial to fully understand the safety and efficacy of MSC applications in tumor treatment. Although MSCs hold significant promise in cancer therapy, their use is not without limitations. For instance, higher doses of MSCs may lead to increased tumor aggressiveness, while the inherent characteristics of MSCs derived from different tissues may contribute to divergent effects on tumor dynamics. In addition, studies have shown that MSCs can inadvertently promote cancerous behaviors in various malignancies, including ovarian (Spaeth et al., 2009), colon (Zhu et al., 2006), and gastric cancer (Okumura et al., 2009). Notably, their influence on tumor progression is complex; MSCs can enhance metastasis, facilitate epithelial–mesenchymal transition, and disrupt immune surveillance, all of which can contribute to tumor advancement.

Given these concerns, it is imperative to critically evaluate the limitations in MSC-based cancer therapies. Rigorous preclinical and clinical investigations are crucial to fully understand the safety and efficacy of MSC applications in oncology. This includes longitudinal studies that explore the long-term effects of MSC administration, as well as assessments of their interactions with the TME. As our understanding of MSC biology continues to evolve, addressing these limitations will be essential for developing effective, safe therapeutic strategies that harness the beneficial properties of MSCs while mitigating their potential adverse effects in cancer therapy.

5.12 Challenges and limitations of mesenchymal stem cell-based therapies

The therapeutic effects of MSCs in cancer are influenced by multiple variables, including the dosage administered, tissue source differentiation status, and the specific tumor type targeted (Cheng et al., 2019). Consequently, the role of MSCs in cancer therapy must be approached with caution. Comprehensive investigations are essential to thoroughly assess the safety and efficacy of MSC-based interventions in oncology. Although MSCs exhibit considerable therapeutic potential, their application is accompanied by notable limitations. For instance, elevated doses of MSCs have been associated with enhanced tumor aggressiveness, and MSCs derived from different tissue sources may exert heterogeneous effects on tumor behavior. Furthermore, emerging evidence indicates that MSCs may unintentionally facilitate tumor progression in various malignancies, including ovarian cancer (Spaeth et al., 2009), colon (Zhu et al., 2006), and gastric cancer (Okumura et al., 2009). Notably, the influence of MSCs on tumor progression is multifaceted; they can promote metastasis, facilitate EMT, and impair immune surveillance, collectively contributing to tumor advancement.

Despite the therapeutic promise of MSC-based approaches in oncology, several limitations must be carefully addressed. Rigorous preclinical and clinical investigations are necessary to establish the safety, therapeutic efficacy, and long-term impact of MSCs administration. Particular attention must be paid to how MSCs interact with the TME, their potential to inadvertently support tumor growth, and their immunoregulatory functions. Long-term follow-up studies are crucial to detect any delayed adverse effects and to evaluate the persistence of therapeutic benefits. As our understanding of MSCs' biology advances, systematically resolving these issues will be essential to harness their regenerative and anticancer properties effectively, while mitigating associated risks in clinical cancer therapy.

5.13 Conclusion

Utilizing MSCs to target the tumor vasculature presents a promising strategy for cancer therapy, capitalizing on their intrinsic tumor-homing properties and ability to influence the TME. Because of their multipotency, self-renewal ability, and low immunogenicity, MSCs are ideal candidates for localized delivery of therapeutic agents such as chemotherapeutics and anti-angiogenic compounds. Recent developments in nanoengineering and exosome modification have further optimized MSC-based delivery systems, enhancing the precision and effectiveness of therapeutic payloads (Kangari et al., 2024). Despite their promise, significant challenges persist, particularly in enhancing the MSC homing efficiency and postengraftment survival, highlighting the need for a deeper understanding of the molecular mechanisms governing these processes. As clinical trials continue to evaluate the safety and therapeutic efficacy of MSC-based therapies, this approach offers the potential to revolutionize elucidating by enabling the targeted delivery with reduced systemic toxicity. Going forward, elucidating the intricate interplay between MSCs and tumor biology will be essential for refining these strategies and improving clinical outcomes.

References

Aghebati-Maleki, L., Dolati, S., Ahmadi, M., Yousefi, M., Majidi, J., Rostamzadeh, D., Shanehbandi, D., Baloo, Z., Mohammadi, M., Ganjalikhani Hakemi, M., Yousefi, B., Baradaran, B., Namdar, A., Shafiee, A., Baradaran, A., … Aghamohammadi, A. (2019). Prospect of mesenchymal stem cells in therapy of osteoporosis: a review. *Journal of Cellular Physiology, 234*(6), 8570–8578.

Ahmed, A. U., & Lesniak, M. S. (2011). Glioblastoma multiforme: Can neural stem cells deliver the therapeutic payload and fulfill the clinical promise? *Expert Review of Neurotherapeutics, 11*(6), 775–777.

Alcayaga-Miranda, F., Cuenca, J., Khoury, M., & Figueroa, F. E. (2016). Prostate tumor-induced angiogenesis is blocked by exosomes derived from menstrual stem cells through the inhibition of reactive oxygen species. *Oncotarget, 7*(28), 44462.

Almeida-Porada, G., Atala, A. J., & Porada, C. D. (2020). Therapeutic mesenchymal stromal cells for immunotherapy and for gene and drug delivery. *Molecular Therapy-Methods & Clinical Development, 16*, 204–224.

Altaner, C. (2008). Prodrug cancer gene therapy. *Cancer Letters, 270*(2), 191–201.

Bartlett, D. L., Liu, Z., Sathaiah, M., Ravindranathan, R., Guo, Z., He, Y., & Guo, Z. S. (2013). Oncolytic viruses as therapeutic cancer vaccines. *Molecular Cancer, 12*, 1–16.

Belete, T. M. (2021). The current status of gene therapy for the treatment of cancer. *Biologics: Targets and Therapy*, 67–77.

Bellon, J. R., Smith, T. L., Brown, P., Jones, M., Davis, K., Taylor, S., Wilson, R., Lee, A., Martin, G., Clark, H., Adams, J., Thompson, L., White, C., Hall, R., King, D., Allen, B., Young, S., Green, P., … Robertson, J. (2005). Sequencing of chemotherapy and radiation therapy in early-stage breast cancer: updated results of a prospective randomized trial. *Journal of Clinical Oncology, 23*(9), 1934–1940.

Borghese, C., Marconi, S., Montagnani, V., Pessina, A., Mauri, P., Ferrari, S., Bonomi, A., Rossino, G., Soncini, M., Pisano, S., Cavicchini, L., Mariani, A., Imberti, L., Ciusani, E., Maccario, R., & Parati, E. (2020). Adipose-derived stem cells primed with paclitaxel inhibit ovarian cancer spheroid growth and overcome paclitaxel resistance. *Pharmaceutics, 12*(5), 401.

Carmeliet, P., & Jain, R. K. (2000). Angiogenesis in cancer and other diseases. *Nature, 407*(6801), 249–257.

Carmeliet, P., & Jain, R. K. (2011). Principles and mechanisms of vessel normalization for cancer and other angiogenic diseases. *Nature Reviews Drug Discovery, 10*(6), 417–427.

Chambers, E., & Mitragotri, S. (2007). Long circulating nanoparticles via adhesion on red blood cells: Mechanism and extended circulation. *Experimental Biology and Medicine, 232*(7), 958–966.

Chastkofsky, M. I., Pituch, K. C., Katagi, H., Zannikou, M., Ilut, L., Xiao, T., Han, Y., Sonabend, A. M., Curiel, D. T., Bonner, E. R., Nazarian, J., Horbinski, C. M., James, C. D., Saratsis, A. M., Hashizume, R., Lesniak, M. S., & Balyasnikova, I. V. (2021). Mesenchymal stem cells successfully deliver oncolytic virotherapy to diffuse intrinsic pontine glioma. *Clinical Cancer Research, 27*(6), 1766–1777.

Chaturvedi, P., Gilkes, D. M., Wong, C. C., Kshitiz, Luo, W., Zhang, H., Wei, H., Takano, N., Schito, L., Levchenko, A., & Semenza, G. L. (2013). Hypoxia-inducible factor-dependent breast cancer-mesenchymal stem cell bidirectional signaling promotes metastasis. *Journal of Clinical Investigation, 123*(1), 189–205.

Chen, Y. C., Chen, H. C., Wu, P. C., Lin, Y. T., Tsai, C. H., Chou, C. H., Hsu, C. W., Lee, C. H., Lin, C. Y., Wang, C. C., Liao, M. H., Wu, C. L., Hsu, H. T., Kuo, Y. L., Li, C. C., Chang, H. Y., Huang, C. F., … Chiu, H. C. (2019). Mesenchymal stem/stromal cell engulfment reveals metastatic advantage in breast cancer. *Cell Reports, 27*(13), 3916–3926.e5.

Cheng, H., Zeidan, R., Shum, P., Poon, Z., Liu, Z., Chow, E. K. H., Lin, Z., Tsai, P., Chan, W. C. W., & Wang, A. Z. (2010). Nanoparticulate cellular patches for cell-mediated tumoritropic delivery. *ACS Nano, 4*(2), 625–631.

Cheng, S., Zhang, Y., Li, X., Wang, H., Chen, J., Liu, R., Sun, L., Zhao, M., Wu, P., Yang, J., Li, H., Zhang, Q., Xu, W., Chen, L., Wang, Y., Liu, S., … Zhang, H. (2019). Engineered mesenchymal stem cells for targeting solid tumors: Therapeutic potential beyond regenerative therapy. *Journal of Pharmacology and Experimental Therapeutics, 370*(2), 231–241.

Cheng, Y., Morshed, R. A., Auffinger, B., Tobias, A. L., Lesniak, M. S., & Zhang, L. (2013). Nanoparticle-programmed self-destructive neural stem cells for glioblastoma targeting and therapy. *Small, 9*(24), 4123–4129.

Choi, M.-R., Stanton-Maxey, K. J., Stanley, J. K., Levin, C. S., Bardhan, R., Akin, D., Li, Z., Feijen, J., Ellenbogen, R. G., & Stantz, K. M. (2007). A cellular Trojan horse for delivery of therapeutic nanoparticles into tumors. *Nano Letters, 7*(12), 3759–3765.

Choi, S., Hong, J. A., Choi, H. J., & Song, J. J. (2022). Enhanced tumor targeting and timely viral release of mesenchymal stem cells/oncolytic virus complex due to GRP78 and inducible E1B55K expressions greatly increase the antitumor effect of systemic treatment. *Molecular Therapy-Oncolytics, 27*, 26–47.

Coccè, V., Bonomi, A., Marchesi, N., Farinella, F., Brini, A. T., Mariani, A., Imberti, L., Ciusani, E., Maccario, R., Pessina, A., & Parati, E. (2017). Drug loaded gingival mesenchymal stromal cells (GinPa-MSCs) inhibit in vitro proliferation of oral squamous cell carcinoma. *Scientific Reports, 7*(1), 9376.

de Araújo Farias, V., Costa, B. M., Teixeira, F. R., da Silva, J. R., Oliveira, A. C., dos Santos, F. L., Lima, R. G., Pereira, L. F., Gomes, M. L., Almeida, P. R., Souza, D. P., Fernandes, C. L., Carvalho, H. C., Martins, A. F., Nunes, T. L., … Teixeira, M. M. (2018). TGF-β and mesenchymal stromal cells in regenerative medicine, autoimmunity and cancer. *Cytokine & Growth Factor Reviews, 43*, 25–37.

Deng, D., Li, X., Wang, Y., Chen, J., Zhang, H., Liu, R., Sun, L., Zhao, M., Wu, P., Yang, J., Li, H., Zhang, Q., Xu, W., Chen, L., Wang, Y., Liu, S., … Zhang, H. (2023). Biotin-avidin system-based delivery enhances the therapeutic performance of MSC-derived exosomes. *ACS Nano, 17*(9), 8530–8550.

Dewhirst, M. W., Cao, Y., & Moeller, B. (1996). Microvascular studies on the origins of perfusion-limited hypoxia. *The British Journal of Cancer. Supplement, 27*, S247.

Du, Y., Li, H., Zhang, Q., Wang, X., Chen, L., Zhou, Y., Liu, J., Sun, M., Zhao, W., Xu, F., Huang, Y., Gao, L., Wu, Y., Lin, H., Yang, Z., Chen, Y., … Zhang, H. (2023). Targeting the hedgehog pathway in MET mutation cancers and its effects on cells associated with cancer development. *Cell Communication and Signaling, 21*(1), 313.

Duebgen, M., Martinez-Quintanilla, J., Tamura, K., Hingtgen, S., Redjal, N., Wakimoto, H., & Shah, K. (2014). Stem cells loaded with multimechanistic oncolytic herpes simplex virus variants for brain tumor therapy. *Journal of the National Cancer Institute, 106*(6), dju090.

Dührsen, L., Hartfuß, S., Hirsch, D., Geiger, S., Maire, C. L., Sedlacik, J., Guenther, C., Westphal, M., Lamszus, K., Hermann, F. G., & Schmidt, N. O. (2019). Preclinical analysis of human mesenchymal stem cells: Tumor tropism and therapeutic efficiency of local HSV-TK suicide gene therapy in glioblastoma. *Oncotarget, 10*(58), 6049–6061.

Durinikova, E., Plava, J., Tyciakova, S., Skvara, P., Vojs Stanova, A., Kozovska, Z., Kucerova, L., & Matuskova, M. (2018). Cytotoxic response of 5-fluorouracil-resistant cells to gene-and cell-directed enzyme/prodrug treatment. *Cancer Gene Therapy, 25*(11), 285–299.

Düzgüneş, N. (2019). Origins of Suicide Gene Therapy. In: Düzgüneş, N. (eds) Suicide Gene Therapy. Methods in Molecular Biology, vol 1895. Humana Press, New York, NY. https://doi.org/10.1007/978-1-4939-8922-5_1.

Ferrara, N., & Kerbel, R. S. (2005). Angiogenesis as a therapeutic target. *Nature, 438*(7070), 967–974.

Folkman, J. (1971). Tumor angiogenesis: Therapeutic implications. *New England Journal of Medicine, 285*(21), 1182–1186.

Folkman, J. (1990). *What is the evidence that tumors are angiogenesis dependent*? (pp. 4–7). Citeseer.

Folkman, J., Merler, E., Abernathy, C., & Williams, G. (1989). Induction of angiogenesis during the transition from hyperplasia to neoplasia. *Nature, 339*(6219), 58–61.

Friedenstein, A., Chailakhyan, R., & Gerasimov, U. (1987). Bone marrow osteogenic stem cells: in vitro cultivation and transplantation in diffusion chambers. *Cell Proliferation, 20*(3), 263–272.

Ghadjar, P., Hayoz, S., Zimmermann, F., Bernhard, J., Mermoud, A., Burger, M., Plasswilm, L., Vassella, E., Studer, U., Glatz, N., Sauter, T., Kestin, L. L., Birkhäuser, F. D., Guckenberger, M., Zilli, T., Moser, A., Hess, C., van den Bogaard, J., … Allal, A. S. (2008). Acute and late toxicity in prostate cancer patients treated by dose escalated intensity modulated radiation therapy and organ tracking. *Radiation Oncology, 3*, 1–8.

Greco, K. A., Franzen, C. A., Foreman, K. E., Flanigan, R. C., Kuo, P. C., & Gupta, G. N. (2016). PLK-1 silencing in bladder cancer by siRNA delivered with exosomes. *Urology, 91*, 241.e1–241.e2417.

Guo, Y., Zhang, Z., Xu, X., Xu, Z., Wang, S., Huang, D., Li, Y., Mou, X., Liu, F., & Xiang, C. (2019). Menstrual blood-derived stem cells as delivery vehicles for oncolytic adenovirus virotherapy for colorectal cancer. *Stem Cells and Development, 28*(13), 882–896.

Heath, V. L., & Bicknell, R. (2009). Anticancer strategies involving the vasculature. *Nature Reviews Clinical Oncology, 6*(7), 395–404.

Hicklin, D. J., & Ellis, L. M. (2005). Role of the vascular endothelial growth factor pathway in tumor growth and angiogenesis. *Journal of Clinical Oncology, 23*(5), 1011–1027.

Holohan, C., Van Schaeybroeck, S., Longley, D. B., & Johnston, P. G. (2013). Cancer drug resistance: An evolving paradigm. *Nature Reviews Cancer, 13*(10), 714–726.

Huang, B., Li, H., Zhang, Y., Yang, M., Liu, X., Wang, Q., Chen, J., Zhao, R., Sun, L., Wu, P., Zhang, H., Xu, W., Li, H., Chen, L., Wang, Y., Liu, S., … Zhang, H. (2015). Active targeting of chemotherapy to disseminated tumors using nanoparticle-carrying T cells. *Science Translational Medicine, 7*(291), 291–294.

Jain, R. K., Tong, R. T., & Munn, L. L. (2007). Effect of vascular normalization by antiangiogenic therapy on interstitial hypertension, peritumor edema, and lymphatic metastasis: Insights from a mathematical model. *Cancer Research, 67*(6), 2729–2735.

Kalimuthu, S., Ramasamy, T. S., Murugan, K., & Sundaram, S. (2018). Migration of mesenchymal stem cells to tumor xenograft models and in vitro drug delivery by doxorubicin. *International Journal of Medical Sciences, 15*(10), 1051.

Kangari, P., Salahlou, R., & Vandghanooni, S. (2024). Harnessing the therapeutic potential of mesenchymal stem cells in cancer treatment. *Advanced Pharmaceutical Bulletin, 14*(3), 574–590.

Ke, X., & Shen, L. (2017). Molecular targeted therapy of cancer: The progress and future prospect. *Frontiers in Laboratory Medicine, 1*(2), 69–75.

Koch, S., Treadwell, J., Sauter, B., Rieber, E. P., & Stuschke, M. (2003). Efficacy of cytotoxic agents used in the treatment of testicular germ cell tumours under normoxic and hypoxic conditions in vitro. *British Journal of Cancer, 89*(11), 2133–2139.

Konerding, M., Fait, E., & Gaumann, A. (2001). 3D microvascular architecture of pre-cancerous lesions and invasive carcinomas of the colon. *British Journal of Cancer, 84*(10), 1354–1362.

Koni, M., Pinnarò, V., & Brizzi, M. F. (2020). The Wnt signalling pathway: A tailored target in cancer. *International Journal of Molecular Sciences, 21*(20), 1–21.

Lan, T., Luo, M., & Wei, X. (2021). Mesenchymal stem/stromal cells in cancer therapy. *Journal of Hematology & Oncology, 14*(1), 195.

Leu, A. J., Berk, D. A., Lymboussaki, A., Alitalo, K., Jain, R. K., & Huang, S. K. (2000). Absence of functional lymphatics within a murine sarcoma: a molecular and functional evaluation. *Cancer research, 60*(16), 4324–4327.

Lin, Z., Chen, X., Li, Q., Wang, H., Zhang, Y., Liu, R., Sun, L., Zhao, M., Wu, P., Yang, J., Li, H., Zhang, Q., Xu, W., Chen, L., Wang, Y., Liu, S., & Zhang, H. (2022). Mesenchymal stem cell-derived exosomes in cancer therapy resistance: Recent advances and therapeutic potential. *Molecular Cancer, 21*(1), 179.

Lyu, T., Li, Q., Chen, X., Zhang, Y., Wang, H., Liu, R., Sun, L., Zhao, M., Wu, P., Yang, J., Li, H., Zhang, Q., Xu, W., Chen, L., Wang, Y., Liu, S., & Zhang, H. (2021). Exosomes from BM-MSCs promote acute myeloid leukemia cell proliferation, invasion and chemoresistance via upregulation of S100A4. *Experimental Hematology & Oncology, 10*, 1–13.

Ma, W. W., & Hidalgo, M. (2007). Exploiting novel molecular targets in gastrointestinal cancers. *World Journal of Gastroenterology: WJG, 13*(44), 5845.

McDonald, D. M., & Choyke, P. L. (2003). Imaging of angiogenesis: From microscope to clinic. *Nature Medicine, 9*(6), 713–725.

Melzer, C., Yang, Y., & Hass, R. (2016). Interaction of MSC with tumor cells. *Cell Commun Signal, 14*(1), 20.

Mohindra, J. K., & Rauth, A. M. (1976). Increased cell killing by metronidazole and nitrofurazone of hypoxic compared to aerobic mammalian cells. *Cancer Research, 36*(3), 930–936.

Müller, F.-J., Snyder, E. Y., & Loring, J. F. (2006). Gene therapy: can neural stem cells deliver? *Nature Reviews Neuroscience, 7*(1), 75–84.

Nakamizo, A., Marini, F., Amano, T., Khan, A., Studeny, M., Gumin, J., Chen, J., Hergert, P., Gianni, D., Hozumi, I., Andreeff, M., & Fine, H. A. (2005). Human bone marrow–derived mesenchymal stem cells in the treatment of gliomas. *Cancer Research, 65*(8), 3307–3318.

O'brien, K., Breyne, K., Ughetto, S., Laurent, L. C., & Breakefield, X. O. (2018). Employing mesenchymal stem cells to support tumor-targeted delivery of extracellular vesicle (EV)-encapsulated microRNA-379. *Oncogene, 37*(16), 2137–2149.

Okumura, T., Wang, S. S. W., Takaishi, S., Tu, S. P., Ng, V., Ericksen, R. E., Rustgi, A. K., & Wang, T. C. (2009). Identification of a bone marrow-derived mesenchymal progenitor cell subset that can contribute to the gastric epithelium. *Laboratory Investigation, 89*(12), 1410–1422.

Oveili, E., Rahmati, S., Maleki, F., Rezaie, J., Zarrabi, A., Maleki, L., & Aghebati-Maleki, A. (2023). The potential use of mesenchymal stem cells-derived exosomes as microRNAs delivery systems in different diseases. *Cell Communication and Signaling, 21*(1), 20.

Padera, T. P., Stoll, B., So, P., Capen, D., Cuttitta, F., Jain, R. K., & Munn, L. L. (2002). Lymphatic metastasis in the absence of functional intratumor lymphatics. *Science, 296*(5574), 1883–1886.

Pascucci, L., Coccè, V., Bonomi, A., Ami, D., Ceccarelli, P., Ciusani, E., Viganò, L., Locatelli, A., Sisto, F., Doglia, S. M., Parati, E., Bernardo, M. E., Muraca, M., Alessandri, G., Bondiolotti, G., & Pessina, A. (2014). Paclitaxel is incorporated by mesenchymal stromal cells and released in exosomes that inhibit in vitro tumor growth: A new approach for drug delivery. *Journal of Controlled Release, 192*, 262–270.

Patel, B., Gaikwad, S., & Prasad, S. (2025). Exploring the significance of extracellular vesicles: Key players in advancing cancer and possible theranostic tools. *Cancer Pathogenesis and Therapy, 3*(2), 109–119.

Pessina, A., Bonomi, A., Rossino, G., Soncini, M., Pisano, S., Cavicchini, L., Mariani, A., Imberti, L., Ciusani, E., Maccario, R., Morigi, M., Camussi, G., & Parati, E. A. (2011). Mesenchymal stromal cells primed with paclitaxel provide a new approach for cancer therapy. *PLoS ONE, 6*(12), e28321.

Petrella, F., Imberti, L., Ciusani, E., Bonomi, A., Maccario, R., Mura, M., Morigi, M., & Camussi, G. (2017). Mesenchymal stromal cells for antineoplastic drug loading and delivery. *Medicines, 4*(4), 87.

Potente, M., Gerhardt, H., & Carmeliet, P. (2011). Basic and therapeutic aspects of angiogenesis. *Cell, 146*(6), 873–887.

Qi, J., Sun, H., Zhang, Y., Wang, Y., Zhang, Z., Li, Z., Xu, W., Liu, Y., Wu, J., Chen, L., Zhang, H., Zhao, J., Li, X., Wang, J., Liu, S., Chen, Y., … Zhao, Y. (2017). Exosomes derived from human bone marrow mesenchymal stem cells promote tumor growth through hedgehog signaling pathway. *Cellular Physiology and Biochemistry, 42*(6), 2242–2254.

Randal, J. (2000). Antiangiogenesis drugs target specific cancers, mechanisms. *JNCI: Journal of the National Cancer Institute, 92*(7), 520–522.

Rezaeian, A., Khatami, F., Heidari Keshel, S., Akbari, M. R., Mirzaei, A., Gholami, K., Mohammadi Farsani, R., & Aghamir, S. M. K. (2022). The effect of mesenchymal stem cells-derived exosomes on the prostate, bladder, and renal cancer cell lines. *Scientific Reports, 12*(1), 20924.

Roger, M., Gondi, C., Meurette, O., Sanson, M., Redon, M. J., Valable, S., Farion, R., Menei, P., & Cohen-Jonathan-Moyal, E. (2010). Mesenchymal stem cells as cellular vehicles for delivery of nanoparticles to brain tumors. *Biomaterials, 31*(32), 8393–8401.

Sadhukha, T., O'Brien, T. D., & Prabha, S. (2014). Nano-engineered mesenchymal stem cells as targeted therapeutic carriers. *Journal of Controlled Release, 196*, 243–251.

Salehi, H., Al-Arag, S., Middendorp, E., Gergely, C., Cuisinier, F., & Orti, V. (2018). Dental pulp stem cells used to deliver the anticancer drug paclitaxel. *Stem Cell Research & Therapy, 9*, 1–10.

Samant, R. S., & Shevde, L. A. (2011). Recent advances in anti-angiogenic therapy of cancer. *Oncotarget, 2*(3), 122.

Schnarr, K., Schreiber, E., Meier, C., Bünz, A., Streich, D., Müller, F., Wang, H., Li, J., Becker, H., & Alexiou, C. (2013). Gold nanoparticle-loaded neural stem cells for photothermal ablation of cancer. *Advanced Healthcare Materials, 2*(7), 976–982.

Selmani, Z., Naji, A., Zidi, I., Favier, B., Gaiffe, E., Obert, L., Borg, C., Saas, P., Tiberghien, P., Rouas-Freiss, N., Carosella, E. D., & Deschaseaux, F. (2008). Human leukocyte antigen-G5 secretion by human mesenchymal stem cells is required to suppress T lymphocyte and natural killer function and to induce CD4+CD25highFOXP3+ regulatory T cells. *Stem Cells, 26*(1), 212–222.

Seo, Y., Kang, M. J., & Kim, H. S. (2021). Strategies to potentiate paracrine therapeutic efficacy of mesenchymal stem cells in inflammatory diseases. *International Journal of Molecular Sciences, 22*(7).

Shi, Q., Li, X., Wang, H., Chen, J., Zhang, Y., Liu, R., Sun, L., Zhao, M., Wu, P., Yang, J., Li, H., Zhang, Q., Xu, W., Chen, L., Wang, Y., Liu, S., … Zhang, H. (2024). Notch signaling pathway in cancer: From mechanistic insights to targeted therapies. *Signal Transduction and Targeted Therapy, 9*(1), 128.

Singh, K., Sharma, A., Patel, R., Verma, S., Kumar, P., Das, A., Mehta, V., Rao, N., Gupta, L., Chatterjee, S., Iyer, R., Nair, P., Joshi, M., Khanna, D., Reddy, H., Banerjee, S., Singh, R., … Mukherjee, T. (2025). Molecular mechanisms and pathways of mesenchymal stem cell-mediated therapy in brain cancer. *Current Stem Cell Research Therapy, 20*(5), 479–493.

Spaeth, E. L., Dembinski, J. L., Sasser, A. K., Watson, K., Klopp, A., Hall, B., Andreeff, M., & Marini, F. (2009). Mesenchymal stem cell transition to tumor-associated fibroblasts contributes to fibrovascular network expansion and tumor progression. *PLoS ONE, 4*(4), e4992.

Sun, Y., Li, X., Ding, L., Wang, J., Chen, J., Liu, R., Zhang, Q., & Xu, H. (2021). Mesenchymal stem cells-derived exosomes for drug delivery. *Stem Cell Research & Therapy, 12*, 1–15.

Sung, H., Ferlay, J., Siegel, R. L., Laversanne, M., Soerjomataram, I., Jemal, A., & Bray, F. (2021). Global cancer statistics 2020: GLOBOCAN estimates of incidence and mortality worldwide for 36 cancers in 185 countries. *CA: A Cancer Journal for Clinicians, 71*(3), 209–249.

Szewc, M., Król, S., Sobierajska, P., Stojko, J., Kubicka-Trząska, A., Słomiński, J., Gola, J., Janicka, I., Nowak, M., Węglarz, T., & Bajek, A. (2022). MSCs as tumor-specific vectors for the delivery of anticancer agents—a potential therapeutic strategy in cancer diseases: Perspectives for quinazoline derivatives. *International Journal of Molecular Sciences, 23*(5), 2745.

Tang, C., Shah, K., Messerli, S. M., Snyder, E. Y., Breakefield, X. O., & Weissleder, R. (2010). Concise review: Nanoparticles and cellular carriers-allies in cancer imaging and cellular gene therapy? *Stem Cells, 28*(9), 1686–1702.

Vaupel, P., Kallinowski, F., & Okunieff, P. (1987). Blood flow, oxygen consumption, and tissue oxygenation of human breast cancer xenografts in nude rats. *Cancer Research, 47*(13), 3496–3503.

Wang, W., Chen, J., Zhang, L., Sun, X., Liu, C., Li, Y., Xu, Y., Zhang, H., Yang, Q., Zhou, J., Zhao, Y., Li, H., Chen, Q., Li, F., Wang, Y., Zhang, Z., … Li, J. (2015). Involvement of Wnt/β-catenin signaling in the mesenchymal stem cells promote metastatic growth and chemoresistance of cholangiocarcinoma. *Oncotarget, 6*(39), 42276–42289.

Wei, H., Chen, J., Wang, S., Fu, F., Zhu, X., Wu, C., Liu, Z., Zhong, G., & Lin, J. (2019). A nanodrug consisting of doxorubicin and exosome derived from mesenchymal stem cells for osteosarcoma treatment in vitro. *International Journal of Nanomedicine*, 8603–8610.

Wei, X., Yang, X., Han, Z., Qu, F., Shao, L., & Shi, Y. (2013). Mesenchymal stem cells: a new trend for cell therapy. *Acta Pharmacologica Sinica, 34*(6), 747–754.

Weng, Z., Zhang, B., Wu, C., Yu, F., Han, B., Li, B., & Li, L. (2021). Therapeutic roles of mesenchymal stem cell-derived extracellular vesicles in cancer. *Journal of Hematology & Oncology, 14*, 1–22.

Wolinsky, J. B., Colson, Y. L., & Grinstaff, M. W. (2012). Local drug delivery strategies for cancer treatment: Gels, nanoparticles, polymeric films, rods, and wafers. *Journal of Controlled Release, 159*(1), 14–26.

Xu, C., Sun, H., Zhao, M., Zhang, Y., Wang, Y., Liu, R., Chen, J., Li, H., Wu, P., Yang, J., Zhang, Q., Xu, W., Chen, L., Wang, Y., Liu, S., … Zhang, H. (2018). A light-triggered mesenchymal stem cell delivery system for photoacoustic imaging and chemo-photothermal therapy of triple negative breast cancer. *Advanced Science, 5*(10), 1800382.

Yoon, A.-R., Hong, J., Li, Y., Shin, H. C., Lee, H., Kim, H. S., & Yun, C.-O (2019). Mesenchymal stem cell–mediated delivery of an oncolytic adenovirus enhances antitumor efficacy in hepatocellular carcinoma. *Cancer Research, 79*(17), 4503–4514.

Yu, Y., Zhang, X., Li, J., Chen, H., Wang, L., Liu, Y., Zhao, Q., Sun, M., Wu, P., Yang, T., Li, H., Zhang, W., Xu, Y., Chen, L., Wang, Y., Liu, S., … Zhang, H. (2025). Targeting the tumor microenvironment with mesenchymal stem cells based delivery approach for efficient delivery of anticancer agents: An updated review. *Biochemical Pharmacology, 232*, 116725.

Zhang, Q., Li, X., Wang, Y., Chen, H., Liu, J., Sun, L., Zhao, R., Wu, P., Yang, M., Guo, J., Huang, L., Xu, T., Li, H., Zhang, W., Wang, F., Liu, Y., Chen, Y., … Zhou, X. (2021). Regulation of pathophysiological and tissue regenerative functions of MSCs mediated via the WNT signaling pathway (Review). *Molecular Medicine Reports, 24*(3), 1–12.

Zhang, Y., Liu, C., Wang, T., Kong, F., Zhang, H., Yi, J., Dong, X., Duan, H., Tao, N., Yang, Y., & Wang, H. (2022). Therapeutic effects of mesenchymal stem cells loaded with oncolytic adenovirus carrying decorin on a breast cancer lung metastatic mouse model. *Molecular Therapy-Oncolytics, 24*, 486–496.

Zheng, L., Zhang, S., Li, Z., Liu, R., Chen, J., Wang, Y., Sun, M., Zhao, Q., Wu, P., Yang, J., Li, H., Zhang, Q., Xu, W., Chen, L., Wang, Y., Liu, S., & Zhang, H. (2012). Antitumor activities of human placenta-derived mesenchymal stem cells expressing endostatin on ovarian cancer. *PLoS ONE, 7*(7), e39119.

Zhou, Y., Li, X., Wang, H., Chen, J., Zhang, Y., Liu, R., Sun, L., Zhao, M., Wu, P., Yang, J., Li, H., Zhang, Q., Xu, W., Chen, L., Wang, Y., Liu, S., … Zhang, H. (2023). The role of mesenchymal stem cells derived exosomes as a novel nanobiotechnology target in the diagnosis and treatment of cancer. *Frontiers in Bioengineering and Biotechnology, 11*, 1214190.

Zhu, W., Xu, W., Jiang, R., Qian, H., Chen, M., Hu, J., Cao, W., Han, C., & Chen, Y. (2006). Mesenchymal stem cells derived from bone marrow favor tumor cell growth in vivo. *Experimental and Molecular Pathology, 80*(3), 267–274.

Zhuang, W.-Z., Chen, H., Zhang, W., Liu, Y., Sun, X., Li, X., Wang, Q., Gu, Y., Li, H., Yang, L., Huang, H., Zhou, L., He, Y., Tang, L., Wang, J., Zhao, H., Yu, J., … Shi, Y. (2021). Mesenchymal stem/stromal cell-based therapy: Mechanism, systemic safety and biodistribution for precision clinical applications. *Journal of Biomedical Science, 28*, 1–38.

CHAPTER

6 Mesenchymal stromal cell-based delivery of oncolytic viruses: A novel strategy for immunotherapy of metastatic tumors

Ubaid Gani[a,b], Awadhesh Kumar Verma[c], Sahar Saleem Bhat[b] and Pranav Kumar Prabhakar[d]

[a]*Department of Zoology, School of Bioengineering and Biosciences, Lovely Professional University, Phagwara, Punjab, India,* [b]*Division of Animal Biotechnology, Faculty of Veterinary Sciences and Animal Husbandry, Sher-e-Kashmir University of Agricultural Sciences and Technology of Kashmir, Srinagar, Jammu and Kashmir, India,* [c]*Department of Biotechnology and Bioinformatics, School of Bioengineering and Biosciences, Lovely Professional University, Phagwara, Punjab, India,* [d]*Department of Biotechnology, School of Engineering and Technology, Nagaland University, Meriema, Kohima, Nagaland, India*

6.1 Introduction

The metastatic spread of cancer cells from a primary tumor to distant organs is the principal cause of morbidity and mortality in oncology, accounting for more than 90% of cancer-related deaths (Siegel et al., 2024). This lethal process involves a cascade of events: Local invasion, intravasation, survival in circulation, extravasation, and colonization of distant sites, culminating in the establishment of therapy-resistant secondary lesions. These metastatic deposits often reside within profoundly immunosuppressive microenvironments, characterized by dense stroma, aberrant vasculature, and a suite of immune-inhibitory cells and signals that together form a formidable barrier to both traditional treatments and modern immunotherapies (Fidler, 2003). The clinical imperative to develop systemic therapies capable of targeting these disseminated, heterogeneous, and immunologically "cold" lesions defines the frontier of contemporary cancer research.

The evolution of cancer immunotherapy, marked by the success of immune checkpoint inhibitors (ICIs) and chimeric antigen receptor (CAR) T cells, has revolutionized outcomes for subsets of patients. However, these strategies frequently fail against metastatic disease because of poor T-cell infiltration, antigenic heterogeneity, and the dominant immunosuppressive tone of the metastatic niche (Sharma et al., 2017). This landscape has catalyzed the development of oncolytic virotherapy, a multimodal approach that uses genetically engineered or naturally occurring viruses to preferentially infect and lyse cancer cells. The first FDA-approved oncolytic virus (OV), talimogene laherparepvec (T-VEC), demonstrated not only local tumor destruction but also the induction of systemic immune responses in a portion of patients with advanced melanoma, validating the core premise of in situ

Advancements in Personalized Cellular Immunotherapy: A Comprehensive Guide to Cancer Treatment.
DOI: https://doi.org/10.1016/B978-0-443-27731-3.00014-0

Copyright © 2026 Elsevier Inc. All rights are reserved, including those for text and data mining, AI training, and similar technologies.

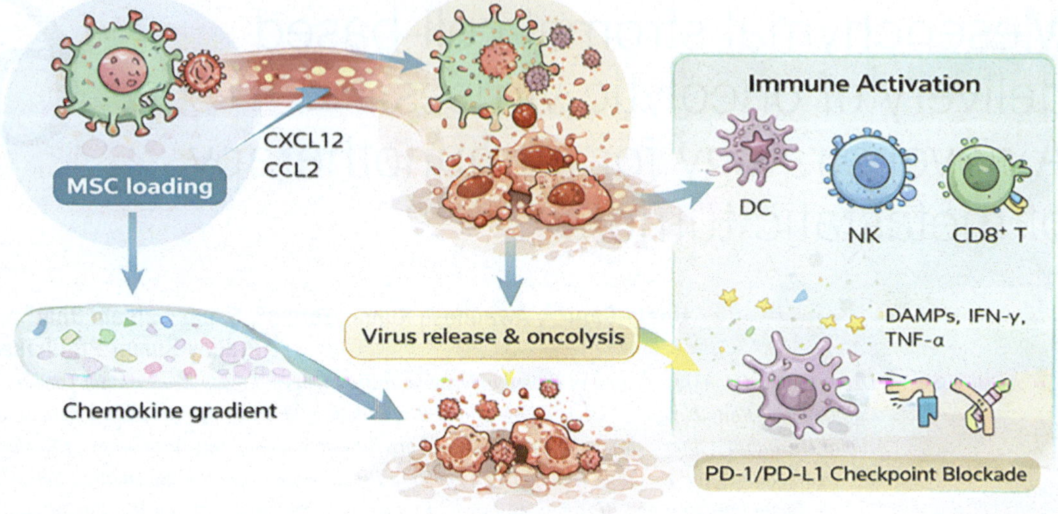

FIGURE 6.1

Schematic representation of MSC-mediated oncolytic virus (MSC–OV) delivery and subsequent antitumor immune activation.

vaccination (Kaufman et al., 2015). OVs act by inducing immunogenic cell death (ICD), releasing tumor-associated antigens and danger-associated molecular patterns (DAMPs), and remodeling the tumor microenvironment (TME) toward a more proinflammatory state.

Despite this promise, the systemic delivery of OVs to treat metastases remains a critical unsolved challenge. Intravenously administered viruses are rapidly neutralized by preexisting antibodies and complement, sequestered by phagocytic cells in the liver and spleen, and face physical and biological barriers to extravasation and intratumoral penetration (Ferguson et al., 2012). To overcome these hurdles, cell-based delivery systems have emerged, with mesenchymal stromal cells (MSCs) standing out as a uniquely qualified candidate. MSCs possess an intrinsic tumor-homing capability, migrating to sites of inflammation, wound healing, and neoplasia, including metastases, in response to chemokine gradients. Furthermore, their immunoprivileged status allows them to evade host immune surveillance, effectively acting as "Trojan horses" that can shield therapeutic OVs from neutralization during transit (Mader et al., 2013). Once within the tumor stroma, MSC lysis releases progeny virions directly into the cancer cell population, initiating a localized cycle of oncolysis and immune activation (Fig. 6.1).

The rationale for combining MSCs with OVs is therefore powerfully synergistic: MSCs provide targeted delivery and immune protection for the virus, while the OV converts the MSC from a potential modulator of the tumor stroma into a targeted, lytic agent that disrupts the metastatic niche and kick-starts a systemic antitumor immune cascade. This chapter will provide a comprehensive examination of this convergent therapeutic platform. Our objectives are to: (1) detail the biological mechanisms underlying MSC tumor-homing and their interactions with OVs; (2) review preclinical evidence demonstrating the efficacy of MSC–OV complexes in models of metastatic disease; (3) analyze the dual role of MSCs in modulating antiviral and antitumor immunity; and (4) discuss the translational challenges, manufacturing considerations, and future directions for clinical development. By synthesizing current knowledge, this chapter aims to illuminate the significant potential of MSC-delivered oncolytic virotherapy as a next-generation strategy for achieving systemic control of metastatic cancer.

6.2 Biological basis of oncolytic virotherapy

Oncolytic virotherapy is a sophisticated and rapidly evolving cancer treatment strategy that repurposes the fundamental nature of viruses, their ability to invade cells, replicate, and cause lysis, and redirects it against malignant tissue. Unlike conventional chemotherapy or radiotherapy, OVs are designed to be selectively destructive to cancer cells while sparing normal tissues. Furthermore, their mechanism of action extends far beyond simple cell killing; they function as potent in situ vaccines, reprogramming the local and systemic immune system to recognize and attack tumors. This dual capacity for direct cytotoxicity and immune stimulation positions OVs as a unique bridge between direct tumor lysis and immunotherapy. This section delves into the core scientific principles that underpin this approach, examining how selectivity is achieved, the major viral platforms in use, the critical immunogenic consequences of viral oncolysis, and the significant challenges that arise when attempting to deliver these agents systemically to treat disseminated metastatic disease.

6.2.1 Mechanisms of tumor selectivity and replication

The absolute requirement for any therapeutic virus is a high degree of specificity for cancerous tissue. This selectivity is not a single feature but a multilayered safety system engineered by exploiting the profound biological differences between a normal cell and a transformed cancer cell.

6.2.1.1 Intrinsic or natural selectivity

Some viruses naturally possess a predilection for replicating in cancer cells because of the very pathways that are dysregulated during oncogenesis. A prime example is the reovirus. In healthy cells, the presence of double-stranded RNA (a hallmark of viral replication) activates an enzyme called protein kinase R (PKR). Activated PKR halts general protein synthesis in the cell, effectively shutting down the viral production line. However, in many cancer cells, the Ras signaling pathway—a common driver of uncontrolled growth—is permanently switched on. A consequence of this activated Ras state is the inhibition of PKR. Therefore when reovirus infects a Ras-activated cancer cell, it finds an environment where its replication cannot be blocked, allowing it to proliferate and eventually lyse the cell. In contrast, in normal cells with functional PKR, reovirus replication is efficiently suppressed. Another common cancer-specific vulnerability is defects in the type I interferon (IFN) response. This antiviral alarm system is frequently compromised in tumors. Viruses such as vesicular stomatitis virus (VSV) and measles virus are exquisitely sensitive to IFN; even low levels can block their replication. In IFN-deficient cancer cells, these viruses replicate uncontrollably, while in normal cells with an intact IFN response, replication is quickly quelled (Guo et al., 2008).

6.2.1.2 Engineered selectivity

Modern genetic engineering allows scientists to build additional layers of safety and specificity into viral platforms. This is achieved through two primary strategies: Transcriptional targeting and transductional targeting.

- **Transcriptional targeting:** This approach places genes that are absolutely essential for viral replication under the control of DNA sequences (promoters or enhancers) that are highly active only in cancer cells. For instance, the promoter for the gene encoding telomerase (hTERT)—an enzyme that allows cells to divide indefinitely and is active in more than 85% of cancers but silent in most healthy adult cells—can be used to drive the expression of key viral genes. This creates a genetic logic gate: The virus can only complete its replicative cycle in a cell that both allows it to enter *and* expresses high levels of telomerase. Other tumor-specific promoters target proteins such as prostate-specific antigen (PSA) for prostate cancer or the alpha-fetoprotein (AFP) promoter for liver cancer (Singh et al., 2012).
- **Transductional targeting:** This strategy focuses on the very first step of infection: viral attachment. The outer proteins of a virus are modified to change which receptors it binds to on a cell's surface. The natural viral receptor-binding domain is often deleted to prevent infection of normal tissues. In its place, genes for new protein "ligands" are inserted. These ligands are designed to bind with high affinity to receptor molecules that are uniquely abundant on the surface of cancer cells (Lin et al., 2018). Examples of such overexpressed tumor receptors include epidermal growth factor receptor (EGFR), human epidermal growth factor receptor 2 (HER2), or specific integrins. This retargeting ensures the virus physically docks only onto cancer cells.

6.2.1.3 Viral arming

Beyond targeting, OVs are frequently "armed" with additional therapeutic genes to enhance their potency. These transgenes are carried within the viral genome and expressed in the infected tumor cell. The most clinically significant example is the immunostimulatory cytokine granulocyte-macrophage colony-stimulating factor (GM-CSF). When the infected tumor cell lyses, it releases a burst of GM-CSF along with new virus particles. GM-CSF acts as a powerful beacon and activator for dendritic cells, the body's premier antigen-presenting cells, thereby dramatically amplifying the immune response against the tumor.

6.2.1.4 The replicative cycle as a therapeutic amplifier

Once a virus successfully infects a permissive cancer cell, it commandeers the cell's machinery to produce hundreds to thousands of new viral progenies. This replication cycle culminates in **oncolysis**—the rupture of the cancer cell membrane. This event does not merely kill one cell; it releases the newly minted virus particles into the immediate TME, where they can proceed to infect and kill adjacent cancer cells. This creates a localized, self-amplifying wave of destruction within the tumor mass, a phenomenon known as the **bystander effect**.

6.2.2 Viral platforms in clinical and preclinical use

The field utilizes a diverse array of viral families, each with distinct biological properties that lend themselves to different therapeutic strategies. The choice of platform involves trade-offs between genome size, immunogenicity, replication speed, and safety profile (Table 6.1).

Table 6.1 Major oncolytic viruses and their properties.

Virus	Genome type	Tumor selectivity mechanism	Immunomodulatory features	Clinical status/notes
Adenovirus	dsDNA	E1A/E1B deletions; tumor-specific promoters	Can carry transgenes encoding cytokines (e.g., GM-CSF)	Widely studied; several ongoing clinical trials
Herpes simplex virus (HSV)	dsDNA	γ34.5 deletion for neuroattenuation; ICP6 deletion for tumor specificity	Strong immunogenicity; can express immune-stimulatory genes	FDA-approved for melanoma (T-VEC)
Vaccinia virus	dsDNA	TK gene deletion; tumor-selective replication	Stimulates robust innate immunity; can carry multiple transgenes	Rapid viral spread; used in solid tumor trials
Reovirus	dsRNA	Exploits activated Ras signaling in tumor cells	Induces type I IFN responses; modulates TME	Naturally occurring; well-tolerated in clinical studies
Measles virus	ssRNA	CD46 overexpression on tumor cells; attenuated strains	Potent induction of immune responses; oncolytic fusion activity	Early-phase clinical trials in multiple solid tumors
Vesicular stomatitis virus (VSV)	ssRNA	Defective interferon signaling in tumor cells	Induces strong innate immunity; can encode cytokines	Highly immunogenic; preclinical and early-phase clinical evaluation
Maraba virus	ssRNA	Oncolytic selectivity in interferon-deficient cells	Strong immunogenic potential; adjuvant in vaccine platforms	Emerging OV; preclinical studies promising

GM-CSF, *Granulocyte-macrophage colony-stimulating factor;* OV, *oncolytic virus;* TME, *tumor microenvironment;* Type I IFN, *type I interferon.*

- **Adenovirus:** A workhorse of gene therapy and virotherapy, adenovirus is a non-enveloped virus with a double-stranded DNA genome. Its advantages include relatively easy genetic manipulation, high efficiency in infecting a wide range of cell types (broad tropism), and a good capacity for carrying foreign genes. A major historical drawback is that a large proportion of the population has preexisting immunity from common cold infections, which can neutralize the therapeutic virus before it reaches the tumor. Newer engineered variants seek to circumvent this by using rare human or animal adenovirus serotypes to which most people have no immunity (Lin et al., 2023).
- **Herpes simplex virus type 1 (HSV-1):** HSV-1 is a large, enveloped, double-stranded DNA virus. Its primary advantage is its enormous genomic capacity. Large segments of its genome are dedicated to neurovirulence and immune evasion in its natural life cycle. These nonessential regions can be deleted to create space for multiple or very large therapeutic transgenes while simultaneously attenuating the virus for safety. The first FDA-approved OV, talimogene laherparepvec (T-VEC), is based on herpes simplex virus-1 (HSV-1). It has deletions in key neurovirulence genes and carries the gene for human GM-CSF (Mondal et al., 2020).
- **Vaccinia virus:** As the virus used in the smallpox vaccine, vaccinia has an unparalleled safety record in humans. It is a large, enveloped, double-stranded DNA poxvirus with a unique life cycle: It replicates entirely in the cytoplasm of the infected cell, meaning its genetic material never enters the nucleus. This eliminates any theoretical risk of integrating into the host genome. Vaccinia has a rapid replication cycle and a remarkable ability to spread directly from cell to cell, evading neutralizing antibodies in the bloodstream. Engineered vaccinia strains, such as JX-594 (Pexa-Vec), are designed to replicate selectively in cancer cells and express immune-stimulating proteins such as GM-CSF (Sahu & Biswas, 2010).
- **Measles virus (Edmonston vaccine strain):** The measles virus is a well-characterized, enveloped, single-stranded RNA virus. It naturally binds to two receptors: CD46 and SLAM. CD46 is a protein that is overexpressed on the surface of many types of cancer cells as a mechanism for evading the immune system. This makes the measles virus inherently tumor-tropic. When the virus infects, it causes infected cells to fuse with their neighbors, forming giant multinucleated cells called syncytia, which rapidly die. The virus is also exceptionally potent at stimulating immune responses (Russell, 2018).
- **Reovirus:** Reovirus (respiratory enteric orphan virus) is unique because it is a naturally occurring oncolytic agent that has not been genetically modified. It is a nonenveloped, double-stranded RNA virus. Its selectivity stems entirely from its dependence on the activated Ras signaling pathway, as described earlier. Its lack of pathogenicity in humans and its natural tumor targeting make it an attractive agent. Pelareorep (Reolysin®) is a formulation of the reovirus that has been tested in numerous clinical trials, often in combination with chemotherapy.
- **Newcastle disease virus (NDV):** NDV is an avian virus that is completely nonpathogenic in humans. It replicates efficiently in human cells that have defective IFN responses—a common trait in cancers. NDV is a potent inducer of type I IFN and ICD in tumor cells, making it a strong immune adjuvant. Its use can be as a direct oncolytic agent or as a vaccine platform (Zamarin & Palese, 2012).

6.2.3 Oncolytic viruses as immune stimulants: Cytokine release, antigen spreading, and T-cell priming

While direct tumor lysis is valuable, the most profound and durable antitumor effects of OVs are mediated by the immune system. OVs are expertly engineered to convert a tumor from an immunologically "cold," suppressed state into a "hot," inflamed site that the body's own defenses can recognize and attack. This process unfolds in a coordinated sequence.

6.2.3.1 Immunogenic cell death

The way an OV kills a cancer cell is critical. It is not a quiet, passive death. Oncolysis is a violent, inflammatory event that releases a specific cocktail of signals, turning the dying tumor cell into a potent vaccine. This cocktail includes:

- **Tumor-associated antigens (TAAs):** A diverse array of the cancer cell's internal proteins, both normal and mutated (neoantigens).
- **Pathogen-associated molecular patterns (PAMPs):** Viral components such as double-stranded RNA or unique glycoproteins that are instantly recognizable as "foreign" by the immune system.
- **Damage-associated molecular patterns (DAMPs):** Molecules from the host cell that are only exposed or released during traumatic cell death, such as calreticulin on the cell surface, ATP, and the nuclear protein HMGBf1. This triad acts as a powerful "danger signal," alerting the immune system that something is profoundly wrong at that location (Galluzzi et al., 2017).

6.2.3.2 Innate immune activation

The released PAMPs and DAMPs are detected by sentinel immune cells (dendritic cells, macrophages) via specialized receptors called pattern-recognition receptors (PRRs). This detection triggers an explosive local **cytokine storm**. Key cytokines released include type I interferons (IFN-α/β), tumor necrosis factor-alpha (TNF-α), and various chemokines, such as CCL5 and CXCL10. This inflammatory soup has several effects: it can help restrict the virus to the tumor site, it activates natural killer (NK) cells for immediate tumor cell killing, and, most importantly, it causes resident dendritic cells to mature and become fully activated.

6.2.3.3 Priming of tumor-specific T cells

Activated dendritic cells are the master coordinators of the adaptive immune response. They gobble up the released tumor antigens, process them, and then travel to the nearest draining lymph node. Here, they present fragments of these tumor antigens on their surface to naive T cells. This is the essential educational step. When a T-cell's receptor recognizes a tumor antigen presented by a mature dendritic cell in the context of strong inflammatory signals (the cytokines), it becomes activated, proliferates massively, and differentiates into an army of effector T cells. These T cells are specifically programmed to hunt down and destroy cells bearing those same tumor antigens.

6.2.3.4 Systemic antigen spreading and memory

This is the hallmark of a successful, durable response. The initial wave of T cells targets the most obvious (immunodominant) antigens, which might be viral proteins or a few abundant tumor proteins. However, as the therapy continues and more tumor cells are destroyed in an inflammatory context, the immune system is exposed to a wider and wider array of tumor antigens. This process, called **epitope spreading or antigen spreading,** diversifies the immune attack. The immune system begins to recognize and attack cancer cells based on many different targets, making it much harder for the tumor to escape by simply down-regulating a single antigen (Gujar et al., 2019). Furthermore, a subset of these activated T cells becomes long-lived **memory T cells**, which can patrol the body for years, providing ongoing surveillance against tumor recurrence.

6.2.4 Limitations of naked viral delivery

Despite their elegant design and potent mechanisms, the clinical translation of OVs, particularly for treating widespread metastases via intravenous injection, has been hampered by significant biological and physical barriers. The term "naked" delivery refers to administering the virus alone, suspended in a simple solution.

6.2.4.1 Humoral immune neutralization

For many viral platforms (especially adenovirus and measles), a large percentage of people have preexisting neutralizing antibodies from prior natural infections or childhood vaccinations. Even in naive patients, the first dose of virus will induce a strong antibody response (Russell et al., 2012). Upon subsequent intravenous administration, these antibodies bind to the virus, tagging it for immediate destruction by the complement system and phagocytic cells (such as macrophages in the liver and spleen). This can clear over 99% of the injected viral dose within minutes, leaving only a tiny fraction to potentially reach the tumor (Alemany et al., 2009).

6.2.4.2 Physical and biological barriers of the tumor

Metastatic tumors, in particular, are often hostile to drug penetration. They can have:

- **High interstitial fluid pressure:** Rapid, chaotic tumor growth compresses blood and lymphatic vessels, creating high internal pressure that opposes the inward flow of therapeutics from the bloodstream.
- **Dense stroma (Desmoplasia):** Many solid tumors, like pancreatic and breast cancers, are encased in a thick, fibrous wall of collagen and other proteins produced by cancer-associated fibroblasts. This acts as a physical mesh that traps viruses and prevents their deep penetration into the tumor mass.
- **Aberrant and heterogeneous vasculature:** Tumor blood vessels are often leaky, twisted, and poorly formed. Blood flow is chaotic and inconsistent, meaning some regions of a tumor may be well-perfused while others are essentially avascular "dead zones," inaccessible to intravenously delivered agents.

6.2.4.3 Off-target sequestration and clearance

Even if they evade neutralizing antibodies, intravenously injected viral particles are quickly filtered from the blood by the body's natural filtration systems. The mononuclear phagocyte system (MPS), primarily the Kupffer cells in the liver and macrophages in the spleen, is highly efficient at recognizing and engulfing viral particles. This not only depletes the therapeutic dose but can also cause dose-limiting liver toxicity.

6.2.4.4 Residual antiviral defenses within the tumor

The TME is not a passive space. Even after a virus arrives, it may face antiviral defenses. Tumor-associated macrophages (TAMs) and fibroblasts can secrete type I IFN. Some cancer cells, despite defective IFN pathways, may retain partial responsiveness. This can create pockets within the tumor where an antiviral state is established, limiting viral replication and spread and protecting a subset of cancer cells from destruction (Nguyen et al., 2008).

These formidable challenges highlight a critical disconnect: The virus is designed to work brilliantly *inside* the tumor, but getting it *to* the tumor, especially to multiple, scattered metastatic sites, is the primary obstacle. This fundamental delivery problem provides the compelling rationale for employing sophisticated delivery vehicles, such as MSCs, which are the focus of the subsequent chapter section. These cellular carriers are engineered to act as protective "Trojan horses," shielding the virus during transit, evading clearance, and actively navigating to the sites of metastatic disease.

6.3 Mesenchymal stromal cells as therapeutic carriers

The quest to overcome the systemic delivery barriers of OVs has catalyzed the exploration of advanced cellular delivery systems. Among these, MSCs have emerged as a uniquely promising "Trojan horse" platform. MSCs are not merely inert delivery bags; they are dynamic, multifunctional cells whose innate biology synergizes powerfully with the requirements of effective virotherapy (Wang et al., 2018). This section details the defining characteristics of MSCs, the mechanisms behind their tumor-homing behavior, their complex immunomodulatory functions, and the compelling rationale for their use as living, targeted vectors for oncolytic agents.

6.3.1 Biological characteristics and sources of mesenchymal stromal cells

MSCs are adult, multipotent progenitor cells first identified in the bone marrow. They are defined by a set of minimal criteria established by the International Society for Cellular Therapy: (1) adherence to plastic under standard culture conditions; (2) specific surface marker expression (positive for CD73, CD90, CD105; negative for hematopoietic markers such as CD34, CD45, CD11b, CD19, and HLA-DR); and (3) in vitro trilineage differentiation potential into osteoblasts (bone), adipocytes (fat), and chondrocytes (cartilage).

A key therapeutic advantage is their availability from multiple tissue sources, each with distinct practical and biological nuances (Table 6.2):

- **Bone marrow (BM-MSCs):** The classic and most studied source. BM-MSCs are obtained via bone marrow aspiration, a somewhat invasive procedure. They have well-characterized properties but may exhibit decreased proliferative capacity with donor age.
- **Adipose tissue (AD-MSCs):** Isolated from lipoaspirate, adipose tissue provides a highly abundant and accessible source of MSCs, often yielding a significantly higher number of cells per gram of tissue compared to bone marrow. Their harvest is associated with a minimally invasive procedure.
- **Umbilical cord (UC-MSCs):** Derived from Wharton's jelly, the perivascular region, or the cord blood of the umbilical cord, a medical waste product. UC-MSCs are considered more primitive, demonstrating superior proliferative capacity, lower immunogenicity, and stronger immunomodulatory potential compared to their adult-derived counterparts. They represent an ethically uncomplicated and potent source.

Regardless of source, MSCs possess two operational characteristics critical for their carrier role: **Immunoprivilege** and **expansion capacity**. They exhibit low immunogenicity due to minimal expression of major histocompatibility complex (MHC) Class II and costimulatory molecules (such as CD80 and CD86), allowing for allogeneic use without triggering a potent host immune rejection. Furthermore,

Table 6.2 MSC sources and characteristics.

MSC source	Tissue origin	Proliferation capacity	Immunomodulatory profile	Tumor homing ability	Clinical advantages/ limitations
Bone marrow (BM-MSCs)	Adult bone marrow	Moderate; decreases with donor age	Strong immunosuppressive effects; secretes cytokines such as TGF-β, IL-10	High; well-studied chemokine receptor profile (CXCR4, CCR2)	Gold standard; limited cell yield in older donors
Adipose tissue (AD-MSCs)	Subcutaneous fat	High; rapid expansion	Moderate immunosuppressive properties; secretes growth factors	Good; robust homing to inflammatory sites	Abundant and easily harvested; variable cytokine secretion
Umbilical cord (UC-MSCs)	Wharton's jelly/cord tissue	Very high; low senescence	Immunologically naïve; potent immunomodulation	Moderate; can be engineered to enhance tumor homing	Suitable for allogeneic "off-the-shelf" therapy; requires GMP expansion
Placenta (PL-MSCs)	Placental tissue	High; rapid proliferation	Strong immunomodulation; secretes IL-6, TGF-β	Moderate; can be genetically modified	Readily available; low ethical concerns; limited long-term data
Dental pulp (DP-MSCs)	Extracted teeth	Moderate	Unique immunomodulatory secretome	Moderate	Novel source; limited large-scale clinical data
Induced MSCs (iMSCs)	Reprogrammed somatic cells	Variable; dependent on reprogramming efficiency	Tunable via genetic engineering	Tunable via chemokine receptor modification	Personalized therapy potential; complex production and validation

GMP, *Good manufacturing practice;* IL-10, *interleukin-10*; TGF-β, *transforming growth factor-beta.*

they can be expanded to clinical-scale numbers in the laboratory, loaded with an OV ex vivo, and cryopreserved for use as an "off-the-shelf" therapeutic product.

6.3.2 Homing capacity to tumor microenvironments

The most critical feature of MSCs for drug delivery is their innate **tumor tropism**—the ability to migrate from the bloodstream or injection site toward primary and metastatic tumors (Spaeth et al., 2008). This homing is not random but a highly directed process driven by the inflammatory and dysregulated nature of the tumor, which mimics a "nonhealing wound."

The homing cascade involves several sequential steps:

1. **Tethering and rolling:** Upon intravenous infusion, MSCs interact with the activated endothelium near tumor sites. They express adhesion molecules (such as VCAM-1) and chemokine receptors that allow them to weakly bind and roll along the vessel walls.
2. **Firm adhesion and transmigration:** The TME secretes a potent cocktail of signaling molecules that creates a chemotactic gradient. MSCs express receptors for these signals, including:
   - **Chemokine receptors:** CXCR4 (binds SDF-1α/CXCL12), CCR2 (binds MCP-1/CCL2), CX3CR1 (binds Fractalkine/CX3CL1) (Kidd et al., 2009).
   - **Growth factor receptors:** Receptors for VEGF, PDGF, and EGF, all abundantly produced by tumors and their associated stroma. The engagement of these receptors triggers activation of integrins on the MSC surface, leading to firm adhesion to the endothelial lining and subsequent migration through the vessel wall (extravasation) toward the source of the signals.
3. **Integration into the tumor stroma:** Once within the tumor, MSCs interact with and become part of the complex tumor stroma. They can differentiate into or influence cancer-associated fibroblasts, pericytes, and other stromal components (Karnoub et al., 2007). This deep integration is what positions them perfectly to serve as local viral factories.

This tropism is remarkably effective for targeting metastases, as these secondary sites also establish inflammatory and chemokine gradients that actively recruit circulating MSCs.

6.3.3 Immunomodulatory and regenerative properties

MSCs are potent modulators of the immune system, a property that is context-dependent and crucial for their therapeutic application with OVs. Their immunomodulation is primarily mediated through the secretion of a vast array of soluble factors and through direct cell–cell contact.

6.3.3.1 Suppressive effects

In inflammatory environments, MSCs can suppress the activity of effector immune cells. They secrete factors such as prostaglandin E2 (PGE2), indoleamine 2,3-dioxygenase (IDO), transforming growth factor-beta (TGF-β), and interleukin-10 (IL-10) (Uccelli et al., 2008). These molecules can:

- Inhibit T-cell proliferation and effector functions.
- Promote the generation of regulatory T cells (Tregs).
- Shift macrophages from a proinflammatory (M1) to an antiinflammatory, tissue-repair (M2) phenotype.

- Suppress NK cell cytotoxicity and dendritic cell maturation. In the context of OV delivery, this immunosuppressive capacity can be initially beneficial by dampening the host's innate antiviral immune response, allowing the loaded OV to establish infection and replicate within the tumor before being cleared.

6.3.3.2 Proinflammatory potential

In certain contexts or when stimulated, MSCs can also secrete proinflammatory cytokines such as IL-6, IL-8, and IFN-γ. This dual nature means their net effect on the TME can be fine-tuned, potentially shifting from initially permissive (for viral replication) to stimulatory (for antitumor immunity) (Waterman et al., 2010).

6.3.3.3 Regenerative and tropic properties

Beyond immunity, MSCs secrete a "secretome" rich in bioactive molecules that promote tissue repair, angiogenesis, and cell survival (e.g., VEGF, HGF, and FGF). While this is beneficial for regenerative medicine, in oncology, it raises a critical paradox: MSCs may inadvertently support tumor growth by enhancing blood supply and stromal resilience. This underscores the importance of using them as short-lived delivery vehicles that are destroyed by the virus they carry, rather than as persistent cellular therapies in cancer.

6.3.3.4 Rationale for mesenchymal stromal cells as delivery vehicles for oncolytic viruses

The convergence of MSC biology and OV therapy creates a rational, synergistic platform designed to overcome the major limitations of standalone virotherapy. The rationale is built on multiple pillars:

1. **Shielding from host immunity:** MSCs provide a physical barrier that protects the enveloped or nonenveloped OV from neutralizing antibodies and complement proteins in the bloodstream. By carrying the virus internally, the MSC–OV complex evades the humoral immune system, dramatically increasing the fraction of the therapeutic dose that survives transit to the tumor.
2. **Active tumor-targeted delivery:** Unlike passive nanoparticle systems, MSCs are living, motile cells that actively seek out tumors and metastases via their homing mechanisms. This transforms systemic delivery from a hope of passive accumulation into an active targeting strategy, increasing the specific localization of the virus to disease sites while reducing off-target exposure.
3. **Penetration of the tumor core:** MSCs do not just accumulate at the tumor periphery. Their ability to extravasate and migrate through the stroma enables them to carry the OV deep into the often poorly vascularized and fibrotic core of the tumor, a region typically inaccessible to freely administered viruses.
4. **Localized viral amplification and production:** Upon reaching the tumor, the OV begins to replicate within its MSC carrier. The ensuing lytic infection of the MSC serves as a localized "burst" release of progeny virions. This delivers a high, concentrated dose of virus directly into the heart of the tumor mass, initiating a powerful, localized infection cycle.
5. **Modulation of the tumor microenvironment:** The MSC carrier can precondition the tumor site. Its initial immunosuppressive secretome may create a temporary "immune-sheltered" niche that allows the OV to establish itself. Subsequently, the immunogenic death of the MSC (caused by the virus) releases viral particles, tumor-homing signals, and MSC-derived immunomodulators in

a coordinated fashion, potentially leading to a more robust and controlled activation of antitumor immunity than the virus alone.

6. **Versatility and safety:** MSCs can be loaded with a wide variety of OV types (adenovirus, measles, vaccinia, etc.). The ex vivo loading process allows for precise control of the viral load per cell. Furthermore, as the OV is designed to lyse its carrier, it provides a built-in safety mechanism to eliminate the MSCs after they have delivered their payload, mitigating long-term concerns about their potential protumor effects.

6.4 Engineering mesenchymal stromal cell–oncolytic virus systems

Mesenchymal stromal/stem cells (MSCs) represent a uniquely advantageous platform for ferrying OVs into otherwise inaccessible tumor compartments. Their intrinsic pathotropism—the ability to migrate preferentially toward sites of inflammation, hypoxia, necrosis, and tumor-derived chemokine gradients—positions them as a biological "delivery vector" capable of overcoming major pharmacological obstacles such as poor tumor perfusion, extracellular matrix barriers, and premature systemic immune inactivation of viral therapeutics (Duebgen et al., 2014).

However, the utility of MSCs as OV carriers depends on sophisticated engineering at multiple biological layers: viral loading, intracellular replication, immune-evasion timing, safety-switch design, and the integration of synthetic regulatory circuits. Each of these variables must be precisely calibrated to ensure that MSCs carry a sufficient viral payload, survive long enough to reach the tumor, and release OVs at a therapeutically meaningful magnitude within the TME, without causing off-target cytolysis or systemic toxicity. The following subsections explore these engineering principles with comprehensive mechanistic detail.

6.4.1 Virus loading, infection efficiency, and replication within mesenchymal stromal cells

Successful OV loading into MSCs is determined by a complex interplay of viral tropism, receptor expression, innate immunity, cell cycle state, and metabolic programming. MSCs derived from bone marrow, adipose tissue, dental pulp, umbilical cord, or placenta differ in receptor density, IFN sensitivity, replicative vigor, and susceptibility to viral transduction, making donor and tissue origin a major variable in engineering outcomes.

6.4.1.1 Receptor-mediated entry and early infection events

OVs rely on specific receptors for binding and entry. Adenovirus requires CAR or integrins; HSV binds nectin-1; measles virus uses CD46 or SLAM; reovirus depends on sialic acid moieties. MSCs do express these receptors, but at levels far lower or more heterogeneous than typical transformed cancer cell lines. In primary MSCs, receptor expression is also dynamic: it can be upregulated by inflammatory cytokines (e.g., TNF-α), hypoxia, culture confluence, and exposure to TME-derived factors.

Once bound, internalization proceeds through clathrin-mediated endocytosis, caveolin-dependent uptake, macropinocytosis, or direct fusion (for enveloped viruses). The efficiency of internalization determines whether viral genomes reach productive replication stages or are instead shuttled toward lysosomal degradation.

6.4.1.2 Innate antiviral barriers within mesenchymal stromal cells

MSCs possess a robust, although intermediate-strength, antiviral machinery. The RIG-I/MDA5 pathway rapidly senses viral RNA, activating MAVS, TBK1, and IRF3 to induce type I IFN. PKR activation leads to phosphorylation of eIF2α, halting translation—even translation of viral proteins. OAS–RNase L and Mx GTPases degrade viral RNA and impede viral polymerase functions. Although these antiviral defenses safeguard MSC stability, they simultaneously restrict OV replication, making infection less productive. Therefore engineering efforts often seek to **transiently attenuate innate immunity**—for example, by hypoxia preconditioning, which naturally reduces IFN responsiveness, or pharmacologic modulation of JAK/STAT signaling, which temporarily silences IFN-induced antiviral genes.

6.4.1.3 Intracellular replication dynamics

Once early viral gene expression begins, replication kinetics depend on viral genome architecture and MSC biology. DNA viruses (adenovirus, HSV) engage nuclear machinery, often forcing MSCs into a pseudo-S-phase environment by modulating pRb, p53, or E2F pathways. RNA viruses (VSV, measles, and reovirus) replicate cytoplasmically and rely heavily on the permissiveness of the cytosolic antiviral state.

Metabolism also plays a central role. MSCs exhibit metabolic plasticity, shifting between glycolysis and oxidative phosphorylation. Viruses typically favor glycolytic states because glycolysis generates abundant nucleotides, biosynthetic intermediates, and redox molecules required for viral assembly. Hypoxic preconditioning, common in MSC clinical manufacturing, enhances glycolysis and often increases OV replication efficiency.

6.4.1.4 Balancing retention and release

For MSC-mediated delivery, viral replication must be **potent but delayed**: MSCs must remain intact until they reach the tumor. Premature lysis in circulation defeats the therapeutic purpose. Thus the viral strain and loading conditions must be calibrated to ensure a controlled latency period, with robust lytic expansion occurring predominantly within the TME.

6.4.2 Genetic modification of mesenchymal stromal cells for enhanced viral delivery

Genetic engineering transforms MSCs from passive viral carriers into programmable therapeutic devices. This involves modulating receptor expression, altering antiviral pathways, enhancing tumor homing, and integrating synthetic gene circuits that control the spatiotemporal dynamics of viral release.

One of the most direct engineering strategies is to overexpress viral entry receptors using lentiviral or transposon-based vectors (Shah, 2012). Enhancing CAR expression dramatically improves adenoviral entry; similarly, augmenting nectin-1 facilitates HSV uptake. Synthetic chimeric receptors, designed to bind capsid-modified or ligand-displaying OVs, further expand infection possibilities. Selective attenuation of antiviral genes—such as PKR, IFITM proteins, or MDA5—can make MSCs more permissive to OVs. CRISPR-mediated knockdown or precise epigenetic modulation may be used to create conditions that permit robust viral replication without compromising genomic stability or introducing tumorigenic risk.

Tumor-directed migration is one of the greatest advantages of MSC-based delivery. Overexpressing chemokine receptors such as CXCR4, CCR2, or CCR5 dramatically increases MSC trafficking toward tumors secreting CXCL12, CCL2, or CCL5, respectively. Tumors with desmoplastic stroma pose physical barriers that MSCs must overcome; engineering MSCs to express matrix-degrading enzymes (e.g., MMPs and heparanase) or integrins enhances intratumoral infiltration (Kang et al., 2012; Hakkarainen et al., 2015). A central challenge in engineering MSC–OV systems is restricting viral replication to the tumor. Synthetic biology provides powerful tools for this. Hypoxia-responsive promoters, acidosis-inducible elements, and inflammation-sensitive transcriptional systems allow viral replication genes (e.g., E1A in adenovirus) to be expressed only under tumor-specific conditions. Advanced logic circuits (AND, OR, NOT gates) can integrate multiple TME features—such as low oxygen, high lactate, and elevated MMP activity—to ensure that replication is tumor-restricted (Nissim & Bar-Ziv, 2010).

MSCs can simultaneously deliver OVs and immunomodulators. Transgenes encoding IL-12, GM-CSF, LIGHT, or IFN-β amplify immune activation following OV-mediated tumor lysis. Similarly, MSCs engineered to produce checkpoint-inhibition nanobodies (anti-PD-L1) or T-cell–recruiting ligands enhance the immunogenic aftermath of OV therapy. This transforms MSC–OV constructs into multiarmed therapeutic systems capable not only of direct oncolysis but also of shaping the systemic antitumor immune response.

6.4.3 Impact on mesenchymal stromal cell viability, differentiation, and immune behavior

MSCs experience profound physiological changes following OV infection. These changes must be understood and modulated to ensure therapeutic predictability and safety. Upon viral infection, MSCs activate stress-response pathways, including ER stress markers, ROS generation, and DNA-damage sensors. Depending on viral load, MSCs may undergo early apoptosis, pyroptosis, or a delayed lytic cycle induced by viral replication. This impacts the duration during which MSCs can home to tumors and determines the magnitude of intratumoral viral release. OV infection may transiently alter lineage-specific transcriptional programs. Adenovirus infection often reduces osteogenic differentiation because of interference with RUNX2, whereas adipogenic differentiation may be suppressed due to perturbation of PPARγ signaling. While these changes do not typically alter therapeutic function when MSCs are used purely as carriers, they are important for regulatory compliance because MSC identity and phenotype must remain measurable and consistent in clinical manufacturing. MSCs normally exert immunosuppressive activity by secreting TGF-β, PGE2, indoleamine 2,3-dioxygenase (IDO), and other modulators. OV infection disrupts this equilibrium. Infected MSCs often transition to a more immunogenic phenotype, upregulating MHC class I, releasing inflammatory cytokines such as IL-6 and CXCL10, and shedding exosomes that contain viral RNA fragments capable of activating innate immune sensors in nearby cells. While undesirable in the bloodstream (where immune evasion is crucial), this immunogenic shift is profoundly beneficial once MSCs arrive within the tumor, as it synergizes with OV-induced ICD to create a potent, in situ vaccine-like effect.

6.4.4 Optimization of virus-to-cell ratios and release kinetics

Determining the appropriate virus-to-MSC ratio—commonly referred to as the multiplicity of infection (MOI)—is one of the most critical engineering steps. The MOI must be high enough to produce a therapeutically robust viral payload but not so high that MSCs lyse prematurely. An MOI that is too low leads to insufficient viral amplification once MSCs reach the tumor. However, an excessively high MOI overwhelms MSC cellular integrity and triggers early innate immune responses, leading to immediate or premature lysis. The optimal MOI varies widely depending on virus type, MSC donor characteristics, culture conditions, and the intended timing of therapeutic peak activity. Fine-tuning viral release involves controlling both replication rate and lytic timing. Genetic delay systems, microenvironment-responsive promoters, and protease-sensitive viral capsids can ensure that MSCs do not release substantial viral loads until they encounter tumor-specific biochemical signatures. Physical approaches—such as encapsulation in hydrogels—can also impose kinetic barriers that slow viral escape into circulation. Advanced technologies, including real-time imaging of reporter viruses, multiomic single-cell profiling, and agent-based mathematical modeling, allow precise characterization of viral amplification inside MSCs. These measurements reveal that even within a homogeneous MSC batch, replication rates vary significantly at the single-cell level. Understanding and accommodating this heterogeneity is essential for clinical translation.

6.4.5 Ex vivo versus in vivo viral loading strategies

The strategy used to load mesenchymal stromal/stem cells (MSCs) with OVs is a decisive determinant of therapeutic potency, biosafety, regulatory feasibility, and translational scalability. Broadly, two conceptual paradigms exist: **Ex vivo loading**, in which MSCs are infected with OVs outside the body under controlled conditions, and **in vivo loading**, in which MSCs acquire OVs after being administered to the patient. Although both approaches aim to achieve efficient cell-mediated viral delivery to tumors, they differ profoundly in their operational workflows, the biological events governing viral uptake, the degree of control over the MSC–OV interface, and the predictability of the final product. Understanding the nuances of each strategy is essential for designing effective MSC–OV therapeutics.

6.4.5.1 Ex vivo viral loading

Ex vivo loading is the currently dominant strategy in preclinical and clinical development because it offers strict control over every step of the MSC infection process (Mader et al., 2013). Under good manufacturing practice (GMP) conditions, isolated MSCs are exposed to a carefully titrated viral inoculum, allowing precise manipulation of MOI, infection timing, culture density, oxygen tension, and cytokine milieu to achieve an optimal balance between viral uptake and MSC viability. Importantly, ex vivo loading ensures that the virus interacts exclusively with MSCs rather than with non-target tissues or systemic immune components.

Mechanistically, ex vivo loading permits modulation of the MSC antiviral landscape. Investigators can transiently prime MSCs with hypoxia, polybrene, or small-molecule IFN inhibitors to enhance viral internalization and early transcription (Hammad et al., 2016). Because these manipulations occur outside the patient's body, they avoid systemic toxicity and allow fine-tuning of intracellular viral replication dynamics. Downstream of viral entry, ex vivo loading enables monitoring of early viral gene expression, quantification of viral genome copy numbers, and adjustment of culture conditions to shape the kinetics of viral amplification and eventual release.

One of the most important advantages of ex vivo loading is quality control. After infection, MSCs can be comprehensively characterized using viability assays, flow cytometry for viral receptors, quantitative PCR for viral genome content, digital droplet PCR for replication intermediates, and transcriptomic profiling to assess cellular responses. Sterility, mycoplasma status, genomic stability, and tumor-homing potency can all be verified before patient infusion, making the final product more predictable and traceable within advanced therapy medicinal product (ATMP) regulatory frameworks.

However, ex vivo loading is not without challenges. Manufacturing patient-specific or donor-derived MSC products under GMP conditions can be complex and costly (Detela et al., 2019). Prolonged culture and viral exposure may induce cellular stress responses or alter cytokine secretion profiles, potentially diminishing MSC homing or immunomodulatory behavior. Furthermore, ex vivo–infected MSCs exhibit a finite latency period before virus-induced lysis begins. This "viable window" must be sufficiently long to permit systemic circulation, immune evasion, and tumor localization. Excessively rapid lysis may result in premature loss of therapeutic cells before they reach the TME.

6.4.5.2 *In vivo viral loading*

In vivo loading represents a fundamentally different paradigm in which MSCs are administered first, followed by systemic delivery of the OV. Circulating MSCs encounter viral particles as they migrate toward tumor-associated chemokine gradients (Yong et al. 2009). This approach reduces ex vivo manipulation and may improve scalability by lowering manufacturing demands. Mechanistically, in vivo loading relies on the biodistribution properties of both MSCs and the virus. MSCs respond to tumor-secreted chemokines such as CXCL12 and CCL2, while systemically delivered viruses undergo partial sequestration by the innate immune system and hepatic filtration before a fraction reaches the tumor (Alemany et al., 2009). However, in vivo loading faces substantial biological variability. Neutralizing antibodies, complement activation, and phagocytic clearance rapidly eliminate circulating viruses, particularly adenoviruses and VSV, thereby limiting the number of infectious particles available for MSC uptake (Alemany et al., 2009). In addition, both MSCs and viruses encounter multiple non-target cell populations in circulation, including endothelial cells, macrophages, hepatocytes, and splenic reticuloendothelial cells, leading to inconsistent and unpredictable MSC-specific viral capture (Yong et al., 2009). A further limitation arises from the temporal mismatch between MSC biodistribution and viral pharmacokinetics. After intravenous infusion, MSCs initially accumulate in the pulmonary vasculature, whereas systemically delivered viruses may follow distinct circulation and clearance kinetics, reducing opportunities for synchronized interaction (Fischer et al., 2009). While engineering strategies to enhance MSC–virus interactions are under investigation, they remain at early developmental stages. From a regulatory perspective, in vivo loading presents additional safety challenges. Because the virus is administered directly to the patient, off-target infection and acute antiviral responses become direct clinical risks. In contrast, ex vivo loading confines viral exposure to the MSC product prior to administration, aligning more closely with current regulatory expectations for cell-based gene therapies (Detela et al., 2019).

6.4.5.3 *Hybrid and next-generation loading approaches*

To reconcile the advantages of both strategies, hybrid loading methods are emerging. One approach uses **sublytic preloading**, in which MSCs are exposed to a low MOI ex vivo, allowing partial viral entry without triggering early cytolysis. After infusion, these partially loaded MSCs encounter additional viral particles in vivo, amplifying their payload in a biologically timed manner. Another approach involves **surface engineering**, whereby MSC membranes are decorated with viral-binding moieties such as heparan sulfate mimetics, penton-base–binding peptides, or synthetic nanoparticle–virus complexes. These modifications enhance the ability of MSCs to selectively capture OVs in vivo without exposing other tissues to excessive viral loads. A third strategy employs **encapsulation technologies** such as biomaterial scaffolds, hydrogel carriers, or extracellular vesicle coatings. These materials can shield MSCs during circulation, modulate viral loading kinetics, and fine-tune the spatial localization of viral release. While these hybrid strategies remain largely experimental, they illustrate the future trajectory of MSC–OV therapy: Fine-grained control over viral capture, delayed lysis, immune evasion, and tumor-targeting precision, delivered through combinations of ex vivo conditioning and in vivo biological reinforcement.

6.5 Mechanisms of antitumor action in mesenchymal stromal cell–oncolytic virus therapy

MSC–OV therapy mediates antitumor activity through a complex, multilayered network of biological mechanisms that extend beyond direct viral oncolysis of malignant cells. Therapeutic efficacy arises from the coordinated interplay between MSC tumor-homing capacity, intratumoral OV replication, immune system modulation, and disruption of the tumor-supportive microenvironment (Wang et al., 2021). In addition to serving as delivery vehicles, MSCs protect OVs from immune neutralization and facilitate their accumulation within primary and metastatic tumor sites, thereby enhancing immune-mediated cytotoxicity and overall therapeutic potency (Alipour et al., 2023).

6.5.1 Tumor targeting and viral propagation in metastatic niches

A defining advantage of MSC–OV systems lies in their ability to reach tumor sites that are inaccessible to free-floating viral particles. Once introduced into the body, MSCs rapidly respond to chemotactic gradients generated by tumor-associated inflammation, hypoxia, and tissue remodeling. Malignant tissues overexpress chemokines such as CXCL12, CCL2, CXCL8, and TGF-β, creating a molecular "breadcrumb trail" that guides MSC migration. These cues are particularly intense in metastatic microenvironments—bone marrow niches, lung parenchyma, hepatic sinusoids, and peritoneal deposits—where chronic inflammation and stromal remodeling occur long before the arrival of circulating tumor cells. Upon reaching the tumor, MSCs navigate the disorganized vasculature and dense extracellular matrix (ECM) via integrin- and metalloproteinase-mediated interactions. Unlike free OVs, which are often neutralized by circulating antibodies or trapped by hepatic filters, MSCs physically shield their viral cargo as they traverse the bloodstream. This cellular escort mechanism ensures that an intact viral payload is delivered to the tumor with minimal systemic loss.

Once within the TME, viral replication is amplified by the unique metabolic and signaling features of malignant cells. Tumors exhibit defective IFN pathways, genomic instability, dysregulated apoptosis networks, and elevated macromolecule synthesis—all of which create a permissive environment for OV expansion. Following controlled lysis of the MSCs, OVs spread radially through tumor cell clusters, exploiting cell–cell junctions, exosomal pathways, and tumor-associated immune suppression to facilitate broad infection. In metastatic niches, where dormant cancer cells or micrometastases are poorly vascularized and often shielded from chemotherapy, MSC–OV delivery provides a rare opportunity for targeted attack. Evidence suggests that MSCs infiltrate marrow endosteal regions, hepatic metastatic foci, and micrometastatic clusters with remarkable precision, delivering OVs to sites where free viral particles seldom accumulate in therapeutically meaningful concentrations. This localized viral propagation initiates successive cycles of tumor cell lysis, releasing neoantigens, DAMPs, and proinflammatory cytokines that dramatically reshape the immunologic landscape of the tumor.

6.5.2 Immune activation cascade

While the direct lytic activity of OVs contributes significantly to tumor debulking, the immune cascade they initiate is now considered one of the most important drivers of long-term antitumor control. Viral infection of tumor cells triggers a robust innate immune response through PRRs such as RIG-I, MDA5, cGAS–STING pathways, and TLR3/7/9 compartments. These sensors detect viral RNA, DNA, capsid proteins, and replication intermediates, stimulating rapid secretion of type I IFNs, TNF-α, IL-6, IL-1β, CXCL10, and GM-CSF (Fig. 6.2).

The early innate phase of the antitumor immune response to OVs recruits a variety of innate effector cells—including neutrophils, NK cells, macrophages, and dendritic cells (DCs)—into the TME. OVs induce inflammation and danger signals that enhance activation and recruitment of NK cells and other innate populations to the tumor site (DePeaux & Delgoffe, 2024). NK cells recognize virally stressed tumor cells, often through altered expression of activating ligands and reduced MHC class I, and contribute to direct cytotoxicity, while macrophage polarization toward a proinflammatory phenotype supports further immune activation. Dying tumor cells release pathogen-associated molecular patterns (PAMPs) and DAMPs, including calreticulin, ATP, and HMGB1, which are hallmarks of ICD and potent stimuli for DC maturation (Galluzzi et al., 2020). Mature DCs engulf tumor antigens and migrate to draining lymph nodes, where they present antigens to naive T cells, bridging innate and adaptive immunity (DePeaux & Delgoffe, 2024). The adaptive phase is characterized by clonal expansion of tumor-specific $CD8^+$ cytotoxic T lymphocytes (CTLs), $CD4^+$ helper T cells, and, in some contexts, B cells that can contribute antibody-mediated antitumor effects. Activated CTLs infiltrate the tumor and eradicate infected and uninfected cancer cells through perforin/granzyme pathways. Furthermore, OV-induced antigen release and presentation can lead to "epitope spreading," whereby immune responses initially directed at viral or dominant tumor antigens broaden to recognize additional tumor antigens that were previously immunologically silent, thereby enhancing overall antitumor immunity MSC–OV systems amplify this immune cascade in several ways. MSCs preferentially release OVs in hypoxic, inflamed zones of the tumor where immune cells are already partially activated, generating synergistic proinflammatory signaling. Some engineered MSCs release immune-enhancing molecules such as IL-12, IFN-β, or ICIs, further strengthening T-cell responses and reversing local immunosuppression. This immunologic reprogramming helps establish durable, systemic antitumor immunity that can suppress recurrent or metastatic disease long after viral clearance.

6.5.3 Disruption of the tumor stroma and vascular networks

In addition to killing tumor cells and stimulating immune responses, MSC–OV therapy dismantles the architectural and metabolic support structures that tumors rely on. The tumor stroma is a complex network composed of cancer-associated fibroblasts (CAFs), matrix components, aberrant vasculature,

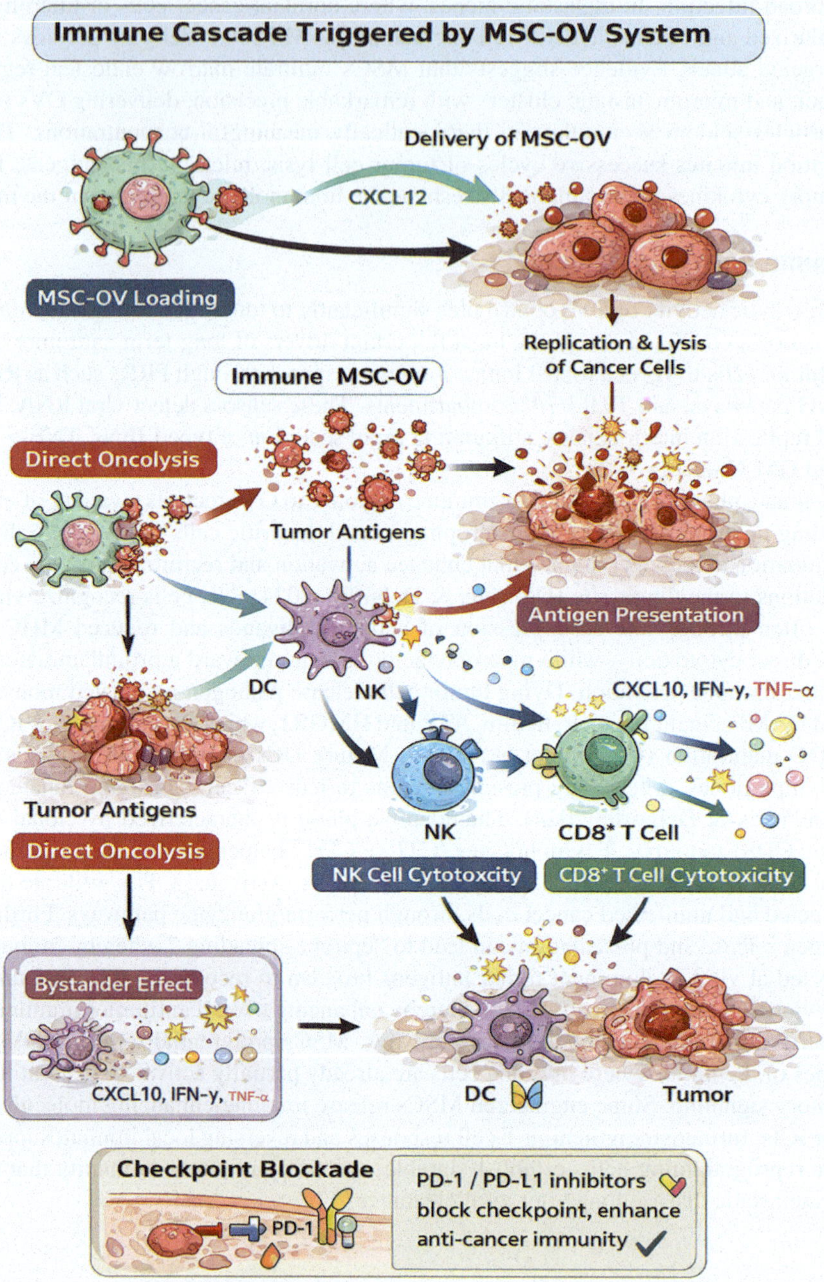

FIGURE 6.2

Immune cascade triggered by MSC-mediated oncolytic virus (MSC–OV) therapy in cancer.

pericytes, immune suppressor cells, and extracellular vesicles. This microenvironment provides growth factors, physical protection, immunosuppressive niches, and resistance pathways that weaken conventional therapies. OVs disrupt the tumor stroma in multiple ways. CAFs are highly susceptible to viral infection due to their proliferative activity and aberrant signaling profiles. Their lysis reduces desmoplastic stiffness, allowing deeper MSC and immune cell penetration. ECM degradation occurs through both viral proteases and the upregulated release of MSC-derived matrix metalloproteinases (MMPs), which remodel dense stromal barriers and increase viral diffusion. Tumor vasculature, characterized by chaotic branching, permeability defects, and hypoxic zones, is also profoundly affected. OVs infect tumor endothelial cells, causing localized vascular damage, increased permeability, and collapse of microvessels. This disrupts nutrient and oxygen supply to the tumor, inducing metabolic stress and secondary tumor cell apoptosis. MSCs contribute by secreting antiangiogenic molecules under inflammatory stimulation, such as thrombospondin-1, TIMPs, and certain microRNAs that regulate VEGF pathways. As the vasculature destabilizes, interstitial pressure decreases, improving OV penetration and immune infiltration.

Moreover, stromal immune-suppressive elements—Treg cells, myeloid-derived suppressor cells (MDSCs), M2-like macrophages—are functionally reprogrammed or reduced following OV infection. Viral replication alters cytokine gradients, disrupts chemokine-mediated recruitment of suppressive cells, and exposes stromal cells to inflammatory conditions that reverse their immunosuppressive phenotypes. This conversion of the TME from immune-cold to immune-hot is a critical mechanism by which MSC–OV therapy enhances responsiveness to immune checkpoint blockade and other immunotherapies. By simultaneously collapsing stromal barriers, damaging vasculature, degrading ECM, and reversing immunosuppression, MSC–OV systems initiate a profound structural collapse of the tumor's physical and immunologic sanctuary. The combined effect is a hostile microenvironment for cancer survival, enabling both viral propagation and immune-mediated elimination to proceed with maximal efficiency.

6.5.4 Modulation of immunosuppressive cells (regulatory T cells, myeloid-derived suppressor cells, and tumor-associated macrophages)

A major barrier to effective antitumor immunity is the presence of immunosuppressive cellular populations within the TME, including Tregs, MDSCs, and TAMs. These populations actively suppress cytotoxic T lymphocyte activity, promote immune tolerance, and secrete cytokines that reinforce tumor immune evasion (Veglia et al., 2018). MSC–OV therapy has emerged as a potent immunotherapeutic strategy capable of reshaping this immunosuppressive network through coordinated mechanisms, including inflammatory reprogramming of myeloid cells, depletion or functional inhibition of suppressive immune subsets, and induction of proinflammatory cytokine milieus that favor effective antitumor immune responses (Russell et al., 2012).

6.5.4.1 *Regulatory T cells*

Tregs accumulate in tumors in response to TGF-β, IL-10, and other immunosuppressive mediators. OV-mediated infection of tumor cells triggers ICD, releasing DAMPs and proinflammatory cytokines that alter chemokine gradients and reduce Treg recruitment. MSCs engineered to secrete IL-12, IFN-β, or IL-15 further inhibit Treg proliferation and function, promoting a shift toward effector T-cell dominance.

Some OV strains have been shown to directly infect Tregs or interfere with FoxP3 transcription, transiently reducing their suppressive phenotype.

6.5.4.2 Myeloid-derived suppressor cells

MDSCs are highly plastic cells recruited to the TME by tumor-secreted factors such as GM-CSF, G-CSF, VEGF, and IL-6. OV infection perturbs these recruitment signals, reducing MDSC trafficking. In addition, viral replication induces type I IFNs and proinflammatory chemokines (e.g., CXCL9 and CXCL10) that promote MDSC differentiation into mature antigen-presenting cells or proinflammatory macrophages. MSC–OV constructs further amplify this effect by delivering immune modulators that bias MDSC differentiation toward a phenotype supportive of antitumor immunity.

6.5.4.3 Tumor-associated macrophages

TAMs often exhibit an M2-like, immunosuppressive phenotype that supports tumor growth, angiogenesis, and metastasis. OV infection of tumor cells and the resulting inflammatory milieu polarizes TAMs toward an M1-like, proinflammatory phenotype. MSCs can be engineered to express molecules such as GM-CSF, LIGHT, or CD40L, which enhance this polarization. The net effect is a repurposing of TAMs from immune suppressors to active participants in cytotoxicity, phagocytosis, antigen presentation, and secretion of proinflammatory cytokines that reinforce both innate and adaptive antitumor responses.

6.5.4.4 Integrated modulation of the immunosuppressive niche

Through these mechanisms, MSC–OV therapy achieves a coordinated remodeling of the TME, diminishing suppressive cell populations while enhancing effector immune cell activity. This not only increases direct tumor killing but also establishes a more immunogenic microenvironment that is receptive to combination therapies such as checkpoint blockade and adoptive T-cell therapy.

6.5.5 Synergistic potential with checkpoint inhibitors and cytokine therapies

The immune reprogramming induced by MSC–OV therapy provides a natural synergy with ICIs and cytokine-based therapies, positioning MSC–OV systems as pivotal components in combination immunotherapy regimens.

6.5.5.1 Checkpoint inhibitors

Checkpoint molecules such as programmed cell death protein-1 (PD-1), programmed death-ligand 1 (PD-L1), and cytotoxic T-lymphocyte-associated protein 4 (CTLA-4) are upregulated in the TME to suppress T-cell activity and maintain tumor tolerance. OV-mediated tumor cell lysis increases antigen presentation and T-cell infiltration, effectively converting “cold” tumors into “hot” tumors. This transformation enhances responsiveness to ICIs. MSCs engineered to express checkpoint-blocking antibodies, nanobodies, or fusion proteins locally amplify this effect, reducing systemic toxicity while maintaining potent T-cell activation at the tumor site.

6.5.5.2 Cytokine therapies

Cytokines such as IL-2, IL-12, IL-15, and GM-CSF play central roles in amplifying immune cell proliferation, activation, and persistence. MSC–OV systems can be genetically engineered to deliver these cytokines directly to the tumor, synergizing with viral lysis-induced ICD to stimulate robust innate and adaptive immune responses. For example, IL-12 expression enhances NK and $CD8^+$ T-cell cytotoxicity, while GM-CSF promotes dendritic cell recruitment and maturation, improving tumor antigen presentation.

6.5.5.3 Mechanistic synergy

The integration of OV-mediated antigen release, mesenchymal stem cell (MSC)–driven tumor homing, immune checkpoint blockade, and localized cytokine delivery establishes a self-reinforcing immunologic network within the TME. In this context, tumor cells are eliminated through direct viral cytotoxicity, immunosuppressive cellular populations are functionally reprogrammed, and effector T cells are expanded, recruited, and sustained within the tumor bed (Russell et al., 2012). By simultaneously enhancing antigen availability, overcoming immune suppression, and improving T-cell infiltration and persistence, MSC–OV–based combination strategies address fundamental barriers such as tumor heterogeneity, immune escape, and poor baseline immunogenicity, positioning MSC–OV therapy as a highly adaptable platform for next-generation combination immunotherapy (Harrington et al., 2019).

6.6 Preclinical and clinical studies

The translation of MSC–OV therapy from concept to clinical application has required extensive preclinical validation, encompassing mechanistic in vitro studies, rigorous in vivo animal models, and carefully designed early-phase clinical trials. Collectively, these studies provide critical insights into viral replication kinetics, MSC viability, tumor targeting, biodistribution, immunologic outcomes, and therapeutic efficacy. The integration of these datasets informs rational design, dosing strategies, and safety considerations for human applications.

6.6.1 In vitro evidence: Tumor cell killing and viral replication profiles

Preclinical evaluation typically begins with in vitro studies, which allow precise control over the cellular and viral environment and provide mechanistic insights into MSC–OV interactions with tumor cells. Tumor cell killing is commonly assessed using cytotoxicity assays, such as MTT, LDH release, or real-time impedance-based measurements. These studies consistently demonstrate that MSCs can act as efficient carriers, delivering OVs to multiple tumor types, including glioblastoma, pancreatic adenocarcinoma, melanoma, and metastatic breast cancer cells.

6.6.1.1 Viral replication kinetics within mesenchymal stromal cells and tumor cells

In vitro models reveal that OVs replicate at variable rates depending on the viral species, tumor cell type, and MSC donor source. Adenoviruses and herpes simplex viruses generally exhibit slower replication kinetics in MSCs compared to permissive tumor cells, which is advantageous for ensuring delivery without premature MSC lysis. RNA viruses such as VSV and measles demonstrate rapid replication but require careful optimization of MOI to balance viral amplification with MSC viability (Iankov et al., 2007).

Reporter systems, including GFP-tagged viral genomes or luciferase constructs, have enabled real-time tracking of viral spread and amplification within cocultures of MSCs and tumor cells. These studies show that MSC-mediated delivery often enhances infection efficiency in tumor cells relative to free viral particles, likely because of MSC-mediated shielding from antiviral neutralizing factors and improved physical proximity to tumor targets (Iankov et al., 2007).

6.6.1.2 *Immunomodulatory effects in vitro*

Coculture systems also allow assessment of immune interactions. MSC–OV–infected tumor cells release DAMPs, cytokines, and chemokines, which can be monitored in conditioned media for their potential to recruit and activate dendritic cells, NK cells, and T lymphocytes in subsequent ex vivo assays. These studies highlight the dual functionality of MSC–OV therapy, combining direct oncolysis with immunogenic remodeling of the tumor milieu.

6.6.2 In vivo animal models: Biodistribution and efficacy studies

Translating in vitro findings into animal models is critical for understanding MSC–OV biodistribution, tumor homing, viral propagation, and systemic safety.

6.6.2.1 *Biodistribution studies*

Preclinical animal studies employ fluorescent or bioluminescent MSCs and viruses to track cell and viral localization. In immunocompromised mice bearing human tumor xenografts, MSCs injected intravenously primarily localize to the lungs initially, then redistribute to tumor sites guided by chemokine gradients. Viral payloads remain largely associated with MSCs until tumor infiltration occurs. Biodistribution studies confirm that MSCs protect the virus from rapid neutralization and hepatic clearance, enabling delivery to poorly vascularized tumor regions that free OVs fail to penetrate.

6.6.2.2 *Efficacy studies in tumor models*

Efficacy studies consistently demonstrate enhanced tumor regression with MSC-mediated OV delivery compared to free-virus administration. In glioblastoma xenografts, MSCs carrying oncolytic adenovirus or herpes virus significantly prolonged survival, reduced tumor burden, and increased intratumoral viral titers. Similar results have been observed in orthotopic pancreatic cancer and metastatic melanoma models, highlighting the broad applicability of MSC–OV strategies across tumor types.

6.6.2.3 *Immune-competent models*

Immunocompetent mouse models allow investigation of immune-mediated mechanisms. Studies demonstrate that MSC–OV therapy increases $CD8^+$ T-cell infiltration, polarizes TAMs toward an M1 phenotype, reduces Treg and MDSC abundance, and generates systemic antitumor immunity. Combination strategies with ICIs in these models further enhance efficacy, demonstrating mechanistic synergy and laying the foundation for clinical translation.

6.6.3 Clinical translation: Current and completed trials of mesenchymal stromal cell–oncolytic viruses

The promising preclinical data have led to early-phase clinical trials investigating MSC–OV therapy in humans. These trials focus primarily on safety, tolerability, biodistribution, and preliminary efficacy. Several early-phase trials have employed MSCs carrying oncolytic adenoviruses, herpes viruses, or

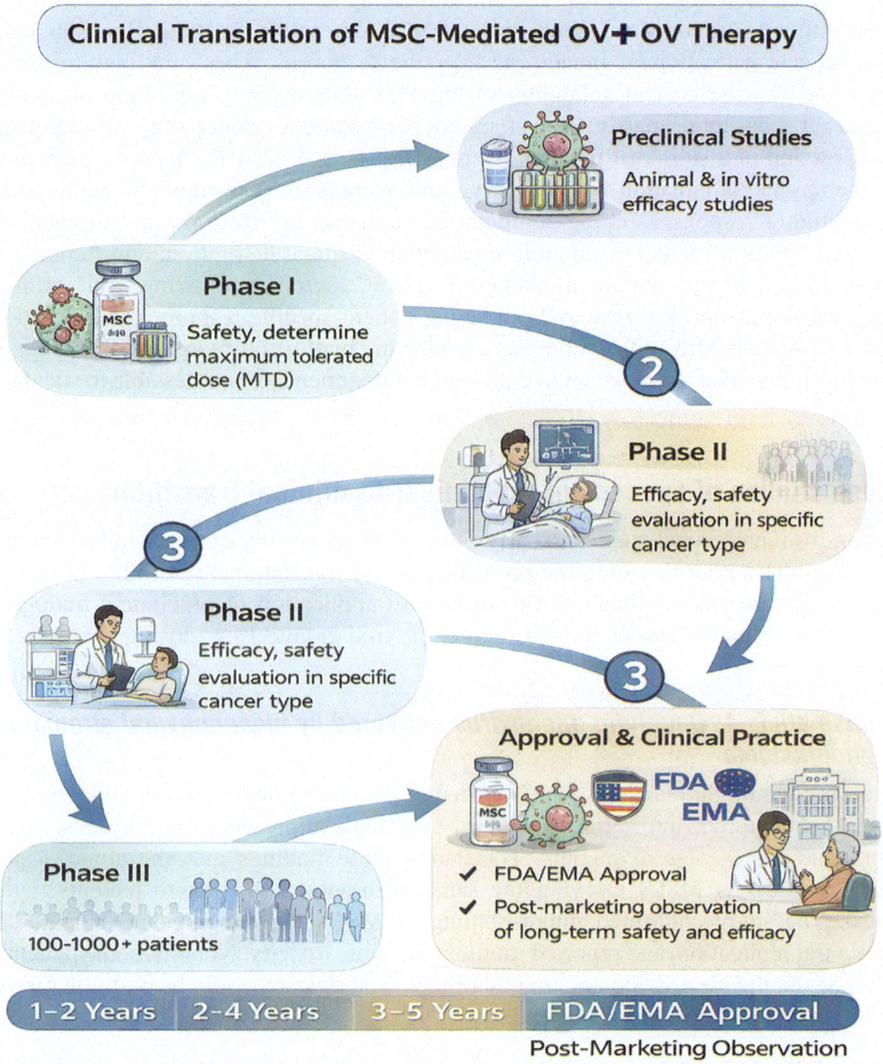

FIGURE 6.3

Clinical translation of MSC-mediated oncolytic virus (MSC-OV) therapy.

measles viruses in patients with advanced solid tumors, including glioblastoma, pancreatic cancer, and metastatic colorectal cancer (Fig. 6.3). Key objectives have included assessing adverse events, viral shedding, MSC viability, and immune responses. Most studies report favorable safety profiles, with transient fever and mild inflammatory reactions being the most common adverse events. Importantly, systemic toxicity is generally lower than free OV administration because of MSC-mediated sequestration and tumor-targeted delivery. Clinical imaging and biomarker studies confirm that MSCs successfully traffic to tumor sites. Viral replication is detectable in tumor biopsies, and immune biomarkers indicate activation of innate and adaptive responses. These findings mirror preclinical observations and provide critical validation of the MSC–OV delivery paradigm in human patients. Although phase I trials are primarily safety-focused, preliminary evidence suggests tumor stabilization or partial regression in a subset of patients (García-Castro et al., 2010). Ongoing trials are exploring MSC–OV therapy in combination with ICIs, cytokine therapy, or conventional chemoradiotherapy to maximize antitumor responses. These combination strategies are informed by mechanistic insights from preclinical studies, including immune modulation, stromal disruption, and tumor-specific viral amplification. Key challenges remain in standardizing MSC sources, optimizing viral loading strategies, controlling viral replication kinetics, and predicting patient-specific responses. Nevertheless, clinical data continue to support MSC–OV therapy as a promising platform for targeted oncolytic virotherapy, particularly for tumors that are resistant to conventional treatments or inaccessible to standard systemic therapies.

6.6.4 Case studies of successful preclinical-to-clinical transition

The transition from preclinical studies to early-phase clinical trials is a critical milestone in MSC–OV therapy, as it validates both the scientific rationale and the translational feasibility of this therapeutic platform. Several case studies illustrate the successful application of preclinical findings to human studies, providing valuable lessons in MSC sourcing, viral engineering, dosing strategies, and safety management.

6.6.4.1 Case study 1: Oncolytic adenovirus delivered by mesenchymal stromal cells in glioblastoma

In preclinical models of glioblastoma, MSCs loaded ex vivo with conditionally replicative adenoviruses demonstrated robust tumor infiltration, enhanced viral replication within tumor cells, and prolonged survival in orthotopic xenograft models. Translating these findings into the clinic, a phase I trial administered autologous MSCs carrying the same adenoviral construct to patients with recurrent glioblastoma. The study confirmed tumor-homing of MSCs via imaging and biopsy, documented intratumoral viral replication, and reported minimal systemic toxicity. Notably, some patients exhibited transient tumor stabilization, demonstrating that preclinical efficacy could be partially reproduced in a human setting (García-Castro et al., 2010).

6.6.4.2 Case study 2: Mesenchymal stromal cells–Herpes simplex virus in pancreatic cancer

Orthotopic pancreatic tumor models showed that MSCs loaded with oncolytic HSV could penetrate the dense stromal barriers, achieve localized viral amplification, and induce immune-mediated tumor regression. Building on these results, a phase I/II trial explored the delivery of HSV-loaded MSCs in patients with unresectable pancreatic adenocarcinoma. Early results indicated that MSCs successfully trafficked to tumor sites, virus-associated cytotoxicity was detectable in biopsies, and systemic cytokine profiles suggested immune activation without dose-limiting toxicity. This case underscores the ability of MSCs to overcome anatomical and physiological barriers that limit the efficacy of free-virus administration (Mader et al., 2009).

6.6.4.3 Case study 3: Oncolytic measles virus in metastatic tumors

Preclinical studies in metastatic melanoma and ovarian cancer models demonstrated that MSCs could protect the oncolytic measles virus from neutralizing antibodies and deliver it efficiently to micrometastatic deposits. A clinical trial subsequently administered allogeneic MSCs loaded with measles virus to patients with metastatic solid tumors. Imaging confirmed tumor localization of MSCs, while biopsies showed intratumoral viral presence. Immune profiling revealed induction of tumor-specific T-cell responses, demonstrating a translation of preclinical immune modulation effects to the clinical setting (Iankov et al. 2007).

6.6.4.4 Key lessons from preclinical-to-clinical transition

These case studies highlight several critical factors for successful translation:

- **MSC source and expansion:** Autologous versus allogeneic MSCs affect immunogenicity and reproducibility; preclinical optimization of culture conditions directly informs clinical protocols.
- **Ex vivo viral loading:** Controlled ex vivo infection ensures predictable viral delivery, a principle reinforced by animal models.
- **Dosing and timing:** Preclinical viral replication kinetics guide clinical MOI selection and infusion schedules.
- **Safety monitoring:** Lessons from animal toxicity studies inform human safety parameters, including cytokine release, off-target infection, and viral shedding.

Collectively, these examples validate the concept that robust preclinical modeling—spanning viral dynamics, tumor homing, immune modulation, and stromal interactions—can guide the successful initiation of human trials with measurable biological and clinical effects.

6.6.5 Comparative outcomes with direct viral delivery

A central rationale for MSC-mediated OV therapy is to overcome the limitations of direct viral delivery, which often suffers from poor tumor targeting, rapid neutralization by antibodies, and systemic toxicity. Comparative studies in preclinical models have elucidated the advantages and challenges of MSC–OV therapy relative to conventional free-virus administration.

- Free viruses administered intravenously are rapidly cleared by the liver, spleen, and circulating neutralizing antibodies, resulting in low tumor accumulation, particularly in poorly vascularized or metastatic sites. MSCs, by contrast, act as cellular vehicles that shield the virus during circulation and actively home to tumors via chemokine gradients (Jittprasong, 2022). Animal studies consistently show higher intratumoral viral titers when OVs are delivered by MSCs compared to free-virus administration, even at equivalent viral doses.
- Direct viral delivery is often limited by the inability of OVs to penetrate dense tumor stroma or reach micrometastatic deposits. MSCs enhance the spatial distribution of OVs, resulting in more extensive tumor infection and improved regression in multiple xenograft and orthotopic models. In metastatic models, MSC–OV therapy has demonstrated superior control of disseminated tumor foci, a setting in which free-virus therapy is largely ineffective (Jittprasong, 2022).
- While free-virus therapy can induce immune activation, the inflammatory response is often acute and systemic, leading to cytokine-mediated toxicity. MSC–OV delivery localizes viral replication

and subsequent immune activation within the TME. This controlled stimulation reduces systemic toxicity while enhancing local recruitment of dendritic cells, NK cells, and tumor-specific T lymphocytes. Comparative studies show that MSC-mediated delivery elicits a stronger and more durable antitumor immune response than free viral injection.

- Direct viral delivery is associated with dose-limiting toxicities, including systemic inflammation, hepatic toxicity, and off-target viral replication. MSCs act as a protective reservoir, limiting exposure to non-target tissues. Preclinical toxicology studies demonstrate that MSC–OV therapy reduces systemic adverse events while maintaining or enhancing antitumor efficacy, making it a safer and more targeted therapeutic approach.

These comparative findings justify the continued development of MSC–OV platforms over free-virus approaches, particularly for solid tumors that are inaccessible, immunosuppressive, or metastatically disseminated. Clinical trials increasingly incorporate MSC-mediated viral delivery as a standard method to maximize efficacy, minimize toxicity, and harness synergistic immune effects.

6.7 Challenges and limitations

Despite the promising preclinical and early clinical outcomes of MSC–OV therapy, the field faces a spectrum of challenges that must be carefully addressed to ensure safety, reproducibility, and clinical success. These limitations arise from the complex biology of MSCs and OVs, the dynamic TME, immune system interactions, and the practical realities of large-scale production under regulatory frameworks. Understanding these obstacles is essential for developing next-generation MSC–OV platforms and designing effective clinical strategies.

6.7.1 Safety concerns

Safety remains the foremost concern in MSC–OV therapy, as the approach inherently involves the combination of live cells and replicating viral vectors. Each component—MSCs and OVs—carries potential risks that can be magnified when combined, necessitating careful preclinical and clinical evaluation.

6.7.1.1 Insertional mutagenesis and genetic stability

Insertional mutagenesis arises when viral vectors integrate transgenes into the host cell genome at sites that disrupt tumor suppressor genes or activate oncogenes. This is particularly relevant for MSCs genetically engineered to enhance viral delivery, express cytokines, or incorporate suicide switches. Although MSCs are generally considered low-risk for malignant transformation, prolonged ex vivo culture increases the likelihood of chromosomal aberrations, telomere shortening, or spontaneous genetic drift (Ben-David et al., 2011; Hacein-Bey-Abina et al., 2003). Preclinical assessments therefore include:

- **Karyotype analysis** to detect gross chromosomal abnormalities.
- **Whole-genome sequencing** to identify small insertions, deletions, or translocations.
- **Soft-agar colony formation assays** to detect anchorage-independent growth.

- **In vivo tumorigenicity assays** in immunodeficient mice to detect rare malignant transformation events.

Even with these measures, the low-frequency nature of insertional events presents a residual risk that must be monitored in long-term clinical follow-up.

6.7.1.2 Systemic inflammation and cytokine release

OVs induce potent innate immune responses, which are central to their antitumor effects but may also precipitate systemic inflammation. Premature lysis of MSCs in circulation or release of high viral loads can trigger:

- **Cytokine release syndrome (CRS):** Rapid elevation of IL-6, TNF-α, and IFN-γ can cause fever, hypotension, vascular leakage, and multiorgan stress.
- **Acute-phase immune toxicity:** Neutrophilia, complement activation, and NK cell hyperactivation may occur, particularly with highly immunogenic viruses.
- **Off-target organ injury:** Although MSCs preferentially home to tumors, some cells may become trapped in lungs, liver, or spleen, leading to localized viral replication and tissue injury.

Mitigation strategies include fine-tuning the MOI during ex vivo loading, engineering MSCs for delayed viral release, incorporating inducible suicide genes, and carefully controlling infusion rates.

6.7.1.3 Viral shedding and environmental safety

MSC-mediated OV delivery does not entirely eliminate the risk of viral shedding. Virus may escape via:

- Circulating lysed MSC debris.
- Excretion in saliva, urine, or feces.
- Localized viral replication in nontarget tissues because of incomplete tumor specificity.

Shedding raises both biosafety concerns for healthcare providers and regulatory hurdles. Clinical trials monitor shedding using sensitive PCR and culture assays, and containment protocols are implemented for both hospital and patient environments. The choice of viral backbone, such as replication-restricted or tumor-specific OVs, directly influences shedding risk and environmental safety.

6.7.2 Immunogenicity and premature clearance

A second major limitation in MSC–OV therapy is the host immune system, which can both enhance antitumor efficacy and simultaneously limit therapeutic delivery. Premature clearance of MSCs or their viral payload reduces tumor-targeted efficacy and complicates dosing strategies.

6.7.2.1 Innate immune clearance

Upon systemic administration, MSCs encounter multiple innate immune defenses:

- **Natural killer (NK) cells:** Recognize and eliminate MSCs displaying stress-induced ligands.
- **Complement system:** May opsonize MSCs, triggering phagocytosis by macrophages.
- **Phagocytic clearance:** Liver Kupffer cells and splenic macrophages rapidly engulf foreign or stressed MSCs, reducing circulation time and tumor-homing efficiency.

Innate clearance is particularly pronounced in allogeneic MSC therapies, where minor histocompatibility mismatches can accelerate recognition.

6.7.2.2 Adaptive immune responses

- **Preexisting antiviral antibodies:** Patients previously exposed to viral vectors (e.g., adenovirus and measles virus) can neutralize OVs before tumor delivery, limiting infection efficiency.
- **T-cell mediated responses:** $CD8^+$ and $CD4^+$ T cells can target viral antigens on MSCs or infected tumor cells, leading to early elimination.
- **Immune memory:** Subsequent doses may be less effective because of rapid recognition and clearance of both virus and carrier cells.

6.7.2.3 Strategies to mitigate premature clearance

Several approaches are being explored to extend MSC survival and viral delivery:

- **Cell surface modification:** PEGylation or cloaking of MSCs to reduce antibody recognition.
- **Transient immunosuppression:** Short-term administration of immune modulators to allow MSCs to reach the tumor.
- **Encapsulation or biomaterial carriers:** Protect MSCs from immune surveillance while permitting controlled viral release.
- **Optimized dosing schedules:** Repeated low-dose administrations to balance immune activation with effective delivery.

Balancing immune evasion with the need for immunogenic tumor killing is a central challenge. Excessive immune suppression may reduce clearance but diminish the therapeutic immunologic amplification, whereas insufficient protection may result in premature clearance and treatment failure.

6.7.3 Heterogeneity of mesenchymal stromal cell sources and patient-specific responses

A central challenge in MSC–OV therapy is the inherent heterogeneity of MSCs, which arises both from their tissue of origin and from inter-individual variability. MSCs are derived from a variety of sources, including bone marrow, adipose tissue, umbilical cord, placenta, and even dental pulp (Phinney et al., 2013). Each of these sources imparts unique phenotypic and functional properties. Bone marrow–derived MSCs, for instance, have been extensively characterized and exhibit robust tumor-homing capacity, but their proliferative potential and cytokine secretion profiles can be significantly influenced by donor age and health status. Adipose-derived MSCs are more abundant and proliferate readily in culture, yet they may display altered immunomodulatory activity compared to bone marrow MSCs (Bourin et al., 2013). Umbilical cord and placental MSCs are highly proliferative and immunologically naïve, making them attractive candidates for allogeneic "off-the-shelf" therapies, though differences in surface markers and chemokine receptor expression may alter their tumor-targeting efficiency (El Omar et al., 2020; Wang et al., 2021).

Beyond the source of MSCs, patient-specific factors profoundly influence therapeutic outcomes. The immune status of a patient, including the presence of preexisting neutralizing antibodies to the viral vector, can significantly reduce the survival and delivery efficiency of MSCs. The TME itself, which may be highly immunosuppressive, hypoxic, or structurally dense, can impede MSC infiltration and OV propagation. In addition, age, comorbidities, prior therapies, and genetic background contribute to variability in both MSC performance and patient response. As a result, two patients receiving ostensibly identical MSC–OV therapy may experience markedly different outcomes, complicating clinical trial design and the establishment of standardized dosing regimens.

These layers of heterogeneity necessitate careful consideration in both preclinical and clinical development. Potency assays, chemokine receptor profiling, and functional tumor-homing assessments are critical for selecting MSC populations capable of reliably delivering viral payloads. Moreover, balancing autologous MSC approaches, which minimize immune rejection but require individualized expansion, with allogeneic approaches, which enable scalability but may face accelerated clearance, is an ongoing challenge. Addressing both MSC source variability and patient-specific factors is essential to achieving consistent, reproducible, and safe MSC–OV therapy (Jittprasong, 2022; Phinney et al., 2013).

6.7.4 Ethical considerations in cellular–viral combinatorial therapies

MSC–OV therapy occupies a unique ethical landscape because it combines two advanced and inherently complex modalities: Living cellular therapy and replicating viral agents. The novelty of this approach introduces uncertainties in both short-term safety and long-term outcomes, raising ethical considerations that extend beyond conventional clinical trials.

First, the balance between potential therapeutic benefit and risk must be carefully communicated to patients. MSC–OV therapies carry risks of off-target viral replication, systemic inflammation, insertional mutagenesis, and unforeseen immune complications. Ethical clinical practice requires transparent informed consent processes in which patients are fully apprised of these uncertainties, as well as the experimental nature of early-phase therapies. Given that many trials enroll patients with refractory or advanced cancers, the ethical imperative is to ensure that patients are not unduly influenced by hope or desperation, but rather make decisions grounded in a clear understanding of potential outcomes (Sugarman et al., 2015). Equity in patient selection also presents an ethical dimension. Autologous MSC–OV therapies are resource-intensive and individualized, potentially restricting access to patients who can afford or tolerate personalized manufacturing. Allogeneic approaches offer broader accessibility but may introduce immunologic variability, raising questions about fairness and optimal allocation of therapy. Institutions and regulatory bodies must consider strategies to ensure equitable access while maintaining rigorous patient selection criteria. Environmental and public health considerations further complicate ethical responsibilities. Viral shedding, whether through excreta or inadvertent release, may pose theoretical risks to family members, healthcare workers, or the community. Ethical stewardship requires rigorous monitoring, containment protocols, and patient education to minimize these risks. Furthermore, the dual-use nature of viral engineering demands careful oversight to prevent misuse, emphasizing the need for institutional safeguards and regulatory compliance.

Finally, ethical responsibility extends to long-term follow-up and transparency in research. Patients may require lifelong monitoring for delayed adverse events, including potential secondary malignancies or immune dysregulation. Equally important is the obligation of researchers to report both positive and negative outcomes, thereby guiding the field responsibly and preventing premature clinical adoption based on incomplete or biased evidence. In sum, MSC–OV therapy challenges traditional ethical frameworks, necessitating a nuanced approach that integrates patient safety, informed consent, equitable access, environmental stewardship, and research integrity (Emanuel et al., 2000; Jittprasong, 2022).

6.8 Innovations and future perspectives

The field of MSC–OV therapy is rapidly evolving, driven by advances in viral engineering, cellular biology, and synthetic biology. While early-generation MSC–OV platforms focused primarily on exploiting the natural tumor-homing capacity of MSCs to deliver conventional OVs, emerging strategies are increasingly sophisticated, aiming to enhance tumor specificity, control viral replication, and program MSC–virus interactions in a precise and predictable manner. These innovations promise to overcome many of the current limitations in safety, efficacy, and scalability, opening new avenues for personalized and combinatorial cancer therapy.

6.8.1 Next-generation viral engineering

Next-generation viral engineering represents a paradigm shift in the design of OVs for MSC-mediated delivery. Traditional OVs rely on relatively broad tumor tropism, often achieving partial tumor selectivity through natural viral receptor expression. However, off-target infection, premature viral release, and systemic immune activation remain major concerns. Advances in molecular virology now allow the incorporation of tumor-specific promoters and microRNA-responsive elements to create viruses that replicate selectively in cancer cells while remaining quiescent in normal tissues.

Tumor-specific promoters enable virus replication to be tightly coupled to molecular signatures characteristic of malignancy. For instance, promoters responsive to telomerase, survivin, or oncogenic pathways such as Ras and Myc can restrict viral gene expression to cancerous cells, minimizing toxicity in healthy organs (Fueyo et al., 2000; Kelly & Russell, 2007). Similarly, incorporation of microRNA target sequences into viral genomes allows posttranscriptional regulation of viral replication. In this approach, viruses are engineered to be suppressed in cells expressing microRNAs abundant in normal tissues but downregulated in tumor cells. This layer of control adds precision to MSC–OV therapy, ensuring that viral propagation occurs predominantly within malignant tissues, even if MSCs traffic through nontarget organs.

Beyond specificity, next-generation viral engineering includes modifications to enhance immunogenicity, evade neutralizing antibodies, and improve intratumoral spread. Incorporating immune-modulatory genes, such as cytokines, chemokines, or checkpoint-blockade molecules, into viral genomes allows the simultaneous induction of direct oncolysis and immune activation. Engineered fusogenic proteins can increase viral dissemination through solid tumor masses, overcoming barriers posed by dense extracellular matrix and hypoxic niches. Together, these strategies create highly refined viral constructs that synergize with MSC-mediated delivery, maximizing therapeutic impact while minimizing systemic toxicity.

6.8.2 Synthetic biology approaches for programmable mesenchymal stromal cell–virus interactions

Synthetic biology offers transformative potential for programming the interactions between MSCs and OVs, creating "smart" cellular vehicles capable of sensing, responding, and adapting to the TME. This approach moves beyond passive MSC delivery toward engineered, responsive systems that can dynamically modulate viral release, cytokine secretion, and immune engagement. One emerging strategy involves integrating synthetic genetic circuits within MSCs to control viral replication. For example, MSCs can be engineered to express inducible promoters that trigger viral release only in response to tumor-associated signals, such as hypoxia, extracellular matrix components, or inflammatory cytokines. This ensures that OVs remain sequestered within MSCs during systemic circulation, reducing off-target effects, and are only activated upon reaching the tumor site.

Another area of innovation is the design of feedback loops in MSCs that sense immune activation and adjust therapy accordingly. Synthetic circuits can be programmed to detect tumor-infiltrating immune cells, cytokine levels, or checkpoint ligand expression, and then modulate viral payloads, immunostimulatory molecule secretion, or MSC survival in real-time (Fedorov et al., 2019;Lim et al., 2018). This level of programmability allows MSC–OV therapy to adapt to the dynamic TME, potentially overcoming resistance mechanisms and enhancing efficacy. Synthetic biology also enables the codelivery of multiple therapeutic modalities within a single MSC carrier. MSCs can be engineered to carry combinations of OVs, immune modulators, or gene-editing constructs, creating multifunctional therapeutic units capable of simultaneously targeting tumor cells, modulating the immune landscape, and reprogramming stromal barriers. These innovations open the possibility of fully personalized, programmable therapies that integrate tumor biology, patient-specific immune responses, and viral engineering into a unified treatment strategy.

6.8.3 Combination therapies with chimeric antigen receptor-T, immune checkpoint blockade, or radiotherapy

The integration of MSC–OV therapy with other cancer treatment modalities represents a major frontier in oncology, aiming to exploit synergistic interactions to overcome the limitations of monotherapies. While MSC–OV therapy alone can induce tumor-selective oncolysis and immune activation, solid tumors often present barriers such as stromal density, immunosuppressive microenvironments, and heterogeneous antigen expression. By combining MSC–OVs with other therapies, these challenges can be strategically addressed to enhance both direct cytotoxicity and antitumor immunity.

6.8.3.1 Synergy with chimeric antigen receptor-T cell therapy

CAR-T cell therapy has revolutionized treatment for hematologic malignancies but faces significant obstacles in solid tumors, including poor T-cell infiltration, immunosuppressive signaling, and antigen heterogeneity (June et al., 2018; Newick et al., 2017). MSC-delivered OVs can ameliorate these challenges by reprogramming the TME. Viral infection of tumor cells triggers ICD, releasing DAMPs and tumor-associated antigens, which enhance CAR-T recognition and activation. In addition, OVs can disrupt the extracellular matrix and stromal barriers, facilitating CAR-T cell infiltration into dense tumor regions. Preclinical studies have demonstrated that combining MSC–OV therapy with CAR-T cells increases tumor cell killing, promotes CAR-T persistence, and reduces tumor escape via antigen loss.

6.8.3.2 Integration with immune checkpoint blockade

ICIs, such as anti-PD-1 or anti-CTLA-4 antibodies, rely on the presence of preexisting T-cell infiltration within the tumor for efficacy. Many solid tumors are immunologically "cold," with minimal cytotoxic T-cell presence. MSC–OV therapy can convert these tumors into "hot" tumors by inducing local inflammation, upregulating MHC expression, and recruiting effector immune cells. OV-mediated release of type I IFNs and proinflammatory cytokines enhances dendritic cell activation and antigen presentation, synergizing with checkpoint blockade to relieve T-cell exhaustion and reinvigorate antitumor responses. The combination of MSC–OVs and checkpoint inhibitors has shown enhanced survival and tumor regression in several preclinical models, highlighting its potential for translation into refractory cancer patients.

6.8.3.3 *Complementarity with radiotherapy*

Radiotherapy remains a cornerstone of cancer treatment but is limited by tumor hypoxia, radioresistance, and systemic toxicity. MSC–OV therapy can complement radiotherapy through multiple mechanisms. Ionizing radiation increases tumor cell susceptibility to viral infection by inducing DNA-damage and stress signaling pathways, which can facilitate viral entry and replication. MSC carriers enable precise localization of OVs to irradiated tumor regions, reducing off-target effects. Moreover, radiotherapy and viral oncolysis can synergistically amplify immunogenic signals, such as DAMPs and neoantigen release, enhancing both local and systemic antitumor immunity. This combinatorial strategy may allow dose reduction of radiation while maintaining efficacy, potentially mitigating long-term toxicity.

6.8.3.4 *Strategic considerations and clinical implications*

While combination therapies hold substantial promise, several strategic considerations are critical. Timing and sequencing of MSC–OV administration relative to CAR-T infusion, checkpoint blockade, or radiotherapy can dramatically influence outcomes. For instance, preconditioning the tumor with MSC–OVs may prime the microenvironment for CAR-T or checkpoint-mediated cytotoxicity, whereas post-treatment viral delivery may be less effective if immune clearance occurs prematurely. Dose optimization, toxicity management, and careful patient selection are also essential to minimize overlapping adverse effects, such as cytokine release syndrome or tissue inflammation.

6.8.4 Personalized oncolytic virotherapy using patient-derived mesenchymal stromal cells

Personalization is emerging as a cornerstone of MSC–OV therapy. Utilizing patient-derived MSCs allows the creation of autologous cellular vehicles that are fully compatible with the host immune system, thereby reducing the risk of immunogenic clearance and improving tumor-homing efficiency. Autologous MSCs can be loaded ex vivo with OVs tailored to the patient's tumor genotype and microenvironment, creating highly individualized therapies. The personalization process involves several key steps. First, MSCs are isolated from bone marrow, adipose tissue, or other accessible sources from the patient. These cells are then expanded under GMP-compliant conditions and loaded with an OV optimized for tumor selectivity and replication kinetics (Du et al., 2020). Additional modifications, such as inducible viral release systems or coexpression of immunomodulatory cytokines, can be incorporated to match the specific immunologic and molecular profile of the patient's tumor. This approach also enables the use of companion diagnostics to guide therapy. Tumor biopsies and genomic profiling inform the selection of viral strains, tumor-specific promoters, or microRNA-regulated circuits, ensuring that the MSC–OV therapy is finely tuned to the patient's disease. While personalized MSC–OV therapies are resource-intensive and require robust manufacturing pipelines, they represent a transformative approach to cancer therapy, potentially maximizing efficacy while minimizing systemic toxicity.

6.8.5 Prospects for artificial intelligence-assisted modeling and precision delivery systems

Artificial intelligence (AI) and computational modeling are poised to revolutionize MSC–OV therapy by enabling predictive, precision-guided strategies. AI can integrate multi-dimensional datasets, including viral kinetics, MSC migration patterns, TME features, and patient immune profiles, to model the dynamics of therapy in silico. These predictive models can optimize viral dosing, MSC infusion schedules, and combination strategies, reducing the reliance on trial-and-error approaches and enhancing translational efficiency (Coley et al., 2021). AI-assisted approaches can also support **precision delivery systems**. For example, computational modeling of MSC trafficking can guide the selection of chemokine receptor modifications or exogenous stimuli that improve tumor homing. Machine learning algorithms can predict patient-specific responses to different OV strains, informing personalized therapy design and minimizing adverse effects (Chung et al., 2020). In the future, integration of AI with real-time imaging and biomarker monitoring may enable adaptive MSC–OV therapies, where viral release, cytokine expression, or MSC survival are dynamically adjusted based on tumor response. Beyond individual patient optimization, AI can accelerate **high-throughput screening** of MSC–OV constructs, predicting the best combinations of viral modifications, MSC sources, and immune modulators. This convergence of biotechnology and computational science promises to make MSC–OV therapy more efficient, precise, and responsive to patient-specific tumor biology, paving the way for the next generation of intelligent, programmable cancer therapeutics.

6.9 Conclusion

MSC–OV therapy represents a convergence of cellular and viral engineering that has the potential to transform the landscape of cancer treatment, particularly for metastatic and refractory malignancies. Across the biological, mechanistic, and translational dimensions, several key insights emerge from the current body of research. Biologically, MSCs provide a versatile and tumor-homing delivery platform capable of protecting OVs from premature immune clearance, enhancing intratumoral viral distribution, and facilitating localized immunomodulation. Mechanistically, the interplay between MSC-mediated viral delivery, tumor-selective viral replication, and immune activation establishes a multifaceted antitumor network, encompassing direct oncolysis, disruption of stromal barriers, and recruitment of innate and adaptive immune responses. Translational studies, ranging from sophisticated in vitro systems to animal models and early-phase clinical trials, provide evidence of feasibility, therapeutic efficacy, and safety, while highlighting the challenges inherent in immunogenicity, heterogeneity, and regulatory compliance.

The promise of MSC–OV systems lies not only in their ability to deliver viral therapies more effectively but also in their potential to synergize with other immunotherapeutic strategies. By combining targeted viral oncolysis with checkpoint blockade, CAR-T therapy, or radiotherapy, these systems can overcome microenvironmental resistance mechanisms and generate robust, systemic antitumor immunity. Innovations in viral engineering, synthetic biology, and precision medicine are further expanding the capabilities of MSC–OV therapies, enabling tumor-specific activation, programmable cellular circuits, and patient-tailored strategies that optimize efficacy while minimizing off-target toxicity. Looking forward, the path toward clinical integration is contingent on addressing the remaining scientific and logistical challenges. Standardization of MSC sources, scalable manufacturing processes, and rigorous long-term safety assessments will be essential to translating these therapies from proof-of-concept studies into widely accessible clinical interventions. Personalized approaches, leveraging patient-derived MSCs and AI-guided treatment optimization, are likely to play a pivotal role in realizing the full potential of MSC–OV therapy as part of a precision immunotherapy paradigm.

Glossary

Adoptive cell therapy (ACT) A cancer treatment strategy in which immune cells are collected from a patient or donor, engineered or expanded ex vivo, and then infused back to attack tumor cells.

Allogeneic MSCs Mesenchymal stromal cells derived from a donor other than the patient, often used in "off-the-shelf" therapies.

Autologous MSCs Mesenchymal stromal cells derived from the same patient, reducing risk of immune rejection and improving compatibility.

Cytokine release syndrome (CRS) A potentially severe systemic inflammatory response caused by rapid immune activation, often observed in cell and viral therapies.

Engineered MSCs MSCs that have been genetically modified to enhance therapeutic properties, such as viral delivery, immunomodulation, or tumor targeting.

Extracellular matrix (ECM) A network of proteins and molecules surrounding cells in tissues, often forming barriers that impede drug or viral delivery in tumors.

Homing The process by which MSCs migrate preferentially to tumor sites, guided by chemokine gradients and adhesion molecules.

Immune checkpoint inhibitors (ICIs) Therapeutic antibodies that block inhibitory immune pathways (e.g., PD-1/PD-L1 and CTLA-4), enhancing T-cell-mediated tumor killing.

Insertional mutagenesis Genetic alterations caused by the integration of foreign DNA, which can disrupt normal genes and potentially induce malignancy.

Mesenchymal stromal cells (MSCs) Multipotent cells capable of differentiating into bone, fat, and cartilage, and are often used as carriers for therapeutic agents due to their tumor-homing ability.

Multiplicity of infection (MOI) The ratio of viral particles to target cells during infection, influencing viral uptake, replication, and cytotoxicity.

Oncolytic virus (OV) A virus engineered or naturally capable of selectively infecting and killing cancer cells while sparing normal tissue.

Programmed viral release A strategy in which MSCs are engineered to release oncolytic viruses in response to specific tumor signals, ensuring targeted viral propagation.

Preclinical models Laboratory-based experiments, including cell cultures and animal studies, are used to evaluate safety, efficacy, and mechanisms before human clinical trials.

Stromal barrier Structural components of the tumor microenvironment, including fibroblasts and ECM, that can restrict immune cell or viral infiltration.

Synthetic biology A field of science that designs and constructs new biological parts, devices, or systems, often used to program MSCs or viral behavior.

Tumor microenvironment (TME) The complex network of cancer cells, stromal cells, immune cells, and extracellular components surrounding a tumor, influencing growth and therapy response.

Tumor-specific promoter A DNA sequence controlling gene expression selectively in tumor cells, used in engineered viruses to restrict replication to malignant tissues.

Viral shedding The release of virus from infected cells into bodily fluids or the environment, which can pose biosafety considerations.

References

Alemany, R., & Cascallo, M. (2009). Oncolytic viruses from the perspective of cancer treatment. *Cancer Letters, 285*(1), 1–13. https://doi.org/10.1016/j.canlet.2009.04.009.

Alipour, M., Baneshi, M., Hosseini, S. M., & Ghaffari, S. (2023). Investigating the potential of oncolytic viruses for cancer treatment via mesenchymal stem cell delivery. *Cell Communication and Signaling, 21*. https://doi.org/10.1186/s12964-023-01232-y. Article 232.

Ben-David, U., Gan, Q. F., Golan-Lev, T., Arad, N., & Benvenisty, N. (2011). Genetic and epigenetic instability in human pluripotent stem cells. *Nature Reviews Genetics, 12*(4), 266–276. https://doi.org/10.1038/nrg2952.

Bourin, P., Bunnell, B. A., Casteilla, L., Dominici, M., Katz, A. J., March, K. L., Redl, H., Rubin, J. P., Yoshimura, K., & Gimble, J. M. (2013). Stromal cells from the adipose tissue-derived stromal vascular fraction and culture expanded adipose tissue-derived stromal/stem cells: A joint statement of the International Federation for Adipose Therapeutics and Science (IFATS) and the International Society for Cellular Therapy (ISCT). *Cytotherapy, 15*(6), 641–648. https://doi.org/10.1016/j.jcyt.2013.02.006.

Chung, E., O'Connor, R. S., & June, C. H. (2020). Engineering multifunctional cellular therapies for cancer. *Nature Reviews Cancer, 20*(12), 685–702. https://doi.org/10.1038/s41568-020-00307-7.

Coley, T., Singh, R., & Dey, S. (2021). Artificial intelligence in oncolytic virotherapy: Predictive modeling for precision medicine. *Frontiers in Oncology, 11*, 650012. https://doi.org/10.3389/fonc.2021.650012.

DePeaux, K., & Delgoffe, G. M. (2024). Integrating innate and adaptive immunity in oncolytic virus therapy. *Trends in Cancer, 10*(2), 135–146. https://doi.org/10.1016/j.trecan.2023.09.012.

Detela, G., & Lodge, A. (2019). EU regulatory pathways for ATMPs: Standard, accelerated and adaptive pathways. *Molecular Therapy – Methods & Clinical Development, 13*, 205–232. https://doi.org/10.1016/j.omtm.2019.01.010.

Du, X., Fang, M., Li, H., & Wang, J. (2020). Mesenchymal stem cells as carriers for systemic delivery of oncolytic viruses. *European Journal of Pharmacology, 874*, 172991. https://doi.org/10.1016/j.ejphar.2020.172991.

Duebgen, M., Martinez-Quintanilla, J., Tamura, K., Hingtgen, S. D., Redjal, N., Wakimoto, H., & Shah, K. (2014). Stem cells loaded with oncolytic herpes simplex virus for brain tumor therapy. *Journal of the National Cancer Institute, 106*(6), dju090. https://doi.org/10.1093/jnci/dju090.

El Omar, R., Dazzi, F., & Krampera, M. (2020). Mesenchymal stromal cells in regenerative medicine: Focus on endogenous repair. *Stem Cells, 38*(6), 646–654. https://doi.org/10.1002/stem.3181.

Emanuel, E. J., Wendler, D., & Grady, C. (2000). What makes clinical research ethical? *JAMA, 283*(20), 2701–2711. https://doi.org/10.1001/jama.283.20.2701.

Fedorov, V. D., Sadelain, M., & Brentjens, R. (2019). Programmable cellular therapeutics: Synthetic biology in cancer immunotherapy. *Nature Reviews Clinical Oncology, 16*(10), 567–580. https://doi.org/10.1038/s41571-019-0215-7.

Ferguson, M. S., Lemoine, N. R., & Wang, Y. (2012). Systemic delivery of oncolytic viruses: Hopes and hurdles. *Advances in Virology, 2012*. https://doi.org/10.1155/2012/805629.

Fidler, I. J. (2003). The pathogenesis of cancer metastasis: The 'seed and soil' hypothesis revisited. *Nature Reviews Cancer, 3*(6), 453–458. https://doi.org/10.1038/nrc1098.

Fischer, U. M., Harting, M. T., Jimenez, F., Monzon-Posadas, W. O., Xue, H., Savitz, S. I., & Cox, C. S. (2009). Pulmonary passage is a major obstacle for intravenous stem cell delivery. *Stem Cells and Development, 18*(5), 683–692. https://doi.org/10.1089/scd.2008.0253.

Fueyo, J., Alemany, R., Gomez-Manzano, C., Fuller, G. N., Khan, A., Conrad, C., & Levin, V. A. (2000). Preclinical characterization of oncolytic adenovirus targeting telomerase-positive tumor cells. *Clinical Cancer Research, 6*(8), 3397–3404. https://doi.org/10.1016/S0304-3835(00)00142-3.

Galluzzi, et al. (2020). *Journal for ImmunoTherapy of Cancer* In press. doi:10.1136/jitc-2019-000337.

Galluzzi, L., Buqué, A., Kepp, O., Zitvogel, L., & Kroemer, G. (2017). Immunogenic cell death in cancer and infectious disease. *Nature Reviews Immunology, 17*(2), 97–111. https://doi.org/10.1038/nri.2016.107.

García-Castro, J., Alemany, R., Cascalló, M., Martínez-Quintanilla, J., Del Mar Arriero, M., Lassaletta, A., Madero, L., & Ramírez, M. (2010). Treatment of metastatic neuroblastoma with systemic oncolytic virotherapy delivered by autologous mesenchymal stem cells: An exploratory study. *Cancer Gene Therapy, 17*(7), 476–483. https://doi.org/10.1038/cgt.2010.4.

Gujar, S., Pol, J. G., Kroemer, G., & Zitvogel, L. (2019). Antitumor benefits of oncolytic viruses. *Nature Reviews Immunology, 19*(6), 355–368. https://doi.org/10.1038/s41577-019-0142-5.

Guo, Z. S., Thorne, S. H., & Bartlett, D. L. (2008). Oncolytic virotherapy: Molecular targets in tumor-selective replication and carrier cell-mediated delivery of oncolytic viruses. *Biochimica et Biophysica Acta - Reviews on Cancer, 1785*(2), 217–231. https://doi.org/10.1016/j.bbcan.2008.02.001.

Hacein-Bey-Abina, S., Von Kalle, C., Schmidt, M., McCormack, M. P., Wulffraat, N., Leboulch, P., Lim, A., Osborne, C. S., Pawliuk, R., Morillon, E., Sorensen, R., Forster, A., Fraser, P., Cohen, J. I., de Saint Basile, G., Alexander, I., Wintergerst, U., Frebourg, T., Aurias, A., … Cavazzana-Calvo, M. (2003). LMO2-associated clonal T cell proliferation in two patients after gene therapy for SCID-X1. *Science, 302*(5644), 415–419. https://doi.org/10.1126/science.1088547.

Hakkarainen, T., Hemminki, A., & Hemminki, O. (2015). Viral entry receptors and intracellular barriers in oncolytic virotherapy. *Current Opinion in Virology, 13*, 49–56. https://doi.org/10.1016/j.coviro.2015.03.006.

Hammad, M., Cornejo, Y. R., & Clements, D. (2016). Enhancing oncolytic virus replication in mesenchymal stem cells through modulation of innate antiviral pathways. *Molecular Therapy – Oncolytics, 3*, 16002. https://doi.org/10.1038/mto.2016.2.

Harrington, K. J., Galanis, E., & Kirn, D. H. (2019). Clinical development of oncolytic viruses: Synergistic immunotherapy strategies. *The Lancet Oncology, 20*(8), e426–e437. https://doi.org/10.1016/S1470-2045(19)30324-2.

Iankov, I. D., Blechacz, B., Liu, C., Schmeckpeper, J. D., Tarara, J. E., Federspiel, M. J., Caplice, N., Russell, S. J., & Galanis, E. (2007). Infected cell carriers for systemic delivery of oncolytic measles virus. *Molecular Therapy, 15*(1), 114–122. https://doi.org/10.1038/sj.mt.6300014.

Jittprasong C., Mesenchymal stem cells as carrier cells to enable effective intratumoral delivery of oncolytic virus for oncolytic virotherapy: A systematic review. Hong Kong arXiv. (2022). doi: https://doi.org/10.48550/arxiv.2204.01238.

June, C. H., O'Connor, R. S., Kawalekar, O. U., Ghassemi, S., & Milone, M. C. (2018). CAR T cell immunotherapy for human cancer. *Science, 359*(6382), 1361–1365. https://doi.org/10.1126/science.aar6711.

Kang, S. K., Shin, I. S., Ko, M. S., Jo, J. Y., & Ra, J. C. (2012). Journey of mesenchymal stem cells for homing: Strategies to enhance efficacy and safety of stem cell therapy. *Stem Cells International*. https://doi.org/10.1155/2012/342968.

Karnoub, A. E., Dash, A. B., Vo, A. P., Sullivan, A., Brooks, M. W., Bell, G. W., Richardson, A. L., Polyak, K., Tubo, R., & Weinberg, R. A. (2007). Mesenchymal stem cells within tumour stroma promote breast cancer metastasis. *Nature, 449*(7162), 557–563. https://doi.org/10.1038/nature06188.

Kaufman, H. L., Kohlhapp, F. J., & Zloza, A. (2015). Oncolytic viruses: A new class of immunotherapy drugs. *Nature Reviews Drug Discovery, 14*(9), 642–662. https://doi.org/10.1038/nrd4663.

Kelly, E., & Russell, S. J. (2007). History of oncolytic viruses: Genesis to genetic engineering. *Molecular Therapy, 15*(4), 651–659. https://doi.org/10.1038/sj.mt.6300108.

Kidd, S., Spaeth, E., Dembinski, J. L., Dietrich, M., Watson, K., Klopp, A., Battula, V. L., Weil, M., Andreeff, M., & Marini, F. C. (2009). Direct evidence of mesenchymal stem cell tropism for tumor and wounding microenvironments using in vivo bioluminescent imaging. *Stem Cells, 27*(10), 2614–2623. https://doi.org/10.1002/stem.187.

Lim, W. A., June, C. H., & Chen, D. S. (2018). Engineering synthetic circuits in immune cells for cancer therapy. *Cell, 175*(2), 313–320. https://doi.org/10.1016/j.cell.2018.09.022.

Lin, C. Z., Xiang, G. L., Zhu, X. H., Xiu, L. L., Sun, J. X., & Zhang, X. Y. (2018). Advances in the mechanisms of action of cancer-targeting oncolytic viruses (review). *Oncology Letters, 15*(4), 4053–4060. https://doi.org/10.3892/ol.2018.7829.

Lin, D., Shen, Y., & Liang, T. (2023). Oncolytic virotherapy: Basic principles, recent advances and future directions. *Signal Transduction and Targeted Therapy, 8*. Article 156 https://doi.org/10.1038/s41392-023-01407-6 .

Mader, E. K., et al. (2009). *Cancer Research*. doi:10.1158/0008-5472.CAN-08-3867.

Mader, E. K., Butler, G., Dowdy, S. C., Mariani, A., Knutson, K. L., Federspiel, M. J., & Dietz, A. B. (2013). Optimizing patient-derived mesenchymal stem cells as virus carriers for a phase I clinical trial in ovarian cancer. *Journal of Translational Medicine, 11*, 20. https://doi.org/10.1186/1479-5876-11-20.

Mader, E. K., Maeyama, Y., Lin, Y., Butler, G. W., Russell, H. M., Galanis, E., & Cattaneo, R. (2013). Mesenchymal stem cell carriers protect oncolytic measles viruses from antibody neutralization in an orthotopic ovarian cancer therapy model. *Clinical Cancer Research, 15*(23), 7246–7257. https://doi.org/10.1158/1078-0432.CCR-13-0197.

Mondal, M., Guo, J., He, P., & Zhou, D. (2020). Recent advances of oncolytic virus in cancer therapy. *Human Vaccines & Immunotherapeutics, 16*(10), 2389–2402. https://doi.org/10.1080/21645515.2020.1723363.

Newick, K., O'Brien, S., Moon, E., & Albelda, S. M. (2017). CAR T cell therapy for solid tumors. *Annual Review of Medicine, 68*, 139–152. https://doi.org/10.1146/annurev-med-062315-120245.

Nguyen, T. L., Abdelbary, H., Arguello, M., Breitbach, C., Leveille, S., Diallo, J.-S., & Bell, J. C. (2008). Tumor cells and the antiviral innate immune response: Implications for oncolytic virus therapy. *Molecular Therapy, 16*(8), 1349–1356. https://doi.org/10.1038/mt.2008.100.

Nissim, L., & Bar-Ziv, R. H. (2010). A tunable dual-promoter integrator for targeting of cancer cells. *Molecular Systems Biology, 6*. https://doi.org/10.1038/msb.2010.99.

Phinney, D. G., & Sensebe, L. (2013). Mesenchymal stromal cells: Skeletal therapeutics and beyond. *Cell Stem Cell, 12*(5), 520–536. https://doi.org/10.1016/j.stem.2013.04.022.

Russell, L. (2018). The emerging role of oncolytic virus therapy against cancer. *Translational Cancer Research, 7*(5), 1081–1095. https://doi.org/10.21037/tcr.2018.08.09.

Russell, S. J., Peng, K. W., & Bell, J. C. (2012). Oncolytic virotherapy. *Nature Biotechnology, 30*(7), 658–670. https://doi.org/10.1038/nbt.2287.

Sahu, A., & Biswas, N. (2010). JX-594: A targeted oncolytic poxvirus for the treatment of cancer. *Journal of Virological Methods, 168*(1–2), 7–17. https://doi.org/10.1016/j.jviromet.2010.07.021.

Shah, K. (2012). Mesenchymal stem cells engineered for cancer therapy. *Advanced Drug Delivery Reviews, 64*(8), 739–748. https://doi.org/10.1016/j.addr.2011.06.010.

Sharma, P., Hu-Lieskovan, S., Wargo, J. A., & Ribas, A. (2017). Primary, adaptive, and acquired resistance to cancer immunotherapy. *Cell, 168*(4), 707–723. https://doi.org/10.1016/j.cell.2017.01.017.

Siegel, R. L., Giaquinto, A. N., & Jemal, A. (2024). Cancer statistics, 2024. *CA Cancer Journal for Clinicians, 74*(1), 12–49. https://doi.org/10.3322/caac.21820.

Singh, P. K., Doley, J., Kumar, G. R., Sahoo, A. P., & Tiwari, A. K. (2012). Oncolytic viruses & their specific targeting to tumour cells. *Indian Journal of Medical Research, 136*(4), 571–584. http://www.icmr.nic.in/ijmr/2012/october/1005.pdf.

Spaeth, E., Klopp, A., Dembinski, J., Andreeff, M., & Marini, F. (2008). Inflammation and tumor microenvironments: Defining the migratory itinerary of mesenchymal stem cells. *Gene Therapy, 15*(10), 730–738. https://doi.org/10.1038/gt.2008.39.

Sugarman, J., & Califf, R. (2015). Ethics and early-phase clinical research: Balancing risk and benefit. *New England Journal of Medicine, 372*(24), 2381–2385. https://doi.org/10.1056/NEJMms1414922.

Uccelli, A., Moretta, L., & Pistoia, V. (2008). Mesenchymal stem cells in health and disease. *Nature Reviews Immunology, 8*(9), 726–736. https://doi.org/10.1038/nri2395.

Veglia, F., Sanseviero, E., & Gabrilovich, D. I. (2018). Myeloid-derived suppressor cells in the era of increasing myeloid cell diversity. *Nature Reviews Immunology, 18*(8), 573–588. https://doi.org/10.1038/s41577-018-0002-4.

Wang, X., Huang, X., Deng, X., Zhang, Y., Zhou, Y., Li, Q., Liu, Z., Chen, J., Liu, Y., Xu, M., Li, H., Wu, J., Sun, S., Zhao, Y., Yang, J., Hu, X., Guo, Z., Lin, X., He, Q., … Chen, L. (2021). Mesenchymal stem cell carriers enhance anti-tumor efficacy of oncolytic virotherapy. *Journal of Cancer Research and Therapeutics*. https://pubmed.ncbi.nlm.nih.gov/33664802.

Wang, Y., Hu, X., An, J., Zhang, A., Wen, L., Chen, S., Liu, H., Zhang, Z., Li, W., Liu, Y., Duan, Y., Zhang, X., Wang, B., Li, X., Zhang, Y., Liu, J., Wang, L., Chen, H., Li, Y., … Zhang, X. (2018). Mesenchymal stem cells for enhanced oncolytic virotherapy: Progress and dilemmas. *Frontiers in Oncology, 8*, 163. https://doi.org/10.3389/fonc.2018.00163.

Waterman, R. S., Tomchuck, S. L., Henkle, S. L., & Betancourt, A. M. (2010). A new mesenchymal stem cell (MSC) paradigm: Polarization into a pro-inflammatory MSC1 or an immunosuppressive MSC2 phenotype. *PLoS ONE, 5*(4). https://doi.org/10.1371/journal.pone.0010088.

Yong, R. L., Shinojima, N., Fueyo, J., & Gumin, J. (2009). Human bone marrow-derived mesenchymal stem cells for intravascular delivery of oncolytic adenovirus. *Cancer Research, 69*(23), 8932–8940. https://doi.org/10.1158/0008-5472.CAN-09-1073.

Zamarin, D., & Palese, P. (2012). Oncolytic Newcastle disease virus for cancer therapy: Old challenges and new directions. *Future Microbiology, 7*(3), 347–367. https://doi.org/10.2217/fmb.12.4.

CHAPTER

7 Cancer treatment advancements in the era of immune checkpoint inhibitors via cellular immunotherapies

Hariharan Thirumalai Vengateswaran[a], Mohammad Habeeb[a], Huay Woon You[b,c], Kiran Balasaheb[d], Girija Balasaheb Bhavar[e] and Govind Sarangdhar Asane[f]

[a]*Department of Pharmaceutics, Crescent School of Pharmacy, B.S. Abdur Rahman Crescent Institute of Science and Technology, Chennai, Tamil Nadu, India,* [b]*Faculty of Science and Technology, Universiti Kebangsaan Malaysia, Bangi, Selangor, Malaysia,* [c]*Pusat PERMATA@Pintar Negara, Universiti Kebangsaan Malaysia, Bangi, Selangor, Malaysia,* [d]*Department of Pharmaceutical Quality Assurance, School of Pharmacy and Technology Management, SVKM NMIMS Global University, Dhule, Maharashtra, India,* [e]*Department of Pharmaceutical Chemistry, School of Pharmacy and Technology Management, SVKM NMIMS Global University, Dhule, Maharashtra, India,* [f]*Department of Pharmaceutics, SNJB Shriman Sureshdada Jain College of Pharmacy, Neminagar, Chandwad, Nashik, Maharashtra, India*

7.1 Introduction

Over the past 10 years, cancer has become an escalating global health concern, marked by rising incidence and mortality rates despite progress in treatment approaches. Spontaneous mutational diversity in most human malignancies presents challenges for therapies targeting specific mutations, as tumors can rapidly adapt and develop resistance. Conventional cancer therapy, such as chemotherapy, radiation treatment, and surgery, often falls short because of limited selectivity, systemic side effects, and inefficacy in advanced or resistant cases, leading to suboptimal outcomes in many patients. However, this diversity simultaneously reveals a variety of tumor-specific antigens that the immune system may be able to recognize and target. These antigens offer a pathway for more personalized and effective therapies, such as immunotherapies, which leverage the body's natural defenses to achieve precision targeting and sustained responses (Garg et al., 2024).

Immunotherapy is a cancer therapy approach that influences the body's immune system to detect and terminate tumor cells. Immunotherapy's great tumor suppression, excellent selectivity, and minimal side effects have made it an immensely appealing cancer treatment approach, offering distinct advantages over conventional modalities. Immune checkpoint inhibitors (ICIs) primarily target inhibitory pathways such as cytotoxic T-lymphocyte-associated protein 4 (CTLA-4) and programmed cell death protein-1 (PD-1) or its ligand programmed death-ligand 1 (PD-L1), which are often exploited by tumors to evade immune surveillance. This approach enhances immune cell activity, particularly of T cells, against tumors. Immune checkpoint blockers (ICBs), such as CTLA-4, PD-1, and its main ligand PD-L1,

Advancements in Personalized Cellular Immunotherapy: A Comprehensive Guide to Cancer Treatment.
DOI: https://doi.org/10.1016/B978-0-443-27731-3.00004-8

Copyright © 2026 Elsevier Inc. All rights are reserved, including those for text and data mining, AI training, and similar technologies.

have shown remarkable potential in restoring immune functionality. These therapies target molecular pathways that prevent the initiation and activity of both innate and adaptive immune cells capable of identifying and combating cancer. Moreover, the success of ICIs in treating malignancies, such as melanoma, non-small cell lung cancer (NSCLC), and RCC, has revolutionized the oncology landscape (Habeeb et al., 2024a). They not only provide durable clinical responses but also provide the way for combination strategies, integrating ICIs with other therapeutic modalities to overcome resistance and improve patient outcomes (Desai et al., 2024).

Engineering autologous T cells to show a chimeric antigen receptor (CAR) that targets CD19 has emerged as another innovative approach, offering a new method for effective cancer treatment. CD19-specific CAR-T cells have demonstrated significant therapeutic effectiveness in individuals with recurrent/refractory large B-cell lymphoma, achieving outcomes that were previously unattainable with traditional therapies. Similarly, these engineered T cells have been extremely active in treating individuals with B-cell acute lymphoblastic leukemia (ALL) who do not respond to conventional treatments. Beyond CAR-T cells, other immunotherapeutic strategies are actively being investigated, including immunostimulatory monoclonal antibodies and small molecules designed to reverse immunosuppression commonly associated with cancer progression. Tumor-specific therapeutic vaccines are under clinical evaluation and are aimed at eliciting a robust and sustained immune response against tumor cells (Zhang & Zhang, 2020). The aim of this chapter is to provide a complete exploration of the mechanisms of action of ICIs and immune system modulators within the field of cancer immunotherapy. In addition, this chapter emphasizes the importance of integrating artificial intelligence (AI) to enhance immunotherapy research, streamline drug development, and optimize clinical applications.

7.2 Tumor microenvironment in cancer

The tumor microenvironment (TME), which is essential in expressing cancer growth and spread, has a substantial influence on the genesis and progression of cancer. Blood arteries, immunosuppressive cells, and a fibrous extracellular matrix are all part of the interactions among tumor cells and their microenvironment, which protects tumor tissue from immune system invasions. These interactions eventually promote tumor growth and metastasis by promoting clonal evolution, multidrug resistance, and cancer cell heterogeneity. The clinical and pathological importance of these interactions in forecasting patient outcomes and treatment efficacy has been shown by research. Cancer growth and progression are greatly influenced by the mutualistic and dynamic connection between cancer cells and the surrounding matrix (Zhang & Zhang, 2020). Cancer patients have been found to differ in the nature and activity of these cell types. By triggering genomic unsteadiness, bolstering cancer stem cells, encouraging metastasis, and inhibiting adaptive immune responses (Fig. 7.1), Tumor-associated macrophages (TAMs), in particular, are essential for the advancement of tumors. Their distinct functional roles within the TME are highlighted by particular surface markers, such as the hemoglobin scavenger receptor (CD163) and macrophage mannose receptor 1 (CD206) (Roy et al., 2023).

7.3 Cancer immunotherapy

Cancer immunotherapy is a therapy that leverages the body's immune system to identify and terminate tumor cells by enhancing its natural ability to recognize tumor-specific antigens.

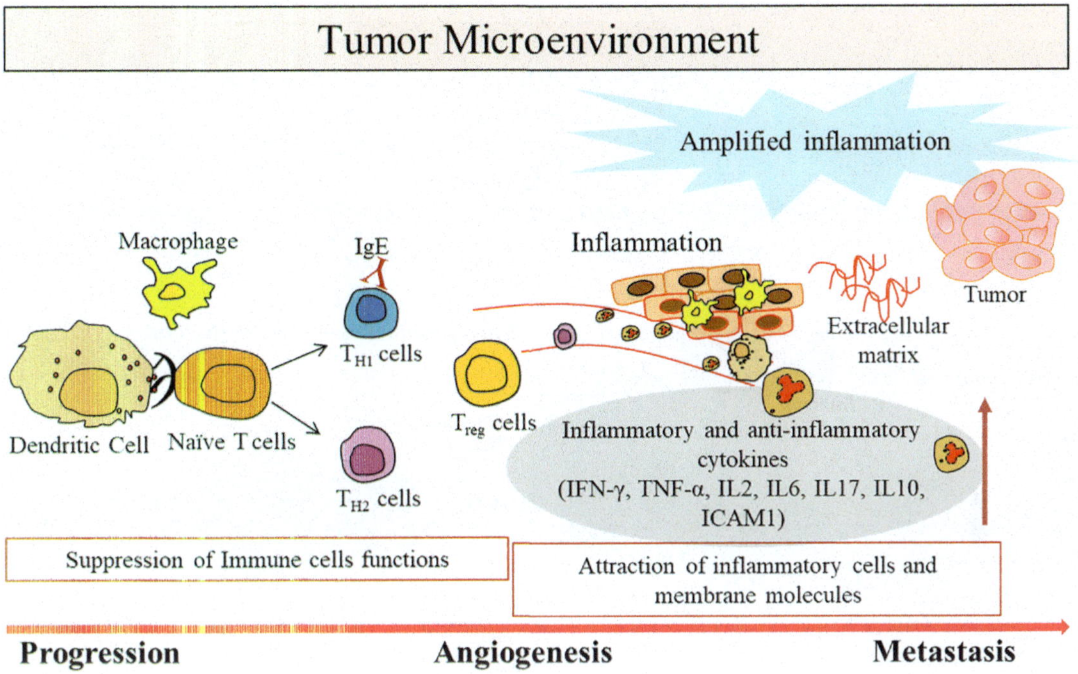

FIGURE 7.1

Illustration of the tumor's progression shows immune function suppression and tumor cell proliferation leading to metastasis.

Reproduced with permission from Roy, R., Singh, S. K., & Misra, S. (2023). Advancements in cancer immunotherapies. Vaccines, 11*(1), 59. https://doi.org/10.3390/vaccines11010059.*

7.3.1 Mechanisms of action in immunotherapy

T-lymphocyte–based cancer immunotherapies have arisen as a key purpose for enlisting the immune system to combat tumors. As an outcome, new investigations, including cancer vaccines, adoptive cellular therapy, and checkpoint inhibition, are ongoing. Cancer vaccines strengthen the immune system to combat tumor cells. When the immune system encounters an unidentified or foreign material, it indicates a risk, but by default, it reacts chemotactically to the recognized or self-substance to function in our system. Tumor-associated antigens (TAAs), neoantigens, or cancer-specific antigens are substances that are shown on the surface of tumor cells but not by healthy cells (Yin et al., 2023). Cytolytic T lymphocytes (CTL) identify these TAAs. The individuals' autologous, tumor-directed, and specific CD8+ CTL detect the melanoma-related antigen, MZ2-E, which is an antagonist antigen. When mouse cells introduced with an HLA-A1 gene presented a nonapeptide, CTL identified it, assuring the connection of the antigen MZ2-E by the HLA-A1 molecule. Therapeutic vaccine-based techniques can benefit from the fact that TAAs are expressed more in tumor cells and less in normal tissues. To stop advanced malignancies from growing further, immune effector mechanisms that aim tumor cells specifically may be developed (Fig. 7.2). Historically, autologous tumor vaccines made from

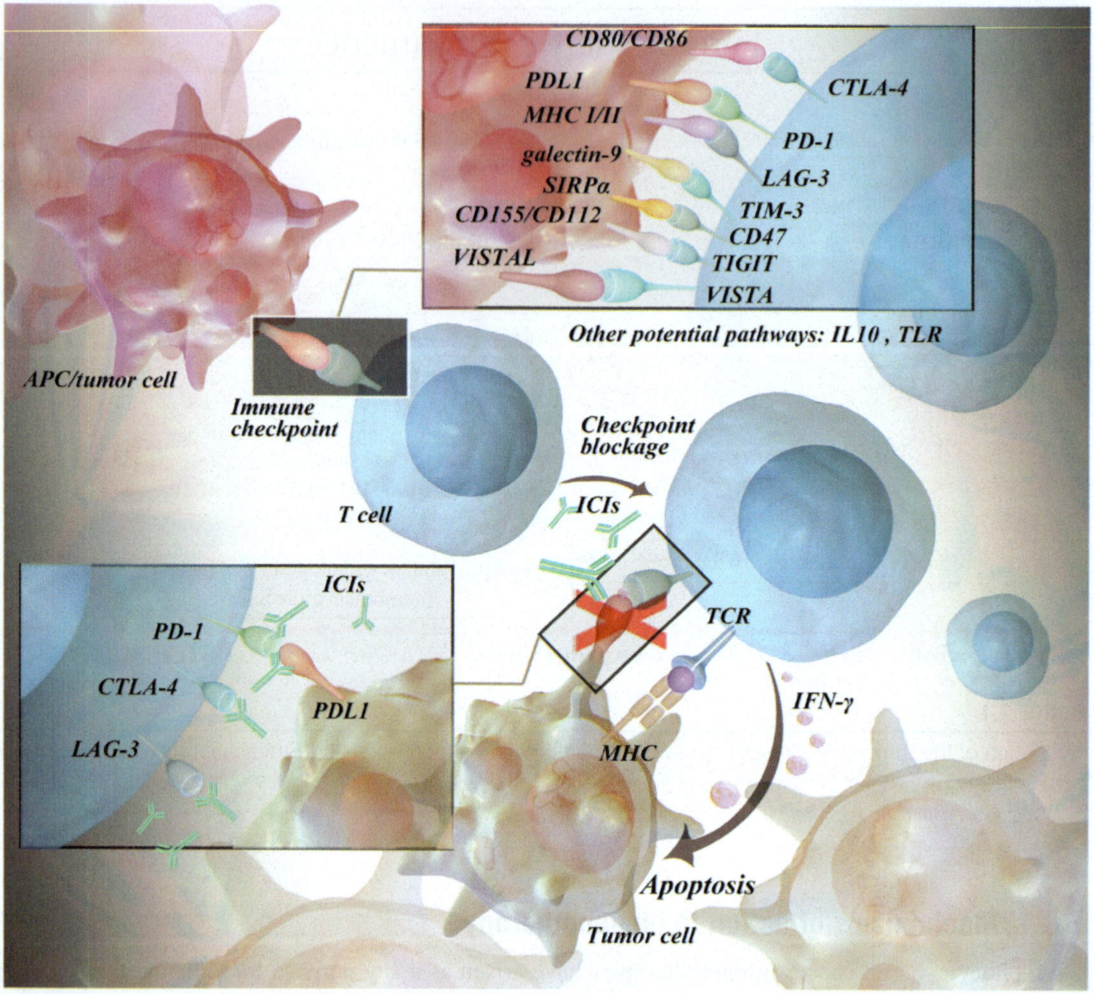

FIGURE 7.2

Immune checkpoint mechanisms of action.

Reproduced with permission from Yin, Q., Wu, L., Han, L., Zheng, X., Tong, R., Li, L., Bai, L., & Bian, Y. (2023). Immune-related adverse events of immune checkpoint inhibitors: A review. Frontiers in Immunology, 14, *1167975. https://doi.org/10.3389/fimmu.2023.1167975.*

patient-derived cancer cells served as the foundation for immunization strategies in the 1970s. To enhance immune responses, these tumor cells were irradiated and then given to the person from whom they were separated, along with a virus or adjuvant. However, this method has numerous drawbacks because tumor specimens are not readily available, particularly in cases of NSCLC (Habeeb et al., 2022). Vaccines must satisfy the standards of inducing higher vigorous immune responses by not producing autoimmune-related harm (Roy et al., 2023).

Table 7.1 Different features of the traditional and emerging ICIs.

Aspect	Traditional ICIs	Emerging ICIs	References
Target diversity	Primarily T cells (e.g., PD-1 and CTLA-4)	Broader immune cells (NK cells, B cells)	Veigas et al. (2021)
Resistance	Limited efficacy in resistant cancers	Potential to overcome resistance by targeting new pathways	Cai et al. (2023)
Mechanism precision	Simple, well-established pathways	Enhanced precision through interaction with diverse immune checkpoints	Joller et al. (2024)
Therapeutic scope	Effective for specific cancers	Promising for refractory and diverse tumor types	Waldman et al. (2020)
Adverse effects	Higher risk of immune-related toxicity	Potentially reduced off-target effects because of selective targeting	Lao et al. (2022)

CTLA-4, *cytotoxic T-lymphocyte-associated protein 4;* ICIs, *immune checkpoint inhibitors;* NK cells, *natural killer cells;* PD-1, *programmed cell death protein-1.*

7.4 Immune checkpoint inhibitors

ICIs are a class of immunotherapies discovered to inhibit particular proteins on immune cells or tumor cells that act as checkpoints, which the body uses to regulate immune responses and prevent autoimmunity. By constraining these checkpoints, ICIs help to restore the immune system's capacity to recognize and attack malignant cells effectively (Table 7.1).

7.4.1 Traditional immune checkpoint inhibitors: Programmed cell death protein 1, programmed death-ligand 1, cytotoxic T-lymphocyte-associated protein 4

The immunological response to an emerging antigen is significantly influenced by T cells, which are accountable for triggering a particular immune response. Relevant ligand molecules called lymphocyte surface receptors are activated upon coming into contact with an antigen-presenting cell (APC). In addition to an indication from costimulators that are activated through the establishment of an immunological synapse, specific recognition of the antigen presented is necessary for cell activation. Cell differentiation antigens belonging to the CD28 family may be costimulators on the layer of lymphocyte cells.

The negative feedback mechanism causes TCR/BCR blocking and a decrease in the number of cytokines generated (in addition to suppressing IL-10) when the PD-1 receptor is activated by ligand indication. In addition, it inhibits the production of antiapoptotic proteins, including Bcl-xL and Bcl-2. It has also been shown how PD-1 activation affects the cell cycle. S-phase kinase-associated protein 2 (SKP2) transcription and the G1 phase transition are inhibited by increased p15 protein expression. This gene codes for the ubiquitin ligase protein component, which breaks down the p27 tumor suppressor. The T-cell phenotype known as "exhausted" is caused by PD-1 overexpression through extended antigenic stimulation, such as carcinogenesis or long-lasting viral infections. This decreases proliferation and the production of interferon-γ (IFN-γ), which in turn causes cytotoxicity. Through this approach, the cancer cells can avoid immune system destruction, only trigger a modest immune response, and establish the conditions required for further carcinogenesis (Zabeti & Vahidi, 2024). Many studies have revealed that PD-L1 is a poor prognostic factor for individuals with skin cancer, renal cell carcinoma, and cancers of the breast, lung, stomach, pancreatic, liver, bladder, and ovary. It has also been established that PD-L1 is stated prominently on the cell external of many types of solid tumors. Given these reports, the investigation revealed a few interesting results that support the idea that tumor surface PD-L1 overexpression is not always linked to a poor prognosis. Antitumor IFN-γ's reaction with neoplastic cells weakened their function, which in turn led to improved PD-L1 expression. This hypothesis can be supported by the significantly longer survival rates observed in melanoma individuals with proven states of PD-L1 lymphocytes. Such judgments must be made cautiously and carefully for the ongoing research in this field. This information is utilized as a marker of resistance to the body's defense activity, though, as it is probable that the PD-1/PD-L1 pathways are involved as an acquired malignance adaption activity (Wojtukiewicz et al., 2021).

From the cytoplasm, the CTLA-4 molecule is drawn to the T-cell effector surface, where it contributes to the immunological synapse. There are two stages to the activation of cytotoxic T cells. Increased sensitivity of CD4 and CD8 receptors results from the initial signal, which is the TCR's acknowledgment of the antigen displayed by histocompatibility molecules on the layer of APC. The reaction of costimulating CD80/CD86 molecules on the exterior of APC with CD28 on the membrane of T cells, which results in lymphocyte initiation and differentiation, is the second signal required for synapse formation. A greater affinity for B7 family ligands, CTLA-4 displaces CD28 from its connection with CD80/86 and thus contests with it for binding to ligands on APC cells. When CTLA-4 binds to ligands (CD80-B7-1, CD86-B7-2) on APC cells, it blocks the T-lymphocyte response, inhibits T-lymphocyte proliferation, inhibits Treg lymphocyte activity, and reduces cytokine secretion, all of which contribute to immunosuppression. Furthermore, T helper lymphocytes are functionally transformed into regulatory T lymphocytes, which have potent immunosuppressive characteristics, when CTLA-4 expression is high. All things considered, CTLA-4 plays a part in the immunological deficit seen in cancer patients. Anti-CTLA-4 antibodies (ipilimumab and tremelimumab) block CTLA-4, causing T-cell activation, inhibition, and reactivation (Habeeb et al., 2024c).

7.4.2 Emerging immune checkpoint inhibitors: TIGIT, LAG-3, TIM-3, and others

T-cell immunoreceptor with immunoglobulin and ITIM Domain (TIGIT) (also referred to as Vsig9, Vstm3, and WUCAM) was discovered as a novel coinhibitory receptor by several research teams in 2009. Regulatory T cells, Tr1 cells, Tfh cells, and dysfunctional CD8+ T cells are among the T-cell types and subsets of natural killer (NK) cells that exhibit stable expression of TIGIT. TIGIT expression is also temporarily upregulated on T cells during TCR stimulation. More recently, B cells, especially those with regulatory roles, have also been shown to express TIGIT. Furthermore, new data suggest that TIGIT may have larger functional roles beyond its traditional involvement with T and NK cells by expressing on innate immune cells, including macrophages and innate lymphoid cells (ILCs), in disease situations. APCs and other nonhematopoietic cells, such as tumor cells, express two ligands, CD155 and CD112, with which TIGIT interacts. The main receptor–ligand interactions in this way are formed by these ligands, which are also shared by the costimulatory receptor CD226 (DNAM-1). Furthermore, TIGIT shares ligands with CD96 and CD112R, which adds to a complicated web of costimulatory and coinhibitory interactions (Fig. 7.3). Particularly, it has been revealed that TIGIT is bound by the FAP2 protein from *Fusobacterium nucleatum*, a microorganism linked to head, neck, and colon malignancies. According to research, despite not binding its murine counterparts, CD114 (Nectin-4) was recently discovered to be a putative ligand for human TIGIT and CD226. However, because of its low binding affinity, the physiological relevance of this interaction is still unknown (Cai et al., 2023).

T-cell initiation triggers the production of LAG-3, a structural homolog of CD4, which endures conditions involving persistent antigenic stimulation, including cancer and persistent infections. The function of LAG-3 has been mostly investigated in T cells, where it functions as an inhibitory receptor, reducing T-cell responses and causing T-cell exhaustion, even though it is also expressed in NK cells, B cells, and plasmacytoid dendritic cells. Furthermore, LAG-3 is expressed in regulatory T-cell populations, such as FOXP3+ Tregs and IL-10-producing Tr1 cells, where it affects immunosuppressive abilities; however, its effects vary depending on the circumstances. Although major histocompatibility complex class II is the typical ligand for LAG-3, alternative ligands have been found, including α-synuclein fibrils, LSECtin, Galectin-3 (GAL-3), Fibrinogen-like protein 1 (FGL1), and the Tcell Receptor–Cluster of Differentiation 3 Complex (TCR-CD3) complex. These various ligands indicate that Lymphocyte activation gene-3 (LAG-3) has a variety of mechanisms to influence immune responses. The precise functional results of these interactions are still unknown, though. For example, LAG-3's binding to MHC class II is essential for controlling cancer and autoimmunity, but further research is needed to ascertain the disease-specific significance of additional ligands, such as FGL1, as well as the TCR-CD3 complex (Joller et al., 2024).

Human and mouse CD4+ as well as CD8+ T cells that produce IFN-γ contain the coinhibitory receptor T-cell immunoglobulin and mucin-domain containing -3 (TIM-3), which regulates type I immunity. Since then, NK cells, macrophages, dendritic cells, mast cells, and B cells have been found to express it. TIM-3's structural components include an IgV domain, a mucin domain, a transmembrane area, and a cytoplasmic tail. The latter has tyrosines that develop phosphorylation when a ligand binds to it, but it lacks traditional inhibitory motifs. GAL-9, phosphatidylserine, HMGB1, and CEACAM-1 are the four known ligands for TIM-3, and they all engage with different parts of its IgV domain. Although phosphatidylserine connects with a conserved pocket framed by the FG and CC loops, which requires calcium coordination, GAL-9 attaches to N-linked carbohydrates on the IgV. The CC′ and FG loops are where CEACAM-1 binds; the same conserved pocket is required for HMGB1 to bind. In particular, therapeutic anti-TIM-3 antibodies exhibit the potential for functional efficacy by blocking interactions with ligands such as CEACAM-1 and phosphatidylserine (Veigas et al., 2021).

7.4.3 Immune checkpoint inhibitors resistance action

ICI-based immunotherapy is a well-established technique that has been applied extensively in clinical environments. Clinical data, however, indicate that some patients receiving ICI treatment still have drug resistance. Some individuals who experience ICI therapy are unsuccessful in responding to antitumor treatment. In immunotherapy, T cells remain the primary lethal killer cells that target malignant cells. Three crucial processes are necessary for their tumor-killing effect: Substantial infiltration, T-cell initiation, and accurate tumor cell acknowledgment (Waldman et al., 2020). Nevertheless, it is challenging to use ICIs to meet all three needs in the intricate TME. Several elements in the intricate

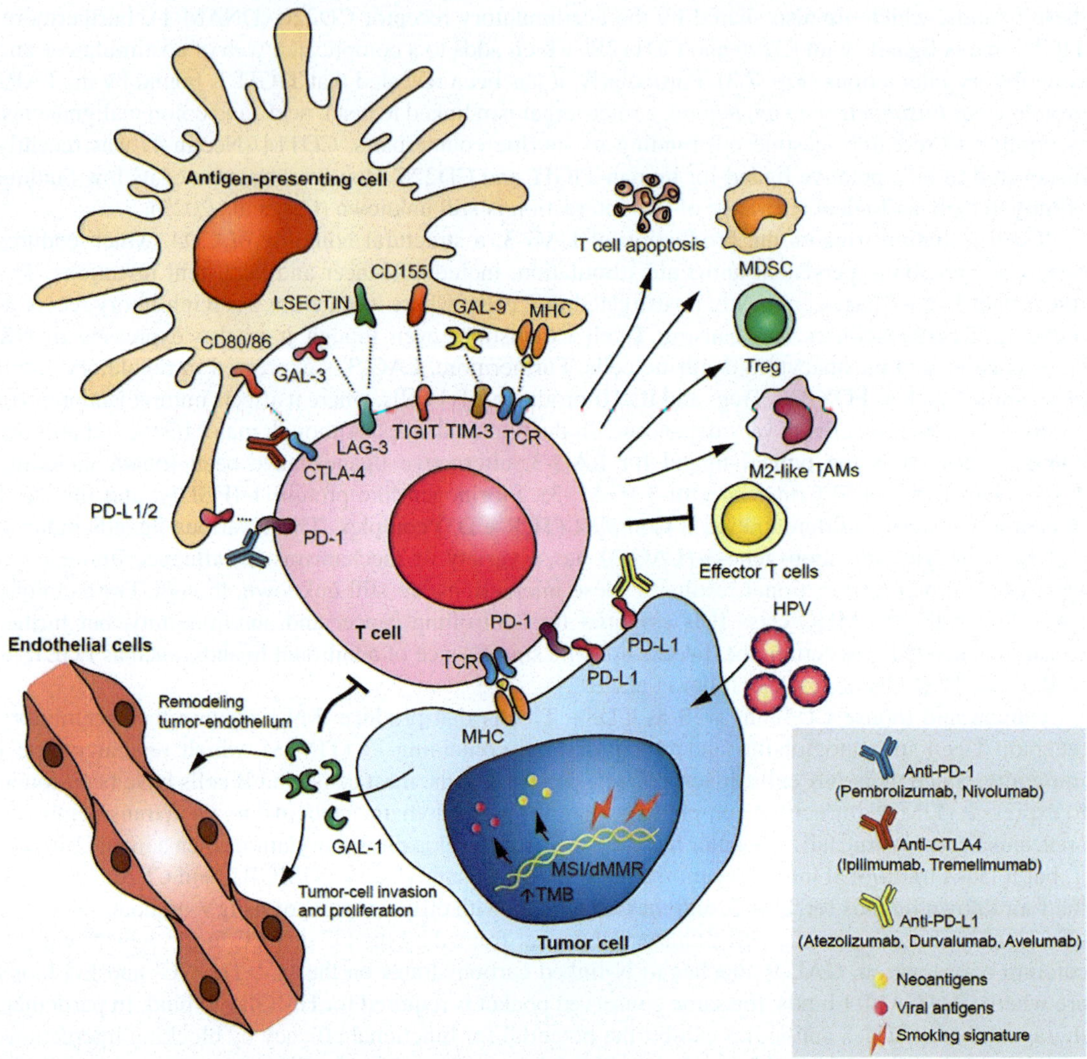

FIGURE 7.3

Immune checkpoint pathways such as PD-1/PD-L1, CTLA-4, TIM-3, LAG-3, TIGIT, and Gal-1 drive immune escape by overwhelming T cells and growing immunosuppressive cells. *PD-L1,* Programmed cell death protein-1; *CTLA-4,* cytotoxic T-lymphocyte-associated protein 4.

Reproduced with permission from Veigas, F., Mahmoud, Y. D., Merlo, J., Rinflerch, A., Rabinovich, G. A., & Girotti, M. R. (2021). Immune checkpoint pathways in head and neck squamous cell carcinoma. Cancers, 13*(5), 1018. https://doi.org/10.3390/cancers13051018.*

TME are linked to resistance to ICIs. Better clinical efficacy from ICIs can only be achieved by reducing barriers to their use through personalized combination therapy (Lao et al., 2022).

7.5 Immune system modulators

Immune system modulators are agents or therapies that improve or stop the action of the immune system to attain a needed beneficial effect.

7.5.1 Agents enhancing immune responses

Immune system modulators act by restoring immune suppression or stimulating immunological cells that boost the body's defenses against cancer. T-cell activity is increased by compounds such as ICIs (anti-PD-1/PD-L1 antibodies), which enable T cells to once again recognize and then destroy tumor cells (Cancer Research Institute, n.d.). To promote high antigen presentation and T-cell activation, a different strategy is to stimulate dendritic cells. To combat tumor immune evasion, new strategies, including small-molecule immune stimulators and toll-like receptor (TLR) agonists, are showing promise. In preclinical and clinical studies, these drugs have shown synergistic effects when combined with medicines such as vaccinations or cytokines. Chen et al. (2022) emphasized the dual function of TLRs in cancer immunotherapy. As modulators of innate and adaptive immunity, the authors examined TLR agonists (TLRa) and antagonists, showcasing their potential in combination therapy and tumor vaccines. While antagonists help in reducing inflammation-induced immune escape, TLRa can stimulate cytotoxic T cells and APCs, fostering antitumor responses. According to the study's findings, clinical translation of TLR-based treatments is hindered by issues such as tumor heterogeneity, immunosuppressive microenvironments, and delivery inefficiencies, regardless of their potential (Chen et al., 2022).

7.5.2 Cytokine therapies in cancer treatment

Signaling proteins such as interleukins (ILs) and interferons (IFNs) are used in cytokine treatments to alter the immune response against cancer (National Cancer Institute, 2024). One important cytokine that boosts anti-tumor immunity is IL-2, which also causes T cells and NK to proliferate. In a similar vein, IFN-alpha suppresses tumor growth while activating NK and dendritic cells. To increase effectiveness and reduce adverse outcomes, new cytokines such as IL-15 and synthetic cytokine variations are being developed. New delivery techniques that improve cytokine stability and lower systemic toxicity, such as PEGylation and nanoparticle encapsulation, are promising technologies in immunotherapy. Huldani et al. (2023) investigated the roles of ILs and IFNs in ovarian cancer, highlighting how they affect tumor growth, immune evasion, and possible treatment applications. According to the findings, IL-6 activation in epithelial ovarian cancer increases metastasis and chemotherapy resistance through pathways such as STAT3, while some ILs, such as IL-27, are identified as tumor suppressors and others, such as IL-6 and IL-33, as tumor promoters. Targeting ILs (e.g., by employing inhibitors for IL-6) and IFNs, which boost immune responses but may potentially accelerate tumor growth in some circumstances, has therapeutic promise in clinical trials. The study concludes that although ILs and IFNs have therapeutic promise, their dual functions necessitate rigorous evaluation to maximize their use in the treatment of ovarian cancer (Huldani et al., 2023).

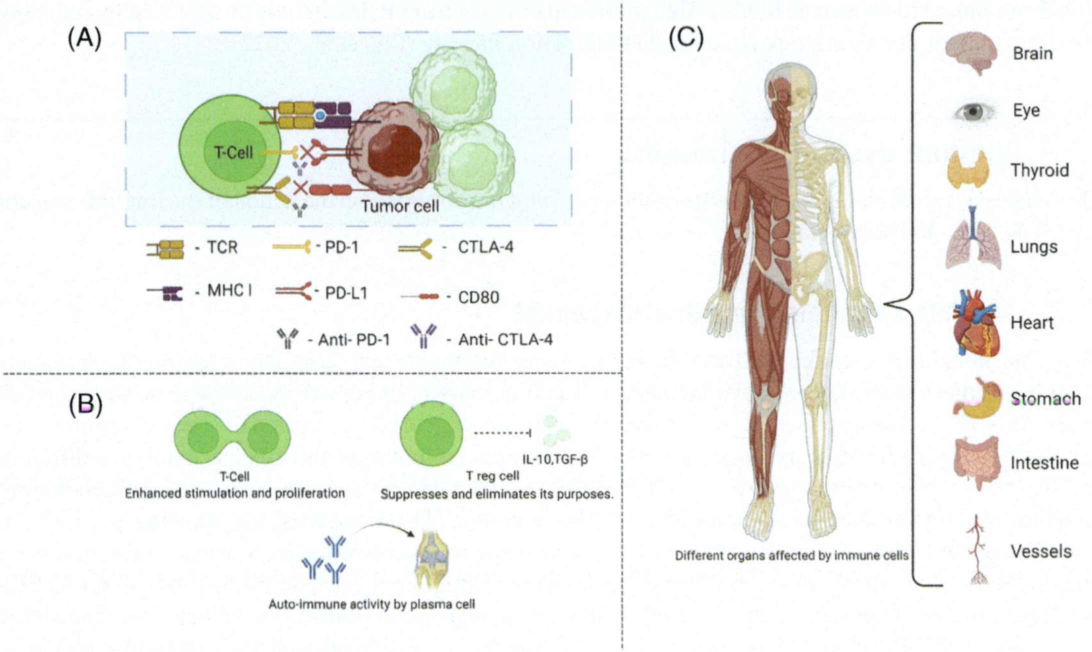

FIGURE 7.4

(A) PD-L1 and CD80 are upregulated by tumor cells to avoid immune detection, (B) immune checkpoint inhibitors may increase the immune system, and (C) immune cells invade several organs as a result of an increase in the immune system, resulting in a variety of immune-related side effects. *PD-L1,* Programmed cell death ligand-1.

7.6 Advantages and disadvantages of immune checkpoint inhibitors

ICIs boost T-cell activation, which allows the immune system to efficiently target tumor cells, offering significant advantages in cancer treatment. In malignancies with significant mutational loads, such as lung cancer and melanoma, where neoantigen presentation elicits strong immune activity, they have demonstrated impressive efficacy in producing long-lasting responses. In addition, ICIs can be used in conjunction with other drugs to enhance therapy results through synergistic effects (Brentjens et al., 2013). In addition, they can strengthen immunological memory, providing tumor control that lasts long after treatment is completed. They are a key strategy for present-day oncology because of their adaptability and capacity to target immune evasion systems. ICIs are not without limitations, either (Postow et al., 2018). They may cause mild to potentially fatal immune-related adverse effects (irAEs), including colitis, hepatitis, and pneumonitis. Because of their underdeveloped immune systems, low mutational loads, and potential for long-term organ damage, pediatric malignancies present unique challenges (Fig. 7.4) (Martins et al., 2019). Their efficacy is further restricted by the TME and resistance mechanisms, including immunosuppressive macrophages and T-cell anergy. To optimize their

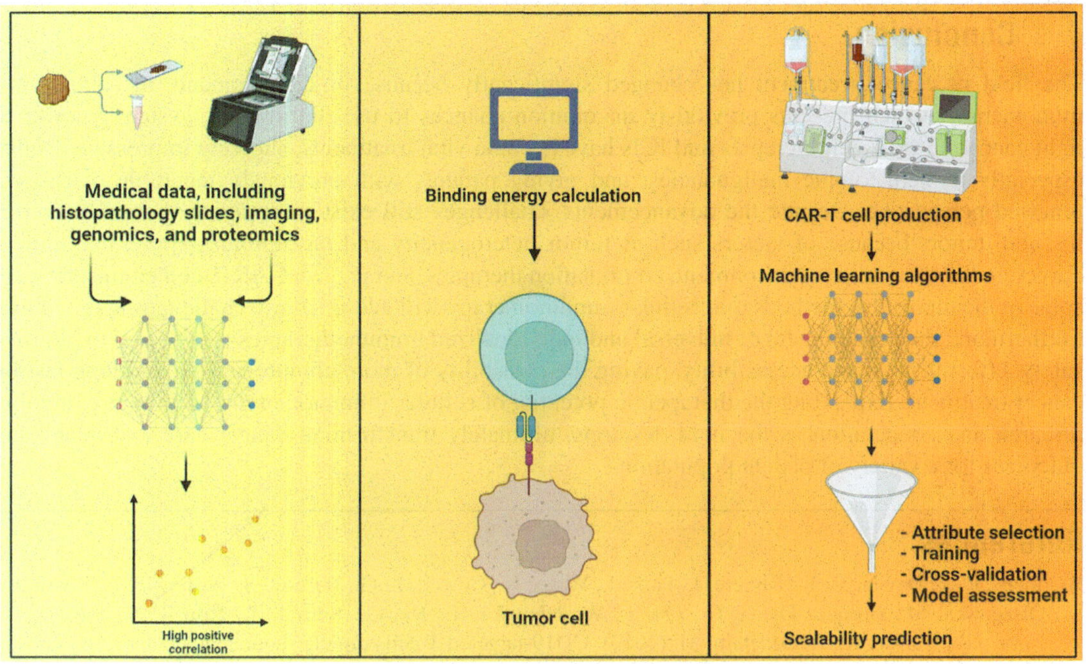

FIGURE 7.5

Role of artificial intelligence in aiding the treatment of immune checkpoint inhibitors.

therapeutic potential, these elements emphasize the necessity of precise patient selection, biomarker-driven approaches, and novel formulations (Park & Cheung, 2017).

7.7 Role of artificial intelligence

Cellular immunotherapy has benefited greatly from AI, which has improved numerous fields of cancer treatment. Various medical data, including histopathology slides, imaging, genomics, and proteomics, are analyzed by AI algorithms to help forecast the effectiveness of immunotherapies, such as ICIs. This makes it easier to identify individuals who will gain advantages from these treatments and to customize care based on each patient's unique characteristics. AI lowers production costs and increases scalability in CAR-T cell therapy by improving manufacturing processes and assuring consistency and quality (Fig. 7.5) (b; Xie et al., 2023). Furthermore, AI facilitates real-time therapy progress tracking, allowing physicians to modify treatments in response to anticipated outcomes. Developing digital twins, or virtual patient models, to mimic treatment reactions is one of the advanced uses of AI that can help with accurate decision-making and lessen dependency on lengthy clinical trials. In addition, immune-related toxicities are predicted using AI-powered models, which enhances safety and results. These developments show that AI might improve efficiency, help with customized treatment in cellular immunotherapy, and address the problems associated with complex drugs (D'Orsi et al., 2024).

7.8 Conclusion

The field of cancer treatment has changed significantly because of developments in cellular immunotherapy, which presents previously uncommon chances to use the immune system's power to fight cancer. CAR-T cell therapies and ICIs have become vital treatments, showing impressive results, especially in hematologic malignancies, and giving patients with previously incurable diseases a renewed perspective. Despite the advancements, challenges still exist in applying these discoveries to solid tumors because of factors such as tumor heterogeneity and the immunosuppressive milieu. Targeting the tumor microenvironment, combination therapies, and precise CAR-T cell engineering are some of the innovative methods that cellular immunotherapy will address these challenges in the future. Furthermore, the development of universal and individualized immunotherapies is expected to improve safety, effectiveness, and accessibility, paving the possibility of more comprehensive and long-lasting cancer treatment. Expanding the therapeutic potential of cellular immunotherapy will require ongoing research and cooperation as the field develops, ultimately transforming cancer care and enhancing outcomes for a variety of patient populations.

References

Brentjens, R. J., Davila, M. L., Riviere, I., Park, J., Wang, X., Cowell, L. G., Bartido, S., Stefanski, J., Taylor, C., Olszewska, M., Borquez-Ojeda, O., Qu, J., Wasielewska, T., He, Q., Bernal, Y., Rijo, I. V., Hedvat, C., Kobos, R., Curran, K., … Sadelain, M. (2013). CD19-targeted T cells rapidly induce molecular remissions in adults with chemotherapy-refractory acute lymphoblastic leukemia. *Science Translational Medicine, 5*(177). doi:10.1126/scitranslmed.3005930.

Cai, L., Li, Y., Tan, J., Xu, L., & Li, Y. (2023). Targeting LAG-3, TIM-3, and TIGIT for cancer immunotherapy. *Journal of Hematology and Oncology, 16*(1). doi:10.1186/s13045-023-01499-1.

Cancer Research Institute (n.d.) Immunomodulators. https://www.cancerresearch.org/treatment-types/immunomodulators.

Chen, X., Zhang, Y., & Fu, Y. (2022). The critical role of Toll-like receptor-mediated signaling in cancer immunotherapy. *Medicine in Drug Discovery, 14*, 100122. doi:10.1016/j.medidd.2022.100122.

Desai, I., Thakur, S., & Pagariya, P. (2024). Current advances in immunotherapy for cancer. *Oral Oncology Reports, 12*, 100652. doi:10.1016/j.oor.2024.100652.

D'Orsi, L., Capasso, B., Lamacchia, G., Pizzichini, P., Ferranti, S., Liverani, A., Fontana, C., Panunzi, S., De Gaetano, A., & Lo Presti, E. (2024). Recent advances in artificial intelligence to improve immunotherapy and the use of digital twins to identify prognosis of patients with solid tumors. *International Journal of Molecular Sciences, 25*(21), 11588. doi:10.3390/ijms252111588.

Garg, P., Pareek, S., Kulkarni, P., Horne, D., Salgia, R., & Singhal, S. S. (2024). Next-generation immunotherapy: Advancing clinical applications in cancer treatment. *Journal of Clinical Medicine, 13*(21). doi:10.3390/jcm13216537.

Habeeb, M., Arsey, S., You, H. W., Kumbhar, S. T., Aher, K. B., Bhavar, G. B., & Vengateswaran, H. T. (2024a). Targeted nanomedicine modulating intercellular communications to arrest renal cell carcinoma progression. *Journal of Drug Delivery Science and Technology, 99*. doi:10.1016/j.jddst.2024.105983.

Habeeb, M., Kareem, T. A., Deepthi, K. L., Khot, V. S., Woon, Y. H., & Pawar, S. S. (2022). Nanomedicine for targeting the lung cancer cells by interpreting the signaling pathways. *Journal of Drug Delivery Science and Technology, 77*. doi:10.1016/j.jddst.2022.103865.

Habeeb, M., Vengateswaran, H. T., You, H. W., Aher, K. B., Bhavar, G. B., & Gaikwad, S. D. (2024b). Development and characterization of carboxylated copper oxide conjugated polymeric nanocomposites and correlating with computational techniques. *Results in Surfaces and Interfaces, 17*. doi:10.1016/j.rsurfi.2024.100323.

Habeeb, M., Vengateswaran, H. T., You, H. W., Saddhono, K., Aher, K. B., & Bhavar, G. B. (2024c). Nanomedicine facilitated cell signaling blockade: Difficulties and strategies to overcome glioblastoma. *Journal of Materials Chemistry B, 12*(7), 1677–1705. doi:10.1039/d3tb02485g.

Huldani, H., Abdul-Jabbar Ali, S., Al-dolaimy, F., Hjazi, A., Andreevich, N. D., Oudaha, K. H., Almulla, A. F., Alsaalamy, A., Kareem Oudah, S., & Mustafa, Y. F. (2023). The potential role of interleukins and interferons in ovarian cancer. *Cytokine, 171*. doi:10.1016/j.cyto.2023.156379.

Joller, N., Anderson, A. C., & Kuchroo, V. K. (2024). LAG-3, TIM-3, and TIGIT: Distinct functions in immune regulation. *Immunity, 57*(2), 206–222. doi:10.1016/j.immuni.2024.01.010.

Lao, Y., Shen, D., Zhang, W., He, R., & Jiang, M. (2022). Immune checkpoint inhibitors in cancer therapy—How to overcome drug resistance? *Cancers, 14*(15), 3575. doi:10.3390/cancers14153575.

Martins, F., Sofiya, L., Sykiotis, G. P., Lamine, F., Maillard, M., Fraga, M., Shabafrouz, K., Ribi, C., Cairoli, A., Guex-Crosier, Y., Kuntzer, T., Michielin, O., Peters, S., Coukos, G., Spertini, F., Thompson, J. A., & Obeid, M. (2019). Adverse effects of immune-checkpoint inhibitors: Epidemiology, management and surveillance. *Nature Reviews Clinical Oncology, 16*(9), 563–580. doi:10.1038/s41571-019-0218-0.

National Cancer Institute. (2024). *Immune system modulators to treat cancer*. https://www.cancer.gov/about-cancer/treatment/types/immunotherapy/immune-system-modulators.

Park, J. A., & Cheung, N. K. V. (2017). Limitations and opportunities for immune checkpoint inhibitors in pediatric malignancies. *Cancer Treatment Reviews, 58*, 22–33. doi:10.1016/j.ctrv.2017.05.006.

Postow, M. A., Sidlow, R., & Hellmann, M. D. (2018). Immune-related adverse events associated with immune checkpoint blockade. *New England Journal of Medicine, 378*(2), 158–168. doi:10.1056/NEJMra1703481.

Roy, R., Singh, S. K., & Misra, S. (2023). Advancements in cancer immunotherapies. *Vaccines, 11*(1). doi:10.3390/vaccines11010059.

Veigas, F., Mahmoud, Y. D., Merlo, J., Rinflerch, A., Rabinovich, G. A., & Girotti, M. R. (2021). Immune checkpoints pathways in head and neck squamous cell carcinoma. *Cancers, 13*(5), 1–31. doi:10.3390/cancers13051018.

Waldman, A. D., Fritz, J. M., & Lenardo, M. J. (2020). A guide to cancer immunotherapy: From T cell basic science to clinical practice. *Nature Reviews Immunology, 20*(11), 651–668. doi:10.1038/s41577-020-0306-5.

Wojtukiewicz, M. Z., Rek, M. M., Karpowicz, K., Górska, M., Polityńska, B., Wojtukiewicz, A. M., Moniuszko, M., Radziwon, P., Tucker, S. C., & Honn, K. V. (2021). Inhibitors of immune checkpoints—PD-1, PD-L1, CTLA-4—New opportunities for cancer patients and a new challenge for internists and general practitioners. *Cancer and Metastasis Reviews, 40*(3), 949–982. doi:10.1007/s10555-021-09976-0.

Xie, J., Luo, X., Deng, X., Tang, Y., Tian, W., Cheng, H., Zhang, J., Zou, Y., Guo, Z., & Xie, X. (2023). Advances in artificial intelligence to predict cancer immunotherapy efficacy. *Frontiers in Immunology, 13*. doi:10.3389/fimmu.2022.1076883.

Yin, Q., Wu, L., Han, L., Zheng, X., Tong, R., Li, L., Bai, L., & Bian, Y. (2023). Immune-related adverse events of immune checkpoint inhibitors: A review. *Frontiers in Immunology, 14*. doi:10.3389/fimmu.2023.1167975.

Zabeti Touchaei, A., & Vahidi, S. (2024). MicroRNAs as regulators of immune checkpoints in cancer immunotherapy: Targeting PD-1/PD-L1 and CTLA-4 pathways. *Cancer Cell International, 24*, 1–32. doi:10.1186/S12935-024-03293-6.

Zhang, Y., & Zhang, Z. (2020). The history and advances in cancer immunotherapy: Understanding the characteristics of tumor-infiltrating immune cells and their therapeutic implications. *Cellular and Molecular Immunology, 17*(8), 807–821. doi:10.1038/s41423-020-0488-6.

CHAPTER

8 The future of cancer immunotherapy: Combining gene and cellular approaches

Balachander Lokeshwari[a], Pazhani Saranraj[a], Karthikeyan Gayathri[a], Alexander Machado Cardoso[b] and Kesavardhini Kesavardhini[a]

[a]*PG and Research Department of Microbiology, Sacred Heart College (Autonomous), Tirupattur, Tamil Nadu, India,* [b]*Department of Biology, Rio de Janeiro State University (UERJ), Rio de Janeiro, Brazil*

8.1 Introduction

Cancer remains a major global health burden, causing millions of deaths annually despite advances in diagnosis and treatment. The conventional approach to treating cancer uses surgery, chemotherapy, and radiotherapy, and alternative therapy methods like immunotherapy (Khalil et al., 2016). Immunotherapy leverages the immune system to identify and eliminate cancer cells; however, tumors can evade immune detection, making it essential to overcome these escape mechanisms. Immune checkpoint inhibitors (ICIs), adoptive cell therapies, cancer vaccines, and cytokine therapies are recent developments in therapeutic treatments (Seliger & Massa, 2021). The emerging areas of gene and cellular therapies can help improve immunotherapy.

The targeted modifications of immune and tumor cells using gene-based treatments are used to improve the response and the adoptive cell therapies, especially chimeric antigen receptor (CAR)-T-cell therapy, and T cell receptor (TCR) engineered cells are able to improve the immune response to the cancer (Bagchi et al., 2020). Molecular biology, genetic engineering, gene delivery systems, and genome editing systems such as Clustered Regularly Interspaced Short Palindromic Repeats (CRISPR) are technologies that have made the integration of gene and cellular therapies more practical and enabled the latter to become more stable and effective (Cheung & Mooney, 2015). Although programmed death-1 (Programmed Death (PD)-1), PD-L1, and Cytotoxic T-Lymphocyte-Associated (CTLA)-4 therapies, genetic diversity in tumors, and immune suppression in the tumor microenvironment (TME) pose a challenge to cellular therapies by enhancing immune targeting of tumors (Iglesias-Escudero et al., 2023).

Adoptive cellular immunotherapy has transformed simple protocols into complicated genetically modified T-cells employing CAR-T cell treatment, which has shown effectiveness with a limited spectrum of blood cancers because of adverse reactions such as cytokine release syndrome (CRS) and antigens. Cancer vaccines, oncolytic viruses, and RNA-based therapeutics use other gene-based techniques that elicit strong immune responses (Guedan et al., 2018). A combination of gene and cellular therapy can improve the functionality of immune cells and overcome the barriers of TME. There are challenges associated with production issues, regulations, and ethical concerns regarding genetic modifications. Current studies aim at the production of safer delivery systems, the enhancement of engineered cellular behavior, and the standardization of treatment systems. Future in vivo cancer

Advancements in Personalized Cellular Immunotherapy: A Comprehensive Guide to Cancer Treatment.
DOI: https://doi.org/10.1016/B978-0-443-27731-3.00012-7
Copyright © 2026 Elsevier Inc. All rights are reserved, including those for text and data mining, AI training, and similar technologies.

immunotherapy techniques combine genetic engineering, cell therapy, and immunology to reduce therapeutic restrictions and maximize their efficacy (Chen et al., 2025).

8.2 Immunological basis of cancer and tumor immune evasion

The immunological basis of cancer is based on the immune system, which is able to distinguish between the self and altered targets that have potentially harmful cells (Liu et al., 2022). Genetic mutations, chromosomal rearrangements as well as alterations in the epigenetics cause the development of new antigens or abnormal expression of self-antigens in cancer cells, rendering them antigenic to the immune system. Antitumor immunity involves both innate and adaptive immune responses through a high network of immune cells, cytokines, and signaling pathways. However, tumors evolve under the pressure of immune response, which results in the appearance of immune-resistant properties (Augustin et al., 2020).

The innate immunity is the initial defensive action by mechanisms independent of any antigens. Natural killer (NK) cells, macrophages, dendritic cells (DCs), and the complement system, which detects signals of tumor cells. On the other hand, adaptive immunity, which is antigen-specific and includes immunological memory, is mostly coordinated by T and B lymphocytes. Cytotoxic T lymphocytes of the Cluster of Differentiation CD^{8+} type directly kill tumor cells that express antigenic peptides through major histocompatibility complex (MHC) class I molecules, and CD^{4+} helper T-cells complement with secretions that mediate immune cell interactions (Cha et al., 2019).

Tumor immune evasion involves adaptations by cancer cells to evade immune recognition, in which the cancer cells use phenotypic or molecular alterations to reduce antigenicity (e.g., loss or modification of tumor antigens), damage MHC antigen presentation, induce immune tolerance, and create an anti-immune local environment (Tie et al., 2022). These processes simultaneously interact and depend on tumor-specific characteristics along with host-determined aspects (Peterson et al., 2022). Immune activation and suppression within the TME are a complex dynamic that determines the dynamics of tumor immune interactions, and it is crucial for the cellular and molecular mechanisms of immune suppression and activation in order to develop effective immunotherapeutic strategies.

8.2.1 Cancer-immune system interactions

The immune system and cancer interact significantly during tumor formation. Stress signals help the innate immune system spot malignancies. Pattern recognition receptors let NK cells and macrophages identify changed cells. MHC class I is reduced by NK cells to kill tumor cells. These cells aid mold-adaptive immunity and early TME. Altvater et al. (2023) found that CD^{4+} helper T-cells and B cells produce CTL-priming and tumor-specific antibodies, whereas CD^{8+} CTLs directly kill tumor cells by causing death. Datta et al. (2019) examined cancer-immune responses as elimination, equilibrium, and escape. Cells of the immune system assault immunogenic tumor cells throughout the elimination process. Signals and cytokines give tumor microenvironment immune cells functional flexibility. Tumor-associated macrophages (TAM)-dominant macrophages that inflame and reduce immunity, while DCs activate or tolerate T-cells. Suppression, tolerance, or weariness can prevent the immune system from responding, allowing tumor growth between host defenses and neoplastic cells. Fig. 8.1 indicates how innate and adaptive immune responses destroy tumor cells to enable tumor immune evasion and tumor progression.

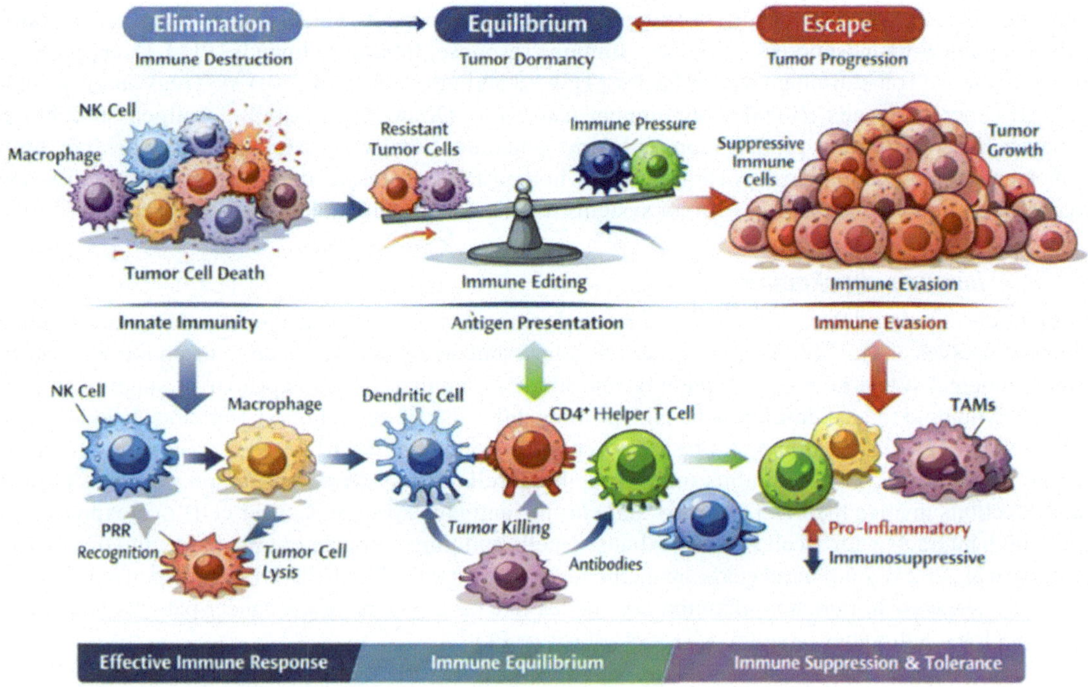

FIGURE 8.1

Interactions of cancer-immune system.

8.2.2 Tumor antigens and immune recognition

Tumor-Specific Antigens (TSAs) and Tumor-Specific Antigens (TAAs) are essential for immune detection and adaptive antitumor immunity. TSAs (neoantigens) are highly immunogenic peptide sequences found in somatic mutations, making them distinct from normal cells. Complicated antigen processing pathways produce tumor antigens. Endogenous antigens are digested by the proteasome, transported through the Transporter Associated with Antigen Processing (TAP) system, and loaded onto MHC class I molecules for CD^{8+} T-cell recognition (Ceja et al., 2024). DCs send foreign antigens on MHC class II molecules to CD^{4^+} T-cells, which then cross-present them onto MHC class I, triggering a CTL response. T lymphocyte activation requires antigen recognition and signaling. CD28-binding B7 molecules are needed to activate the TCR when it detects its 3D peptide-MHC complex. Synthetic T cells become effector and memory cells that kill tumor cells (Liu et al., 2024). Central and peripheral tolerance reduce immune response to naturally generated TAAs, delivering tumor antigen.

8.2.3 Mechanisms of immune response

Tumor cells are utilized in order to evade detection and removal by the immune system. Complex tumor cell and microenvironment changes weaken antitumor immunity. Sun et al. (2016) found that cancer progression increases evasion mechanisms, increasing growth and metastasis. Decreased MHC class I molecules and TAP peptide-loading mechanism components hinder CD^{8+} tumor cell identification. Tumor immunosuppression reduces immune response. Inhibitory ligands (PD-L1), suppressive cytokines (e.g., Transforming Growth Factor (TGF)-β and interleukin (IL-10)), or Tregs and myeloid-derived suppressor cells (MDSC) recruitment cause this (Xiang et al., 2022). Antigen stimulation exhausts T-cells, limiting proliferation, cytokine generation, and cytotoxicity, which does not limit tumor growth. Immunological evasion involves homeostatic processes that avoid autoimmunity and maintain immunological balance. Tumor systems regulate local immunological tolerance.

8.2.3.1 *Immune checkpoints*

Immune checkpoints decrease T-cell activation and produce immunological tolerance. Negative immune control involves CTLA-4, PD-1, and PD-L1 to prevent autoreactivity. These mechanisms disturb immunological homeostasis by bypassing antitumor immunity. CTLA-4 primarily suppresses secondary lymphoid organ development and priming after T-cell activation (Crompton et al., 2014). CTLA-4 strongly inhibits antigen-presenting cells by competing with the CD28 signaling complex for B7 molecules. CTLA-4 inhibits tumor-specific T-cells and systemic antitumor response. Cancer and infections involve the PD-1 pathway and chronic antigen exposure. Cancer cells overexpress PD-L1, which interacts with T-cell PD-1 to exhaust T-cells and limit cytokine output and cytotoxic activity (Chang et al., 2023). Immune checkpoint medicines can disrupt the PD-1/PD-L1 axis, which upregulates factors in response to immune-inflammatory signals like cancer cells. Some cancer patients responded to monoclonal antibodies targeting CTLA-4, PD-1, or PD-L1.

8.2.3.2 *Immunosuppressive tumor microenvironment*

Cancer cells, stromal cells, blood vessels, and immune cells impede antitumor immunity in the TME. Regulatory T-cells (Tregs) reduce effector T-cells via direct contact and cytokine-mediated pathways (IL-10, TGF-β), while MDSCs control immunological response via amino acids or reactive oxygen/nitrogen species (Rodríguez-García et al., 2020). TAMs release cytokines and chemokines that promote tumor development and immune evasion, weakening cytotoxic T and NK cells. Tumor cell metabolic reprogramming causes dietary restriction, hypoxia, and immune cell dysfunction (Libutti et al., 2018). Lactic acid and adenosine stimulate suppressive T-cells, stabilize Tregs, and induce lymphocytes that invade tumors. Immune suppression reduces antitumor immunity and treatment resistance (Henke et al., 2020). ICIs and other drugs can boost antitumor responses and prevent cancer.

8.3 Cancer immunotherapy modalities

8.3.1 Immune checkpoint inhibitors

Oncology has been transformed by ICIs that alter immune responses and prevent autoimmunity (Wellhausen et al., 2024). These drugs promote immune identification of cancer cells, which involve T-cell-mediated malignancies' tumor evasion mechanisms and checkpoint molecules. ICIs have revolutionized immune-system-based cancer treatment in melanoma, non-small cell lung cancer, and renal cell carcinoma by improving late-stage survival. Future ICI therapy depends on optimizing the efficacy and safety of personalized and combinational immunotherapeutic methods (Qin et al., 2025). Table 8.1

Table 8.1 Immune checkpoint regulatory pathways and their clinical relevance.

Pathway	Key molecules	Primary site of action	Mechanisms of immune regulation	Therapeutic agents	Significance	Limitations	References
Immune checkpoint inhibitors (ICIs)	PD-1, PD-L1, CTLA-4	Tumor microenvironment and lymphoid organs	Block inhibitory immune signals to restore T cell–mediated antitumor responses	Nivolumab, pembrolizumab, ipilimumab, atezolizumab	Durable responses and improved survival in multiple cancers	Variable response rates; immune-related adverse events	Amer (2014)
PD-1/PD-L1 pathway	PD-1, PD-L1, PD-L2	Peripheral tissues and tumor microenvironment	Inhibits T cell proliferation and cytokine release, promoting T cell exhaustion	Nivolumab, pembrolizumab, atezolizumab, durvalumab	Broad efficacy across solid tumors; predictive biomarker potential	irAEs (dermatitis, colitis, pneumonitis); resistance mechanisms	Cogdill et al. (2017)
CTLA-4 pathway	CTLA-4, CD80, CD86	Secondary lymphoid organs	Suppresses early T cell activation by competing with CD28	Ipilimumab	Enhances T cell priming; synergistic with PD-1 blockade	Higher incidence of severe irAEs; narrow therapeutic window	Catanzaro et al. (2024)

irAEs, *immune-related adverse effects;* PD-1, *programmed death-1.*

summarizes significant immune checkpoint regulation pathways, biological targets, sites of action, immune modulation techniques, and cancer immunotherapy medicines.

8.3.1.1 *Programmed death-1/Programmed death-L1 pathway*

The most extensively employed immune checkpoint system, the PD-1 pathway (PD-1, PD-L1, and PD-L2), is used in cancer immunotherapy. PD-1 is expressed by B and NK cells and activated by T-cells. Tumor cells and antigen-presenting cells in the TME upregulate the protein PD-L1. Tumor immune escape occurs when PD-L1 and PD-1 suppress T cell proliferation, cytokine production, and induction. In PD-L1-deficient malignancies, antibody profiling predicts tumor response, showing complex host immune interactions (Dias et al., 2024). Although PD-1/PD-L1 inhibitors are clinically effective, they may cause immune-related adverse effects like dermatitis, colitis, and pneumonitis due to increased immunological reaction and decreased peripheral tolerance. Improve efficacy, reduce adverse effects, and improve patient outcomes.

8.3.1.2 *CTLA-4 pathway*

CTLA-4 controls T cell activation in secondary lymphoid organs, both on activated conventional T-cells and Tregs. CLTA-4 binds CD80 and CD86 more strongly than the receptor, and CD28 inhibits immune system overstimulation. Blocking the CTLA-4 pathway with monoclonal antibodies like ipilimumab increases the immune response during T cell priming and growth in lymph nodes, resulting in antitumor immunity against melanoma. PD-1/PD-L1 inhibitors and CTLA-4 inhibition against early T cell activation improve clinical outcomes in patients with a variety of cancers. However, this treatment may lead to immune-related side effects due to systemic immune activation during clinical care. Diorio et al. (2024) found that inhibiting the CTLA-4 pathway regulates Tregs for immunological tolerance, a two-pronged effect. Additional research on dose, biomarkers, and combinations is necessary to maximize the therapeutic potential of CTLA-4-targeted therapies.

8.3.2 Cancer vaccines

Cancer vaccines are designed to reactivate the immune system against tumor-associated antigens in order to induce a specific and antitumor response. Unlike prophylactic vaccines for infectious diseases, these vaccines are used to overcome immune tolerance against self-antigens expressed in the tumor. Recent progress in the discovery of antigens, immunogens/design of the delivery has been used in cancer vaccines as a feasible strategy in immunotherapy (Mi et al., 2019). The efficacy of these vaccines is determined by several factors that include the antigen, vaccines, adjuvants used, and immune responses. Recent advances in personalized neoantigen vaccines and approaches to integrate the vaccines with ICIs demonstrate great potential for cancer vaccines that are divided into peptide-based, nucleic acid-based, and cell-based tumor antigen vaccines.

8.3.2.1 *Peptide vaccines*

Peptide vaccines are composed of short synthetic peptides of tumor-associated or tumor-specific antigens. Peptide-based cancer vaccines are developed to the T cell responses. These peptides bind to MHC molecules on antigen-presenting cells and promote the activation of CD^{8+} cytolytic T lymphocytes and CD^{4+} helper T-cells. Additionally, it also improves efficacy, approaches of long peptides, multiepitope vaccines, and combinations with immune checkpoint blockage are used to overcome tumor-induced immunosuppression (Corrigan, 2008).

8.3.2.2 DNA vaccines

DNA vaccines are composed of tumor antigen-expressing plasmid DNA, which could be translated by the host cells to generate tumor antigens endogenously to the immune system. This mechanism induces both humoral and cellular immunity, which provides advantages including production, stability of expression, and the ability to include several antigens together with immunostimulatory molecules. However, the use of DNA vaccines in clinics has drawbacks including but not limited to poor transfection efficiency and immunogenicity in human beings. Different approaches to enhance immune responses, such as electroporation, molecular adjuvants, and optimization (Gao et al., 2025). Furthermore, the combined use of DNA vaccines with other immunotherapies has also been studied as a strategy to enhance antitumor responses.

8.3.2.3 RNA vaccines

Cancer vaccines made of RNA, particularly messenger RNA (mRNA) vaccines, are appealing due to their scalability and high immunogenicity. Such vaccines carry the genes for tumor antigens that are translated in the cytoplasm of host cells, without the genotoxicity associated with DNA vaccines. Advances in RNA stabilization and delivery on lipid nanoparticles (LNPs) have improved their clinical feasibility. mRNA vaccines can also be tailored to express neoantigens for the induction of specific antitumor immunity. Clinical trials found that mRNA vaccines can effectively induce strong T cell responses when administered with ICIs (Garcia-Guerra et al., 2018). The therapeutic potential of treating infectious diseases, delivery system optimization challenges remain, which can minimize innate immune overstimulation and induce a durable immune response.

8.3.2.4 Dendritic cell vaccines

Cell-based immunotherapy uses patient-derived DC vaccines loaded with tumor antigens and incorporated into the patients. Antigen-presenting DCs initiate and regulate adaptive immune response, priming T cells and enhancing antigens. Clinical experiments have shown that DC vaccinations are safe and immunogenic, especially against prostate cancer (Garnipudi, 2024).

8.3.3 Adoptive cell therapy

Adoptive cell therapy (ACT) involves the expansion of ex vivo or genetic modification of immune cells. This treatment exploits the specificity and killing potential of immune effector cells and has shown remarkable efficacy for hematologic malignancies. Tumor lymphocyte therapy, CAR-engineered T and NK cells, and TCR-engineered immune cells are a part of ACT approaches (Gonzalez et al., 2025).

8.3.3.1 Tumor-infiltrating lymphocyte therapy

Tumor-infiltrating lymphocyte (TIL) therapy employs autologous T lymphocytes to generate tumor-reactive T-cells that penetrate the TME. T-cells demonstrate notable antitumor effects, especially in metastatic melanoma, and are being used for application in other solid tumors. Key challenges associated with this therapy include the need for tumor resection, variability in both the yields of TILs.

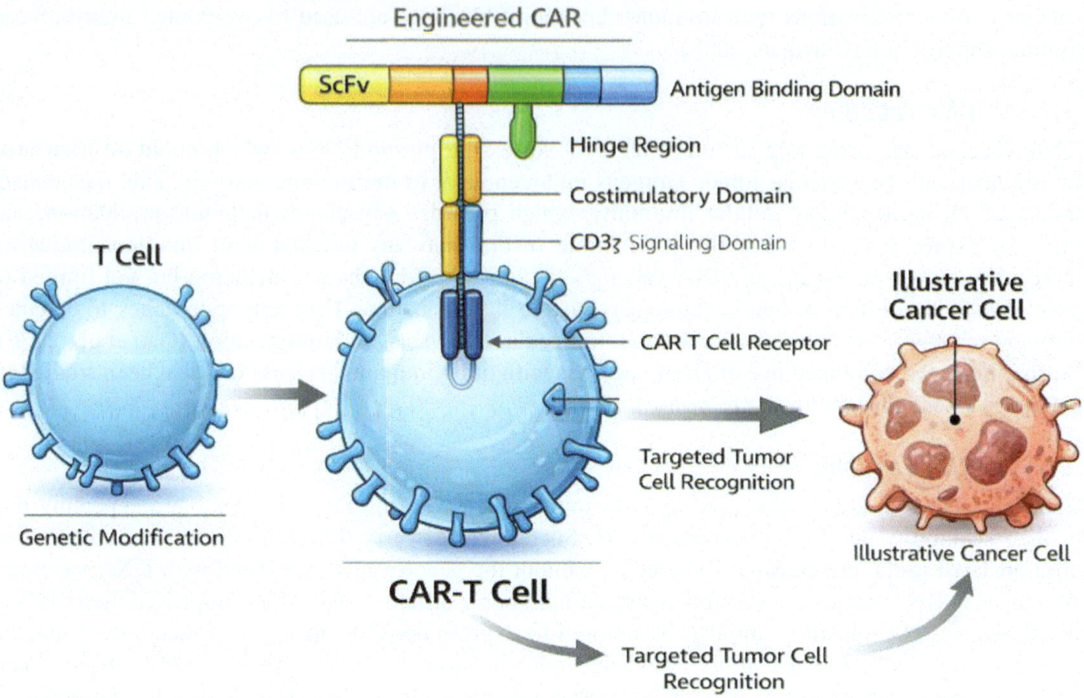

FIGURE 8.2

Structure and functional components of chimeric antigen receptor (CAR) T-cells.

Nevertheless, improvements in culture techniques and patient selection are improving the viability and effectiveness of TIL therapy (Goyal et al., 2023)

8.3.3.2 CAR-T cell therapy

CAR has been used to genetically modify T-cells to produce tumor antigens that can be recognized without MHC. The combination of intracellular signaling and antibody binding domains increases T cell activation when exposed to antigens, which is highly beneficial for B-cell malignancies. However, CAR-T treatment has severe toxicities like neurotoxicity and CRS, and it cannot treat solid tumors because of antigen heterogeneity and the immunosuppressive TME (Tokarew et al., 2018). Fig. 8.2 illustrates the CAR-T cell's structural design, featuring components such as the antigen-binding domain, hinge region, and CD3 signaling. These engineered CAR-T cells effectively target tumor antigens on cancer cells, facilitating immune-mediated destruction of the tumor cells.

8.3.3.3 CAR-NK cell therapy

CAR-NK cell therapy is a promising alternative to CAR-T cell therapy, utilizing the innate cytotoxic abilities of NK cells (Gujjar et al., 2025). This approach has advantages such as a lower risk of host disease and reduced incidence of CRS, making it suitable for allogeneic applications. Preclinical and early clinical studies have shown encouraging antitumor activity with safety profiles.

8.3.3.4 *T cell receptor engineered immune cells*

TCR-engineered immune cell therapy is a strategy to introduce tumor-specific TCR into patients' lymphocytes that are capable of recognizing intracellular antigens presented by MHC molecules (Hargadon, 2022). Nevertheless, TCR-engineered strategies are restricted by MHC restriction and the possibility of off-target toxicity as a result of cross-reactivity.

8.4 Gene therapy in cancer immunotherapy

A major development in cancer immunotherapy is gene therapy, which targets the TME and immune and tumor cells genetically. Gene editing techniques have promoted the integration of gene therapy with immuno-oncology by improving antitumor immunity and thwarting immune evasion mechanisms. In vivo techniques directly transfer transgenes to tumor or immune cells to affect cytokine production and immune checkpoint activities, while ex vivo techniques modify immune cells like T or NK cells for patient reinfusion (e.g., CAR-T and TCR therapies). These include the risks of immunological reactions and off-target effects, as well as gene therapy, gene delivery efficiency, specificity, safety, and long-term expression control (Huayamares et al., 2025).

8.4.1 Gene editing technologies

Genome editing technologies are an important platform for cancer immunotherapy. They facilitate the promotion of immune cell function and disruption of tumor-mediated suppressive pathways. This capability has enabled the generation of engineered immune cells and therapeutic gene constructs. CRISPR-Cas systems, transcription activator-like effector nucleases (TALENs), and zinc finger nucleases (ZFNs) are the main gene editing platforms; they have varying molecular structures and utilize different strategies for recognition of DNA substrate, targeted DNA double-strand breaks repaired by cellular DNA machinery. Gene editing technology depends on editing efficiency, targeting specificity, system design simplicity and clinical safety. In immuno-oncology, these technologies allow the knockout of immune checkpoint genes, insertion of synthetic receptors, potentiation in cytokine signaling, and improvement of immune cell persistence (Janji & Chouaïb, 2022). Table 8.2 provides significant gene editing methods, molecular tools, mechanisms of action, and therapeutic uses in immune cell engineering for cancer immunotherapies.

8.4.1.1 *CRISPR-Cas systems*

CRISPR-Cas technology is a pivotal gene editing tool renowned for its simplicity, flexibility, and efficiency, originating from bacterial immune systems. Cas endonucleases are guided by RNA to create sequence-specific double-strand breaks, enabling economical gene targeting. In cancer immunotherapy, CRISPR-Cas is utilized to modify immune cells, enhancing their tumor destruction capabilities by editing inhibitory receptors and allowing for modifications. Cas variants and techniques like base and prime editing are enhancing precision and safety for clinical applications of CRISPR-based immunotherapies (Jiao et al., 2024)

8.4.1.2 *Transcription activator-like effector nucleases*

TALENs are a precise gene targeting technology characterized by high specificity and minimal off-target effects. It is composed of a DNA-binding domain from transcription activator-like effectors and the

Table 8.2 Overview of gene editing technologies in immunotherapy.

Gene editing technology	Components	Mechanisms	Applications in immunotherapy	Advantages	Limitations	References
CRISPR-Cas systems	Guide RNA, Cas nuclease	RNA-guided DNA cleavage causing targeted double-strand breaks	Immune checkpoint knockout (e.g., PD-1), CAR insertion, multiplex gene editing	High efficiency, ease of design, multiplex capability, and cost-effectiveness.	Off-target effects, dependence on Protospacer Adjacent Motif (PAM) sequences, delivery challenges, and potential immunogenicity of Cas proteins.	Dong et al. (2021)
TALENs	DNA-binding domains	Paired nucleases induce site-specific DNA breaks	Universal CAR-T cell generation, TCR knockout	High specificity, lower off-target activity compared to early CRISPR systems.	Complex protein engineering, large construct size complicating delivery, and limited multiplexing capability.	Wang et al. (2022)
Zinc finger nucleases (ZFNs)	Zinc finger motifs, nuclease	Sequence-specific DNA cleavage via modular DNA recognition	Early engineered T cell therapies, immune gene disruption	High specificity when well-designed; clinically validated in early trials.	Labor-intensive design, limited flexibility, higher cost, and reduced scalability compared to CRISPR-based approaches.	Kopecka et al. (2021)

CAR, *chimeric antigen receptor;* PD-1, *programmed death-1;* TALENs, *transcription activator-like effector nucleases;* TCR, *T cell receptor.*

nuclease domain; they enable targeted modifications of genome loci. In immuno-oncology, TALENs are used to modify immune cells, particularly by knocking out endogenous T-cell receptors to create CAR-T cells with lowered graft-versus-host disease risks (Kardani, 2024). Nonetheless, TALENs present significant therapeutic potential for established gene editing methods.

8.4.1.3 Zinc finger nucleases

ZFNs are the advanced genome editing tools for site-specific genome modification, consisting of zinc finger DNA-binding motifs attached to a FokI nuclease domain that induces sequence-dependent DNA breakage (Khatami et al., 2022). Each zinc finger recognition code comprises a DNA sequence, and a modular collection of zinc finger motifs can be used to generate DNA-binding domains with their specificity. ZFNs have been used in early-stage clinical trials to engineer immune cells for therapeutic purposes, including inactivation of genes associated with immune regulation and viral entry. These include engineered T cell therapies for cancer and targeted genome editing in humans. ZFNs are being substituted by CRISPR technology for their safety profiles in the development of next-generation gene editing tools.

8.4.2 Viral and non-viral gene delivery systems

The effective cancer immunotherapy has been categorized into two systems such as the viral and non-viral vectors. Viral vectors leverage virus characteristics for efficient transfection, while non-viral systems enhance safety by reducing immunogenicity (Klichinsky et al., 2020). Key factors influencing the delivery system choice include target cell type, gene expression duration, genetic size, and safety profiles. In immuno-oncology, high transduction efficiency in immune cells or tumor tissue with minimal toxicity is crucial.

8.4.2.1 Lentiviral vectors

Lentiviral vectors are widely used in cancer immunotherapy, as they can transduce dividing and non-dividing cells, which results in stable and long-term expression of the genes (Chouaib et al., 2016). These vectors are based on human immunodeficiency virus and are rationally designed for safety. CAR-T and TCR-engineered cell therapies, because stable genomic integration allows therapeutic gene expression to persist over time for engineered immune cells to be effective (Li et al., 2024c; Lin et al., 2024). However, the risk of insertional mutagenesis for the vector design and integration in vectors and targeted integration methods is improving the safety of lentiviral systems.

8.4.2.2 Retroviral vectors

Retroviral vectors derived from gamma-retroviruses have been influential in the development of gene and immunotherapy, notably early CAR-T cell therapies (Mai et al., 2024). Such vectors are efficient in mediating integration into the host genome, but only transduce the tumor cells. Although they allow stable gene expression, to integrate transcriptional sites that present a risk of insertional oncogenesis and are used in the lentiviral systems.

8.4.2.3 Adenoviral vectors

Adenoviruses are modified to serve as viral vectors, which enable high-level transient gene expression in multiple cell types because they can accommodate large loads (Maksymova et al., 2025). They have been widely exploited as vectors for the delivery of immunostimulatory genes (e.g., cytokines) to the sites of tumors, which is a critical factor for oncolytic therapy and in situ vaccination against cancer. However, cancer immunotherapy by increasing regional immune activation in the TME.

8.4.2.4 *Adeno-associated virus vectors*

Adeno-associated virus (AAV) vectors are immunogenic; they may be ideal for in vivo gene delivery (McCann et al., 2018). These vectors allow sustained gene expression as an episome for a very long period but without the genes being inserted into the host genome, thus reducing the risk of insertional mutagenesis. In the field of cancer immunotherapy, AAV mechanisms can regulate immune reactions or provide a suitable environment for cancer destruction by the immune system. Vector engineering is being actively used to overcome these challenges and improve the potency of AAV-based immunotherapies.

8.4.2.5 *Lipid nanoparticles*

LNPs have been used as a major non-viral gene delivery system for nucleic acid therapeutics. Such nanoparticles can stabilize genetic materials from their degradation, improve cellular uptake, and efficiently deliver mRNA, siRNA, or gene editing agents. The development of mRNA vaccines has enhanced cancer immunotherapy. LNPs offer many benefits such as scalability of production, transient modifications and minimal risk of genomic integration. In cancer immunotherapy, LNPs are used to carry into immune cells or tumors mRNAs that encode tumor antigens, cytokines, or CRISPR tools (Mitra et al., 2024).

8.4.2.6 *Polymeric systems*

Polymeric gene delivery vectors are formulated to encapsulate and transport nucleic acids to the cells with the aid of synthetic or natural polymers. They have tunable physicochemical properties that may be designed toward drug release, targeting, and biodegradability. Polymeric systems in comparison to viral vectors generally have reduced immunogenicity and can be formulated with various nucleic acid forms (Mo et al., 2023). In cancer immunotherapy, polymeric nanoparticles (NPs) are being investigated for use in localized gene delivery and combination therapies.

8.4.3 Therapeutic genes in immuno-oncology

8.4.3.1 *Cytokine genes*

Cytokine genes that code for small secreted proteins are the key signaling molecules in the immune system. Their expression is controlled at multiple levels to tightly control appropriate immune responses. Activation by pathogens stimulates signaling cascades, such as NF-κB, Mitogen-Activated Protein Kinase (MAPK), and Janus Kinase–Signal Transducer and Activator of Transcription (JAK-STAT) pathways that induce the expression of critical cytokine genes, mainly interleukins, interferons, chemokines, and tumor necrosis factors. These proteins are involved in critical functions that attract, activate, expand, and differentiate leukocytes to coordinate both innate and adaptive immunity. Cytokine gene expression dysregulation underlies a broad range of diseases; excess transcription can cause inflammation and autoimmune disease, and a lack of sufficient expression may result in immunodeficiency (Murciano-Goroff et al., 2020).

8.4.3.1.1 Interleukin-2

IL-2 is a critical cytokine that mediates T cell proliferation, survival, and differentiation. While recombinant IL-2 has some efficacy in treating cancer for selected tumors, its clinical utility is limited by high levels of systemic toxicity. Gene therapy ploys have been developed to solve these issues through local and controlled IL-2 expression. Studies have explored IL-2 gene transfer into tumors and immune cells to enhance antitumor immunity. IL-2 expression can be localized to the tumor environment to improve therapeutic efficacy and minimize systemic toxicity (Noreen et al., 2024).

8.4.3.1.2 Interleukin-12

IL-12 is a critical immune stimulatory cytokine that promotes Th1 immune responses and enhances cytotoxic activities of T and NK cells (Otchere et al., 2023). Although IL-12 has demonstrated potent antitumor effects in preclinical studies, systemic administration of this cytokine is associated with potential toxicity. Viral as well as non-viral vectors have been investigated in IL-12 gene therapy or combined with ICIs to convert the TME to immunotherapy.

8.4.3.1.3 Interferon-γ

Interferon-γ (IFN-γ) is an essential cytokine for the induction of anticancer immunity and functions to improve tumor antigen presentation, activate macrophages, and suppress tumor progression (Pacheco & Tai, 2025). IFN-γ gene therapy is used to enhance such immune responses in the TME and thereby improve immune-mediated tumor control. Preliminary data suggest that the gene-mediated delivery can be used with other immunotherapeutic modalities. However, prolonged IFN-γ signaling could result in adaptive immune resistance to be controlled for improving gene therapy (Leonard et al., 2024).

8.5 Cellular immunotherapy approaches

Cellular immunotherapy, which targets and kills cancer cells, is an emerging oncology treatment (Qin et al., 2025). Hematologic malignancies have improved with ACT, increasing their application in solid tumors. For *ex vivo* efficacy studies, patient or donor immune effector cells are separated, manipulated, and targeted (Ranjan et al., 2024). TILs, TCR-engineered T-cells, and CAR-engineered cells, which detect antigens independent of the MHC and have great cytotoxicity, are the most viable techniques. Antigen heterogeneity, immune responses, and off-tumor toxicity hinder cellular immunotherapies. Optimizing cell design, enhancing injected cell permanence, and producing allogeneic resources. Multimodal therapy includes cellular immunotherapy. Immune cells and the TME aid long-term tumor management (Ruan & Wang, 2025).

8.5.1 Chimeric antigen receptor-based therapies

CAR treatment is a novel immunotherapeutic method that uses synthetic receptors to drive immune cells to recognize tumor-associated antigens. Clinical efficacy of CAR-T cell therapy, targeting CD19 in B-cell malignancies, has led to regulatory approval of products for leukemia and lymphoma, resulting in significant and long-lasting response rates (Sarvepalli et al., 2023). Since CAR engineering has advanced, newer generations have been developed with enhanced efficacy and safety. CAR technology has also changed from T-cells to NK and macrophages. Next-generation CAR-based therapeutics address problems in solid tumors with dense stroma and immunosuppressive factors (e.g., CD70 or IDO), demonstrating synthetic immunological potential for cancer treatment (Wang et al., 2024).

8.5.1.1 CAR-T cell design and generations

The five generations of CAR-T treatment have improved therapeutic outcomes. First-generation CARs exhibited limited in vivo expansion due to their signaling domains. Second-generation CARs with a costimulatory domain increased T-cell growth and responsiveness, especially in CD19-targeted treatments (Sloas et al., 2021). However, the third generation incorporated two costimulatory domains but did not consistently enhance efficacy over the second generation, raising toxicity concerns. Fourth-generation CARs, or TRUCKs, could induce transgenic payloads to modify the TME. Fifth-generation CARs have added signaling domains to boost T-cell proliferation for prolonged, effective cancer immunotherapies for diverse malignancies (Zhang et al., 2022a).

8.5.1.2 Chimeric antigen receptor natural killer cell therapies

CAR-NK cell immunotherapies enhance the natural ability of NK cells to fight tumors through genetic modification, an alternative to T-cell therapies. These cells can kill tumor cells directly without sensitization and are being developed for specific antigen reactivity while maintaining natural killing capabilities (Wang et al., 2021). These strong antitumor activities against hematologic cancers have a safety profile that suggests lower rates of severe CRS and neurotoxicity compared to CAR-T cells (Sun et al., 2025). Strategies to improve CAR-NK cell efficacy include cytokine support, modifications to CAR signaling, and gene editing techniques. CAR-NK cells are anticipated to play a significant role in cancer treatment or replace CAR-T cells.

8.5.1.3 CAR-macrophage therapies

CAR-macrophage (CAR-M) therapies, which use macrophages to become TAMs in solid tumors and decrease cancer growth, are a developing immunotherapy. CAR-M therapies aim to retrain macrophages to recognize and phagocytose tumor-associated antigens, resulting in pro-inflammatory and anticancer effects. In addition to targeting tumor cells, CAR-M cells positively impact the TME by activating cytokine release, antigen-presenting, and adaptive immune system activation (Niculescu-Duvaz, 2000). Preclinical data suggest CAR-M therapies can increase immune cell function and modify tumor stroma with T cell therapies that are ineffective with CAR-T therapy. However, challenges remain, such as achieving efficient gene transduction in macrophages and persistent CAR expression. Early clinical development of CAR-M treatments could transform immunotherapy for malignant cancers.

8.6 Synergistic integration of gene and cellular therapies

8.6.1 Gene-edited immune cells

Gene-modified immune cells are key in modern cellular immunotherapies with unprecedented control of immune cell properties and effector function (Wang et al., 2025). Unlike conventional medications that are transient and may cause systemic toxicity, this type of genetic manipulation induces stable changes in the epigenetic profile of these modified cells that last for the life span of the cell. This long-term stability is particularly attractive for ACT therapies in which sustained efficacy and durability are critical for optimal therapeutic applications. As a result, gene editing has emerged as a cornerstone for enhancing the safety of immune therapies (Yuan et al., 2024). CRISPR/Cas9 has transformed stem cell genome editing because of its ease of use and effectiveness, which permits several genetic targets. Consequently, many clinical trials have the ability to use gene-edited T cells and NK cells to tackle various cancers. In addition to their potential in cancer therapy, engineered immune cells for effective therapies to enable immune tolerance and autoimmunity (Jiang et al., 2025).

8.6.1.1 Inhibitory receptor targeting

Inhibitory receptors are essential for maintaining the balance of the immune system, avoiding hyperactivation and autoimmunity. But in TME, these receptors are hijacked to shut down antitumor immunity. TILs encounter antigens chronically, under conditions of chronic antigen stimulation and stress experience, which in turn, results in the sustained expression of inhibitory receptors by TILs, leading to dysfunction as a consequence of decreased cytokine production and proliferation. Gene editing provides another approach to permanently knocking down immune cell inhibitory receptors so that the specificity can be increased and off-target effects minimized. Thus, by the multiple inhibitory receptors such as PD-1 and lymphocyte activation gene-3 (LAG-3), it is able to block redundancy in immune regulation, thereby enhancing immune attack on tumors (Zafar et al., 2024).

8.6.1.1.1 Programmed death-1

The programmed cell death protein 1 (PD-1) is an inhibitory marker of activated immune cells and controls immune tolerance. Its association with its ligands PD-L1 and PD-L2 suppresses the activity of T cells and is involved in tumor immune escape that many cancers exhibit. PD-1/PD-L1 inhibition has already been clinically effective, and gene editing PD-1 in CAR-T cells demonstrates the potential to better immunotherapy by increasing cell survival and cytotoxicity (Zhang et al., 2022a). In contrast to antibodies, PD-1 genetic deletion has the potential of lasting protection against PD-L1 inhibition and has the risk of overactivation of the immune system.

8.6.1.1.2 Lymphocyte activation gene 3

LAG-3 is an inhibitory receptor present on activated T-cells, Regulatory T-cells, and NK cells, which play an important role in regulating the activation of immunity (Fesnak et al., 2016). It binds to the MHC II, preventing TCR signaling and effector actions. Co-expression of LAG-3 with PD-1 in the TME invites resistance to immunotherapy and depletion of T-cells. Gene editing targeting LAG-3 is potentially useful in the rejuvenation of depleted immune cells and in the adoption of cell therapies, as preclinical testing has shown increased proliferation and cytotoxicity in LAG-3-driven T-cells, particularly in association with PD-1 deficiency. Nevertheless, there are difficulties with targeting LAG-3 because it is a complex protein that interacts with different ligands, and its complete deletion may be detrimental to immune regulation (Zhou et al., 2025).

8.6.1.2 Enhancement of immune cell persistence and cytotoxicity

The efficacy of cancer immunotherapy depends on the persistence and cytotoxic capacity of adoptively transferred immune cells. Gene editing strategies are being employed to enhance the longevity and effectiveness of these immune cells by targeting the metabolic pathways (Canella & Rajappa, 2023).

Techniques include promoting anti-apoptotic B-cell Lymphoma (BCL-2) family proteins and limiting pro-apoptotic signals, alongside supporting cytokines like IL-7 and IL-15 to improve proliferation and memory development without increasing toxicity. Furthermore, gene editing can boost the expression of effector molecules, enhance mitochondrial function, and adapt metabolic processes to tumor responses. These advancements aim to ensure immune cells persist longer and possess greater cytotoxic capabilities, enhancing cellular immunotherapy efficacy (Baena et al., 2025).

8.7 Tumor microenvironment modulation

8.7.1 Gene-modified cells targeting tumor microenvironment barriers

Genetically modified immune cells, such as CAR-engineered lymphocytes, are being developed to improve CAR-T therapy in solid tumors (Tufail et al., 2025). Engineering CAR-T cells to produce chemokine receptors for tumor targeting and altering cells to secrete ECM breakdown enzymes for immune cell penetration are major efforts. Synthetic promoters and microRNA-controlled circuits reduce adverse effects by controlling enzyme expression (Brancewicz & Kucharzewska, 2025). Also, immunosuppressive factors in the TME are targeted by engineering immune cells using dominant receptors to retain their functionality and CAR-T cells to release pro-inflammatory cytokines, which have the potential to reprogram the immune environment.

8.7.2 Hypoxia, fibrosis, and metabolic suppression

Hypoxia in solid tumors results from vasculature and rapid tumor proliferation, leading to immunosuppression. Hypoxia-Inducible Factors (HIFs) increase in these conditions, promoting angiogenesis, immune evasion, and inhibiting CTLs and NK cells. This milieu promotes regulatory T-cells and MDSC, as well as the expression of immunological checkpoints such as PD-L1, resulting in local immunotolerance (Cheng et al., 2025). Cancer-Associated Fibroblasts (CAFs)-mediated fibrosis has a stroma-rich structure that blocks immune cell access and promotes immunosuppression via TGF-β and Vascular Endothelial Growth Factor (VEGF). Immune cells can be directed against fibrotic tissues to improve tumor entry. Additionally, tumor metabolism inhibits T cell activity by suppressing immunological function through glucose, amino acids, and metabolic waste accumulation (Li et al., 2025).

8.7.3 Immunosuppressive cells

8.7.3.1 Regulatory T-cells

Tregs play a crucial role in maintaining immune balance, but their presence in tumors for effective antitumor responses. They inhibit effector T-cells through mechanisms such as cytokine release (IL-10 and TGF-β), expression of the inhibitory receptor CTLA-4, and IL-2 absorption (Li et al., 2024a). Immunotherapy approaches targeting Tregs include selective depletion, functional reprogramming, or engineering immune cells to resist Treg suppression (Morva et al., 2025). Preclinical studies on genetically engineered CAR-T cells have demonstrated effectiveness in Treg inhibition or in producing cytokines that target Tregs.

8.7.3.2 Myeloid-derived suppressor cells

MDSCs are the myeloid cells that increase in cancer, resulting in severe immunosuppression by blocking the activation of T cells due to the effects of arginase and the synthesis of nitric oxide and reactive oxygen species. They also mediate the generation of regulatory T-cells and differentiation of TAMs via inflammatory cytokines and growth factors of cancer and stromal cells (Li et al., 2024b). Some of the therapeutic approaches against MDSCs are the prevention of their recruitment, inhibition of their suppressive enzymes, to non-suppressive forms. Furthermore, genetically modified immune cells that inhibit the activity of MDSC are also used to counter the myeloid-derived immunosuppression within the TME.

8.7.3.3 Tumor-associated macrophages

The immune cells of solid tumors are called TAMs, which have been shown to induce an antitumor effect. TAMs are likely to assume an M2-like phenotype in the TME, which promotes tumor growth, angiogenesis, and immune evasion. They also set free growth factors, matrix remodeling enzymes, and immunosuppressive cytokines, which help tumors to develop (Guedan et al., 2018). TAMS is employed to induce an antitumor phenotype by inhibiting the stimulating factors, by the toll-like receptors, or the immune cells to activate macrophages for TAMs into tumor facilitators into antitumor cells.

8.8 Emerging technologies

Genome and cell-based therapy with targeted and sustainable techniques is advancing cancer immunotherapy (Riva & Petrini, 2019). Gene editing, especially using CRISPR-Cas, improves immune cell performance by boosting tumor identification and immunosuppressive resistance. Multiplex gene editing of CAR-T, CAR-NK, and TCR-engineered treatment to improve specificity and reduce toxicity. Synthetic biology may produce immune systems that release cytokines to reduce adverse impacts (Wang et al., 2024; Wellhausen et al., 2024). RNA-based immune response control is dynamic without genetic alterations. AI and single-cell omics enable customized therapy using new tumor antigens and biomarkers.

8.8.1 Synthetic biology and immune cells

Synthetic biology is also transforming the cancer immunotherapy approach by improving the specificity of immune cells (Xiang et al., 2022). It uses modular genetic circuits, unlike traditional genetic engineering, to allow immune cells to successfully cross complex TMEs to differentiate tumor and tissues to overcome the obstacles of off-tumor toxicity and therapy resistance. The most significant progress has been made in the form of CAR-T and CAR-NK, where several tumor antigens are recognized, eliminating the problem of antigen escape and enhancing the selectivity of the treatment (Zhang et al., 2022b). Artificial receptors promote real-time immune cell response with respect to tumor-specific applications. Also, synthetic biology enhances the persistence and metabolic pathways of immune cells, which allows them to survive under adverse environments such as hypoxia. Safety measures play a very important role in controlling the engineered immune cells in case of toxicity.

Artificial intelligence and the genome editing into these therapies, which will lead to a new phase of adaptive cancer therapy (Wang et al., 2025).

8.8.2 Nanotechnology-assisted gene and cell delivery

Nanotechnology has also enhanced the precision and noninvasive delivery of therapeutic genes and cells into immune cells and tumors in cancer immunotherapy (Yang, 2015). Nanoparticles may be modified in terms of size, charge on the surface, functionalization, biological barriers, including enzymatic degradation, and low cellular uptake. They are used to transfer nucleic acids, proteins as well as small molecules that modulate immune response and TME. Nanoparticles made of lipids are less in modification of the genome and polymer nanoparticles and inorganic nanoparticles have their own gains in terms of prolonged release and stability.

Nanocarriers are important to improve the efficacy of adoptively transferred immune cells by allowing them to be specifically delivered to tumor cells, which leads to better antitumor effects with limited systemic toxicity. These tumor stimuli can also be considered a design innovation aimed at attacking immune cells (Yuan et al., 2024). Nonetheless, there are still some challenges, such as mass-manufacturing, reproducibility, and long-term safety, which require paying specific attention to the design of materials. Synthetic biology and immunotherapy are major breakthroughs in the multipotent immune therapies as opposed to the traditional vector delivery techniques (Zhou et al., 2025).

8.9 Clinical translation and regulatory challenges

8.9.1 Manufacturing scalability and standardization

Scalability and standardization of clinical production processes of cell-based immunotherapy is one of the main obstacles to clinical translation since it includes multiple complicated procedures such as cell isolation, genetic modification, expansion, quality testing, and cryopreservation. These complexities are dangerous to product and patient safety. Other therapies, such as CAR-T cells, are their terms of scalability because of patient specificity through patient quality systems (Ceja et al., 2024). In order to overcome these challenges, manufacturers are embracing automation, closed-system production, and digital process controls. Therefore, to enhance bioengineering, regulatory frameworks, and global manufacturing with significant improvements to achieve scalable and standardized production.

8.9.2 Safety concerns

Safety concerns pose significant challenges for novel immunotherapies, as robust immune activation can lead to unpredictable and severe adverse effects (Wang et al., 2021). Unlike predictable chemotherapy toxicities, immunotherapy-related toxicities are heterogeneous and specific to individuals, necessitating safety assessments and clinical monitoring. The complexity of genetically engineered immune cells increases risks, such as overgrowth, for monitoring toxicities and potential malignancies (Dias et al., 2024). Ongoing developments in synthetic biology aim to create safer immunotherapy systems through switchable controls and adjusted gene expression, though their clinical applications.

8.9.2.1 *Cytokine release syndrome*

CAR-T cell therapies of CRS are characterized by a rapid immune response and the excessive release of pro-inflammatory cytokines like IL-6, leading to symptoms ranging from flu-like effects to severe systemic inflammation and multi-organ dysfunction. The pathophysiology involves interactions between engineered immune cells and immune cells, resulting in a cycle of inflammation (Gujjar et al., 2025). High tumor burden and quick immune cell expansion are key risk factors, emphasizing the need for careful patient selection and risk assessment. Management strategies include IL-6 receptors, necessitating meticulous treatment to ensure their effectiveness. Standardized treatment techniques have improved outcomes, yet the clinical application of new immunotherapies is challenged by CRS with lower cytokine production to enhance therapeutic implementations.

8.9.2.2 *Neurotoxicity*

Neurotoxicity, specifically immune effector cell-associated neurotoxicity syndrome, is a significant adverse effect of cellular immunotherapy, varying from mild cognitive impairments to severe conditions such as encephalopathy, seizures, cerebral edema, and coma. Unlike CRS, neurotoxicity's occurrence does not consistently correlate with cytokine levels and can happen independently or post-resolution of systemic inflammation (Wang et al., 2025). Blood-brain barrier disruption, endothelial activation, and cytokines in the central nervous system, leading to neural dysfunction and inflammation, are influenced by genetic characteristics of modified cells and patient susceptibility.

8.9.3 Regulatory frameworks and ethical considerations

Advanced immunotherapies face regulatory issues, including the regulation of organizations, such as the Food and Drug Administration and European Medicines Agency, under such categories as advanced therapy medicinal products (Garnipudi, 2024). These medicines are tested preclinically and clinically to ensure patient safety and quality due to their rapid advancement. Due to long-term adverse effects, initial ethical issues are greater.

8.10 Future perspectives

The future of immunotherapy lies in the integration of synthetic biology, systems immunology, and precision medicine. Genome editing technologies like CRISPR and programmable immune cells that address the issues of specificity, potency, and safety, and adjust to the microenvironments of the tumor to enhance its response to clinical advantages over time (Qin et al., 2025). Also, therapeutic development and personalized treatments can be optimized with the use of machine learning models based on clinical and genomic data on a large scale, serve patients during their selection, and create combination regimens (Kardani, 2024). The immunotherapy uses to the autoimmune diseases and chronic infections (Diorio et al., 2024). These are the translational obstacles, such as safety oversight, regulation, and ethical issues, to transforming the healthcare provision with multidisciplinary cooperation.

8.11 Conclusion

The future of cancer immunotherapy integrates gene and cellular therapies, indicating a new precision oncology age. CRISPR-Cas, TALENs, and ZFNs are used to design immune cells, including CAR-T, CAR-NK, and TCR-modified cells, to improve their specificity and persistence in the immunosuppressive TME. These modified cells address immune evasion, antigen heterogeneity, and TME-mediated suppression by targeting inhibitory checkpoints (PD-1 and LAG-3), optimizing metabolic pathways, and permitting localized cytokine production. Lentivirus, AAV, and LNPs gene delivery technologies enable in vivo and ex vivo changes, extending therapeutic possibilities. New synthetic biology and nanotechnology technologies create smart therapies that activate and modulate the microenvironment. The clinical translation includes manufacturing scalability, standardization, acute toxicities such as CRS and neurotoxicity, and complex regulatory frameworks. Interdisciplinary collaboration, automated bioprocessing, and AI/omics-integrated predictive biomarkers are needed to overcome these obstacles. The combination of genetic and cellular mechanisms in immuno-oncology could improve patient outcomes and worldwide treatment.

References

Altvater, B., Kailayangiri, S., Spurny, C., Flügge, M., Meltzer, J., Greune, L., Urban, K., Schwöppe, C., Brand, C., Schliemann, C., Hintelmann, H., Harrach, S., Hartmann, W., Abken, H., Kuehle, J., Schambach, A., Görlich, D., Berdel, W. E., & Rössig, C. (2023). CAR T cells as micro pharmacies against solid cancers: Combining effector T-cell mediated cell death with vascular targeting in a one-step engineering process. *Cancer Gene Therapy, 30*(10), 1355–1368.

Amer, M. H. (2014). Gene therapy for cancer: Present status and future perspective. *Mol. Cell. Ther., 2*, 27.

Augustin, R. C., Delgoffe, G. M., & Najjar, Y. G. (2020). Characteristics of the tumor microenvironment that influence immune cell functions: Hypoxia, oxidative stress, metabolic alterations. *Cancers, 12*(12), 3802.

Baena, J. C., Victoria, J., Toro-Pedroza, A., Aragón, C. C., Ortiz-Guzman, J., García-Robledo, J. E., Torres, D., Ríos-Serna, L. J., Albornóz, L. L., Rosales, J., Cañas, C. A., Cruz, G., Osorio, F. O., Fleitas, T., Laponogov, I., Loukanov, A., & Veselkov, K. (2025). Smart CAR-T Nano symbionts: Archetypes and proto-models. *Frontiers in Immunology, 16*, 1635159.

Bagchi, S., Yuan, R., & Engleman, E. G. (2020). Immune checkpoint inhibitors for the treatment of cancer: Clinical impact and mechanisms of response and resistance. *Annual Review of Pathology: Mechanisms of Diseases, 16*, 223–249.

Brancewicz, J., & Kucharzewska, P. (2025). Emerging macrophage-based therapies for cancer: A review of preclinical and clinical advances. *Frontiers in Immunology, 16*, 1679271.

Canella, A., & Rajappa, P. (2023). Therapeutic utility of engineered myeloid cells in the tumor microenvironment. *Cancer Gene Therapy, 30*(7), 964–972.

Catanzaro, E., Beltrán-Visiedo, M., Galluzzi, L., & Krysko, D. V. (2024). Immunogenicity of cell death and cancer immunotherapy with immune checkpoint inhibitors. *Cellular & Molecular Immunolgy, 22*(1), 24–38.

Ceja, M. A., Khericha, M., Harris, C., Puig-Saus, C., & Chen, Y. Y. (2024). CAR-T cell manufacturing: Major process parameters and next-generation strategies. *Journal of Experimental Medicine, 221*(2), e20230903.

Cha, J., Chan, L.-C., Song, M. S., & Hung, M. (2019). New approaches on cancer immunotherapy. *Cold Spring Harbor Perspectives in Medicine, 10*(8), a036863.

Chang, H. L., Schwettmann, B., McArthur, H. L., & Chan, I. S. (2023). Antibody-drug conjugates in breast cancer: overcoming resistance and boosting immune response. *Journal of Clinical Investigation, 133*(18), e172156.

Chen, A. X. Y., Yap, K. M., Kim, J. S., Sek, K., Huang, Y., Dunbar, P., Wiebking, V., Armitage, J., Munoz, I., Todd, K. L., Derrick, E. B., Nguyen, D. Q., Tong, J., Chan, C. W., Hoang, T. X., Audsley, K. M., van Elsas, M. J., Middelburg, J., Lee, J. N., & Beavis, P. A. (2025). Rewiring endogenous genes in CAR T cells for tumour-restricted payload delivery. *Nature, 644*(8075), 241–249.

Cheng, N., Yang, X., Chen, D., & Xu, C. F. (2025). Editorial: Vector-based gene delivery in cancer immunotherapy. *Frontiers in Immunology, 16*, 1587359.

Cheung, A. S., & Mooney, D. (2015). Engineered materials for cancer immunotherapy. *Nano Today, 10*(4), 511–531.

Chouaib, S., Noman, M. Z., Kosmatopoulos, K., & Curran, M. A. (2016). Hypoxic stress: Obstacles and opportunities for innovative immunotherapy of cancer. *Oncogene, 36*(4), 439–445.

Cogdill, A. P., Andrews, M. C., & Wargo, J. A. (2017). Hallmarks of response to immune checkpoint blockade. *British Journal of Cancer, 117*(1), 1–7.

Corrigan, C. J. (2008). Immunosuppression and Immunomodulation. In N. F. Adkinson, B. S. Bochner, A. W. Burks, W. W. Busse, S. T. Holgate, R. F. Lemanske, & R. E. O'Hehir (Eds.), *Middleton's allergy: Principles and practice* (pp. 287–302). Elsevier.

Crompton, J. G., Sukumar, M., Roychoudhuri, R., Clever, D., Gros, A., Eil, R., Tran, E., Hanada, K., Yu, Z., Palmer, D. C., Kerkar, S. P., Michalek, R. D., Upham, T., Leonardi, A., Acquavella, N., Wang, E., Marincola, F. M., Gattinoni, L., Muranski, P., & Restifo, N. P. (2014). Akt inhibition enhances expansion of potent tumor-specific lymphocytes with memory cell characteristics. *Cancer Research, 75*(2), 296–305.

Datta, M., Coussens, L. M., Nishikawa, H., Hodi, F. S., & Jain, R. K. (2019). Reprogramming the tumor microenvironment to improve immunotherapy: Emerging strategies and combination therapies. *American Society of Clinical Oncology Educational Book, 39*, 165–174.

Dias, J., García, J., Agliardi, G., & Roddie, C. (2024). CAR-T cell manufacturing landscape—Lessons from the past decade and considerations for early clinical development. *Molecular Therapy Methods & Clinical Development, 32*(2), 101250.

Diorio, C., Teachey, D. T., & Grupp, S. A. (2024). Allogeneic chimeric antigen receptor cell therapies for cancer: Progress made and remaining roadblocks. *Nature Reviews Clinical Oncology, 22*(1), 10–22.

Dong, Y., Wan, Z., Gao, X., Yang, G., & Liu, L. (2021). Reprogramming immune cells for enhanced cancer immunotherapy: Targets and strategies. *Frontiers in Immunology, 12*, 609762.

Fesnak, A. D., June, C. H., & Levine, B. L. (2016). Engineered T cells: The promise and challenges of cancer immunotherapy. *Nature Reviews Cancer, 16*(9), 566–581.

Gao, L., Wang, J., & Bi, Y. (2025). Nanotechnology for neurodegenerative diseases: Recent progress in brain-targeted delivery, stimuli-responsive platforms, and organelle-specific therapeutics. *International Journal of Nanomedicine, 20*, 11015–11045.

Garcia-Guerra, A., Dunwell, T. L., & Trigueros, S. (2018). Nano-scale gene delivery systems: Current technology, obstacles, and future directions. *Current Medicinal Chemistry, 25*(21), 2448–2464.

Garnipudi, D. K. (2024). Gene therapy delivery systems: Evaluating the latest methods for effective gene therapy delivery. *African Journal of Biomedical Research, 27*(4s), 501–508.

González, J., Park, K. W., Evans, D., Sharma, R., Sahaym, O., Gopalakrishnan, S., Dar, A. I., Valdez, T. A., & Sharma, A. (2025). Nano approaches to nucleic acid delivery: Barriers, solutions, and current landscape. *Wiley Interdisciplinary Reviews: Nanomedicine and Nanobiotechnology, 17*(2), e70010.

Goyal, V. K., Rani, D., Arora, R., Mehrotra, S., Deng, C., & Wang, Y. (2023). Unlocking the potential of nano-enabled precision agriculture for efficient and sustainable farming. *Plants, 12*(21), 3744.

Guedan, S., Ruella, M., & June, C. H. (2018). Emerging cellular therapies for cancer. *Annual Review of Immunology, 37*, 145–171.

Gujjar, S., Kukal, S., Prakash, J., Balaji, N., Sainger, S., Roy, S., Rallapalli, S., Mahadevappa, R., Minocha, S., Kumar, S., & Mathapati, S. (2025). Nanomaterials for biomedical applications: Addressing regulatory hurdles and strategic solutions. *Nano Trends, 11*, 100127.

Hargadon, K. M. (2022). Genetic dysregulation of immunologic and oncogenic signaling pathways associated with tumor-intrinsic immune resistance: A molecular basis for combination targeted therapy-immunotherapy for cancer. *Cellular and Molecular Life Sciences, 80*, 40.

Henke, E., Nandigama, R., & Ergün, S. (2020). Extracellular matrix in the tumor microenvironment and its impact on cancer therapy. *Frontiers in Molecular Biosciences, 6*, 160.

Huayamares, S. G., Lian, L., Rab, R., Hou, Y., Radmand, A., Kim, H., Zenhausern, R., Achyut, B. R., Ross, M. G., Lokugamage, M. P., Loughrey, D., Peck, H. E., Echeverri, E. S., Da Silva Sanchez, A. J., Shajii, A., Li, A., Tiegreen, K. E., Santangelo, P. J., Sorscher, E. J., & Dahlman, J. E. (2025). Nanoparticle delivery of a prodrug-activating bacterial enzyme leads to anti-tumor responses. *Nature Communications, 16*, 3490.

Iglesias-Escudero, M., Arias-González, N., & Martínez-Cáceres, E. (2023). Regulatory cells and the effect of cancer immunotherapy. *Molecular Cancer, 22*, 26.

Janji, B., & Chouaïb, S. (2022). The promise of targeting hypoxia to improve cancer immunotherapy. *Frontiers in Immunology, 13*, 880810.

Jiang, M., Zhang, K., Wang, Z., Gao, M., Su, S., He, J., Xu, H., Bo, Z., Jiang, Z., Zhang, C., Hui, J. H. P., & Wei, R. (2025). Nanomaterials in gene therapy and genome editing: Challenges and emerging directions. *Journal of Nanbiotechnology, 23*, 169.

Jiao, L., Sun, Z., Sun, Z., Liu, J., Deng, G., & Wang, X. (2024). Nanotechnology-based non-viral vectors for gene delivery in cardiovascular diseases. *Frontiers in Bioengineering and Biotechnology, 12*, 1349077.

Kardani, S. L. (2024). Nanocarrier-based formulations: Regulatory challenges, ethical and safety considerations in pharmaceuticals. *Asian Journal of Pharmaceutics, 18*(2), 476–484.

Khalil, D. N., Smith, E. L., Brentjens, R. J., & Wolchok, J. D. (2016). The future of cancer treatment: Immunomodulation, CARs and combination immunotherapy. *Nature Reviews Clinical Oncology, 13*(5), 273–290.

Khatami, F., Aghamir, Z. S., Jahanshahi, F., Feiz-Abadi, S. A., Birang, F., Khoshchehreh, M., Shabestari, A. N., & Aghamir, S. M. K. (2022). The gene manipulation and cellular immunotherapy combination in the treatment of cancer. *Iranian Journal of Biotechnology, 20*(2), e3094.

Klichinsky, M., Ruella, M., Shestova, O., Lu, X. M., Best, A., Zeeman, M. E., Schmierer, M., Gabrusiewicz, K., Anderson, N. R., Petty, N. E., Cummins, K. D., Shen, F., Shan, X., Veliz, K., Blouch, K., Yashiro–Ohtani, Y., Kenderian, S. S., Kim, M., O'Connor, R. S., & Gill, S. (2020). Human chimeric antigen receptor macrophages for cancer immunotherapy. *Nature Biotechnology, 38*(8), 947–953.

Kopecka, J., Salaroglio, I. C., Pérez-Ruíz, E., Sarmento-Ribeiro, A. B., Saponara, S., de las Rivas, J., & Riganti, C. (2021). Hypoxia as a driver of resistance to immunotherapy. *Drug Resistance Updates, 59*, 100787.

Leonard, E. K., Tomala, J., Gould, J. R., Leff, M. I., Lin, J.-X., Li, P., Porter, M. J., Johansen, E., Thompson, L., Cao, S. D., Hou, S., Henclová, T., Huličiak, M., Sargunas, P. R., Fabilane, C. S., Vaněk, O., Kovář, M., Schneider, B., Raimondi, G., & Spangler, J. B. (2024). Engineered cytokine/antibody fusion proteins improve IL-2 delivery to pro-inflammatory cells and promote antitumor activity. *JCI Insight, 9*(18), e173469.

Li, H., Zhou, Q., Cao, N., Hu, C., Wang, J., He, Y., Jiang, S., Li, Q., Chen, M., Gong, L., Luo, M., Deng, X., & Luo, Z. (2025). Nanobodies and their derivatives: Pioneering the future of cancer immunotherapy. *Cell Communication and Signaling, 23*, 271.

Li, J., Chen, P., & Ma, W. (2024a). The next frontier in immunotherapy: Potential and challenges of CAR-macrophages. *Experimental Hematology & Oncology, 13*, 73.

Li, N., Geng, S., Dong, Z., Jin, Y., Ying, H., Li, H., & Shi, L. (2024b). A new era of cancer immunotherapy: Combining revolutionary technologies for enhanced CAR-M therapy. *Molecular Cancer, 23*, 119.

Li, Y.-R., Halladay, T., & Yang, L. (2024c). Immune evasion in cell-based immunotherapy: Unraveling challenges and novel strategies. *Journal of Biomedical Science, 31*, 49.

Libutti, S. K., Tamarkin, L., & Nilubol, N. (2018). Targeting the invincible barrier for drug delivery in solid cancers: Interstitial fluid pressure. *Oncotarget, 9*(87), 35723–35725.

Lin, H., Li, C., Zhang, W., Wu, B., Wang, Y., Wang, S., Wang, D., Li, X., & Huang, H. (2024). Synthetic cells and molecules in cellular immunotherapy. *International Journal of Biological Sciences, 20*(8), 2833–2855.

Liu, B., Zhou, H., Tan, L., Siu, K. T. H., & Guan, X. (2024). Exploring treatment options in cancer: Tumor treatment strategies. *Signal Transduction and Targeted Therapy, 9*, 175.

Liu, Z., Zhou, Z., Dang, Q., Xu, H., Lv, J., Li, H., & Han, X. (2022). Immunosuppression in tumor immune microenvironment and its optimization from CAR-T cell therapy. *Theranostics, 12*(14), 6273–6290.

Mai, Z., Lin, Y., Lin, P., Zhao, X., & Cui, L. (2024). Modulating extracellular matrix stiffness: A strategic approach to boost cancer immunotherapy. *Cell Death & Disease, 15*, 307.

Maksymova, L., Pilger, Y. A., Nuhn, L., & Van Ginderachter, J. A. (2025). Nanobodies targeting the Tumor Microenvironment and their formulation as nanomedicines. *Molecular Cancer, 24*, 65.

McCann, J. V., Null, J. L., & Dudley, A. C. (2018). Deadly DAaRTS destroy cancer cells via a Tumor Microenvironment–mediated trigger. *Journal of Clinical Investigation, 128*(7), 2750–2752.

Mi, Y., Hagan, C. T., Vincent, B. G., & Wang, A. Z. (2019). Emerging nano-/microapproaches for cancer immunotherapy. *Advances in Therapy, 2*(6), 1900128.

Mitra, A., Kumar, A., Amdare, N., & Pathak, R. (2024). Current landscape of cancer immunotherapy: Harnessing the immune arsenal to overcome immune evasion. *Biology, 13*(5), 307.

Mo, F., Liu, J., Chen, G., Gong, N., & Li, Z. (2023). Editorial: Nanotechnology and bioengineering platforms for drug and gene delivery. *Frontiers in Bioengineering and Biotechnology, 11*, 1328961.

Morva, A., Arroyo, A. B., Andreeva, L., Tapia-Abellán, A., & Luengo-Gil, G. (2025). Unleashing the power of CAR-M therapy in solid tumors: A comprehensive review. *Frontiers in Immunology, 16*, 1615760.

Murciano-Goroff, Y. R., Warner, A. B., & Wolchok, J. D. (2020). The future of cancer immunotherapy: Microenvironment-targeting combinations. *Cell Research, 30*(6), 507–519.

Niculescu-Duvaz, I. (2000). Prodrug-activating systems in suicide gene therapy. *Journal of Clinical Investigation, 105*(9), 1161–1167.

Noreen, S., Saleem, M. H., Ali, B., Khan, K. A., Hafeez, A., & Javed, M. A. (2024). Advancing plant resilience against microplastics and metals through nanotechnology. *BioNanoScience, 14*(2), 2065–2080.

Otchere, E. K., McKay, B. M., English, M., & Aryee, A. N. A. (2023). Current trends in nano-delivery systems for functional foods: a systematic review. *PeerJ, 11*, e14980.

Pacheco, M. F. M., & Tai, L. (2025). Reprogramming the Tumor Microenvironment to boost adoptive T cell therapy. *Frontiers in Immunology, 16*, 1677548.

Peterson, C., Denlinger, N., & Yang, Y. (2022). Recent advances and challenges in cancer immunotherapy. *Cancers, 14*(16), 3972.

Qin, H., Zhou, Z., Shi, R., Mai, Y., Xu, Y., Fu, P., Cheng, G., Zhang, Z., Chen, W., Chen, Y., Chen, Y., Xu, R., & Lu, Q. (2025). Insights into next-generation immunotherapy designs and tools: molecular mechanisms and therapeutic prospects. *Journal of Hematology & Oncology, 18*, 62.

Qin, L., Li, Y., Liu, J., & An, X. (2025). Advancements in cellular immunotherapy: Overcoming resistance in lung and colorectal cancer. *Frontiers in Immunology, 16*, 1554256.

Ranjan, K., Dube, P., & Mishra, S. K. (2024). Editorial: Engineered medicines to mitigate resistance to cancer immunotherapy. *Frontiers in Medicine, 11*, 1452812.

Riva, L., & Petrini, C. (2019). A few ethical issues in translational research for gene and cell therapy. *Journal of Translational Medicine, 17*, 395.

Rodríguez-García, A., Palazón, A., Noguera-Ortega, E., Powell, D. J., & Guedan, S. (2020). CAR-T cells hit the tumor microenvironment: Strategies to overcome tumor escape. *Frontiers in Immunology, 11*, 1109.

Ruan, L., & Wang, L. (2025). Adoptive cell therapy against tumor immune evasion: Mechanisms, innovations, and future directions. *Frontiers in Oncology, 15*, 1530541.

Sarvepalli, S., Pasika, S. R., Verma, V., Thumma, A., Bolla, S., Nukala, P. K., Butreddy, A., & Bolla, P. K. (2025). A review on the stability challenges of advanced biologic therapeutics. *Pharmaceutics, 17*(5), 550.

Seliger, B., & Massa, C. (2021). Immune therapy resistance and immune escape of tumors. *Cancers, 13*(3), 551.

Sloas, C., Gill, S., & Klichinsky, M. (2021). Engineered CAR-macrophages as adoptive immunotherapies for solid tumors. *Frontiers in Immunology, 12*, 783305.

Sun, C., Dotti, G., & Savoldo, B. (2016). Utilizing cell-based therapeutics to overcome immune evasion in hematologic malignancies. *Blood, 128*(23), 2929–2936.

Sun, M., Białasek, M., Mayoux, M., Lin, M.-S., Buck, A., Marszałek, I., Taciak, B., Bühler, M., Górczak, M., Kucharzewska, P., Kurpiel, D., Siemińska, M., Nowakowska, J., Brancewicz, J., Gabrusiewicz, K., Lipiński, W., Pfammatter, S., Pascolo, S., Hütter, G., & Weiß, T. (2025). Adoptive cell therapy with macrophage-drug conjugates facilitates cytotoxic drug transfer and immune activation in glioblastoma models. *Science Translational Medicine, 17*(803), eadr4058.

Tie, Y., Tang, F., Wei, Y., & Wei, X. (2022). Immunosuppressive cells in cancer: Mechanisms and potential therapeutic targets. *Journal of Hematology & Oncology, 15*, 61.

Tokarew, N., Ogonek, J., Endres, S., von Bergwelt-Baildon, M., & Kobold, S. (2018). Teaching an old dog new tricks: next-generation CAR T cells. *British Journal of Cancer, 120*(1), 26–37.

Tufail, M., Jiang, C., & Li, N. (2025). Immune evasion in cancer: mechanisms and cutting-edge therapeutic approaches. *Signal Transduction and Targeted Therapy, 10*, 227.

Wang, B., Zhao, Q., Zhang, Y., Liu, Z., Zheng, Z., Liu, S., Meng, L., Xin, Y., & Jiang, X. (2021). Targeting hypoxia in the Tumor Microenvironment: A potential strategy to improve cancer immunotherapy. *Journal of Experimental & Clinical Cancer Research, 40*, 24.

Wang, D., Wu, X., & Sun, Y. (2022). Therapeutic targets and biomarkers of tumor immunotherapy: Response versus non-response. *Signal Transduction and Targeted Therapy, 7*, 331.

Wang, Y., Liu, C., Ren, Y., Song, J., Fan, K., Gao, L., Ji, X., Chen, X., & Zhao, H. (2024). Nanomaterial-based strategies for attenuating T-cell-mediated immunodepression in stroke patients: Advancing research perspectives. *International Journal of Nanomedicine, 19*, 5793–5812.

Wang, Z., Li, P., Zeng, X., Ji, J., Zhang, C., Fan, Z., Wang, Z., Zhu, P., & Chen, Z. (2025). CAR-T therapy dilemma and innovative design strategies for next generation. *Cell Death & Disease, 16*, 211.

Wellhausen, N., Baek, J. S., Gill, S., & June, C. H. (2024). Enhancing cellular immunotherapies in cancer by engineering selective therapeutic resistance. *Nature Reviews Cancer, 24*(9), 614–626.

Xiang, X., Niu, Y., Wang, Z.-H., Ye, L., Peng, W., & Zhou, Q. (2022). Cancer-associated fibroblasts: Vital suppressors of the immune response in the Tumor Microenvironment. *Cytokine & Growth Factor Reviews, 67*, 35–48.

Yang, Y. (2015). Cancer immunotherapy: Harnessing the immune system to battle cancer. *Journal of Clinical Investigation, 125*(9), 3335–3337.

Yuan, C., Wang, Y., & Guo, Z. S. (2024). Editorial: Recent advances in gene modified immune cells and oncolytic virus for cancer immunotherapy. *Frontiers in Immunology, 15*, 1454183.

Zafar, A., Khan, M. J., Abu, J., & Naeem, A. (2024). Revolutionizing cancer care strategies: Immunotherapy, gene therapy, and molecular targeted therapy. *Molecular Biology Reports, 51*, 379.

Zhang, H., Feng, L., de Mello, P. A., Mao, C., Near, R. I., Csizmadia, E., Chan, L. L., Enjyoji, K., Gao, W., Zhao, H., & Robson, S. C. (2022a). Glycoengineered anti-CD39 promotes anticancer responses by depleting suppressive cells and inhibiting angiogenesis in tumor models. *Journal of Clinical Investigation, 132*(13), e157431.

Zhang, Y., Liu, Z., Wei, W., & Li, Y. (2022b). TCR engineered T cells for solid tumor immunotherapy. *Experimental Hematology and Oncology, 11*, 38.

Zhou, M., Huang, Z., Ma, Z., Chen, J., Lin, S., Yang, X., Gong, Q., Braunstein, Z., Wei, Y., Rao, X., & Zhong, J. (2025). The next frontier in antibody-drug conjugates: challenges and opportunities in cancer and autoimmune therapy. *Cancer Drug Resistance, 8*, 34.

CHAPTER

Platelet shielding of cancer cells: Implications for immunotherapy strategies

9

Vivek P. Chavda[a], Disha Joshi[b], Arya Vyas[a,c] and Manish Patel[a]

[a]*Department of Pharmaceutics and Pharmaceutical Technology, L. M. College of Pharmacy, Ahmedabad, Gujarat, India,* [b]*Department of Pharmaceutical Technology, L. J. Institute of Pharmacy, Lok Jagruti Kendra (LJ) University, Ahmedabad, Gujarat, India,* [c]*PhD Section, Gujarat Technological University, Ahmedabad, Gujarat, India*

9.1 Introduction

In recent years, the traditional role of platelets has expanded beyond their well-known function in homeostasis. These cellular components are now recognized as important contributors to the immune system, highlighting their broader significance in physiological processes. With around 200,000 platelets per microliter, platelets are the most prevalent immune-modulating cells in circulation. They release inflammatory mediators such as platelet factor 4/C-X-C motif chemokine ligand 4 (PF4/CXCL4), pro-platelet basic protein (PPBP), regulated upon activation, normal T cell expressed and secreted (RANTES), and interleukin-1 β (IL-1β), which recruit and activate leukocytes locally and systemically (Flad & Brandt, 2010).

In the context of pathogen invasion and sterile tissue injury, platelets traditionally serve as essential components of the body's response mechanisms (Nicolai et al., 2024). These cellular elements exhibit multifaceted functions, including the direct ingestion and elimination of bacteria, the release of antimicrobial agents, and the orchestration of both innate and adaptive immune cell differentiation, polarization, and migration (Li et al., 2024; Morrell et al., 2019). Beyond their involvement in pathological conditions, platelets are also essential in regulating immune cell functions and development under normal physiological circumstances. This regulatory role is exemplified by various experimental findings, as shown by elevated helper T cells 17 subset CD4^{+} cells in mice with depleted platelet counts (Shi et al., 2014), an increased population of reparative monocytes in mice lacking platelet-derived β2-microglobulin (β2M) under normal conditions (Hilt et al., 2019; Hilt et al., 2021), and compromised vascular integrity in thrombocytopenic mice (Ho-Tin-Noé et al., 2018). These studies indicate that platelets serve as an immunological barometer, calibrating the inflammatory milieu. While activated platelets stimulate inflammatory responses to tissue damage or pathogenic invasion, quiescent platelets contribute to the maintenance of immune homeostasis under normal physiological conditions.

Immunotherapy has revolutionized cancer (Ghemrawi et al., 2024) and autoimmune disease treatment (Jung & Kim, 2022) through immune checkpoint inhibitors (Shiravand et al., 2022), chimeric antigen receptor T-cell therapy (Singh et al., 2022), and bispecific antibodies (Klein et al., 2024). Nonetheless, challenges persist in managing immune-related adverse effects because of intense immune

Advancements in Personalized Cellular Immunotherapy: A Comprehensive Guide to Cancer Treatment.
DOI: https://doi.org/10.1016/B978-0-443-27731-3.00005-X
Copyright © 2026 Elsevier Inc. All rights are reserved, including those for text and data mining, AI training, and similar technologies.

system activation, emphasizing the potential role of platelet interaction in shielding immune cells from overactivation. Utilizing platelets as a protective shield could mitigate certain risks linked to immunotherapy, such as overstimulation of the immune system (Schmied et al., 2021a). Research have demonstrated that platelets possess the ability to engage in intimate interactions with the immune cells, providing them with protection under stressful conditions and regulating the immune response (Senchenkova et al., 2019).

This chapter seeks to uncover and elaborate on the innovative concept of platelet shielding as a novel strategy in immunotherapy. The primary aim is to investigate how platelets, beyond their traditional roles in hemostasis and immune modulation, can act as protective agents for immune cells, potentially mitigating the adverse effects of intense immune activation during immunotherapy. By delving into the mechanisms of platelet-immune cell interactions, the chapter aims to present new insights into how platelet shielding can balance efficacy and safety in cancer and autoimmune disease treatments. Additionally, the chapter will explore the potential translational applications of these findings, proposing platelet shielding as a cutting-edge approach to enhance the therapeutic index of current immunotherapy protocols. Through this comprehensive examination, the objective is to contribute to the evolving landscape of immunotherapy by highlighting platelet shielding's role in achieving more effective and safer treatment outcomes.

9.2 Platelet biology

9.2.1 Structure and function of platelets

Platelets are diminutive, anucleated cellular components originating from megakaryocytes within the bone marrow (BM). These components play a crucial role in the mechanisms of hemostasis and thrombosis. Recent investigations have revealed that platelets contribute to a broad spectrum of biological mechanisms, surpassing their conventional roles. These newly identified functions encompass inflammatory processes, angiogenesis, stem cell translocation, and cellular reproduction (Martinod & Deppermann, 2021). Despite lacking a nucleus, they remain metabolically active, as they possess essential organelles, including the mitochondria, endoplasmic reticulum, and Golgi apparatus to synthesize various proteins from mRNA (Yadav & Storrie, 2017).

In contrast to other cellular organelles, mitochondria exhibit a broader range of functions within platelets. Healthy platelets generally contain 5–8 mitochondria, and for optimal platelet functionality, it is essential that the majority of these structures remain intact. Platelet mitochondria are integral to essential functions, including activation and apoptosis, underscoring their heightened importance in these anucleated cells (Melchinger et al., 2019). The platelet secretome, a collection of diverse granules, is housed within a specialized organelle zone in these platelets. This assortment encompasses alpha granules, lysosomal granules, dense granules, and glycogen granules, all of which contribute to the platelet's unique composition (Heijnen & van der Sluijs, 2015). Comprising more than 1500 proteins, including 262 phosphoproteins and 190 membrane proteins, this secretome is essential in modulating the functional responses of platelets. The multifaceted roles of platelets are enabled by their capacity to react to a broad spectrum of stimuli and their abundant granular reservoir of growth factors, cytokines, and other biological regulators (Chaudhary et al., 2022).

9.2.2 Platelet activation and signaling pathways

Platelets are indispensable participants in the intricate process of homeostasis, wherein they orchestrate the formation of blood clots to prevent excessive bleeding following vascular injury. These nonnucleated cells rapidly react to injuries in blood vessels, triggering a sequence of platelet activation and clumping. When encountering damaged blood vessel lining or exposed extracellular matrix (ECM), platelets undergo a series of physical and chemical alterations that allow them to stick, expand, and cluster at the injury site (Scridon, 2022). Key signaling pathways are initiated upon activation of platelet surface receptors, such as glycoprotein VI (GPVI) and integrin αIIbβ3. GPVI binds to exposed collagen at injury sites, triggering a cascade that leads to platelet adhesion, while αIIbβ3 integrins facilitate platelet aggregation through fibrinogen bridging (Rayes et al., 2019).

Activation process also triggers the discharge of intracellular signaling molecules such as calcium ions (Janus-Bell & Mangin, 2023), adenosine diphosphate (ADP) (Lu et al., 2022), and thromboxane A2 (Hong et al., 2023), which amplify the response by recruiting additional platelets to the injury site. These signaling processes not only contribute to the stability of the forming clot but also trigger the activation of PI3K and PKC signaling cascades. These pathways play an important role in regulating the alteration of platelet shape and their ability to spread (Janus-Bell & Mangin, 2023). The activation of platelets has implications that extend beyond blood clotting. The activation of platelets triggers the release of diverse substances, encompassing cytokines, chemokines, and growth factors. These released compounds exert a considerable influence on the regulation of inflammatory responses and facilitate the recruitment of immune cells (Scherlinger et al., 2023). Consequently, the signaling pathways in platelets are essential not only in the formation of blood clots but also in connecting hemostatic functions with immune and inflammatory responses. The sequence of events for platelet activation and signaling is shown in Figs. 9.1 and 9.2.

9.2.3 Platelets in hemostasis and thrombosis

Contemporary research has underscored the dual role of platelets in both physiological and pathological conditions (Ling et al., 2023). In hemostasis, platelets act as "first responders," adhering to damaged blood vessels and forming clots (Menter et al., 2017). The preceding section outlined how the activation of platelets triggers a cascade of events. These include alterations in platelet shape, the release of granular contents, and the presentation of surface receptors such as GPIIb/IIIa. These receptors play a crucial role in binding to fibrinogen and von Willebrand factor (vWF). The initial phase, referred to as primary hemostasis, culminates in the establishment of a provisional platelet aggregate. The subsequent stage, known as secondary hemostasis, involves the coagulation cascade. During this process, activated platelets offer a phospholipid surface for the organization of clotting factors. This facilitates the production of thrombin and the reinforcement of the platelet aggregate through the incorporation of fibrin filaments (Periayah et al., 2017).

Recent research has enhanced our understanding of platelet biology in hemostasis, showing that platelet-derived extracellular vesicles (PEVs) are crucial for facilitating communication between cells, coagulation enhancement, and immune response modulation. Furthermore, advanced imaging techniques have revealed the dynamic interactions between platelets and endothelial cells during hemostasis (Das et al., 2024).

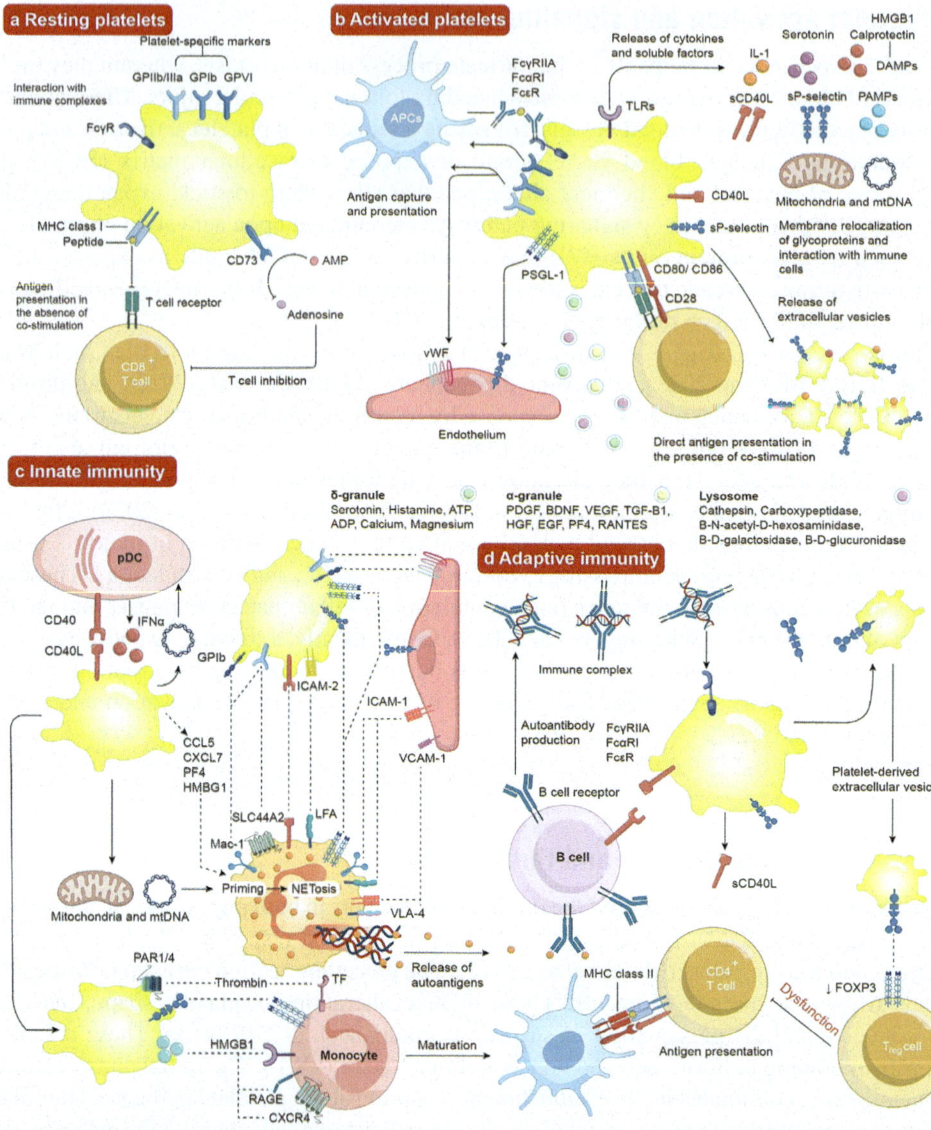

FIGURE 9.1 Platelets in the immune system.

Platelets act as immune helpers. At rest, they keep immune responses balanced and help clear immune complexes. When activated, they release signals and interact with immune cells, promoting inflammation and antigen presentation. They stimulate innate immunity by activating neutrophils and monocytes, and shape adaptive immunity by influencing T cells and B cells, including antibody production and immune regulation.

From Yan, M., Wang, Z., Qiu, Z. et al. (2024). Platelet signaling in immune landscape: comprehensive mechanism and clinical therapy. Biomarker Research 12, 164. https://doi.org/10.1186/s40364-024-00700-y.

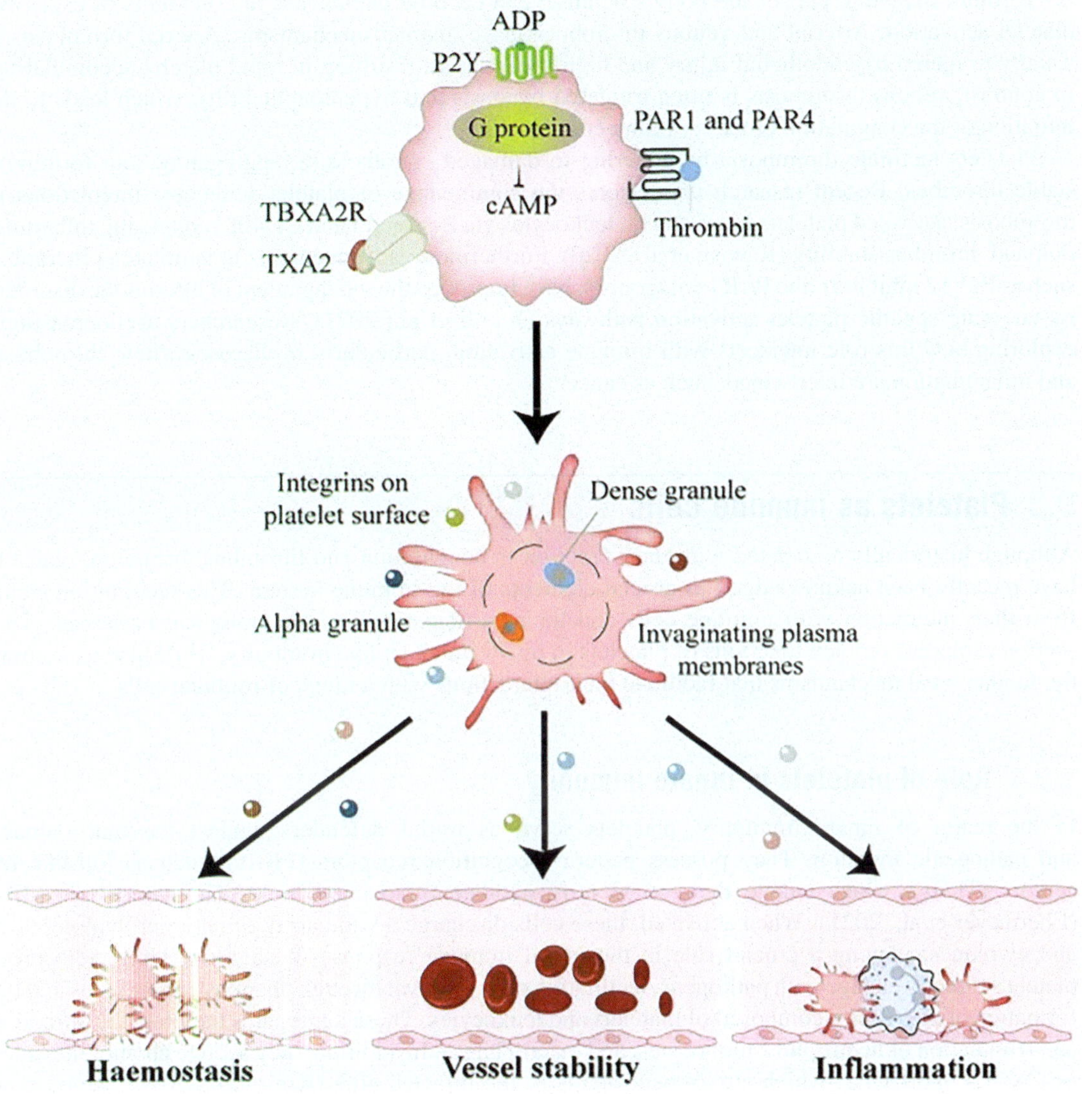

FIGURE 9.2 Platelet activation and function.

When a blood vessel is injured, platelets quickly "switch on." Signals like thrombin, ADP, collagen, and TXA_2 activate receptors on their surface, triggering internal changes that make them release their stored contents. This helps platelets stick together and respond rapidly. Beyond forming clots, platelets also support inflammation, healing, immune defense, and blood vessel stability.

From Wang L, Wang X, Guo E, Mao X and Miao S (2022) Emerging roles of platelets in cancer biology and their potential as therapeutic targets. Frontiers in oncology:939089. doi:https://doi.org/10.3389/fonc.2022.939089.

Thrombosis, while part of the body's defense, can become pathogenic in conditions of excessive platelet activation. Arterial and venous thrombosis have distinct mechanisms. Arterial thrombosis is usually triggered by endothelial injury and high shear stress, resulting in rapid platelet accumulation. In contrast, venous thrombosis is often triggered by stasis and hypercoagulability, which leads to the initiation of the coagulation process (Yeung et al., 2018).

Platelets facilitate thrombosis by adhering to damaged endothelium, aggregating, and forming a stable thrombus. Recent research underscores the significance of platelet–leukocyte interactions in thrombosis; activated platelets interact with leukocytes via P-selectin and CD40L, enhancing inflammation and thrombus stability (Rawish et al., 2020). Furthermore, advancements in antiplatelet therapies, such as P2Y12 inhibitors and PAR1 antagonists, have improved the management of thrombotic disorders by targeting specific platelet activation pathways (Jourdi et al., 2021). Researchers are increasingly exploring how this role intersects with immune activation, particularly in diseases where thrombosis and inflammation are intertwined, such as cancer.

9.3 Platelets as immune cells

Although historically associated with their function in coagulation and thrombus formation, platelets have recently been acknowledged as crucial elements of the immune system. This recognition stems from their interaction with immune cells and their participation in immunological processes. This section explores complex functions of platelets in innate and adaptive immunity, while also examining the fundamental mechanisms that facilitate their interactions with a range of immune cells.

9.3.1 Role of platelets in innate immunity

In the realm of innate immunity, platelets serve as initial defenders against vascular damage and pathogenic invasion. They possess pattern recognition receptors (PRRs), such as toll-like receptors (TLRs), which enable them to identify pathogen-associated molecular patterns (PAMPs) (Ebermeyer et al., 2021). When activated, these cells discharge a wide array of antimicrobial peptides and cytokines, playing a crucial role in the initial immune response mechanism. Upon activation, platelets directly engage with pathogens, facilitating their removal through phagocytic processes and the formation of aggregates composed of platelets and leukocytes. These aggregates facilitate the attraction and stimulation of neutrophils, monocytes, and macrophages. In addition, they secrete chemokines such as CXCL4 and CCL5, which attract neutrophils to the infected area (Kral et al., 2016). Moreosver, platelets are pivotal in the generation of neutrophil extracellular traps (NETs), which serve to trap and eliminate pathogens, preventing their dissemination (Carestia et al., 2016).

Contemporary investigations have underscored the pivotal role of platelets in affecting the complement system, an essential component of the body's innate immune defense mechanisms. Platelet-derived complement proteins have the capacity to augment the process of opsonization and elimination of pathogens, thereby serving as a bridge between innate and adaptive immune responses (Nicolai et al., 2024). Furthermore, platelets secrete danger-associated molecular patterns (DAMPs), establishing a connection between hemostasis and the immune system, while underscoring their involvement in inflammatory processes (Scherlinger et al., 2023).

9.3.2 Platelets in adaptive immune responses

Platelets play a crucial role in shaping adaptive immune responses by influencing the activation and regulation of T and B lymphocytes. The antigen-capturing and presenting capabilities of platelets contribute to the enhanced activation of T cells by dendritic cells. This function is facilitated by specific molecules on the platelet surface, namely P-selectin, CD154, and CD40L, which enable the antigen-presenting process (Maouia et al., 2020). In addition, platelets release an array of cytokines and chemokines that modulate immune responses. As an example, IL-1β derived from these anucleate cells stimulates the growth and differentiation of T helper cells, whereas platelet-derived CXCL4 modulates B cell activity and immunoglobulin synthesis (Mehdi-Alamdarlou et al., 2023).

Yu et al. (2021) have noted that platelets and lymphocytes exhibit a reciprocal interaction. T cells, when activated, can trigger the activation and clumping of platelets, which facilitates the development of immunological synapses. This interaction plays a vital role in controlling immune responses and preserving immune hemostasis. Emerging research has also shown that blood platelets can emit tiny vesicles that transport immune-related molecules, such as proteins and RNA, to remote immune cells, thereby affecting their operation and conduct. These EVs serve as messengers, influencing the behavior of distant immune cells (Puricelli et al., 2023). These findings underscore the significance of platelets as versatile cellular components that connect hemostasis with both innate and adaptive immune processes in the body's defense system.

9.3.3 Mechanisms of platelet-immune cell interactions

Platelet and immune cells communicate through multiple pathways, which include direct cellular contact, secretion of immune-modulating substances, and discharge of EVs (Rolla et al., 2021). The initial mechanism involves direct cellular contact, wherein platelets display surface proteins such as P-selectin, GPIIb/IIIa, and CD40L, enabling immediate interactions with immune cells. These connections play a pivotal role in the development of platelet–leukocyte aggregates and in the regulation of immune cell functions, as elaborated in the preceding section (Becker et al., 2018). Concerning the second mechanism of immune mediator secretion, platelets, upon activation, discharge a vast array of bioactive compounds from their alpha and dense granules. This diverse collection of substances, comprising chemokines, cytokines, and growth factors, subsequently modulates the activities of numerous immune cells (Golebiewska & Poole, 2015). As an illustration, TGF-β derived from platelets affects the development of regulatory T cells, while CCL5 originating from platelets attracts monocytes and T cells to areas of inflammation (Maouia et al., 2020).

Platelets employ an additional mechanism involving the discharge of EVs, including microparticles (MPs) and exosomes. These EVs serve as carriers for immune-modulating molecules such as RNA, lipids, and proteins. Analogous to the adaptive immune response, these vesicles transmit signals to remote immune cells, thus modifying their function and behavior (Ferreira et al., 2020). Ongoing research have revealed that platelet-derived EVs play a crucial role in the immune system by influencing the maturation of dendritic cells, the activation of T cells, and the production of antibodies by B cells, thereby acting as key messengers within the immune network (Qiao et al., 2023). The transfer of RNA to immune cells has the potential to modify their genetic expression and operational capabilities. This mechanism of genetic material transmission represents a recently uncovered means by which platelets influence immune responses, enabling swift interaction between platelets and immune cells

(Kanth Manne et al., 2020). Continued investigations into the immunological functions of platelets hold promise for creating innovative therapeutic strategies for infectious and inflammatory disorders.

9.4 The concept of platelet shielding

9.4.1 Definition and mechanisms of platelet shielding

The BM is the major site of creation for platelets, which are tiny, anucleate cells generated from megakaryocytes. Megakaryocytes produce proplatelets, which are lengthy projections that are broken up by the forces of blood flow in the blood arteries during platelet production. These proplatelets are then further divided into smaller platelets as they circulate through the bloodstream (Selvadurai & Hamilton, 2018). Platelets have historically been mostly recognized for their function in thrombosis and hemostasis. Nevertheless, there is mounting evidence that platelets serve a variety of other purposes. For instance, platelets can release cytokines that are both pro- and antiinflammatory and express a number of functional TLRs. Platelets' impact on immune system modulation is highlighted by ongoing research into the methods by which they balance the production and suppression of inflammation (Zhou et al., 2023). Platelets have been demonstrated to enable several stages of tumor growth in the setting of cancer, including local growth, bloodstream migration, and metastatic establishment. Platelets are crucial for immune system evasion during a number of those processes. Cytotoxic lymphocytes may directly or indirectly impede tumor cell interaction, providing protection from immunosurveillance (Schmied et al., 2021b).

All of the main elements of the hemostatic system, including platelets, are impacted by tumor cells. Important stages in the development of cancer have been connected to platelets and platelet activation. Platelets can shield tumor cells from physical pressures in the vasculature by coating them and allowing them to evade immune system effector cells. This is one of the two ways that platelets can protect tumor cells. Platelets, for instance, have been demonstrated to shield tumors from TNF-α cytotoxicity and natural killer (NK) cells (Philippe et al., 1993; Shau et al., 1993). Crucial phases of the development of cancer have been linked to platelets and their activation, as outlined in Fig. 9.3 (Bambace & Holmes, 2011).

Historically, platelets' primary role has been to aid in thrombosis and hemostasis. Nevertheless, there is growing evidence that platelets can perform a variety of additional functions. For instance, platelets can create proinflammatory and antiinflammatory cytokines and have many functioning TLR to influence immune system activity; yet, it is still unknown how platelets decide whether to initiate or suppress inflammation (Holinstat, 2017; Li et al., 2012).

Platelets are known to be activated by tumor cells, which causes them to aggregate and form complexes that promote tumor growth and metastasis, among other advantages. Circulating tumor cells (CTCs) are shielded from shear stress and immune surveillance by platelets that cluster around them. At metastatic sites, this helps the CTCs adhere to the vascular endothelium. Furthermore, growth factors such as transforming growth factor beta (TGFβ), vascular endothelial growth factor (VEGF), and platelet-derived growth factor (PDGF) are released by activated platelets and aid in angiogenesis and the epithelial–mesenchymal transition (EMT), two processes that are critical for the spread of tumors (Zhou et al., 2023). Through direct contacts between tumor cells and platelets or the release of soluble platelet agonists into the bloodstream, tumors can directly cause platelet activation. ADP and other

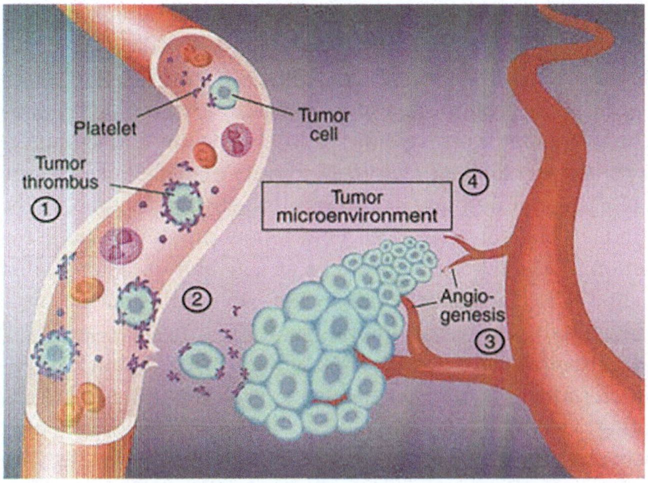

FIGURE 9.3

Platelets play a crucial role in various stages of cancer progression. "The current understanding of how platelets contribute to malignancy includes: (A platelet aggregation induced by tumor cells, which occurs after tumor cells intravasate into the bloodstream, helping to 'shield' circulating tumor cells from immune detection and physical removal, (B) platelets assist in the arrest of tumor cells within blood vessels, promote endothelial cell retraction, and facilitate tissue invasion, (C) platelets stimulate endothelial cell proliferation and the formation of new blood vessels, essential for tumor angiogenesis and growth, and (Ds) interactions between platelets, tumor cells, and stromal cells in the tumor microenvironment are influenced by platelet activation and the release of platelet proteins, contributing to the inflammatory response. Furthermore, platelet-derived proteins and metabolites that promote proteolysis and tissue remodeling also support tumor growth and metastasis, including in bone metastases."

From Bambace, N. M. & Holmes, C. E. (2011). The platelet contribution to cancer progression. Journal of Thrombosis and Haemostasis, *9(2), 237–249. https://doi.org/10.1111/j.1538-7836.2010.04131.x.*

traditional soluble agents can cause platelet activation (Abdol Razak et al., 2018). Numerous tumor cells contain the transmembrane protein podoplanin, which has a direct interaction with the platelet activation receptor CLEC-2. Other tumor-environmental cells that can express podoplanin and trigger platelet activation include vascular endothelial cells (VECs) and macrophages close to leaky blood vessels (Suzuki-Inoue, 2019). Furthermore, the G protein-coupled receptor (GPCR) CD97 is expressed by a number of cancer cell types, and its extracellular domain has the ability to attach to and activate platelets (Ward et al., 2018). In addition, tumor cells have the ability to emit molecular patterns linked to damage, such as the alarmin protein HMGB1, which attaches to platelets' TLR4 and initiates activation (Yu et al., 2014). In addition, type I collagen is deposited in the ECM around breast cancer CTCs during EMT, promoting platelet contact and effective colonization, according to a recent study by Xiong et al. Platelets usually come into contact with collagen following vascular damage, which triggers platelet aggregation. The expression of the ECM-regulating gene Hsp47 in CTCs controls the deposition of collagen (Xiong et al., 2020).

By triggering the release or expression of traditional platelet agonists involved in hemostasis and coagulation, tumor cells might indirectly activate platelets. Since the 20th century, it has been known that platelet aggregating activity (PAA) and procoagulant activity (PCA) of tumor cells are related. The ability of tumor cells to change prothrombin into thrombin, a vital platelet agonist, is known as PCA (Zhou et al., 2023). Thrombin inhibitors have the ability to prevent tumor-induced platelet aggregation, indicating that tumor cells use coagulant factors to activate platelets, which in turn produce thrombin. Cancer procoagulant (CP), a protein found in tumor cells, activates factor X in the extrinsic coagulation pathway, which can induce platelet activation. Few follow-up studies have focused on CP's role in tumor–platelet interactions, though it is thought to be an activator of factor X, distinct from the intrinsic pathway involving factor VII (Cavanaugh et al., 1988). Furthermore, tumors mimic vascular injury to cause platelet aggregation by expressing tissue factor (TF) on their surface. Normally, TF triggers the extrinsic coagulation pathway following vascular injury. Thrombin can be directly produced by pancreatic tumors, which further encourages platelet activation (Abdol Razak et al., 2018).

Through circulating innate immune cells, tumor cells can also activate platelets. Shao et al. (2011) discovered, for instance, that cancer mucins bind to platelet P-selectins and neutrophil L-selectins, bringing the two cell types together. Both platelets and neutrophils are activated by this contact in conjunction with bidirectional signaling through P-selectin and P-selectin glycoprotein ligand-1 (PSGL-1). Cathepsin G, which is released when neutrophils are activated, activates platelets through the protease-activated receptor-4 (PAR4) (Shao et al., 2011). Moreover, NETs, which are made up of neutrophil proteases and DNA, have the ability to activate platelets and encourage thrombosis. Intriguingly, platelets themselves may also trigger NET formation, indicating a self-sustaining loop of interaction between platelets and neutrophils. Tumor-induced NET creation is important for platelet activation (Cedervall et al., 2018).

Certain tumor cells acquire the capacity to infiltrate, move into the surrounding stroma, and enter the blood or lymphatic system when cancer spreads. By promoting the EMT, which increases the invasive and resistant characteristics of cancer cells, platelets contribute to this process. Tumor cell motility and invasion capacity are increased by EMT, which includes alterations such as cytoskeletal rearrangement, loss of epithelial polarity, and decreased cell–cell adhesion (Massagué & Obenauf, 2016; Tsubakihara & Moustakas, 2018). N-cadherin levels are higher, and E-cadherin levels are lower in cancer cells undergoing EMT, which facilitates the cells' separation from the epithelium. In addition, ECM remodeling brought about by EMT creates a microenvironment that promotes tumor growth and metastasis (Tsubakihara & Moustakas, 2018; Xiong et al., 2020).

TGFβ, which platelets retain and emit, is one of the signals that platelets release to cause EMT in cancer cells. Breast, colon, and ovarian cancer are among the tumors for which TGFβ stimulates EMT and metastasis. Metastatic spread is greatly decreased when platelets' TGFβ is wiped down. A mesenchymal, prometastatic phenotype is caused by TGFβ activating the TGFβ/Smad pathway in tumor cells and platelet–tumor cell contacts activating the NF-κB pathway (Guo et al., 2019; Labelle et al., 2011). EMT is also induced by the interaction of cancer cell podoplanin with platelet CLEC-2. By recruiting ezrin, radixin, and moesin (ERM) proteins and destabilizing cadherin junctions, podoplanin connects to the actin cytoskeleton and increases cell invasiveness. By modifying EMT gene expression via the microRNA miR-939, decreasing E-cadherin levels, and raising mesenchymal markers such as vimentin and fibronectin, platelet-secreted MPs can also cause EMT in ovarian cancer cells (Lowe et al., 2012; Tang et al., 2017).

Both primary tumor cells and CTCs are susceptible to EMT induction by platelets. In primary cancers, the platelet marker CD42b correlates with EMT markers, such as decreased E-cadherin and elevated Snail1, suggesting platelet-induced EMT (Miyashita et al., 2015). In addition, platelets control vascular maturation and shape, which promotes tumor growth in the primary tumor microenvironment (Chen et al., 2023). In CTCs, some show mesenchymal features, and platelet aggregation is observed in clusters of CTCs with mesenchymal traits, which correlates with disease progression. These results propose that platelets may contribute to EMT activation and metastasis in both primary tumor cells and CTCs (Yu et al., 2013).

Tumor metastasis is inefficient, with most CTCs dying because of stressors such as shear stress and immune attacks. However, platelets help CTCs survive by forming a protective layer, shielding them from NK cells. Depleting platelets or fibrinogen has no effect on metastasis when NK cell activities are impaired, indicating that platelets help tumor cells avoid immune attack (Nieswandt et al., 1999; Palumbo et al., 2005).

Platelets protect CTCs by altering NK cell recognition. They transfer surface molecules, such as glycoprotein IIb and MHC class I, to tumor cells, helping them evade NK cell detection (Placke et al., 2012). In addition, platelets secrete TGF-β, which lowers NK cell activity and facilitates immune evasion. Furthermore, platelets impede NK cell identification by causing tumor cells to lose NKG2D ligands (Kopp et al., 2009; Maurer et al., 2018).

Platelets also support neutrophil-mediated tumor growth. CTCs resistant to neutrophil cytotoxicity form clusters with neutrophils, promoting tumor cell proliferation. Platelets may help neutrophils migrate to the tumor. They also facilitate CTC trapping by NETs, promoting metastasis. Finally, platelets prevent anoikis in CTCs by activating RhoA and YAP, which leads to resistance to apoptosis. In addition, there is evidence that platelets increase the survival of CTCs by shielding them from shear stress (Chen et al., 2023; Zhou et al., 2023).

Although CTCs' bigger size generally traps them in tight capillaries, they can also arrest in wider arteries and actively adhere to the endothelium, which facilitates extravasation and metastasis (Labelle & Hynes, 2012). Platelets help CTC adhere to the vasculature, which is a crucial part of this process. To facilitate connections between platelets, tumor cells, and the vascular wall, they express a variety of adhesion molecules, including integrins, P-selectin, and PECAM-1 (Chen et al., 2023; Li, 2016; Xu et al., 2016). When platelets coat CTCs, they stabilize interactions with endothelial cells, improving adhesion. Inducing platelet aggregation can enhance CTC retention and metastasis development in experimental models (Gong et al., 2012; Kunita et al., 2007; Plantureux et al., 2020). Platelets also facilitate "tethering/rolling" of tumor cells on the endothelium, enhancing stable adhesion through mechanisms involving P-selectin and αIIbβ3 integrins (Burdick & Konstantopoulos, 2004; Chen et al., 2023).

In addition, platelets increase adhesion molecule expression on tumor cells. For example, platelet–tumor cell interactions can stimulate acid sphingomyelinase (ASM), leading to ceramide production on tumor cells, which clusters integrins and signaling molecules (Carpinteiro et al., 2015). Platelets also release integrin β3 and microvesicles carrying CD41, enhancing tumor cell adhesion (Janowska-Wieczorek et al., 2005; Plantureux et al., 2020;).

Endothelial cells in metastatic locations may be stimulated to express adhesion molecules such as E-selectin and ICAM-1, which aid CTCs in binding, even in the absence of vascular damage (Labelle & Hynes, 2012). Through CD154 and MPs, platelets can directly activate endothelial cells, enhancing adhesion molecule production and boosting contacts between tumor cells and endothelium

(Plantureux et al., 2020; Semple et al., 2011). Coculture of platelets, leukocytes, and tumor cells is necessary for full activation of endothelial cells, as seen in the expression of E-selectin (Läubli et al., 2009).

Platelets continue to support the metastatic process by encouraging trans endothelial migration, invasion, and tumor cell survival after CTCs attach to the endothelium at the metastatic location. The significance of platelet–tumor cell interaction in extravasation was highlighted by a study that revealed thrombocytopenic mice had fewer extravasated tumor cells than wild-type mice (Weber et al., 2016). Platelets disrupt endothelial integrity by releasing ATP, which binds to P2Y2 receptors on endothelial cells, causing gap formation (Schumacher et al., 2013).

In addition, platelets promote tumor cell invasion via the ECM. Increased ECM breakdown was seen in tumor cells treated with activated platelet membranes. Inhibition of platelet binding reduced this degradation (Pang et al., 2015). Platelet-derived microvesicles can also upregulate matrix metalloproteinases (MMPs), crucial for ECM degradation and metastasis, especially in highly metastatic cancer cell lines (Janowska-Wieczorek et al., 2005).

Another important factor in metastasis is the interaction between leukocytes and platelets. Leukocytes promote tumor cell survival in early metastatic stages, and their recruitment correlates with metastasis. However, in some cases, leukocytes can have inhibitory effects on metastasis, such as tumor-entrained neutrophils that can kill tumor cells via reactive oxygen species (Granot et al., 2011). The type of leukocytes and their activation play a key part in either promoting or inhibiting metastasis. Platelets attract leukocytes to the metastatic site by releasing chemokines such as CXCL5 and CXCL7, and further studies are needed to fully understand how platelets influence leukocyte activity in metastasis (Labelle & Hynes, 2012; Labelle et al., 2014).

9.4.2 Pathological contexts: Inflammation and infection

Although platelets' primary function is to maintain hemostasis, they are also essential for inflammation and the immune system. In addition to their role in viral infections, human platelets have α-granules that release antibacterial proteins called thrombocidins when the platelet is activated. These proteins, which are C-terminal truncation variations of the platelet-derived chemokines CXCL7, CXCL4, and CCL5, have a wide range of antibacterial action, but they work especially well against Staphylococcus aureus and *E. coli* (Yeaman et al., 2007). By interacting directly or indirectly with particular receptors, bacteria can activate platelets. For instance, *Staphylococcus aureus* quickly activates platelets by building fibrinogen/fibronectin bridges between platelet GPIIb/IIIa and bacterial fibronectin-binding protein A. In addition, it activates platelets by causing the activation signal to be triggered by IgG binding to FcεRIIa. Furthermore, functional TLRs, of which TLR4 is the most prevalent, are expressed by platelets. Depending on the platelets' level of activation, TLR4 expression can be changed (Cognasse et al., 2005). In addition, platelets have PRRs, which activate platelets when they come into contact with bacterial components. These receptors include complement and immunoglobulin. These PRRs use damage-associated molecular patterns, such as those from the Siglec (sialic acid-binding Ig-like lectins) family, to identify danger signals (Crocker et al., 2007; Land, 2013).

Neutrophils, which release extracellular DNA traps, or NETs, to capture and destroy germs, can be further stimulated by activated platelets (Clark et al., 2007; Fitzgerald et al., 2006). Furthermore, the immune response to bacterial infections in vivo is improved by the interaction of complement C3 and GPIb.

Platelets that stick to the endothelium or sub-endothelial region after vascular injury or inflammation aid in attracting inflammatory cells and coordinating the inflammatory response (Humdelshausen et al., 2009; Von Hundelshausen & Weber, 2007). Platelets interact with inflammatory cells such as neutrophils and monocytes to help them adhere to the artery wall. These adhesive interactions include: (i) platelets' P-selectin binding to leukocytes' P-selectin glycoprotein ligand-1 (PSGL-1); (ii) platelets' GPIb or junctional adhesion molecule-C (JAM-3) binding to leukocytes' β2-integrin Mac-1 under low shear stress; and (iii) fibrinogen attached to platelets' GPIIb/IIIa interacting with leukocytes' Mac-1 (Ehlers et al., 2003; Santoso et al., 2002; Wagner & Frenette, 2008; Weber & Springer, 1997). Platelets are an important source of proinflammatory chemokines and cytokines in addition to attracting inflammatory cells (Von Hundelshausen & Weber, 2007) and can increase inflammation by releasing MPs (Mause & Weber, 2010; Vasina et al., 2010). According to recent research, different subclasses of α-granules found in platelets release diverse subsets of chemokines in response to pathological or physiological situations (Italiano et al., 2008; Van Nispen Tot Pannerden et al., 2010).

Megakaryocytes are identified by platelet-derived chemokines, such as CXCL4 (PF4) and CXCL7, which are the most prevalent. Platelets also release other chemokines, such as CCL5, MIF, CXCL12, and CXCL5 (Zlotnik & Yoshie, 2000). In addition, platelets trigger the acute phase response, mostly by the cytoplasmic synthesis of IL-1β from premessenger RNA. During both sterile and infectious inflammation, platelets express components of inflammasomes and build functional inflammasomes, releasing IL-1β in MPs, particularly during thrombus formation following vascular injury. Through IL-1 receptors, IL-1β increases its own production in platelets by initiating a self-maintaining autocrine loop (Brown et al., 2013).

The pathophysiology of several disorders, including atherosclerosis, arthritis, vasculitis, sepsis, neuroinflammation, and inflammatory bowel diseases, is influenced by platelets because of their proinflammatory function (Boilard et al., 2010; Clark et al., 2007; Langer et al., 2012). In neuroinflammation, TNF-induced brain inflammation is facilitated by interactions between platelets, endothelial cells, and microglia in the brain microvasculature (Zhou et al., 2009). Leukocyte adhesion and rolling in TNF-induced brain inflammation are decreased by platelet reduction (Carvalho-Tavares et al., 2000). In diseases such as cerebral malaria, platelets also cause brain endothelial cells to produce proinflammatory reactions. Platelet–leukocyte interactions may impact the course of disorders such as multiple sclerosis, stroke, and Alzheimer's, particularly when protein misfolding causes platelet aggregation and the release of proinflammatory mediators (Barbier et al., 2011; Thornton et al., 2010).

9.5 Implications for immunotherapy

The study of cancer biology is well aware of the part platelets play in the development of cancer. The additional roles and extent to which platelets contribute to the development of malignancies are being further explored by new research. In addition to releasing substances that promote tumor growth, platelets can form heterotypic aggregates with tumor cells, giving them an edge in evading the immune system. The suppression of immune cell activity, which might impair the body's capacity to stop tumor formation and progression, may be their most important function (Kanikala-Marie et al., 2018).

In addition, there is proof that platelets hide the CTCs from immune surveillance and destruction (Meikle et al., 2017). The selective survival benefits that tumor cells receive from interacting with platelets allow them to evade or resist immune responses and gain more access to growth substances generated by platelets. The relatively simple passage of platelets facilitates these interactions as well because of their tiny size and active cytoskeleton, migrate into neighboring tissues without being hindered by a cell nucleus (Kanikala-Marie et al., 2018; Sharma et al., 2014). Tumor cells are then stimulated to home, extravasate, and metastasize in the perivascular gaps by substances secreted by platelets. Therefore there is a good chance that focusing on or inhibiting these interactions between tumor cells and platelets will have positive effects. As anticipated, it has been shown that inhibiting this tumor cell-platelet interaction affects clinical trials and lowers tumor cell survival, proliferation, and metastasis in cancer models used in experiments (Kanikala-Marie et al., 2018).

9.5.1 Enhancing efficacy of immunotherapeutic agents

Platelets, through their interactions with immune cells, contribute to both innate and adaptive immunity, which is crucial for immunological control (Kanikala-Marie et al., 2018). They influence immune responses through vesicle-mediated protein transfer to immune cells during stimulation (Laffont et al., 2016; Sprague et al., 2008;). TGF-β, an immunosuppressive factor that controls T regulatory cell homeostasis, is abundant in platelets (Assoian et al., 1983; Li et al., 2006). It has been demonstrated that blocking the TGF-β1 receptor can stop the progression of ovarian cancer brought on by platelet-derived TGF-β1 (Hu et al., 2017). A protumorigenic microenvironment is produced when immune recognition pathways malfunction, and platelets may inhibit immune cells that attack cancer cells. For example, platelets can hide tumor cells from NK cells by transferring MHC-I molecules onto tumor surfaces or inhibiting CD4+ and CD8+ T cells via TGF-β and lactate (Nieswandt et al., 1999; Placke et al., 2012).

Platelets significantly influence tumor immunity by modulating immune cell responses and masking tumor cells, enabling them to evade immune surveillance. They also suppress CD8+ T cells, critical for cancer immunotherapy, which may affect disease progression and treatment outcomes (Kanikala-Marie et al., 2018). In colorectal cancer patients, the lack of CD8+ T cells in tumors correlates with higher recurrence rates, while their presence predicts long-term disease-free survival (Curtis et al., 2012; Mlecnik et al., 2011). By restoring CD8+ T-cell activity, targeting tumor-educated platelets may enhance cancer immunotherapy.

Programmed death-ligand 1 (PD-L1) is another important component of immunotherapy; it binds to the PD-1 receptors on CD8+ T cells to inhibit them (Chen & Mellman, 2013; Pardoll, 2012). Despite their potential, PD-L1 inhibitors' effectiveness varies because of factors such as immune escape mechanisms, few tumor-infiltrating cells, and variations in PD-L1 expression. Notably, by expressing platelet proteins or joining platelet aggregates, CTCs can avoid immune monitoring. Research shows that in patients with metastatic non-small cell lung cancer (NSCLC), higher platelet counts are associated with worse responses to PD-L1 therapy and a lower survival rate (Diem et al., 2017).

Inhibiting platelets could enhance immunotherapy by boosting T-cell activity. Preclinical models using genetically modified mice with impaired platelets showed that platelet inhibition improves adoptive T-cell therapy responses, highlighting the potential for integrating platelet-targeted approaches into cancer immunotherapy (Nieswandt et al., 1999).

Checkpoint inhibitors have completely changed the way solid tumors are treated, allowing for long-lasting remissions in conditions that were thought to be incurable before. These therapies aim to disrupt T-cell checkpoint signaling, thereby stimulating anticancer immune responses. Recent advancements have explored the use of platelets as carriers for monoclonal antibodies (mAbs) to enhance the efficacy of this therapeutic approach (Cacic et al., 2023).

To treat breast cancer and melanoma, Wang et al. (2017) created a platelet-based medication delivery system in which platelets were coupled with antibodies that target the programmed death-1 (PD-1) protein. This strategy makes use of platelets' capacity to enter tumors and deliver the medication there through PEVs. When compared to free anti-PD-1 therapy, anti-PD-1-conjugated platelets performed better, providing a longer circulation half-life, more robust localized immune responses, and improved tumor reduction. By employing antibody-crosslinked nanogel backpacks that disintegrate following platelet activation in response to surface reduction activity, Lu et al. (2022) presented a unique method for attaching a noticeably greater quantity of anti-PD-1 antibodies to each platelet.

To improve BM infiltration, Hu et al. (2018) used anti-PD-1-conjugated platelets in conjunction with hematopoietic stem cells (HSCs) to treat acute myeloid leukemia (AML) in mice. Although proper BM delivery required HSC conjugation, adding platelets significantly increased the antileukemic inflammatory response and prolonged survival. In addition, the same researchers implanted chimeric antigen receptor (CAR)-T cells in biodegradable hydrogel into melanoma resection cavities together with anti-PD-1-conjugated platelets (Hu et al., 2021). Anti-PD-1 antibodies were released by in situ activation, increasing the effectiveness of CAR-T cell treatment. Following partial melanoma resection, Zhang et al. (2018) created mouse PD-1-expressing platelets from megakaryocyte progenitors that gathered in tumor beds. When loaded with cyclophosphamide, these platelets decreased the number of regulatory T cells in the tumor microenvironment and increased the recruitment of cytotoxic CD8+ T cells to the tumor, further amplifying their anticancer effects.

By reducing tumor-associated macrophages via CSF1R signaling blockage, the anti-PD-1-conjugated platelets' inflammatory anticancer impact was enhanced when they were added to a hydrogel containing the tyrosine kinase inhibitor pexidartinib (Li et al., 2022). When used in conjunction with this treatment, vadimezan increased platelet recruitment to tumors through vascular disruption, which greatly increased survival in a number of cancer types in mice (Li et al., 2021). Furthermore, by "wounding" tumors, boosting platelet infiltration, and encouraging the release of tumor antigens, photothermal treatment (PTT) was coupled with anti-PD-1-conjugated platelets to enhance immune responses (Gao et al., 2023; Han et al., 2019).

These results demonstrate that, whereas platelets may increase resistance to specific immunotherapies, they may also be modified to become very potent immunomodulatory agents that can be used with currently available medications to treat cancer.

9.5.2 Role of platelet shielding in tumor microenvironment

The microenvironment of the majority of solid tumors, especially carcinomas, is made up of the parenchyma, or tumor cells, and the stroma, which is made up of connective tissue and other noncancerous mesenchymal cells that help the tumor grow and survive. Importantly, the different cell types in the tumor microenvironment communicate with themselves and with one another to encourage tumor growth. As a result, the microenvironment and tumor properties are interdependent. Platelets that pass through the circulatory network become activated, attach, aggregate, and may wind up in the tumor's microenvironment, where they may have an impact on the parenchyma and tumor-associated stroma. This is because malignancies require this network to proliferate. It has been suggested that cells in the tumor microenvironment, including fibroblasts, immune cells, pericytes, endothelial cells, and neoplastic epithelial cells, may contribute to the development of cancer's traits

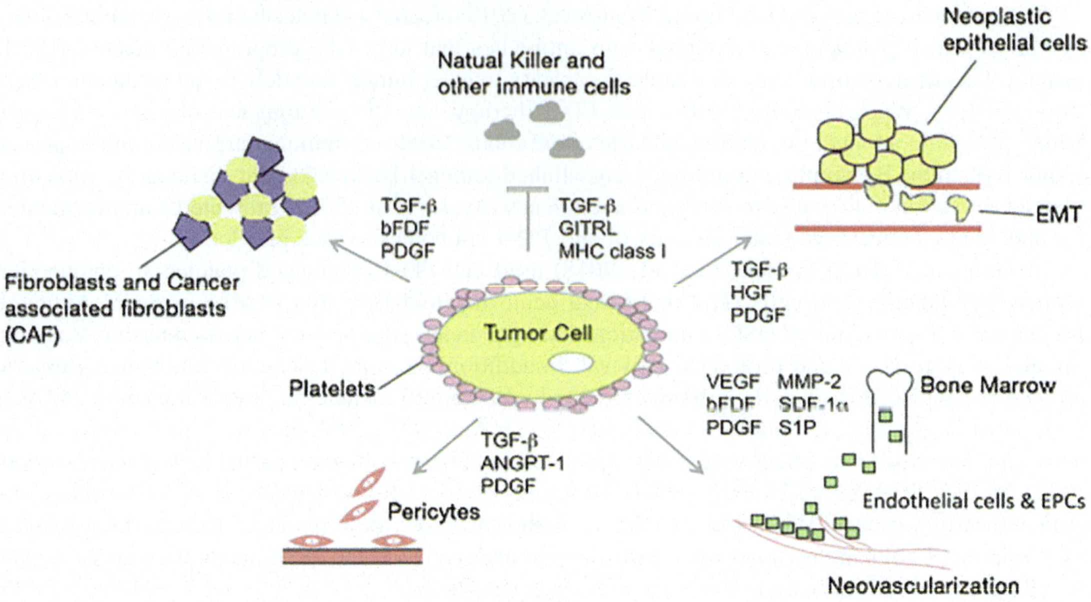

FIGURE 9.4 Interaction of platelets and tumor microenvironment.

"The microenvironment of solid tumors, particularly carcinomas, comprises both malignant tumor cells (parenchyma) and nonmalignant mesenchymal cells within the stroma, supported by connective tissue. This microenvironment sustains tumor growth through complex interactions between its cellular components. Tumors depend on a vascular network, where activated platelets can adhere, aggregate, and integrate into the tumor microenvironment. These platelets may influence both the parenchyma and the tumor-associated stroma. Cells within this microenvironment—such as neoplastic epithelial cells, endothelial cells, pericytes, fibroblasts, and immune cells—play a critical role in acquiring cancer hallmarks."

From Yan, MengJie & Jurasz, Paul. (2016). The role of platelets in the tumor microenvironment: From solid tumors to leukemia. Biochimica et Biophysica Acta (BBA) - Molecular Cell Research, *1863(3), 392–400. https://doi.org/10.1016/j.bbamcr.2015.07.008.*

(Hanahan & Weinberg, 2011). Interaction of platelets with the various components of tumor microenvironment has been illustrated in Fig. 9.4.

Through vascular leakage, platelets themselves, as well as their release, can control the TME network. Antiplatelet antibodies that deplete platelets can drastically lower the number of microvessels (De Palma et al., 2017; Stone et al., 2012). In addition, platelets in the primary TME can change the morphology of tumor arteries by inducing vascular permeability. Because platelets assist in vascular maturation and change the ECM, which is connected to the tumor cells' capacity to spread out from the original site, it is evident that they have a major effect on the initial invasion of tumor cells. Meanwhile, research has been done on platelet extravasation. The tumor microenvironment's platelet-related compounds enable platelets to carry out a number of functions in cancer biology (Chen et al., 2023).

The network of the tumor microenvironment has become the focus of recent studies in the field of cancer biology, rather than the tumor cells themselves. To gain a better knowledge of the biological significance of platelets in the development of cancer, it is necessary to pinpoint the precise sites of platelet–tumor interactions. Research has shown that platelets invaded the tumor stroma and had predictive value in patients with pancreatic and colorectal cancer (Miao et al., 2022; Xu et al., 2019; Zhang et al., 2018).

9.5.3 Case studies: Successful integration of platelet shielding in therapy

Immunotherapy has advanced significantly in the clinical treatment of tumors. Low antigen presentation efficacy, insufficient immune cell infiltration, and tumor cell immune evasion are still problems, though. The clinical success of immunotherapy was severely limited by these problems, which led to less than ideal efficacy and safety. Additionally, the presence of an immune escape mechanism and immunosuppressive microenvironment makes it more difficult for nanoagents to activate immunity (Ngwa et al., 2018; Zhang et al., 2023; Zhang et al., 2024).

Since tumor exosomes are transmitters for tumor cells, they are essential to the formation, growth, and dissemination of malignancies. Tumor metastasis can be effectively decreased by controlling the release of tumor exosomes. Furthermore, the host's anticancer immune response is critically regulated negatively by immunological checkpoints. By promoting immune evasion of tumor cells, this vital function limits the effectiveness of tumor immunotherapy (Morrissey et al., 2021; Sun et al., 2018).

Yan et al. (2022) created self-adaptive platelet pharmacytes called siRab@aPDL1/PLTs, which were made up of surface aPD-L1 nanogel particles and within Rab27 siRNA (siRab) formed multicationic nanocomplexes. Even after loading the therapeutic chemicals, the platelets retain their fundamental structure and morphology. Platelet pharmacytes exhibit good immunotherapeutic effects and a remarkably long circulatory half-life of 16.9 hours following intravenous administration. Furthermore, intratumoral enrichment of siRab@aPDL1/PLTs has been demonstrated in mice models. When siRab@aPDL1/PLTs are activated within tumor tissue, positively charged siRab NCs may be released. These NCs have the ability to effectively transfect tumor cells, mute Rab27a to stop the production of tumor-derived exosomes, decrease tumor growth, and enhance tumor invasion of T lymphocytes. This study offered fresh insights into the development of modified platelets for tumor immunotherapy by demonstrating, based on animal tests, the capacity of these cells to serve as carriers for gene therapy and immunosuppressive elimination (Yan et al., 2022).

In a different study, Fan et al. (2021) found that intravenous administration of *E. coli* TOP10 can significantly increase the recruitment of macrophages in tumors by causing quick and strong clotting in tumor tissues in a mouse tumor model. Furthermore, lipopolysaccharide on the bacterial wall primarily polarized the recruited macrophages into an anticancer M1 phenotype. After that, platelets modified with the CD47 antibody (anti-CD47) (P@aCD47) were injected. Platelets carried anti-CD47 to the tumor site by obstructing the "don't eat me" signal sent by the tumor cells. This encouraged M1 macrophage phagocytosis and tumor cell destruction. Together with the P@aCD47 treatment approach, this bacterium dramatically reduced tumor growth in the xenograft CT26 tumor-bearing mouse model. A possible method for the targeted distribution of immune checkpoint inhibitors for immunotherapeutic systems is the tumor local coagulation recruitment of macrophages. The long-term safety of the bacteria following injection, whether it results in a series of clotting events, increased platelet counts, and the potential for thrombotic problems in tumors, are some of the issues that still need to be resolved before clinical translation can take place. Based on the photoresponsively modified platelet, immunostimulatory drug loading may increase the therapeutic efficacy of tumor therapy. A possible method for the targeted distribution of immune checkpoint inhibitors for immunotherapeutic systems is the tumor local coagulation recruitment of macrophages. The long-term safety of the bacteria following injection, whether it results in a series of clotting events, increased platelet counts, and the potential for thrombotic problems in tumors, are some of the issues that still need to be resolved before clinical translation can take place. The immunostimulatory drug loading may improve the therapeutic efficacy of tumor therapy based on the photoresponsive engineered platelet (Fan et al., 2021)

Platelet liposomes (Pt@PL–IgG) containing IgG antibodies were created by Yang et al. (2023). IgG was attached to the surface of tumor cells by targeting and fusing PM's membrane, which improved NK cells' ability to identify and eliminate tumor cells. To further improve tumor killing, cisplatin encapsulated in platelet liposomes was released into tumor cells during membrane fusion. In a mouse model with 4T1 tumors, Pt@PL–IgG demonstrated a potent capacity to stop tumor growth. The flow cytometry analysis of tumor tissues revealed a considerable improvement in the infiltration of CD69+ NK cells and CD8+ T cells. In addition, in the B16-F10 melanoma tumor model, Pt@PL–IgG showed a higher ability to inhibit lung metastases. By overcoming tumor heterogeneity and demonstrating remarkable immune efficiency in several tumor models, this approach offers fresh perspectives on how to deal with tumor heterogeneity's problems (Yang et al., 2023).

9.6 Challenges and limitations

9.6.1 Potential adverse effects of platelet activation

The utilization of platelets in immunotherapies has shown immense promise in the treatment of various diseases, yet it is not without its challenges and limitations. One of the primary concerns is the potential for adverse effects stemming from platelet activation, which is a crucial component of hemostasis and wound healing. A significant danger linked to unregulated platelet activation is the discharge of proinflammatory substances, which could potentially worsen inflammatory disorders or induce atypical immune reactions. This is particularly worrisome in chronic inflammatory conditions such as atherosclerosis, where persistent platelet activation may result in impaired endothelial function and plaque formation (Medina-Leyte et al., 2021). Thrombotic complications, such as deep vein thrombosis, pulmonary embolism, and myocardial infarction, are significantly influenced by heightened platelet activation, which plays a pivotal role in their development. Such thrombotic incidents are potentially fatal and pose substantial challenges in clinical management, particularly for individuals with coexisting medical conditions (Pastori et al., 2023).

Despite their life-saving potential, platelet transfusions can provoke immune reactions in recipients. These responses may manifest as mild complications, such as febrile nonhemolytic transfusion reactions (Goel et al., 2019), allergic manifestations, or escalate to more severe immune reactions such as transfusion-related acute lung injury (TRALI) (Tung et al., 2022). The risk of these adverse outcomes highlights the necessity for meticulous observation and judicious management of patients receiving platelet transfusions.

9.6.2 Variability in patient response to immunotherapy

As a revolutionary cancer treatment, immunotherapy has shown remarkable potential. However, individual patient outcomes can differ considerably, highlighting the relevance of genetic factors that significantly. This highlights the necessity of creating customized treatment approaches. Increasingly, biomarkers with predictive capabilities are employed to personalize immunotherapy regimens for individual patients. Variations in factors such as tumor mutational burden (TMB) and the expression levels of immune checkpoint proteins such as PD-L1 have been identified as key biomarkers that correlate with variable patient outcomes. Elevated levels of TMB and PD-L1 are frequently associated with better efficacy of immune checkpoint inhibitors. Nevertheless, these biomarkers do not provide absolute predictive accuracy, as some patients with low TMB or PD-L1 expression also respond favorably to the treatment, underscoring the complexity of immune system responses (Sankar et al., 2022).

In addition, immunotherapy can provoke severe autoimmune reactions in a subset of patients. These immune-related adverse reactions (irAEs) have the potential to affect any organ system, with severity ranging from mild to life-threatening. Managing irAEs requires a coordinated effort from various medical disciplines and often necessitates the suspension of immunotherapy, potentially compromising the treatment efficacy (Blum et al., 2023).

9.6.3 Technical challenges in targeting platelets

The therapeutic use of platelets as targets poses numerous technical obstacles, owing to the complexity of platelet biology and their involvement in various physiological processes. The multifaceted involvement of platelets in thrombosis, hemostasis, inflammation, and immune responses makes it challenging to develop targeted therapies without altering their essential functions. The development of sophisticated platelet function testing methods, including flow cytometry and aggregometry, has enhanced our understanding of disorders related to platelet function. However, the interpretation of intricate data produced by these technologies remains problematic. Furthermore, the high costs liked to cutting-edge platelet function testing hinder their widespread adoption in clinical settings (Yoon et al., 2024).

The paucity of targetable antigens within the tumor microenvironment can hinder the efficacy of active targeting approaches. To address this challenge, researchers have explored leveraging therapeutic responses to disease, such as the stimulation of vascular antigens, to enhance the number of binding sites and improve targeting of nanoparticles to tumors (Wang et al., 2023). However, these methods are often confined to specific clinical scenarios where the disease sites are known and accessible, precluding the treatment of disseminated disease, which remains the primary cause of mortality in cancer.

Developing drugs that selectively target platelet functions without causing adverse effects is another significant challenge. While effective in preventing thrombotic events, antiplatelet medications such as P2Y12 inhibitors and GP IIb/IIIa inhibitors can elevate the risk of hemorrhage. Striking a balance between the advantages of antiplatelet therapy and the potential for adverse effects necessitates careful deliberation and individualized patient management strategies (Sim et al., 2023). Furthermore, the diverse platelet responses observed across individuals pose challenges in developing universally effective treatments. To address this, researchers are investigating personalized medicine strategies, including the application of pharmacogenomics, to customize platelet-targeted therapies according to each patient's unique genetic makeup. This approach aims to optimize therapeutic outcomes while minimizing adverse effects (Kumar et al., 2024). Notwithstanding these obstacles, research in platelet-focused immunotherapies continues to progress, with scientists investigating novel approaches to improve the precision and effectiveness of these treatments. Novel strategies, including genetic modification, chemical alteration of cell surfaces, and the utilization of external physical forces, show potential in tackling the safety and effectiveness issues linked to living intelligent therapeutic agents. As researchers gain a deeper insight into the intricate relationship between platelets and the immune system, it is anticipated that the creation of more potent and tailored platelet-focused immunotherapies will gain momentum. This progress is expected to usher in a new phase in cancer treatment.

9.7 Conclusion

The evolving understanding of platelets as integral components of the immune system has significant implications for both physiology and clinical practice. This chapter has elucidated the multifaceted roles of platelets, emphasizing their dual function in hemostasis and immune modulation. By releasing a variety of cytokines and chemokines, platelets not only facilitate clot formation but also orchestrate immune responses, enhancing the recruitment and activation of leukocytes during infections and tissue injuries. Furthermore, their interactions with T and B cells underscore their importance in adaptive immunity, showcasing their ability to influence lymphocyte differentiation and function. Platelets contribute to cancer progression by promoting tumor growth, aiding in immune suppression, and facilitating metastasis through their interactions with tumor cells. They secrete growth factors and cytokines that enhance tumor cell survival and dissemination. In addition, platelet-mediated suppression of immune cells, such as CD8+ T cells, hinders the efficacy of cancer immunotherapies. By shielding tumor cells from immune surveillance, platelets play a pivotal role in immune evasion mechanisms, which are critical in the development of therapeutic resistance.

The potential for utilizing platelets in immunotherapy presents an exciting avenue for future research. Strategies such as "platelet shielding" could mitigate adverse effects associated with immune overactivation, thereby enhancing therapeutic outcomes. As we continue to uncover the complex signaling pathways and mechanisms governing platelet-immune cell interactions, it becomes increasingly clear that these anucleate cells are not merely passive players in hemostasis but active participants in maintaining immune homeostasis and responding to pathological challenges. This knowledge paves the way for innovative therapeutic approaches targeting platelet functions in various diseases. Furthermore, the integration of platelets in innovative therapeutic strategies has demonstrated potential to overcome some of the limitations associated with cancer immunotherapy. Techniques such as platelet-based drug delivery systems, antibody-conjugated platelets, and the modification of platelets to carry therapeutic agents have shown promise in enhancing therapeutic outcomes. These strategies exploit platelets' unique properties, such as their capacity for targeted localization within tumor tissues, to improve drug delivery, boost immune responses, and mitigate tumor progression.

In conclusion, platelets serve as a critical nexus in cancer biology, acting as both facilitators of tumor progression and vehicles for innovative therapeutic advancements. The comprehensive understanding of their roles and interactions within the tumor microenvironment is essential for advancing cancer treatment, offering hope for more targeted, effective, and personalized approaches to combat cancer in the future.

References

Abdol Razak, N., Jones, G., Bhandari, M., Berndt, M., & Metharom, P. (2018). Cancer-associated thrombosis: An overview of mechanisms, risk factors, and treatment. *Cancers, 10*(10), 380. doi:10.3390/cancers10100380.

Assoian, R. K., Komoriya, A., Meyers, C. A., Miller, D. M., & Sporn, M. B. (1983). Transforming growth factor-β in human platelets. Identification of a major storage site, purification, and characterization. *Journal of Biological Chemistry, 258*(11), 7155–7160.

Bambace, N. M., & Holmes, C. E. (2011). The platelet contribution to cancer progression. *Journal of Thrombosis and Haemostasis, 9*(2), 237–249. doi:10.1111/j.1538-7836.2010.04131.x.

Barbier, M., Faille, D., Loriod, B., Textoris, J., Camus, C., Puthier, D., Flori, L., Wassmer, S. C., Victorero, G., Alessi, M. C., Fusaï, T., Nguyen, C., Grau, G. E., & Rihet, P. (2011). Platelets alter gene expression profile in human brain endothelial cells in an in vitro model of cerebral malaria. *PLoS ONE, 6*(5). doi:10.1371/journal.pone.0019651.

Becker, R. C., Sexton, T., & Smyth, S. S. (2018). Translational implications of platelets as vascular first responders. *Circulation Research, 122*(3), 506–522. doi:10.1161/CIRCRESAHA.117.310939.

Blum S. M., Rouhani S. J., & Sullivan R. J., Effects of immune-related adverse events (irAEs) and their treatment on antitumor immune responses. Immunological Reviews. doi: 10.1111/imr.13262.

Boilard, E., Nigrovic, P. A., Larabee, K., Watts, G. F. M., Coblyn, J. S., Weinblatt, M. E., Massarotti, E. M., Remold-O'Donnell, E., Farndale, R. W., Ware, J., & Lee, D. M. (2010). Platelets amplify inflammation in arthritis via collagen-dependent microparticle production. *Science, 327*(5965), 580–583. doi:10.1126/science.1181928.

Brown, G. T., Narayanan, P., Li, W., Silverstein, R. L., & McIntyre, T. M. (2013). Lipopolysaccharide stimulates platelets through an IL-1β autocrine loop. *Journal of Immunology, 191*(10), 5196–5203. doi:10.4049/jimmunol.1300354.

Burdick, M. M., & Konstantopoulos, K. (2004). Platelet-induced enhancement of LS174T colon carcinoma and THP-1 monocytoid cell adhesion to vascular endothelium under flow. *American Journal of Physiology - Cell Physiology, 287*(2), C539. doi:10.1152/ajpcell.00450.2003.

Cacic, D., Hervig, T., & Reikvam, H. å (2023). Platelets for advanced drug delivery in cancer. *Expert Opinion on Drug Delivery, 20*(5), 673–688. doi:10.1080/17425247.2023.2217378.

Carestia, A., Kaufman, T., & Schattner, M. (2016). Platelets: New bricks in the building of neutrophil extracellular traps. *Frontiers in Immunology, 7*. doi:10.3389/fimmu.2016.00271.

Carpinteiro, A., Becker, K. A., Japtok, L., Hessler, G., Keitsch, S., Požgajovà, M., Schmid, K. W., Adams, C., Müller, S., Kleuser, B., Edwards, M. J., Grassmé, H., Helfrich, I., & Gulbins, E. (2015). Regulation of hematogenous tumor metastasis by acid sphingomyelinase. *EMBO Molecular Medicine, 7*(6), 714–734. doi:10.15252/emmm.201404571.

Carvalho-Tavares, J., Hickey, M. J., Hutchison, J., Michaud, J., Sutcliffe, I. T., & Kubes, P. (2000). A role for platelets and endothelial selectins in tumor necrosis factor-α-induced leukocyte recruitment in the brain microvasculature. *Circulation Research, 87*(12), 1141–1148. doi:10.1161/01.RES.87.12.1141.

Cavanaugh, P. G., Sloane, B. F., & Honn, K. V. (1988). Role of the coagulation system in tumor-cell-induced platelet aggregation and metastasis. *Haemostasis, 18*(1), 37–46. doi:10.1159/000215781.

Cedervall, J., Hamidi, A., & Olsson, A. K. (2018). Platelets, NETs and cancer. *Thrombosis Research, 164*, S148. doi:10.1016/j.thromres.2018.01.049.

Chaudhary, P. K., Kim, S., & Kim, S. (2022). An insight into recent advances on platelet function in health and disease. *International Journal of Molecular Sciences, 23*(11). doi:10.3390/ijms23116022.

Chen, D. S., & Mellman, I. (2013). Oncology meets immunology: The cancer-immunity cycle. *Immunity, 39*(1), 1–10. doi:10.1016/j.immuni.2013.07.012.

Chen, L., Zhu, C., Pan, F., Chen, Y., Xiong, L., Li, Y., Chu, X., & Huang, G. (2023). Platelets in the tumor microenvironment and their biological effects on cancer hallmarks. *Frontiers in Oncology, 13*. doi:10.3389/fonc.2023.1121401.

Clark, S. R., Ma, A. C., Tavener, S. A., McDonald, B., Goodarzi, Z., Kelly, M. M., Patel, K. D., Chakrabarti, S., McAvoy, E., Sinclair, G. D., Keys, E. M., Allen-Vercoe, E., DeVinney, R., Doig, C. J., Green, F. H. Y., & Kubes, P. (2007). Platelet TLR4 activates neutrophil extracellular traps to ensnare bacteria in septic blood. *Nature Medicine, 13*(4), 463–469. doi:10.1038/nm1565.

Cognasse, F., Hamzeh, H., Chavarin, P., Acquart, S., Genin, C., & Garraud, O. (2005). Evidence of toll-like receptor molecules on human platelets. *Immunology & Cell Biology, 83*(2), 196–198. doi:10.1111/j.1440-1711.2005.01314.x.

Crocker, P. R., Paulson, J. C., & Varki, A. (2007). Siglecs and their roles in the immune system. *Nature Reviews Immunology, 7*(4), 255–266. doi:10.1038/nri2056.

Curtis, N. J., Primrose, J. N., Thomas, G. J., Mirnezami, A. H., & Ottensmeier, C. H. (2012). The adaptive immune response to colorectal cancer: From the laboratory to clinical practice. *European Journal of Surgical Oncology (EJSO), 38*(10), 889–896. doi:10.1016/j.ejso.2012.05.011.

Das, D., Jothimani, G., Banerjee, A., Dey, A., Duttaroy, A. K., & Pathak, S. (2024). A brief review on recent advances in diagnostic and therapeutic applications of extracellular vesicles in cardiovascular disease. *The International Journal of Biochemistry & Cell Biology, 173*, 106616. doi:10.1016/j.biocel.2024.106616.

De Palma, M., Biziato, D., & Petrova, T. V. (2017). Microenvironmental regulation of tumour angiogenesis. *Nature Reviews Cancer, 17*(8), 457–474. doi:10.1038/nrc.2017.51.

Diem, S., Schmid, S., Krapf, M., Flatz, L., Born, D., Jochum, W., Templeton, A. J., & Früh, M. (2017). Neutrophil-to-lymphocyte ratio (NLR) and platelet-to-lymphocyte ratio (PLR) as prognostic markers in patients with non-small cell lung cancer (NSCLC) treated with nivolumab. *Lung Cancer, 111*, 176–181. doi:10.1016/j.lungcan.2017.07.024.

Ebermeyer, T. héo, Cognasse, F., Berthelot, P., Mismetti, P., Garraud, O., & Hamzeh-Cognasse, H. (2021). Platelet innate immune receptors and TLRs: A double-edged sword. *International Journal of Molecular Sciences, 22*(15), 7894. doi:10.3390/ijms22157894.

Ehlers, R., Ustinov, V., Chen, Z., Zhang, X., Rao, R., Luscinskas, F. W., Lopez, J., Plow, E., & Simon, D. I. (2003). Targeting platelet–leukocyte interactions. *The Journal of Experimental Medicine, 198*(7), 1077–1088. doi:10.1084/jem.20022181.

Fan, J.-X., Liu, X.-H., Wang, X.-N., Niu, M.-T., Chen, Q. i-W., Zheng, D. i-W., Wei, J.-S., Yang, X.-Q., Zeng, X., & Zhang, X.-Z. (2021). Antibody engineered platelets attracted by bacteria-induced tumor-specific blood coagulation for checkpoint inhibitor immunotherapy. *Advanced Functional Materials, 31*(22). doi:10.1002/adfm.202009744.

Ferreira, P. M., Bozbas, E., Tannetta, S. D., Alroqaiba, N., Zhou, R., Crawley, J. T. B., Gibbins, J. M., Jones, C. I., Ahnström, J., & Yaqoob, P. (2020). Mode of induction of platelet-derived extracellular vesicles is a critical determinant of their phenotype and function. *Scientific Reports, 10*(1). doi:10.1038/s41598-020-73005-3.

Fitzgerald, J. R., Loughman, A., Keane, F., Brennan, M., Knobel, M., Higgins, J., Visai, L., Speziale, P., Cox, D., & Foster, T. J. (2006). Fibronectin-binding proteins of Staphylococcus aureus mediate activation of human platelets via fibrinogen and fibronectin bridges to integrin GPIIb/IIIa and IgG binding to the FcγRIIa receptor. *Molecular Microbiology, 59*(1), 212–230. doi:10.1111/j.1365-2958.2005.04922.x.

Flad, H. D., & Brandt, E. (2010). Platelet-derived chemokines: Pathophysiology and therapeutic aspects. *Cellular and Molecular Life Sciences, 67*(14), 2363–2386. doi:10.1007/s00018-010-0306-x.

Gao, Y. u, Chen, X., Wang, B. o, Wang, S., Wang, J., Ren, L., Jin, W.-, Han, H., & Wang, L. (2023). Engineering platelets with PDL1 antibodies and iron oxide nanoparticles for postsurgical cancer immunotherapy. *ACS Applied Bio Materials, 6*(1), 257–266. doi:10.1021/acsabm.2c00869.

Ghemrawi, R., Abuamer, L., Kremesh, S., Hussien, G., Ahmed, R., Mousa, W., Khoder, G., & Khair, M. (2024). Revolutionizing cancer treatment: Recent advances in immunotherapy. *Biomedicines, 12*(9), 2158. doi:10.3390/biomedicines12092158.

Goel, R., Tobian, A. A. R., & Shaz, B. H. (2019). Noninfectious transfusion-associated adverse events and their mitigation strategies. *Blood, 133*(17), 1831–1839. doi:10.1182/blood-2018-10-833988.

Golebiewska, E. M., & Poole, A. W. (2015). Platelet secretion: From haemostasis to wound healing and beyond. *Blood Reviews, 29*(3), 153–162. doi:10.1016/j.blre.2014.10.003.

Gong, L., Mi, H. J., Zhu, H., Zhou, X., & Yang, H. (2012). P-selectin-mediated platelet activation promotes adhesion of non-small cell lung carcinoma cells on vascular endothelial cells under flow. *Molecular Medicine Reports, 5*(4), 935–942. doi:10.3892/mmr.2012.760.

Granot, Z., Henke, E., Comen, E. A., King, T. A., Norton, L., & Benezra, R. (2011). Tumor entrained neutrophils inhibit seeding in the premetastatic lung. *Cancer Cell, 20*(3), 300–314. doi:10.1016/j.ccr.2011.08.012.

Guo, Y. i, Cui, W., Pei, Y., & Xu, D. (2019). Platelets promote invasion and induce epithelial to mesenchymal transition in ovarian cancer cells by TGF-β signaling pathway. *Gynecologic Oncology, 153*(3), 639–650. doi:10.1016/j.ygyno.2019.02.026.

Han, X., Chen, J., Chu, J., Liang, C., Ma, Q., Fan, Q., Liu, Z., & Wang, C. (2019). Platelets as platforms for inhibition of tumor recurrence post-physical therapy by delivery of anti-PD-L1 checkpoint antibody. *Journal of Controlled Release, 304*, 233–241. doi:10.1016/j.jconrel.2019.05.008.

Hanahan, D., & Weinberg, R. A. (2011). Hallmarks of cancer: The next generation. *Cell, 144*(5), 646–674. doi:10.1016/j.cell.2011.02.013.

Heijnen, H., & van der Sluijs, P. (2015). Platelet secretory behaviour: As diverse as the granules … or not? *Journal of Thrombosis and Haemostasis, 13*(12), 2141–2151. doi:10.1111/jth.13147.

Hilt, Z. T., Maurya, P., Tesoro, L., Pariser, D. N., Ture, S. K., Cleary, S. J., Looney, M. R., McGrath, K. E., & Morrell, C. N. (2021). β2m signals monocytes through non-canonical tgfβ receptor signal transduction. *Circulation Research, 128*(5), 655–669. doi:10.1161/CIRCRESAHA.120.317119.

Hilt, Z. T., Pariser, D. N., Ture, S. K., Mohan, A., Quijada, P., Asante, A. A., Cameron, S. J., Sterling, J. A., Merkel, A. R., Johanson, A. L., Jenkins, J. L., Small, E. M., McGrath, K. E., Palis, J., Elliott, M. R., & Morrell, C. N. (2019). Platelet-derived β2M regulates monocyte inflammatory responses. *JCI Insight, 4*(5). doi:10.1172/jci.insight.122943.

Holinstat, M. (2017). Normal platelet function. *Cancer and Metastasis Reviews, 36*(2), 195–198. doi:10.1007/s10555-017-9677-x.

Hong, H. J., Nam, G. S., & Nam, K. S. (2023). Daidzein inhibits human platelet activation by downregulating thromboxane A2 production and granule release, regardless of COX-1 activity. *International Journal of Molecular Sciences, 24*(15). doi:10.3390/ijms241511985.

Ho-Tin-Noé, B., Boulaftali, Y., & Camerer, E. (2018). Platelets and vascular integrity: How platelets prevent bleeding in inflammation. *Blood, 131*(3), 277–288. doi:10.1182/blood-2017-06-742676.

Hu, Q., Hisamatsu, T., Haemmerle, M., Cho, M. S., Pradeep, S., Rupaimoole, R., Rodriguez-Aguayo, C., Lopez-Berestein, G., Wong, S. T. C., Sood, A. K., & Afshar-Kharghan, V. (2017). Role of platelet-derived Tgfβ1 in the progression of ovarian cancer. *Clinical Cancer Research, 23*(18), 5611–5621. doi:10.1158/1078-0432.CCR-16-3272.

Hu, Q., Li, H., Archibong, E., Chen, Q., Ruan, H., Ahn, S., Dukhovlinova, E., Kang, Y., Wen, D. i, Dotti, G., & Gu, Z. (2021). Inhibition of post-surgery tumour recurrence via a hydrogel releasing CAR-T cells and anti-PDL1-conjugated platelets. *Nature Biomedical Engineering, 5*(9), 1038–1047. doi:10.1038/s41551-021-00712-1.

Hu, Q., Sun, W., Wang, J., Ruan, H., Zhang, X., Ye, Y., Shen, S., Wang, C., Lu, W., Cheng, K., Dotti, G., Zeidner, J. F., Wang, J., & Gu, Z. (2018). Conjugation of haematopoietic stem cells and platelets decorated with anti-PD-1 antibodies augments anti-leukaemia efficacy. *Nature Biomedical Engineering, 2*(11), 831–840. doi:10.1038/s41551-018-0310-2.

Italiano, J. E., Richardson, J. L., Patel-Hett, S., Battinelli, E., Zaslavsky, A., Short, S., Ryeom, S., Folkman, J., & Klement, G. L. (2008). Angiogenesis is regulated by a novel mechanism: Pro- and antiangiogenic proteins are organized into separate platelet α granules and differentially released. *Blood, 111*(3), 1227–1233. doi:10.1182/blood-2007-09-113837.

Janowska-Wieczorek, A., Wysoczynski, M., Kijowski, J., Marquez-Curtis, L., Machalinski, B., Ratajczak, J., & Ratajczak, M. Z. (2005). Microvesicles derived from activated platelets induce metastasis and angiogenesis in lung cancer. *International Journal of Cancer, 113*(5), 752–760. doi:10.1002/ijc.20657.

Janus-Bell, E., & Mangin, P. H. (2023). The relative importance of platelet integrins in hemostasis, thrombosis and beyond. *Haematologica, 108*(7), 1734–1747. doi:10.3324/haematol.2022.282136.

Jourdi, G., Lordkipanidzé, M., Philippe, A., Bachelot-Loza, C., & Gaussem, P. (2021). Current and novel antiplatelet therapies for the treatment of cardiovascular diseases. *International Journal of Molecular Sciences, 22*(23), 13079. doi:10.3390/ijms222313079.

Jung, S. M., & Kim, W. U. (2022). Targeted immunotherapy for autoimmune disease. *Immune Network, 22*(1). doi:10.4110/in.2022.22.e9.

Kanikala-Marie, P., Lam, M., Sorokin, A. V., Overman, M. J., Kopetz, S., & Menter, D. G. (2018). Platelet metabolism and other targeted drugs: Potential impact on immunotherapy. *Frontiers in Oncology, 8*. doi:10.3389/fonc.2018.00107.

Kanth Manne, B., Denorme, F., Middleton, E. A., Portier, I., Rowley, J. W., Stubben, C., Petrey, A. C., Tolley, N. D., Guo, L., Cody, M., Weyrich, A. S., Yost, C. C., Rondina, M. T., & Campbell, R. A. (2020). Platelet gene expression and function in patients with COVID-19. *Blood, 136*(11), 1317–1329. doi:10.1182/blood.2020007214.

Klein, C., Brinkmann, U., Reichert, J. M., & Kontermann, R. E. (2024). The present and future of bispecific antibodies for cancer therapy. *Nature Reviews Drug Discovery, 23*(4), 301–319. doi:10.1038/s41573-024-00896-6.

Kopp, H. G., Placke, T., & Salih, H. R. (2009). Platelet-derived transforming growth factor-β down-regulates NKG2D thereby inhibiting natural killer cell antitumor reactivity. *Cancer Research, 69*(19), 7775–7783. doi:10.1158/0008-5472.CAN-09-2123.

Kral, J. B., Schrottmaier, W. C., Salzmann, M., & Assinger, A. (2016). Platelet interaction with innate immune cells. *Transfusion Medicine and Hemotherapy, 43*(2), 78–88. doi:10.1159/000444807.

Kumar, S., Schroeder, J. A., & Shi, Q. (2024). Platelet-targeted gene therapy induces immune tolerance in hemophilia and beyond. *Journal of Thrombosis and Haemostasis, 22*(1), 23–34. doi:10.1016/j.jtha.2023.07.025.

Kunita, A., Kashima, T. G., Morishita, Y., Fukayama, M., Kato, Y., Tsuruo, T., & Fujita, N. (2007). The platelet aggregation-inducing factor aggrus/podoplanin promotes pulmonary metastasis. *American Journal of Pathology, 170*(4), 1337–1347. doi:10.2353/ajpath.2007.060790.

Labelle, M., Begum, S., & Hynes, R. (2011). Direct signaling between platelets and cancer cells induces an epithelial-mesenchymal-like transition and promotes metastasis. *Cancer Cell, 20*(5), 576–590. doi:10.1016/j.ccr.2011.09.009.

Labelle, M., Begum, S., & Hynes, R. O. (2014). Platelets guide the formation of early metastatic niches. *Proceedings of the National Academy of Sciences of the United States of America, 111*(30), E3053. doi:10.1073/pnas.1411082111.

Labelle, M., & Hynes, R. O. (2012). The initial hours of metastasis: The importance of cooperative host-tumor cell interactions during hematogenous dissemination. *Cancer Discovery, 2*(12), 1091–1099. doi:10.1158/2159-8290.CD-12-0329.

Laffont, B., Corduan, A., Rousseau, M., Duchez, A. C., Lee, C. H. C., Boilard, E., & Provost, P. (2016). Platelet microparticles reprogram macrophage gene expression and function. *Thrombosis and Haemostasis, 115*(2), 311–323. doi:10.1160/TH15-05-0389.

Land, W. G. (2013). Transfusion-related acute lung injury: The work of DAMPs. *Transfusion Medicine and Hemotherapy, 40*(1), 3–13. doi:10.1159/000345688.

Langer, H. F., Choi, E. Y., Zhou, H., Schleicher, R., Chung, K. J., Tang, Z., Göbel, K., Bdeir, K., Chatzigeorgiou, A., Wong, C., Bhatia, S., Kruhlak, M. J., Rose, J. W., Burns, J. B., Hill, K. E., Qu, H., Zhang, Y., Lehrmann, E., Becker, K. G., & Wang, Y. (2012). Platelets contribute to the pathogenesis of experimental autoimmune encephalomyelitis. *Circulation Research, 110*(9), 1202–1210. doi:10.1161/CIRCRESAHA.111.256370.

Läubli, H., Spanaus, K. S., & Borsig, L. (2009). Selectin-mediated activation of endothelial cells induces expression of CCL5 and promotes metastasis through recruitment of monocytes. *Blood, 114*(20), 4583–4591. doi:10.1182/blood-2008-10-186585.

Li, C., Li, J., Li, Y., Lang, S., Yougbare, I., Zhu, G., Chen, P., & Ni, H. (2012). Crosstalk between platelets and the immune system: Old systems with new discoveries. *Advances in Hematology, 2012*, 1–14. doi:10.1155/2012/384685.

Li, C., Ture, S. K., Nieves-Lopez, B., Blick-Nitko, S. K., Maurya, P., Livada, A. C., Stahl, T. J., Kim, M., Pietropaoli, A. P., & Morrell, C. N. (2024). Thrombocytopenia independently leads to changes in monocyte immune function. *Circulation Research, 134*(8), 970–986. doi:10.1161/CIRCRESAHA.123.323662.

Li, H., Wang, Z., Chen, Z., Ci, T., Chen, G., Wen, D., Li, R., Wang, J., Meng, H., Bell, R. B., Gu, Z., Dotti, G., & Gu, Z. (2021). Disrupting tumour vasculature and recruitment of aPDL1-loaded platelets control tumour metastasis. *Nature Communications, 12*(1). doi:10.1038/s41467-021-22674-3.

Li, M. O., Wan, Y. Y., Sanjabi, S., Robertson, A. K. L., & Flavell, R. A. (2006). Transforming growth factor-β regulation of immune responses. *Annual Review of Immunology, 24*, 99–146. doi:10.1146/annurev.immunol.24.021605.090737.

Li, N. (2016). Platelets in cancer metastasis: To help the "villain" to do evil. *International Journal of Cancer, 138*(9), 2078–2087. doi:10.1002/ijc.29847.

Li, Z., Ding, Y., Liu, J., Wang, J., Mo, F., Wang, Y., Chen-Mayfield, T. J., Sondel, P. M., Hong, S., & Hu, Q. (2022). Depletion of tumor associated macrophages enhances local and systemic platelet-mediated anti-PD-1 delivery for post-surgery tumor recurrence treatment. *Nature Communications, 13*(1). doi:10.1038/s41467-022-29388-0.

Ling, L., Zhang, J., Li, Y., Liu, C., Du, L., & Zhou, J. (2023). Platelets play a dual role in the pathophysiology of transfusion-related acute lung injury. *Respiratory Physiology & Neurobiology, 309*, 104004. doi:10.1016/j.resp.2022.104004.

Lowe, K. L., Navarro-Nunez, L., & Watson, S. P. (2012). PL-06 platelet CLEC-2 and podoplanin in cancer metastasis. *Thrombosis Research*, 129. doi:10.1016/S0049-3848(12)70013-0.

Lu, P. H., Liao, T. H., Chen, Y. H., Hsu, Y. L., Kuo, C. Y., Chan, C. C., Wang, L. K., Chern, C. Y., & Tsai, F. M. (2022). Coumarin derivatives inhibit ADP-induced platelet activation and aggregation. *Molecules, 27*(13). doi:10.3390/molecules27134054.

Lu, Q., Ye, H., Wang, K., Zhao, J., Wang, H., Song, J., Fan, X., Lu, Y., Cao, L., Wan, B., Zhang, H., He, Z., & Sun, J. (2022). Bioengineered platelets combining chemotherapy and immunotherapy for postsurgical melanoma treatment: Internal core-loaded doxorubicin and external surface-anchored anti-PD-L1 antibody backpacks. *Nano Letters, 22*(7), 3141–3150. doi:10.1021/acs.nanolett.2c00907.

Maouia, A., Rebetz, J., Kapur, R., & Semple, J. W. (2020). The immune nature of platelets revisited. *Transfusion Medicine Reviews, 34*(4), 209–220. doi:10.1016/j.tmrv.2020.09.005.

Martinod, K., & Deppermann, C. (2021). Immunothrombosis and thromboinflammation in host defense and disease. *Platelets, 32*(3), 314–324. doi:10.1080/09537104.2020.1817360.

Massagué, J., & Obenauf, A. C. (2016). Metastatic colonization by circulating tumour cells. *Nature, 529*(7586), 298–306. doi:10.1038/nature17038.

Maurer, S., Kropp, K. N., Klein, G., Steinle, A., Haen, S. P., Walz, J. S., Hinterleitner, C., Märklin, M., Kopp, H. G., & Salih, H. R. (2018). Platelet-mediated shedding of NKG2D ligands impairs NK cell immune-surveillance of tumor cells. *OncoImmunology, 7*(2). doi:10.1080/2162402X.2017.1364827.

Mause, S. F., & Weber, C. (2010). Microparticles: Protagonists of a novel communication network for intercellular information exchange. *Circulation Research, 107*(9), 1047–1057. doi:10.1161/CIRCRESAHA.110.226456.

Medina-Leyte, D. J., Zepeda-García, O., Domínguez-Pérez, M., González-Garrido, A., Villarreal-Molina, T., & Jacobo-Albavera, L. (2021). Endothelial dysfunction, inflammation and coronary artery disease: Potential biomarkers and promising therapeutical approaches. *International Journal of Molecular Sciences, 22*(8). doi:10.3390/ijms22083850.

Mehdi-Alamdarlou, S., Ahmadi, F., Shahbazi, M. A., Azadi, A., & Ashrafi, H. (2023). Platelets and platelet-derived vesicles as an innovative cellular and subcellular platform for managing multiple sclerosis. *Molecular Biology Reports, 50*(5), 4675–4686. doi:10.1007/s11033-023-08322-7.

Meikle, C. K. S., Kelly, C. A., Garg, P., Wuescher, L. M., Ali, R. A., & Worth, R. G. (2017). Cancer and thrombosis: The platelet perspective. *Frontiers in Cell and Developmental Biology, 4*. doi:10.3389/fcell.2016.00147.

Melchinger, H., Jain, K., Tyagi, T., & Hwa, J. (2019). Role of platelet mitochondria: Life in a nucleus-free zone. *Frontiers in Cardiovascular Medicine, 6*. doi:10.3389/fcvm.2019.00153.

Menter, D. G., Kopetz, S., Hawk, E., Sood, A. K., Loree, J. M., Gresele, P., & Honn, K. V. (2017). Platelet "first responders" in wound response, cancer, and metastasis. *Cancer and Metastasis Reviews, 36*(2), 199–213. doi:10.1007/s10555-017-9682-0.

Miao, Y., Xu, Z., Feng, W., Zheng, M., Xu, Z., Gao, H., Li, W., Zhang, Y., Zong, Y., Lu, A., & Zhao, J. (2022). Platelet infiltration predicts survival in postsurgical colorectal cancer patients. *International Journal of Cancer, 150*(3), 509–520. doi:10.1002/ijc.33816.

Miyashita, T., Tajima, H., Makino, I., Nakagawara, H., Kitagawa, H., Fushida, S., Harmon, J. W., & Ohta, T. (2015). Metastasis-promoting role of extravasated platelet activation in tumor. *Journal of Surgical Research, 193*(1), 289–294. doi:10.1016/j.jss.2014.07.037.

Mlecnik, B., Tosolini, M., Kirilovsky, A., Berger, A., Bindea, G., Meatchi, T., Bruneval, P., Trajanoski, Z., Fridman, W. H., Pagès, F., & Galon, J. (2011). Histopathologic-based prognostic factors of colorectal cancers are associated with the state of the local immune reaction. *Journal of Clinical Oncology, 29*(6), 610–618. doi:10.1200/JCO.2010.30.5425.

Morrell, C. N., Pariser, D. N., Hilt, Z. T., & Vega Ocasio, D. (2019). The platelet Napoleon complex-small cells, but big immune regulatory functions. *Annual Review of Immunology, 37*, 125–144. doi:10.1146/annurev-immunol-042718-041607.

Morrissey, S. M., Zhang, F., Ding, C., Montoya-Durango, D. E., Hu, X., Yang, C., Wang, Z., Yuan, F., Fox, M., Zhang, H. G., Guo, H., Tieri, D., Kong, M., Watson, C. T., Mitchell, R. A., Zhang, X., McMasters, K. M., Huang, J., & Yan, J. (2021). Tumor-derived exosomes drive immunosuppressive macrophages in a pre-metastatic niche through glycolytic dominant metabolic reprogramming. *Cell Metabolism, 33*(10), 2040. doi:10.1016/j.cmet.2021.09.002.

Ngwa, W., Irabor, O. C., Schoenfeld, J. D., Hesser, J., Demaria, S., & Formenti, S. C. (2018). Using immunotherapy to boost the abscopal effect. *Nature Reviews Cancer, 18*(5), 313–322. doi:10.1038/nrc.2018.6.

Nicolai, L., Pekayvaz, K., & Massberg, S. (2024). Platelets: Orchestrators of immunity in host defense and beyond. *Immunity, 57*(5), 957–972. doi:10.1016/j.immuni.2024.04.008.

Nieswandt, B., Hafner, M., Echtenacher, B., & Männel, D. N. (1999). Lysis of tumor cells by natural killer cells in mice is impeded by platelets. *Cancer Research, 59*(6), 1295–1300.

Palumbo, J. S., Talmage, K. E., Massari, J. V., La Jeunesse, C. M., Flick, M. J., Kombrinck, K. W., Jirousková, M., & Degen, J. L. (2005). Platelets and fibrin(ogen) increase metastatic potential by impeding natural killer cell-mediated elimination of tumor cells. *Blood, 105*(1), 178–185. doi:10.1182/blood-2004-06-2272.

Pang, J. H., Coupland, L. A., Freeman, C., Chong, B. H., & Parish, C. R. (2015). Activation of tumour cell ECM degradation by thrombin-activated platelet membranes: Potentially a P-selectin and GPIIb/IIIa-dependent process. *Clinical & Experimental Metastasis, 32*(5), 495–505. doi:10.1007/s10585-015-9722-5.

Pardoll, D. M. (2012). The blockade of immune checkpoints in cancer immunotherapy. *Nature Reviews Cancer, 12*(4), 252–264. doi:10.1038/nrc3239.

Pastori, D., Cormaci, V. M., Marucci, S., Franchino, G., Del Sole, F., Capozza, A., Fallarino, A., Corso, C., Valeriani, E., Menichelli, D., & Pignatelli, P. (2023). A comprehensive review of risk factors for venous thromboembolism: From epidemiology to pathophysiology. *International Journal of Molecular Sciences, 24*(4). doi:10.3390/ijms24043169.

Periayah, M. H., Halim, A. S., & Saad, A. Z. M. (2017). Mechanism action of platelets and crucial blood coagulation pathways in Hemostasis. *International Journal of Hematology-Oncology and Stem Cell Research, 11*(4), 319–327. http://ijhoscr.tums.ac.ir/index.php/ijhoscr/article/download/573/585.

Philippe, C., Philippe, B., Fouqueray, B., Perez, J., Lebret, M., & Baud, L. (1993). Protection from tumor necrosis factor-mediated cytolysis by platelets. *American Journal of Pathology, 143*(6), 1713–1723. http://ajp.amjpathol.org/.

Placke, T., Örgel, M., Schaller, M., Jung, G., Rammensee, H. G., Kopp, H. G., & Salih, H. R. (2012). Platelet-derived MHC class I confers a pseudonormal phenotype to cancer cells that subverts the antitumor reactivity of natural killer immune cells. *Cancer Research, 72*(2), 440–448. doi:10.1158/0008-5472.CAN-11-1872Germany.

Plantureux, L., Mège, D., Crescence, L., Carminita, E., Robert, S., Cointe, S., Brouilly, N., Ezzedine, W., Dignat-George, F., Dubois, C., & Panicot-Dubois, L. (2020). The interaction of platelets with colorectal cancer cells inhibits tumor growth but promotes metastasis. *Cancer Research, 80*(2), 291–303. doi:10.1158/0008-5472.can-19-1181.

Puricelli, C., Boggio, E., Gigliotti, C. L., Stoppa, I., Sutti, S., Giordano, M., Dianzani, U., & Rolla, R. (2023). Platelets, protean cells with all-around functions and multifaceted pharmacological applications. *International Journal of Molecular Sciences, 24*(5). doi:10.3390/ijms24054565.

Qiao, K., Cui, X., Gao, J., Yu, F., Liu, H., Dai, Y., Liu, J., Yang, Y., Xian, X., Hu, J., Tang, J., & Cui, X. (2023). Roles of extracellular vesicles derived from immune cells in atherosclerosis. *Extracellular Vesicle, 2*, 100028. doi:10.1016/j.vesic.2023.100028.

Rawish, E., Nording, H., Münte, T., & Langer, H. F. (2020). Platelets as mediators of neuroinflammation and thrombosis. *Frontiers in Immunology, 11*. doi:10.3389/fimmu.2020.548631.

Rayes, J., Watson, S. P., & Nieswandt, B. (2019). Functional significance of the platelet immune receptors GPVI and CLEC-2. *Journal of Clinical Investigation, 129*(1), 12–23. doi:10.1172/jci122955.

Rolla, R., Puricelli, C., Bertoni, A., Boggio, E., Gigliotti, C. L., Chiocchetti, A., Cappellano, G., & Dianzani, U. (2021). Platelets: "Multiple choice" effectors in the immune response and their implication in COVID-19 thromboinflammatory process. *International Journal of Laboratory Hematology, 43*(5), 895–906. doi:10.1111/ijlh.13516.

Sankar, K., Ye, J. C., Li, Z., Zheng, L., Song, W., & Hu-Lieskovan, S. (2022). The role of biomarkers in personalized immunotherapy. *Biomarker Research, 10*(1). doi:10.1186/s40364-022-00378-0.

Santoso, S., Sachs, U. J. H., Kroll, H., Linder, M., Ruf, A., Preissner, K. T., & Chavakis, T. (2002). The junctional adhesion molecule 3 (JAM-3) on human platelets is a counterreceptor for the leukocyte integrin Mac-1. *Journal of Experimental Medicine, 196*(5), 679–691. doi:10.1084/jem.20020267.

Scherlinger, M., Richez, C., Tsokos, G. C., Boilard, E., & Blanco, P. (2023). The role of platelets in immune-mediated inflammatory diseases. *Nature Reviews Immunology, 23*(8), 495–510. doi:10.1038/s41577-023-00834-4.

Schmied, L., Höglund, P., & Meinke, S. (2021a). Platelet-mediated protection of cancer cells from immune surveillance – Possible implications for cancer immunotherapy. *Frontiers in Immunology, 12*. doi:10.3389/fimmu.2021.640578.

Schumacher, D., Strilic, B., Sivaraj, K. K., Wettschureck, N., & Offermanns, S. (2013). Platelet-derived nucleotides promote tumor-cell transendothelial migration and metastasis via P2Y2 receptor. *Cancer Cell, 24*(1), 130–137. doi:10.1016/j.ccr.2013.05.008.

Scridon, A. (2022). Platelets and their role in hemostasis and thrombosis—From physiology to pathophysiology and therapeutic implications. *International Journal of Molecular Sciences, 23*(21), 12772. doi:10.3390/ijms232112772.

Selvadurai, M. V., & Hamilton, J. R. (2018). Structure and function of the open canalicular system–The platelet's specialized internal membrane network. *Platelets, 29*(4), 319–325. doi:10.1080/09537104.2018.1431388.

Semple, J. W., Italiano, J. E., & Freedman, J. (2011). Platelets and the immune continuum. *Nature Reviews Immunology, 11*(4), 264–274. doi:10.1038/nri2956.

Senchenkova, E. Y., Ansari, J., Becker, F., Vital, S. A., Al-Yafeai, Z., Sparkenbaugh, E. M., Pawlinski, R., Stokes, K. Y., Carroll, J. L., Dragoi, A. M., Qin, C. X., Ritchie, R. H., Sun, H., Cuellar-Saenz, H. H., Rubinstein, M. R., Han, Y. W., Orr, A. W., Perretti, M., Granger, D. N., & Gavins, F. N. E. (2019). Novel role for the AnxA1-Fpr2/ALX signaling axis as a key regulator of platelet function to promote resolution of inflammation. *Circulation, 140*(4), 319–335. doi:10.1161/CIRCULATIONAHA.118.039345.

Shao, B., Wahrenbrock, M. G., Yao, L., David, T., Coughlin, S. R., Xia, L., Varki, A., & McEver, R. P. (2011). Carcinoma mucins trigger reciprocal activation of platelets and neutrophils in a murine model of Trousseau syndrome. *Blood, 118*(15), 4015–4023. doi:10.1182/blood-2011-07-368514.

Sharma, D., Brummel-Ziedins, K. E., Bouchard, B. A., & Holmes, C. E. (2014). Platelets in tumor progression: A host factor that offers multiple potential targets in the treatment of cancer. *Journal of Cellular Physiology, 229*(8), 1005–1015. doi:10.1002/jcp.24539.

Shau, H., Roth, M. D., & Golub, S. H. (1993). Regulation of natural killer function by nonlymphoid cells. *Natural Immunity, 12*(4-5), 235–249.

Shi, G., Field, D. J., Ko, K. A., Ture, S., Srivastava, K., Levy, S., Kowalska, M. A., Poncz, M., Fowell, D. J., & Morrell, C. N. (2014). Platelet factor 4 limits Th17 differentiation and cardiac allograft rejection. *Journal of Clinical Investigation, 124*(2), 543–552. doi:10.1172/JCI71858.

Shiravand, Y., Khodadadi, F., Kashani, S. M. A., Hosseini-Fard, S. R., Hosseini, S., Sadeghirad, H., Ladwa, R., O'byrne, K., & Kulasinghe, A. (2022). Immune checkpoint inhibitors in cancer therapy. *Current Oncology, 29*(5), 3044–3060. doi:10.3390/curroncol29050247.

Sim, M. M. S., Shiferawe, S., & Wood, J. P. (2023). Novel strategies in antithrombotic therapy: Targeting thrombosis while preserving hemostasis. *Frontiers in Cardiovascular Medicine, 10.* doi:10.3389/fcvm.2023.1272971.

Singh, S., Khasbage, S., Kaur, R. J., Sidhu, J. K., & Bhandari, B. (2022). Chimeric antigen receptor T cell: A cancer immunotherapy. *Indian Journal of Pharmacology, 54*(3), 226–233. doi:10.4103/ijp.ijp_531_20.

Sprague, D. L., Elzey, B. D., Crist, S. A., Waldschmidt, T. J., Jensen, R. J., & Ratliff, T. L. (2008). Platelet-mediated modulation of adaptive immunity: Unique delivery of CD154 signal by platelet-derived membrane vesicles. *Blood, 111*(10), 5028–5036. doi:10.1182/blood-2007-06-097410.

Stone, R. L., Nick, A. M., McNeish, I. A., Balkwill, F., Han, H. D., Bottsford-Miller, J., Rupaimoole, R., Armaiz-Pena, G. N., Pecot, C. V., Coward, J., Deavers, M. T., Vasquez, H. G., Urbauer, D., Landen, C. N., Hu, W., Gershenson, H., Matsuo, K., Shahzad, M. M. K., King, E. R., & Tekedereli, I. (2012). Paraneoplastic thrombocytosis in ovarian cancer. *New England Journal of Medicine, 366*(7), 610–618. doi:10.1056/NEJMoa1110352.

Sun, C., Mezzadra, R., & Schumacher, T. N. (2018). Regulation and function of the PD-L1 checkpoint. *Immunity, 48*(3), 434–452. doi:10.1016/j.immuni.2018.03.014.

Suzuki-Inoue, K. (2019). Platelets and cancer-associated thrombosis: focusing on the platelet activation receptor CLEC-2 and podoplanin. *Blood, 134*(22), 1912–1918. doi:10.1182/blood.2019001388.

Tang, M., Jiang, L. U, Lin, Y., Wu, X., Wang, K., He, Q., Wang, X., & Li, W. (2017). Platelet microparticle-mediated transfer of miR-939 to epithelial ovarian cancer cells promotes epithelial to mesenchymal transition. *Oncotarget, 8*(57), 97464–97475. doi:10.18632/oncotarget.22136.

Thornton, P., McColl, B. W., Greenhalgh, A., Denes, A., Allan, S. M., & Rothwell, N. J. (2010). Platelet interleukin-1α drives cerebrovascular inflammation. *Blood, 115*(17), 3632–3639. doi:10.1182/blood-2009-11-252643.

Tsubakihara, Y., & Moustakas, A. (2018). Epithelial-mesenchymal transition and metastasis under the control of transforming growth factor β. *International Journal of Molecular Sciences, 19*(11), 3672. doi:10.3390/ijms19113672.

Tung, J. P., Chiaretti, S., Dean, M. M., Sultana, A. J., Reade, M. C., & Fung, Y. L. (2022). Transfusion-related acute lung injury (TRALI): Potential pathways of development, strategies for prevention and treatment, and future research directions. *Blood Reviews, 53*. doi:10.1016/j.blre.2021.100926.

Van Nispen Tot Pannerden, H., De Haas, F., Geerts, W., Posthuma, G., Van Dijk, S., & Heijnen, H. F. G. (2010). The platelet interior revisited: Electron tomography reveals tubular α-granule subtypes. *Blood, 116*(7), 1147–1156. doi:10.1182/blood-2010-02-268680.

Vasina, E., Heemskerk, J. W. M., Weber, C., & Koenen, R. R. (2010). Platelets and platelet-derived microparticles in vascular inflammatory disease. *Inflammation & Allergy - Drug Targets, 9*(5), 346–354. doi:10.2174/187152810793938008.

Von Hundelshausen, P., Koenen, R. R., & Weber, C. (2009). Platelet-mediated enhancement of leukocyte adhesion. *Microcirculation, 16*(1), 84–96. doi:10.1080/10739680802564787.

Von Hundelshausen, P., & Weber, C. (2007). Platelets as immune cells: Bridging inflammation and cardiovascular disease. *Circulation Research, 100*(1), 27–40. doi:10.1161/01.RES.0000252802.25497.b7.

Wagner, D. D., & Frenette, P. S. (2008). The vessel wall and its interactions. *Blood, 111*(11), 5271–5281. doi:10.1182/blood-2008-01-078204.

Wang, C., Sun, W., Ye, Y., Hu, Q., Bomba, H. N., & Gu, Z. (2017). In situ activation of platelets with checkpoint inhibitors for post-surgical cancer immunotherapy. *Nature Biomedical Engineering, 1*(2). doi:10.1038/s41551-016-0011.

Wang, Q., Shao, X., Zhang, Y., Zhu, M., Wang, F. X. C., Mu, J., Li, J., Yao, H., & Chen, K. (2023). Role of tumor microenvironment in cancer progression and therapeutic strategy. *Cancer Medicine, 12*(10), 11149–11165. doi:10.1002/cam4.5698.

Ward, Y., Lake, R., Faraji, F., Sperger, J., Martin, P., Gilliard, C., Ku, K. P., Rodems, T., Niles, D., Tillman, H., Yin, J. J., Hunter, K., Sowalsky, A. G., Lang, J., & Kelly, K. (2018). Platelets promote metastasis via binding tumor CD97 leading to bidirectional signaling that coordinates transendothelial migration. *Cell Reports, 23*(3), 808–822. doi:10.1016/j.celrep.2018.03.092.

Weber, C., & Springer, T. A (1997). Neutrophil accumulation on activated, surface-adherent platelets in flow is mediated by interaction of Mac-1 with fibrinogen bound to alphaIIbbeta3 and stimulated by platelet-activating factor. *Journal of Clinical Investigation, 100*(8), 2085–2093. doi:10.1172/JCI119742.

Weber, M. R., Zuka, M., Lorger, M., Tschan, M., Torbett, B. E., Zijlstra, A., Quigley, J. P., Staflin, K., Eliceiri, B. P., Krueger, J. S., Marchese, P., Ruggeri, Z. M., & Felding, B. H. (2016). Activated tumor cell integrin $\alpha v\beta 3$ cooperates with platelets to promote extravasation and metastasis from the blood stream. *Thrombosis Research, 140*, S27. doi:10.1016/S0049-3848(16)30095-0.

Xiong, G., Chen, J., Zhang, G., Wang, S., Kawasaki, K., Zhu, J., Zhang, Y., Nagata, K., Li, Z., Zhou, B. P., & Xu, R. (2020). Hsp47 promotes cancer metastasis by enhancing collagen-dependent cancer cell-platelet interaction. *Proceedings of the National Academy of Sciences of the United States of America, 117*(7), 3748–3758. doi:10.1073/pnas.1911951117.

Xu, S. S., Xu, H. X., Wang, W. Q., Li, S., Li, H., Li, T. J., Zhang, W. H., Liu, L., & Yu, X. J. (2019). Tumor-infiltrating platelets predict postoperative recurrence and survival in resectable pancreatic neuroendocrine tumor. *World Journal of Gastroenterology, 25*(41), 6248–6257. doi:10.3748/wjg.v25.i41.6248.

Xu, X. R., Carrim, N., Neves, M. A. D., McKeown, T., Stratton, T. W., Coelho, R. M. P., Lei, X., Chen, P., Xu, J., Dai, X., Li, B. X., & Ni, H. (2016). Platelets and platelet adhesion molecules: Novel mechanisms of thrombosis and anti-thrombotic therapies. *Thrombosis Journal, 14*. doi:10.1186/s12959-016-0100-6.

Yadav, S., & Storrie, B. (2017). The cellular basis of platelet secretion: Emerging structure/function relationships. *Platelets, 28*(2), 108–118. doi:10.1080/09537104.2016.1257786.

Yan, J., Liu, X., Wu, F., Ge, C., Ye, H., Chen, X., Wei, Y., Zhou, R., Duan, S., Zhu, R., Zheng, Y., & Yin, L. (2022). Platelet pharmacytes for the hierarchical amplification of antitumor immunity in response to self-generated immune signals. *Advanced Materials, 34*(23). doi:10.1002/adma.202109517.

Yang, X., Bian, J., Wang, Z., He, M., Yang, Y., Li, Q., Luo, X., Zhou, Z., Li, J., Ju, S., & Sun, M. (2023). A bio-liposome activating natural killer cell by illuminating tumor homogenization antigen properties. *Advanced Science, 10*(12). doi:10.1002/advs.202205449.

Yeaman, M. R., Yount, N. Y., Waring, A. J., Gank, K. D., Kupferwasser, D., Wiese, R., Bayer, A. S., & Welch, W. H. (2007). Modular determinants of antimicrobial activity in platelet factor-4 family kinocidins. *Biochimica et Biophysica Acta - Biomembranes, 1768*(3), 609–619. doi:10.1016/j.bbamem.2006.11.010.

Yeung, J., Li, W., Holinstat, M., & Isom, L. L. (2018). Platelet signaling and disease: Targeted therapy for thrombosis and other related diseases. *Pharmacological Reviews, 70*(3), 526–548. doi:10.1124/pr.117.014530.

Yoon, I., Han, J. H., & Jeon, H. J. (2024). Advances in platelet-dysfunction diagnostic technologies. *Biomolecules, 14*(6). doi:10.3390/biom14060714.

Yu, L., Guo, Y., Chang, Z., Zhang, D., Zhang, S., Pei, H., Pang, J., Zhao, Z. J., & Chen, Y. (2021). Bidirectional interaction between cancer cells and platelets provides potential strategies for cancer therapies. *Frontiers in Oncology, 11*. doi:10.3389/fonc.2021.764119.

Yu, L. X., Yan, L., Yang, W., Wu, F. Q., Ling, Y., Chen, S. Z., Tang, L., Tan, Y. X., Cao, D., Wu, M. C., Yan, H. X., & Wang, H. Y. (2014). Platelets promote tumour metastasis via interaction between TLR4 and tumour cell-released high-mobility group protein. *Nature Communications, 5*. doi:10.1038/ncomms6256.

Yu, M., Bardia, A., Wittner, B. S., Stott, S. L., Smas, M. E., Ting, D. T., Isakoff, S. J., Ciciliano, J. C., Wells, M. N., Shah, A. M., Concannon, K. F., Donaldson, M. C., Sequist, L. V., Brachtel, E., Sgroi, D., Baselga, J., Ramaswamy, S., Toner, M., Haber, D. A., & Maheswaran, S. (2013). Circulating breast tumor cells exhibit dynamic changes in epithelial and mesenchymal composition. *Science, 339*(6119), 580–584. doi:10.1126/science.1228522.

Zhang, K., Li, H., Ma, Z., Zhong, W., Yu, Y., & Zhao, Y. (2024). Engineered platelets for cancer therapy. *Aggregate*. doi:10.1002/agt2.704.

Zhang, S. R., Yao, L., Wang, W. Q., Xu, J. Z., Xu, H. X., Jin, W., Gao, H. L., Wu, C. T., Qi, Z. H., Li, H., Li, S., Ni, Q. X., Yu, X. J., Fu, D. L., & Liu, L. (2018). Tumor-infiltrating platelets predict postsurgical survival in patients with pancreatic ductal adenocarcinoma. *Annals of Surgical Oncology, 25*(13), 3984–3993. doi:10.1245/s10434-018-6727-8.

Zhang, W., Li, S., Liu, Y., Xing, R., Jin, Z., Yan, X., & Xue, H. (2023). Immunosuppressive microenvironment improvement and treatment of aggressive malignancy pancreatic ductal adenocarcinoma based on local administration of injectable hydrogel. *Nano Today, 50*, 101832. doi:10.1016/j.nantod.2023.101832.

Zhang, X., Wang, J., Chen, Z., Hu, Q., Wang, C., Yan, J., Dotti, G., Huang, P., & Gu, Z. (2018). Engineering PD-1-presenting platelets for cancer immunotherapy. *Nano Letters, 18*(9), 5716–5725. doi:10.1021/acs.nanolett.8b02321.

Zhou, H., Andonegui, G., Wong, C. H. Y., & Kubes, P. (2009). Role of endothelial TLR4 for neutrophil recruitment into central nervous system microvessels in systemic inflammation. *Journal of Immunology, 183*(8), 5244–5250. doi:10.4049/jimmunol.0901309.

Zhou, L., Zhang, Z., Tian, Y., Li, Z., Liu, Z., & Zhu, S. (2023). The critical role of platelet in cancer progression and metastasis. *European Journal of Medical Research, 28*(1). doi:10.1186/s40001-023-01342-w.

Zlotnik, A., & Yoshie, O. (2000). Chemokines. *Immunity, 12*(2), 121–127. doi:10.1016/s1074-7613(00)80165-x.

CHAPTER

10

Cancer cell-platelet interactions: Harnessing therapeutic opportunities in cancer treatment

Sanya Jain and Gudhanti Siva Naga Koteswara Rao
Shobhaben Pratapbhai Patel School of Pharmacy & Technology Management, SVKM's NMIMS (Deemed to be University), Mumbai, Maharashtra, India

10.1 Introduction

10.1.1 Overview of cancer and platelet interactions

Platelets are anucleated cells, circulating in the bloodstream that clump together to form a plug at the injury site where von Willebrand factor (vWF) is expressed to prevent blood loss. Their main function is to maintain homeostasis and regulate inflammation, adaptive, and innate immune responses. It mediates platelet tethering on the thrombogenic surface. Inhibition of platelet GPIbα interaction with vWF leads to a reduction in the platelet-tumor cell contact and metastasis. The secretion of vWF fiber from endothelial cells is mediated by tumor cell-generated vascular endothelial growth factor-A (VEGF-A) (Castanho & &Garnes, 2019; Goto et al., 2015; Morris et al., 2022; Zaid & Merhi, 2022). Platelets play a pivotal role in maintaining homeostasis and thrombosis in several pathological events. These cells are essential for maintaining vascular homeostasis through the activation of coagulation factors in the absence or presence of injury (Goto et al., 2015; Gupta et al., 2020). They are activated by tumor cells through various stimuli, promoting tumor cell proliferation, intracellular signaling, and EMT. The interaction between platelets and tumor cells plays a key role in tumor growth, thrombosis, angiogenesis, dissemination, metastasis, and immune escape. Platelets have several key functions in the tumor microenvironment: Contributing to tumor growth, facilitating the thrombosis of the tumor, inducing tumor angiogenesis, and promoting tumor cell invasion and metastasis. They also play a part in immune surveillance evasion, with PD-1–PD-L1 ligand being of high expression in cancer patients but not in normal platelets. Platelets increase PD-L1 expression in cancer cells through different signaling pathways and thus facilitate invasion by tumor cells and assist in the detachment of cancer cells from their primary site (Li et al., 2023). (Fig. 10.1)

10.1.2 Significance of platelet interactions in cancer progression

Platelets play a significant role in cancer progression by influencing several tumor development stages. They protect the circulating tumor cells from immune surveillance and shear in the bloodstream, ensuring their survival and further dispersion. They induce pathways in cancer cells, causing a more invasive mesenchymal-like phenotype and enhancing metastasis. Platelets assist in arresting the Circulating Tumor Cells (CTCs) on the vascular wall and supporting extravasation into the distant organs through the secretion of growth factors (Zhou et al., 2023).

Advancements in Personalized Cellular Immunotherapy: A Comprehensive Guide to Cancer Treatment.
DOI: https://doi.org/10.1016/B978-0-443-27731-3.00013-9

Copyright © 2026 Elsevier Inc. All rights are reserved, including those for text and data mining, AI training, and similar technologies.

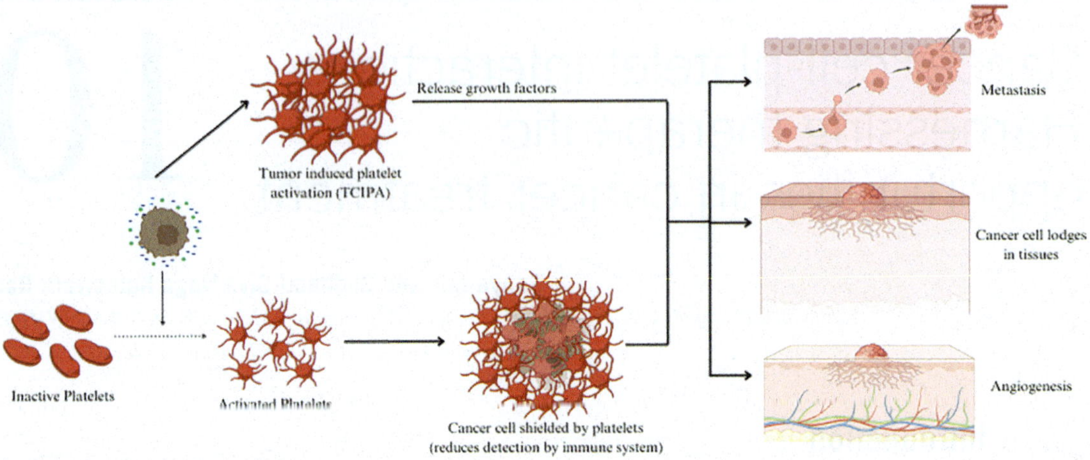

FIGURE 10.1

Cancer cell–platelet interactions.

Platelets aid in the tumorigenic process by providing physical and mechanical support to cancer cells, evading the immune system, and lead to metastasis by forming "early metastatic niches" (Labelle et al., 2014; Repsold & Joubert, 2021). They can interact either directly or indirectly, mediated by cancer cells and microvesicles, respectively (Lazar & Goldfinger, 2021). These interactions aid cancer cells to bypass the immunological monitoring, leading to an increase in the tumor invasiveness, neovascularization, releasing growth factors and cytokines, which promote angiogenesis and metastasis (Battinelli, 2017; Li, 2016). Calcium is also required to maintain the structural integrity and ligand-binding ability of integrins such as GPIIb/IIIa and $\alpha v\beta 3$ as a result of the platelet-cell interactions (Lazar & Goldfinger, 2021). GPIIb/IIIa can modulate cancer either directly or indirectly (Morris et al., 2022). These cells facilitate metastasis by interacting physically and functionally with cancer cells and are involved in the intercellular communication by receptor–ligand interactions with monocytes, leukocytes, and granulocytes (Castanho & &Garnes, 2019; Braun et al., 2021) (Fig. 10.2).

Altering the platelets and fibrin in the TME also leads to the spread of the tumor (Obermann et al., 2021). Platelets interact with leukocytes to manipulate immune responses, as well as change the relationship between innate and adaptive immunity. They can attract immune cells into tumor sites, modify their functions, and even protect tumor cells from natural killer cell cytotoxicity (Meikle et al., 2017; Stoiber & Assinger, 2020). On interaction with leucocytes, they aid in cancer development and regulate inflammation by attracting leucocytes to the tumor site (Stoiber & Assinger, 2020). Granulocyte recruitment is dependent on platelet-derived CXCL5/7 chemokines, which prevent metastasis by blocking platelet-to-granulocyte communication (Labelle et al., 2014). The interaction between platelets and cancer cells leads to a time-dependent increase in the cyclin-dependent kinase inhibitor and a significant decrease in Rb phosphorylation (Plantureux et al., 2020). These interactions activate the TGFβ/Smad and NF-β B pathways in cancer cells, predisposing tumor cells for further metastasis (Labelle et al., 2011).

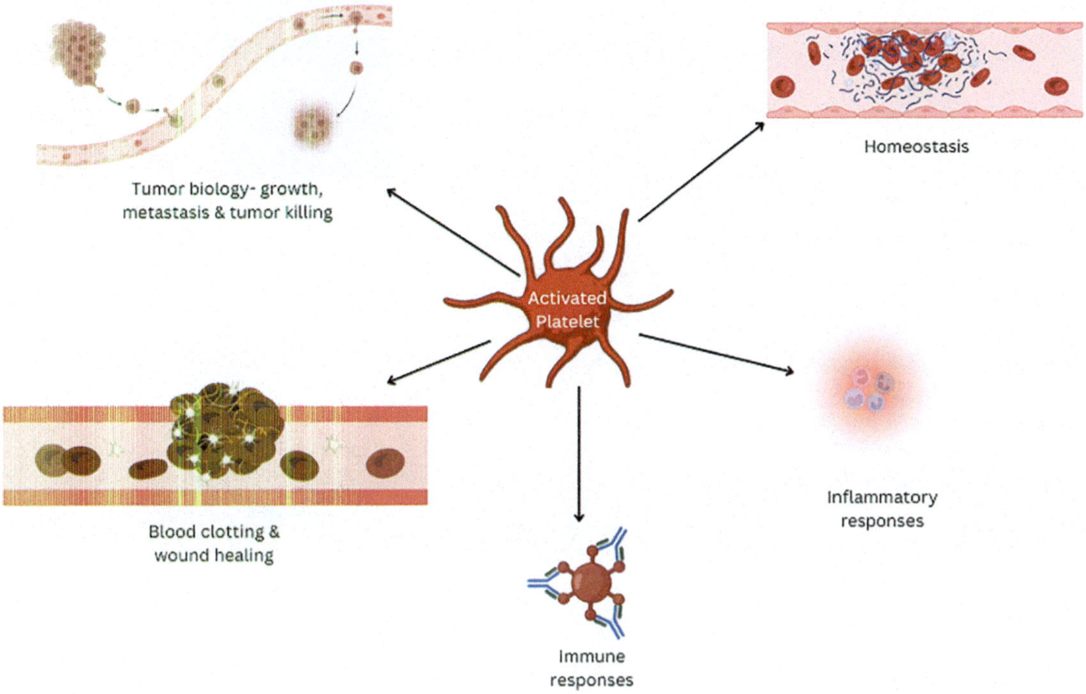

FIGURE 10.2

Functions of activated platelets.

Platelets enable metastatic processes by acting as a shield for immune elimination and stimulate their arrest in the endothelium (Gay & Felding-Habermann, 2011). They also recruit myeloid cells and stimulate tumor cell proliferation to micrometastatic foci, enhancing the cellular plasticity and survival of tumor cells during their progression (Schlesinger, 2018; Wang et al., 2018).

Tumor-educated platelets contain extremely high levels of active biomolecules that are released following platelet activation and play a role in tumor progression. Targeting platelet activation, adhesive proteins responsible for cancer cell-platelet interaction, and platelet agonists is likely to inhibit cancer-associated thrombosis and tumor progression (Plantureux et al., 2018). They can also be used for cancer detection, diagnosis, and treatment effects (In 'T Veld & Wurdinger, 2019) (Fig. 10.3).

10.2 Receptors involved in platelet–cell interactions

Several receptors are involved in platelet-cell interactions and play a significant role in maintaining homeostasis and cancer metastasis. Table 10.1 represents a detailed overview of the different receptors involved in cancer-platelet cell interactions (Fig. 10.4).

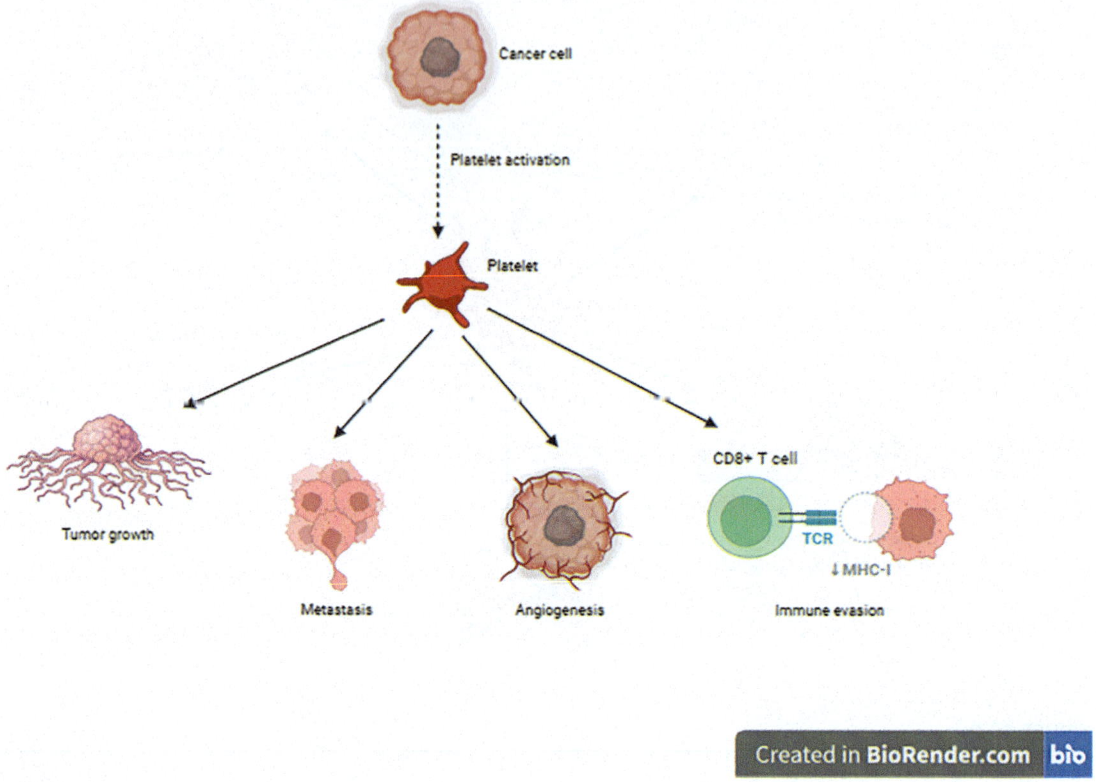

FIGURE 10.3

Role of platelets in cancer development.

10.2.1 Adhesion receptors

The first category of receptors involved in platelet–cell interactions is adhesion receptors. These receptors facilitate platelet interaction with damaged endothelial surfaces, subendothelial extracellular matrix components, and leucocytes (Prabhu et al., 2011; Ruggeri, 2003; Yip et al., 2005). This category includes PECAM-1, GPIIb/IIIa, GPVI, and integrins (αIIbβ3), which play a significant role in platelet adhesion and aggregation.

Cancer platelet–cell interactions involve GPIb-IX-V and GPVI receptors present on the platelets. The GPIbα component of the complex and GPVI enhance CTC arrest and extravasation, leading to metastatic progression. Although mice with dysfunctional GPIb-IX had reduced metastatic foci, GPIbα inhibition, reliant on P-selectin expression, enhanced metastasis in an experimental mice model (Morris et al., 2022). During homeostasis, GP IIb/IIIa recognizes the Arg-Gly-Asp sequence and results in platelet adhesion and aggregation. GPIIb/IIIa demonstrates conformational changes with fibrinogen binding, speeding up the aggregation process (Fullard, 2004). Under conditions of shear stress, GPIb-IX-V receptors on the platelets interact with adhesive receptors such as VWF, which slows down the

Table 10.1 Receptors involved in platelet–cell interactions.

Receptor family	Example(s)	References
Adhesion receptors	(GP) Ib-IX-V, (GP) VI, (GP) IIb/IIIa (αIIbβ3), and PECAM-1	Jurasz et al. (2004)
Toll-like receptors	TLR1, TLR2, TLR4, TLR6 (extracellular); TLR-3, TLR-7, TLR-8, TLR-9 (intracellular)	Dib et al. (2020); Tokarz-Deptuła et al. (2024)
Nod-like receptors	NOD2, NLRP1, NLRP3, NLRC4	Tokarz-Deptuła et al. (2024); Vogel et al. (2018); Zhang et al. (2015)
C-type lectin receptors families	CLEC-2 and DC SIGN	Qiao et al. (2015); Tokarz-Deptuła et al. (2024)
GPCRs	TPα and TPβ, PAR-1 and PAR-4, P2Y receptors (P2Y1 and P2Y12), 5-HT2A, prostacyclin receptor (IP), GRK3, GRK6	
ITAM and hemITAM-associated receptors	FcγRIIa	Boylan et al. (2008)
P-selectins	CD62P	Tokarz-Deptuła et al. (2024)
Other extracellular receptors	CD116, CD18, CD40, FcγRIIa, MHC class I, GARP, $LPA_{1,}$ $LPA_{2,}$ LPA_3, $P2Y_{1,}$ $P2Y_{12}$, PAR-1,3,4, ADP, TxA, IL1β, IL1R1, IL18Ra, TNF, IFNγ, TGFβR, CCL-1, 3, 4, 5, CXCL-1, 2, 3, 4, 6, 7, 8, 12, 16, 17, 22, 24, 26, ICAM-1, ICAM-2, JAM-A, JAM-C, PECAM1, C1q, C8, C9, C3a, C3aR, and ACE-2	Tokarz-Deptuła et al. (2024)

ICAM, *Intercellular adhesion molecule;* IFNγ, *interferon gamma;* JAM, *junctional adhesion molecule;* MHC, *major histocompatibility complex;* NLRP, *nucleotide-binding oligomerization domain leucine-rich repeat and pyrin domain binding;* NOD, *nucleotide-binding and oligomerization domain;* PAR, *protease-activated receptor;* PECAM1, *platelet endothelial cell adhesion molecule 1;* TNF, *tumor necrosis factor.;* TLR, *toll-like receptor; CLEC-2, C-type lectin-like receptor 2.*

platelet movement, allowing further adhesion (Jurasz et al., 2004; Saboor et al., 2013; Varga-Szabo et al., 2008).

Integrins such as αIIbβ3 and α6β1 are involved in platelet adhesion and play a significant role in altering the platelet morphology within the signaling pathways (Hosseini et al., 2019; Nuyttens et al., 2011; Saboor et al., 2013). Platelet aggregation leads to the formation of an "adhesion monolayer," which is resistant to the disintegrating stimuli and requires some changes in the platelet structure. This cascade results in platelet integrin receptor activation (αIIbβ3), enabling fibrinogen binding (Jurasz et al., 2004).

PECAM-1 acts by modulating integrin function in the platelets during the adhesion and aggregation process, which also acts as a leucocyte trafficking regulator (Aljohani et al., 2024). PECAM-1 can restrict collagen and CRP-induced platelet aggregation when activated by a wild-type immunoglobulin

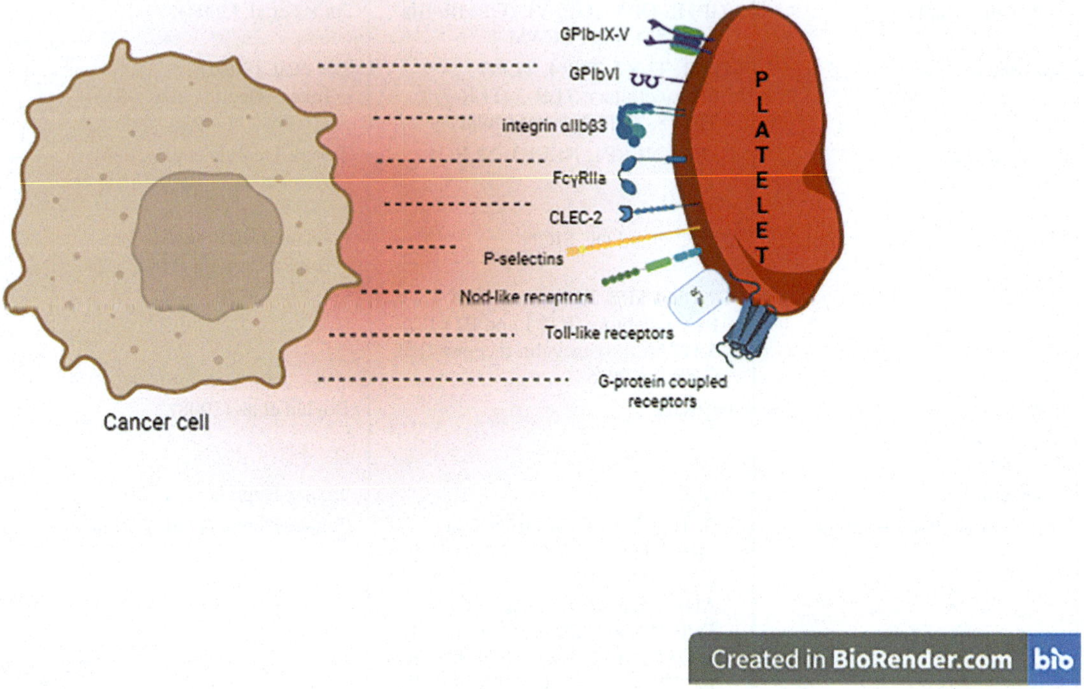

FIGURE 10.4

Therapeutic targets on platelets for cancer–platelet cell interactions.

chimera, whereas PECAM-1–deficient platelets enhanced aggregation and secretion responses, indicating PECAM-1 as a negative physiological regulator (Jones et al., 2001).

10.2.2 Immunoreceptor tyrosine-based activation motif and hemITAM-associated receptors

Immune tyrosine-based activation motif (ITAM) and hemITAM are platelet adhesion-type receptors that promote tumor cell proliferation by attracting myeloid cells and shielding tumor cells (Schlesinger, 2018). Major ITAM receptors involved in platelet-cell interactions include FcγRIIa, CLEC-2, and GPVI.

FcγRIIa activates the downstream signaling pathways involving Syk and phospholipase Cγ2, leading to thrombus formation (Boylan et al., 2008). GPVI and CLEC-2 follow a shared downstream signaling pathway of hemITAM (Haining et al., 2019). CLEC-2 is a hemITAM C-type lectin receptor (Qiao et al., 2015). It recognizes podoplanin and contains a single ITAM domain. It assists in the activation of platelets and aids in thrombus stabilization, although its role in homeostasis is less understood (Boulaftali et al., 2013). GPVI activates platelet aggregation by triggering the intercellular signaling and is dependent on the FcRγ chain for signaling activities (Boulaftali et al., 2013; Borst & Gawaz, 2021; Kehrel et al., 1998). GPVI, coupled with integrin $\alpha 2\beta 1$, acts as a crucial receptor for collagen type I and is responsible for stabilizing platelet adhesion under higher shear rates. When GPVI binds to collagen, it causes the receptor to cluster, which increases the affinity toward collagen and brings it closer to the signaling molecules associated with that receptor (Schlesinger, 2018). GPVI forms clusters and causes ITAM phosphorylation by Fyn and Lyn kinases by binding to galectin 3 on colon cancer cells. Syk causes platelet activation and results in the release of growth factors. PDGF-activated DGF receptor causes COX-2 stabilization and PGE2 release. This PGE2 binds to the EP receptor on the tumor cells, causing $p21^{waf}$ downregulation and cyclin B1 upregulation.

10.2.3 C-type lectin receptor families

CLEC-2 is an important C-type lectin receptor present on the platelet surface and is involved in cancer-associated thrombosis (Sun et al., 2024). It promotes platelet adhesion and interactions by modulating signaling pathways, encompassing Syk and Src tyrosine kinases by facilitating tyrosine phosphorylation (Pollitt et al., 2014; Suzuki-Inoue, 2006). It undergoes phosphorylation on binding to PDPN, which activates Syk. This cascade leads to Cγ2 phosphorylation, resulting in increased intracellular calcium levels; thereby promoting platelet activation (Sun et al., 2024).

10.2.4 P-selectins

P-selectin-P-selectin glycoprotein ligand-1 and GP Ibalpha-alphaMbeta2 integrin play a crucial role in neutrophil-platelet interactions (Li et al., 2014). Along with the above-mentioned receptors, activated platelets express P-selectin, which is responsible for platelet-leucocyte interactions. It traverses rapidly to the surface during the platelet-activation process and stimulates leucocyte activation by interacting with PSGL-1 (Jurasz et al., 2004). Sulfated glycosphingolipids are key ligands in platelet adhesion and aggregation, and also stabilize platelet aggregates (Merten & Thiagarajan, 2001). They facilitate platelet aggregation by stabilizing the platelet-to-platelet contact after initial binding via integrins such as GP IIb/IIIa, which later form larger and stable platelet aggregates for adequate homeostasis (Merten & Thiagarajan, 2000). Platelet–endothelium and leucocyte–endothelium interactions were observed in mice models after 60 min of ischemia. And these interactions were dependent on the P-selectin expression (Massberg et al., 1998). mAbs against P-selectin failed to detect an increase in the circulating platelets, but significantly reduced adherence to the intracellular matrix and enhanced its expression in the intestinal vasculature (Massberg et al., 1998).

Hence, understanding the relationship between P-selectin and its ligands may lead to the identification of novel strategies for developing cancer therapy while minimizing the adverse effects.

10.2.5 Toll-like receptors

Under the effect of certain conditions and stimuli, such as toll-like receptor-7 (TLR-7) activation can lead to the interaction of platelets with the circulating leucocytes by modifying the CD40 or P-selectin surface expression. These eventually give rise to platelet-leucocyte aggregates. The molecular pattern recognition receptors, TLRs ranging from TLR1 to TLR10 mRNA transcripts have been identified in the platelets. These TLRs can alter neutrophil platelet-leucocyte interactions (Koupenova et al., 2018).

Platelet-leukocyte interactions occur when P-selectin on the activated platelet membrane binds to the P-selectin glycoprotein ligand 1 (PSGL1) present on the leukocytes (Repsold & Joubert, 2021).

10.2.6 Nod-like receptors

Nod-like receptors (NLRs) represent a subtype of PPRs that mediate the upregulation expression of IL-1β and NLRP3 to mediate platelet-leukocyte aggregation and have been implicated in initiating cancer-associated platelet-leukocyte interactions (Dib et al., 2020). Hence, they can be a potential target to reduce platelet activation and improve disease outcomes.

10.2.7 G protein-coupled receptors

G protein-coupled receptors (GPCRs) tend to regulate platelet function by influencing cancer cell behavior and metastasis. They influence cancer progression via AKT/mTOR, Hippo, and MAPKs pathways. The cancer cell invasion is aided by Rho GTPases, which induce cytoskeletal changes; thereby maintaining the tumor-permissive environment, which is also caused by interactions with the extracellular matrix (Liu et al., 2016; O'Hayre et al., 2014; Wu et al., 2019). An adhesion GPCR, CD97, is overexpressed in several cancer types and can indirectly interact with the platelets, leading to bidirectional expansion of cancer. This interaction stimulates platelet activation and granule secretion, further facilitating cancer cell migration and metastasis by disrupting the endothelial junctions through ATP release (Ward et al., 2018). Several mediators, such as TXA2, ADP, and PGE2, released by the platelets activate additional platelets and drive the metastatic signaling pathways in cancer cells. Such interactions are vital for EMT and cancer cell dissemination (Rovati et al., 2022). Receptors involved in Platelet-Cell Interactions are shown in Table 10.1.

This complex interplay among such receptors thus opens avenues for the development of targeted therapies for abnormal platelet activation conditions.

10.3 Role of platelet–cancer cell interactions in cancer progression

10.3.1 Cancer cell survival and growth

Platelets release growth factors from their granules that support the growth and survival of tumors (Anvari et al., 2021; Goubran et al., 2014). Moreover, platelets protect the tumor cells from immune attacks and hence facilitate the survival of circulating cancer cells in the bloodstream (Anvari et al., 2021; Bambace & Holmes, 2011; Wang et al., 2018). The presence of platelets also leads to the generation of a metastatic niche in which TME becomes amenable for the recruitment of fibroblasts and immune cells that aid in colonization by tumor cells (Braun et al., 2021; Xu et al., 2018). This crosstalk between platelets and cancer cells leads to plasticity in cancer cells to support survival and extravasation in the dissemination process (Wang et al., 2018).

10.3.2 Angiogenesis and tumor vascularization

Platelets support tumor angiogenesis through the secretion of proangiogenic factors that help develop new blood vessels (Bambace and Holmes, 2011; Goubran et al., 2014; Plantureux et al., 2020). VEGF, matrix metalloproteinases (MMPs), and basic fibroblast growth factor (bFGF) are also involved with proangiogenic factors promoting the formation of blood vessels. Platelets can adhere via receptors such as P-selectin and integrins to the tumor cell surface, which the latter then further bind to the blood vessel vasculature (Italiano et al., 2008; Wojtukiewicz et al., 2017). They can also modify the TME by attracting immune cells and by creating an inflammatory environment, which promotes angiogenesis (Schlesinger, 2018).

10.3.3 Metastatic cascade and platelet support

Platelets facilitate several stages of metastasis, which include protection of circulating tumor cells from shear forces and immune attacks, and help in their arrest at the vascular level (Anvari et al., 2021; Schlesinger, 2018). The key stages involved in metastasis are platelet adhesion to the endothelium, tumor cell emboli adhesion to the blood vessel vasculature, tumor cell extravasation, immune cell evasion by NK cells, and recruitment of myeloid cells to the metastatic site, which facilitates tumor growth and proliferation (Bambace & Holmes, 2011; Gay & Felding-Habermann, 2011; Schlesinger, 2018). The molecular mechanisms involved are integrin-mediated adhesion, selectin binding, growth factor release, and protease activation (Bambace & Holmes, 2011; Gay & Felding-Habermann, 2011; Liao et al., 2023; Liu et al., 2021). They induce EMT in tumor cells, thus improving their invasive characteristics (Labelle et al., 2011; Plantureux et al., 2020). PDGFs also help to open the capillary endothelium to facilitate extravasation of tumor cells and secondary tumor formation (Schlesinger, 2018).

10.4 Mechanisms of platelet–cancer cell interactions

10.4.1 Platelet activation by cancer

Cancer cells can activate platelets by a variety of mechanisms, such as direct cell-to-cell contact, secretion of growth factors, platelet receptor modulation, and release of mediators such as HMGB1, CD97, Podoplanin, and extracellular vesicles. Carcinoma mucins and vWF are also involved in the activation of cancer cells and exhibit a vital role in the activation process (Bekendam & Ravid, 2023; Braun et al., 2021; Schlesinger, 2018; Strasenburg et al., 2022; Yu et al., 2021). Platelet activation starts with either direct contact with cancer cells or with the release of soluble factors. This is followed by the release of growth factors such as VEGF, TGFβ, and PDGF; thereby stimulating tumor growth and metastasis (Zhou et al., 2023). This shields the cancer cells from shear stress and immune attacks so that they can adhere to the metastatic site (Wang et al., 2018; Zhou et al., 2023). These platelets that have been accumulated can be used to deliver chemotherapeutic and imaging agents (Morris et al., 2022) (Table 10.2).

10.4.2 Cancer cell-induced platelet aggregation

Cancer cell-induced platelet aggregation (CIPA) promotes metastasis by supporting tumor cell survival and metastasis. Several signaling molecules are used in this process: MMPs, adhesion receptors, selectins, CAMs, LPA, and ATX. MMP is regulated by NO, which can suppress CIPA by reducing the production of a specific type of MMP, MMP-2 (Jurasz et al., 2001). CAMs can also facilitate the interaction between CTCs and platelets to promote platelet aggregation and protect CTCs (Liu et al., 2021). Similarly, LPA, ATX, and selectins promote platelet aggregation and metastasis (Leblanc et al., 2018) (Fig. 10.5).

Table 10.2 Mechanisms of platelet activation in cancer progression.

Mechanism	Description
Tumor cell proliferation	Platelets release growth factors that support tumor growth
Immune evasion	Platelets cloak tumor cells and prevent immune detection
Extravasation	Platelets enable CTCs' metastasis to the distant organs through the blood endothelium
Angiogenesis	Angiogenic factors secreted by platelets promote blood vessel formation

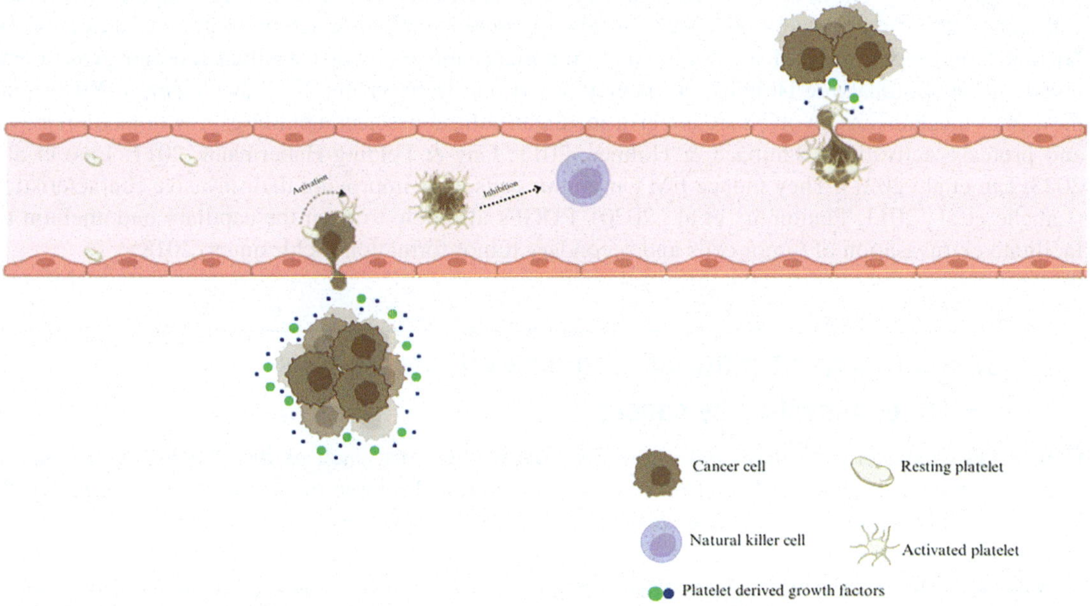

FIGURE 10.5

Role of platelets in cancer immune evasion.

From Puricelli, C., Boggio, E., Gigliotti, C.L., Stoppa, I., Sutti, S., Giordano, M., ... Rolla, R. (2023). Platelets, Protean cells with all-around functions and multifaceted pharmacological applications. International Journal of Molecular Sciences, 24(5).

CIPA significantly contributes to metastasis by protecting tumor cells, promoting their extravasation, and inducing tumor growth. This understanding can provide potential therapeutic opportunities by disrupting the platelet-tumor cycle and thereby minimizing metastasis.

10.4.3 Signaling pathways

Inside-out and outside-in signaling pathways are two essential processes through which cancer cells communicate with platelets, hence altering the progression and metastasis of cancer. Several mechanisms have been reported to enable such communication between intracellular and extracellular environments that impact the behaviour of cancer cells and the functioning of platelets in significant ways.

10.4.3.1 Inside-out signaling

This pathway refers to the signaling pathway inside the platelets, which undergoes a conformational change by αIIbβ3 integrin, which, on activation by several stimuli, leads to an increase in binding forces with the cell surface of cancer cells (Janus-Bell & Mangin, 2023; Huang et al., 2019. This improves the potential to bind to the ligands attached to cancer cells, making cancer cells bind to the cancer cells. The internal integrin activation brings about a concomitant increase in the affinity for ligands such as fibrinogen. This is vital to initiate adhesion and aggregation in order to eventually form and stabilize the thrombus (Durrant et al., 2017; Huang et al., 2019; Zou et al., 2007). This process depends on proteins such as talin or kindlin, binding to the presented cytoplasmic domains through the integrins. It stabilizes the transition in conformation towards a high-affinity state (Chen et al., n.d.; Huang et al., 2019; Suzuki-Inoue et al., 2007).

10.4.3.2 Outside-in signaling

Outside-in signaling initiates a downstream sequence of events between the activated integrins on the platelets and their ligands (Huang et al., 2019; Janus-Bell and Mangin, 2023). This signaling event triggers cell responses once integrins bind with their ligands. These processes are associated with spreading, blood retraction, and thrombi stabilization. It also includes a complex network of enzymes and cytoskeletal elements and results in EMT, LPA release, and increased ROS production (Huang et al., 2019; Durrant et al., 2017; Liu et al., 2023; Xue et al., 2024; Zhang et al., 2023; Zhou et al., 2023). Therefore this mechanism is significant in cancer progression.

10.4.4 Release of platelet-derived growth factors

Platelet-derived growth factors and cytokines are pivotal in tumor growth regulation. PDGFs, particularly PDGF-B and their receptors, are involved in tumorigenesis and are dynamic pro-angiogenic factors (Raica & Cimpean, 2010). They have the ability to drive tumor cell proliferation and survival by genetic as well as epigenetic alterations, which are detected in some cases, such as skin and GIT stromal tumors (Heldin et al., 2018). PDGF signaling is also implicated in autocrine stimulation and malignant cells in tumor progression (Andrae et al., 2008; Board & Jayson, 2005). PDGFs initiate angiogenesis by stimulating VEGF expression and recruiting pericytes to the newly formed blood vessels, enhancing glioma angiogenesis (Andrae et al., 2008; Guo et al., 2003). These pericytes stabilize the newly formed blood vessels, improving the integrity of tumor vasculature (Geindreau et al., 2022). Such angiogenic growth factors interact with immune cells present in the TME (Ge et al., 2023). Platelets are blood cells with proangiogenic factors. These factors released by the platelets further contribute to angiogenesis by releasing proteinases, which are responsible for denaturing proteins present in the ECM and also aid in the migration of proangiogenic cells to the tumor site (Sierko & Wojtukiewicz, 2004). Tumor cells activate platelets, which then release growth factors that stimulate proliferation in the tumor cells, giving the tumor cells a mesenchymal-like phenotype and aiding in metastasis (Schlesinger, 2018).

10.4.5 Immune modulation and evasion

The primary mechanism of immune evasion via platelet interactions is when the platelets protect the CTCs from the immune system by forming a protective clot. (Ali et al., 2015; Cox et al., 2011). They also assist NK cell-mediated killing by upregulating immune checkpoints such as CD55, reducing the cytotoxicity. (Cluxton et al., 2019; Sun et al., 2025). This hinders the immune system diminishes immune cell access and their ability to recognize and attack the targets. (Cox et al., 2011). Since platelets are rich in Transforming Growth Factor (TGF)-β, they can suppress immune responses by establishing an immunosuppressive environment. (Metelli et al., 2020; Schmied et al., 2021). Another mechanism is when the NKG2D ligand is shed from the tumor cells on platelet activation. This leads to the recognition and killing of cancer cells, impairing the ability of NK cells to conduct immune surveillance (Cluxton et al., 2019; Maurer et al., 2018). Platelet-derived microparticles can be used to regulate gene expression and promote immune evasion and metastasis, and this is done by transferring miRNAs to the tumor cells (Lazar & Goldfinger, 2018).

10.5 Types of cancer–cell platelet interactions

Cancer cells enter the bloodstream during metastasis and are shielded by other platelets, prolonging their survival (Palacios-Acedo et al., 2019). P-selectin serves as a cancer cell adhesion receptor in platelets by upregulating tissue factors in monocytes by interacting with PSGL-1 and GPIb (Furie et al., 2001). This receptor binds with cancer cells via TCIPA and forms clusters that cannot be recognized by NK cells.

On neutrophil extracellular traps (NET) extrusion and release, platelets interact with neutrophils. Platelets reprogram with monocytes by IL-1β, TNF-α, IL-8, and monocyte chemoattractant protein (MCP-1) trigger responses. Monocytes trigger cytokine secretion mediated via GPIBα and phosphatidylserine recognition. Platelets also interact with dendritic cells (DC) by CD40L-CD40, subsequently leading to phagocytosis (Dib et al., 2020).

Direct receptor-receptor binding occurs when a specific receptor present on the cancer cell surface binds to its corresponding receptor on the platelet, initiating a signaling cascade and promoting tumor spread. Initially, platelets bind and roll over the cancer-cell surface as a result of the interactions between PSGL on the platelets and P-selectin on the cancer cells. CADH-6 and integrins ($\alpha6\beta1$) get activated, which further strengthens the link between $\alpha6\beta1$ and ADAM9, which is generated from cancer cells. GPIIb/IIIa on the platelets and Cadherin-6 on cancer cells. Cadherin-6, being a GPIIb-IIIa ligand, induces platelet activation and promotes metastasis by delivering angiogenic modulators. The interaction between Simultaneously, CLEC-2 on the platelets binds to podoplanin on cancer cells, which leads to the activation and aggregation of platelets. Furthermore, CD44 on the platelets interacts with hyaluronic acid in the TME, keeping the platelets anchored and supporting tumor growth and dissemination (Morris et al., 2022). Similarly, adhesion GPCR CD97 is found to directly activate the platelets by coordinating tumor invasion and endothelial cell retraction. This interaction activates the platelets and leads to granule secretion and ATP release. Hence, targeting CD97 can be a potential approach to prevent metastatic spread (Ward et al., 2018).

The extracellular matrix proteins play a key role in facilitating complex interactions between cancer cells and platelets. Platelets, having a variety of receptors, can bind with vWF, collagen, fibrinogen, and fibronectin being specific extracellular adhesive proteins. Both these molecules can bridge platelet and tumor cell receptors and may associate even when they have a lower direct binding affinity (Jurasz et al., 2004; Morris et al., 2022). This association results in improved adhesion of cancer cells to the blood endothelium, which is a significant step in tumor development out of the body and metastasis (Ding et al., 2023). Upon binding to the proteins, these platelets activate their pathways that lead to tumor growth, metastasis, and evasion of the immune system through the release of chemokines and growth factors (Palacios-Acedo et al., 2019). Activation of protease-activated receptors (PARs) present on the platelets has been linked to increased chances of interaction with cancer cells (Morris et al., 2022).

Platelet membrane GPs, integrins, and lectins enable the direct interaction of platelets with tumor cells. It leads to a change in platelet shape and promotes the release of factors necessary for activating other platelets (Trivanović et al., 2024). Besides, in addition to facilitating the cancer-related coagulation process, platelet interactions with other mechanisms can protect cancer cells from immune responses and also create procoagulant surfaces (Bambace & Holmes, 2011). DAMPs too can stimulate platelets, which enhance the survival and proliferation of cancer cells (Braun et al., 2021; Wang et al., 2018). Lysophosphatidic acid (LPA) released from the platelets has been shown to enhance the motility and survival of cancer cells (Braun et al., 2021; Lazar & Goldfinger, 2021).

Surgical removal of nonmetastatic cancer increases the risk of metastasis associated with the activation of platelets during surgery. Enhanced platelet-CTC aggregates tend to capture CTCs by using NETs. Knocking down TLR or blocking the activation of the platelet brings down the levels of metastasis. These NETs thicken the blood and enhance tumor propagation, promoting the activation of neutrophils as well as growing the tumor size (Morris et al., 2022).

These interactions differ among different cancer types and define the thrombotic risk and metastatic potential. With an understanding of these mechanisms, novel therapies can be developed that are specific to mitigate cancer progression and associated thrombosis.

10.6 Signaling pathways involved in cancer cell–platelet interactions

Platelets interact with tumor cells with several receptors and signaling pathways, which promote mesenchymal-like phenotype in cancer cells, leading to metastasis. They modulate immune responses by recruiting immune cells at the tumor site and by interacting with leukocytes to influence and alter the tumor-associated inflammation. Several mechanisms are involved in cancer platelet-cell interactions; the primary types are discussed in the following sections.

10.6.1 TGF-β/Smad and NF-κB pathway

The TGF-β/Smad and NF-κB pathways are closely connected and play pivotal roles in various cellular processes, such as EMT and cancer progression by cross-signaling mechanisms (Javelaud & Mauviel, 2005). These pathways are activated due to platelet-tumor cell interactions. (Labelle et al., 2011). The TGF-β pathway is activated by platelet-derived TGF-β, which further activates the NF-κB pathway, transitioning to increase the invasiveness of mesenchymal-like phenotype and promote metastasis. This leads to the inhibition of NF-κB signaling, which can block TGF-β induced EMT (Labelle et al., 2011; Wu et al., 2018). The noncoding RNA NKILA is upregulated by TGF-β and plays a vital role in negative feedback regulation of the NF-κB pathway, thus decreasing the TGF-β-induced tumor metastasis. The junB promoter contains NF- κB sites, which are TGF-β–responsive regions that require an intact NF-κB pathway for their transactivation by TGF-β and are mediated by members of the Smad family. Furthermore, Smad3 was found to interact in vivo with NF-κB subunit p52 and induce tissue factor expression (Giarratana et al., 2024; López-Rovira et al., 2000).

10.6.2 Integrin signaling (particularly $\alpha v\beta 3$)

Integrins are glycoproteins that aim to regulate and support the physical interactions between platelets and tumor cells (Lavergne et al., 2017). Platelets in interaction with integrins have an indirect effect. For instance, $\alpha\beta 51$ on cancer cells can be activated on the release of acid sphingomyelinase (Asm). This Asm activation results in cancer metastasis. In addition, integrin $\alpha 2\beta 1$ and heat shock protein (Hsp47) are involved in cancer platelet-cell interactions, where $\alpha 2\beta 1$ further promotes EMT by activating the Wntβ-catenin pathway and protein expression. Hsp47, too, is an EMT initiator and has been reported to promote collagen-dependent platelet-cancer cell interactions. (Morris et al., 2022)The downregulation of EMT-associated marker proteins is detected in vimentin upregulation and E-cadherin downregulation (Schlesinger, 2018).

10.6.3 Lysophosphatidic acid-lysophosphatidic acid receptor pathway

The LPAR2 subtype of the LPA-LPAR pathway can regulate cancer cell migration and can be used as a potential therapeutic target for gastric cancer (Yang et al., 2013). The receptors involved in this modulate tumor growth and metastasis, and target the TME cells (Cao, 2013). Cancer cell-secreted exosomes contain an enzyme, autotaxin (ATX), which produces a signaling molecule, LPA from lysophosphatidylcholine. This product then activates the LPA-LPAR pathway (Khasabova et al., 2023; Ward et al., 2011). Since cancer cells overexpress LPARs such as LPAR1, tumor cells can respond readily to LPA signaling. LPA on binding to the LPARs present in the tumor cells activates the signaling pathways, promoting angiogenesis, invasion, and metastasis (Wang et al., 2022). They can contribute to cancer-related pain. (Khasabova et al., 2023). Hence, these receptors could be used as a potential target for new treatment approaches (Shih & Holland, 2006). PDGF overactivation occurs in the case of malignancy, suggesting it as a therapeutic target for treating certain tumors (Heldin et al., 2018).

10.6.4 CLEC-2-podoplanin pathway

The transmembrane sialoglycoprotein podoplanin (aggress) plays a role in tumor cell-induced platelet aggregation and tumor metastasis (Kato et al., 2008). An endogenous receptor of podoplanin on the platelets, the interaction between it and CLEC-2 is dependent upon the sialic acid moiety on the O-glycans of podoplanin (Kato et al., 2008; Sun et al., 2024; Suzuki-Inoue et al., 2007). The PLAG4 domain mediates the interaction of podoplanin with CLEC-2. This interaction can be antagonized by either PLAG3 or PLAG4 domains of podoplanin, thereby suppressing podoplanin-derived tumor metastasis. A monoclonal antibody that targets the PLGA4 domain of podoplanin can inhibit the interaction of podoplanin with CLEC-2 (Sekiguchi et al., 2016).

10.6.5 JAK/STAT pathway

Cancer cells overexpress cytokines, which activate the Janus kinase/signal transducer and activator of transcription (JAK/STAT) pathway for adhesion molecule expression on the cell surface

(Brooks and Putoczki, 2020). These molecules bind to the platelet receptors and contribute to the adhesion of platelets and the transition of epithelial to mesenchymal cell type (Jin, 2020). The activated platelets further release pro-inflammatory mediators and growth factors, which promote metastasis in tumors and involve STAT3 (Jin et al., 2020).

10.6.6 P38MAPK-MMP-9 pathway

Platelets can promote the invasive capabilities of tumor cells by up-regulating the activity and expression of MMP-9 through the p38MAPK pathway (Radziwon-Balicka et al., 2014). This is majorly because of the stress factors mediated by cancer cells that activate the platelets and facilitate their adhesion to the cancer cell surface (Abraham et al., 2024; Corre et al., 2017; Manikanta et al., 2020). This initiates the p39MAPK pathway, which results in the production of inflammatory cytokines. These cytokines thus produced can attract immune cells to the affected tumor sites (García-Hernández et al., 2021; Grave et al., 2022). Platelets can suppress mRNA transcription as well as translation and upregulate MMP-9 at the biomolecular levels. MAPK proteins and MMP-9 interact with each other to release proteins such as clusterin and TSP1. This influences the cancer cell invasiveness and hence could be used as a viable treatment option (Radziwon-Balicka et al., 2014).

10.6.7 Protease-activated receptor pathway

The PAR pathway is initiated by PAR-1 receptor activation by tissue factors, thrombin, and other growth factors (Liu et al., 2017; Wojtukiewicz et al., 2015). This leads to a cascade of events that ultimately promote tumor cell adhesion (Liu et al., 2017; Srivastava et al., 2024). Whereas some cancer cells express PAR-1, which allows them to directly interact with thrombin, further promoting their invasive behavior (Srivastava et al., 2024; Wojtukiewicz et al., 2015). A high concentration of thrombin found in the TME is also responsible for activating PARs present on both cancer cells and platelets, which further advances tumor progression (Liu et al., 2017; Zigler et al., 2011). Human platelets desensitized to PAR1 signaling are counteracted by PAR4 signaling, which allows the reestablishment of aggregation through PKC-induced fibrinogen and ADP release (Fälker et al., 2011).

10.6.8 Cathepsin K signaling

Cathepsin K is a cysteine protease that is overexpressed in several cancer subtypes (Littlewood-Evans et al., 1997). This enzyme is released by cancer cells into the TME, which directly activates PAR3 and PAR4 via their proteolytic activity (Qian et al., 2022; Wang et al., 2018). This eventually results in platelet aggregation and migration to distant tissues that enhance metastasis (Morris et al., 2022). This further causes EMT induction in cancer cells (Braun et al., 2021). Cathepsin B has also been shown to contribute to the interaction between platelets and cancer cells

10.6.9 Lysophosphatidic acid signaling

LPA, derived from the activated platelets, acts as a signaling molecule in cancer cell-platelet interactions. This molecule then binds to targeted receptors present in the cancer cells, promoting metastasis

(Leblanc et al., 2018; Ikeda et al., 2024). LPA results in a cascade of events on binding to LPARs that trigger intercellular signaling either directly or indirectly and increase cancer cell motility (Ikeda et al., 2024; Qi et al., 2024). It also activates its own GPCR by the Ras pathway and protein tyrosine phosphorylation (Moolenaar, 1994). LPA is also responsible for increasing tumor invasiveness by proximal CD97-LPAR heterodimer signaling and increasing vascular permeability (Ward et al., 2018). LPA signaling is mainly involved in bone cancer. Bone metastasis can be reduced by silencing the LPAR1 receptor; hence, it can be used as a therapeutic target to improve the quality of life of such patients (Boucharaba et al., 2006).

The above-discussed receptors involved in their respective cancer signaling pathways could be considered potential targets for cancer therapy.

10.7 Therapeutic targeting of platelet–cancer cell interactions

10.7.1 Anti-P-selectin antibodies

P-selectins facilitate platelet adhesion to tumor cells in neuroblastoma and small-cell lung cancer via carbohydrate structures such as sialyl-Lewis X. Such interactions can be inhibited by anti-P-selectin antibodies, an opening avenue for a potential therapeutic strategy to prevent metastasis. (Aigner et al., 1997; Stone & Wagner, 1993). CD24 is a glycoprotein present in the tumor cells and acts as a ligand for P-selectin. This helps in the attachment of cancer cells to platelets. This interaction, although crucial for the dispersal of tumor cells, can be blocked by inhibiting CD24; this makes another therapeutic approach viable for research (Aigner et al., 1997). Similarly, inhibiting GPIb-IX-V or vWF function along with blocking antibodies reduces platelet-cancer cell interactions (Jurasz et al., 2004). Similarly, targeting $\alpha v\beta 3$ integrins can also reduce platelet-tumor cell contact. (Lavergne et al., 2017).

10.7.2 Antibodies targeting cancer cell surface markers

GPIIb/IIIa receptors withstand conformational changes on the release of platelet agonists such as ADP, collagen, and thrombin, having an increased affinity for vWF and fibrinogen, which may lead to cancer cell arrest. Abciximab and XV454 belong to the class of GPIIb/IIIa antagonists and are potent TCIPA inhibitors. They act by reducing tumor cell adhesion to the vascular endothelium and inhibiting angiogenesis. Abcximab acts by blocking the $\alpha V\beta 3$ integrin present in cancer and endothelial cells, which play a major role in platelet–cancer cell interactions. By blocking $\alpha V\beta 3$, platelet aggregation and platelet-cancer cell interactions can be reduced (Morris et al., 2022). One of the most commonly used antiplatelet drugs, aspirin, inhibits platelet aggregation by COX inhibition, followed by the production of TXA_2 by the platelets, and induces tumor cell-induced platelet aggregation (Jurasz et al., 2004). Similarly, aspirin and the P2Y12 blocker, 2MeSAMp, can disrupt a crucial step in platelet activity inhibition (Morris et al., 2022). NF-κB pathway within the TME can be proposed as a potential new target for antiplatelet therapy in inflammatory diseases such as cancer (Cao et al., 2024; Fuentes et al., 2016). Also, mAbs such as LYP7 can block the ristocetin-induced platelet aggregation and inhibit metastasis (Boukerche et al., 1996). Targeting P-selectin also protects cancer cells from immune response since it shields them from NK cell-mediated cytotoxicity. Anti-P-selectin antibodies

Table 10.3 Comparison of therapeutic strategies targeting platelet interactions.

Therapeutic strategy	Mechanism of action	Potential benefits
Antiplatelet drugs	Inhibit platelet aggregation and adhesion	Reduction in metastasis
Monoclonal antibodies	Target specific receptors involved in platelet–tumor interactions	Prevent CTC adhesion and immune evasion
Small molecule inhibitors	Disrupt signaling pathways activated by several platelet interactions	Control cell growth and metastasis

may disrupt this interaction, which could be beneficial for the immune system to target cancer cells (Natoni et al., 2022).

10.7.3 Small molecule inhibitors targeting cancer cell signaling pathways

Small molecule inhibitors are being developed as promising approaches to target pathways of cancer cell signaling. The advantages of these molecules over other therapies include selectivity, efficacy, and penetration to the target site. Thus PD-1/PD-L1 pathway inhibitors are being designed to inhibit PD-L1 transcription and translation and speed up the degradation of the PD-L1 protein for enhancing tumor immunotherapy (Baglini et al., 2022; Wu et al., 2021). Even small molecules that could inhibit the CSC progression, important for self-renewal and tumorigenesis, are being investigated. This mainly occurs by targeting signaling pathways such as Wnt, Notch, Hippo, Hedgehog, and autophagy (Clara et al., 2020; Fang et al., 2016; Yang et al., 2020). Other research areas include small-molecule inhibitors against tyrosine kinases, which include PDGFR and the fibroblast growth factor receptor, that will control aberrant cellular growth in cancers (Showalter & Kraker, 1997).

10.7.4 Platelet inhibition and immunotherapy

Combining antiplatelet drugs with immunotherapy is on the rise to further improve the efficacy of cancer treatment. This approach of using platelets to inhibit the TGFβ pathway is involved in immune suppression and tumor progression. Inhibitors such as alisertib can reverse this suppression and enhance T-cell proliferation and immunity against tumors. Combining TGFβ inhibitors with PD-1/PD-L1 blockade has proven to show synergistic effects in preclinical colon carcinoma models (Holmgaard et al., 2018) (Table 10.3).

10.7.5 Emerging therapies

New advancements in cancer treatment involve developing inhibitors of platelet-cancer cell interactions that allow the survival and dissemination of cancer cells within the bloodstream. Such strategies involve the inhibition of platelet receptors, the release of platelet particles, and the inhibition of platelet-specific enzymes. Moreover, a future direction in this regard would be using nano-platelets as a targeted delivery system (Wang et al., 2022). Studies have demonstrated that Phallusia nigra heparan sulfate can reduce metastasis in colon cancer models. To suppress podoplanin-CLEC-2 binding and thereby prevent podoplanin-induced aggregation of platelets, a new anti-human Aggrus (podoplanin) mAb called MS-1 has been developed (Morris et al., 2022). PDGFR inhibitors can disrupt signaling pathways that allow tumor growth and metastasis. Such inhibitors may interfere with the protumorigenic activity of platelets in TME (Showalter & Kraker, 1997). Nanoparticle-mediated delivery systems can be designed to deliver therapeutic agents such as gene therapy, RNAi, and biologics directly to the tumor cells. They are applied to codeliver chemotherapeutic drugs and P-gp inhibitors to significantly improve drug accumulation within cancer-resistant cells (Pan et al., 2021; Patil et al., 2009, 2010). This approach appears most effective in overcoming MDR in CSCs and other tumor-resistant cells. (Pan et al., 2021; Patil et al., 2009).

These strategies could be employed to improve the efficacy of existing therapies, overcome resistance, and improve patient outcomes via targeted interventions.

10.8 Overcoming drug resistance through targeting platelets

Mechanisms of drug resistance in cancer include genetic mutations, efflux pumps, and altered signaling pathways. Platelets participate in these processes by allowing metastasis and protecting tumor cells in circulation. Interference with the interaction between platelets and tumor cells can disrupt protective mechanisms that may improve the effectiveness of chemotherapy (Garg et al., 2024; Wang et al., 2022). IL-6 may also promote chemotherapeutic resistance by inhibiting the activation of proteases that are involved in apoptosis (Huong et al., 2019). These factors negate the antiproliferative effects of the chemotherapeutic drugs and can play a role in rendering chemotherapy ineffective. The effect of platelets on the efficacy of chemotherapeutic drugs varies, but they can decrease the efficacy of platinum-based treatments for NSCLC (Huong et al., 2019). The rebound growth following antiangiogenic therapy is also contributed to by platelets (Haemmerle et al., 2018). The studies with NP-based drug delivery systems, targeting the drug resistance models with increased cytotoxicity and platelet-induced inhibition of drug-to-cell growth, have shown a captivating trend in the clinical evidence, which aims to target the platelet interaction for improving chemotherapeutic response (Jing et al., 2018; Li et al., 2022; Patil et al., 2010).

The clinical evidence suggests that the efficacy of chemotherapy can be improved by preventing platelet-to-cancer cell adhesion (Elaskalani et al., 2017; Wang et al., 2018). Thus the index of platelet activation can be indicated as a predictive biomarker for resistance to chemotherapeutic drugs (Huong et al., 2019). Antiplatelet therapies have been reported to result in better treatments because they block the crosstalk between cancer cells and platelets; thus it is established for ovarian and pancreatic cancers, which are usually associated with thrombocytosis (Elaskalani et al., 2017; Haemmerle et al., 2018).

10.9 Challenges, limitations, and future directions

Despite promising advancements, some problems persist in targeting such types of interactions. Challenges include the complexity of platelet biology, potential off-target effects, and strict requirements for precise targeting without interference with the normal platelet function. Long-term safety and efficacy need to be investigated for these novel therapies (Shapira et al., 2011; Wang et al., 2022). In addition, current studies lack insight into the molecular mechanisms, TME, and heterogeneity, particularly in antiplatelet therapies.

There are significant gaps in the relationship between cancer types and platelet roles in the TME (Braun et al., 2021). Different platelet subpopulations can potentially have different roles during tumor development, and hence, the study about specific roles of PDGFs is necessary (Braun et al., 2021; Ghosh & Jain, 2023). The immune and stromal cells that form the TME are also involved in platelet-cancer cell interactions. The translational potential of antiplatelet therapies is limited because of bleeding complications, and therefore more research is required to find optimal strategies (Braun et al., 2021). Clinical biomarkers of these interactions may be of significant importance in early detection, monitoring the treatment response, and personalized medicine approaches (Yu et al., 2021). Advanced in vitro and in vivo models, such as organ-on-chip systems, are required to study the complexity of the interactions between platelets and cancer cells (Ghosh & Jain, 2023).

Integrin signaling involves multiple interacting proteins and pathways. Hence, it poses a significant challenge to control it. A better understanding of the pathways and the signaling molecules involved remains a future perspective. Elucidation of specific ligands and mechanisms is required to advance treatment options. Balancing safety and efficacy remains a challenge; hence, future research should focus on optimizing therapies to reduce side effects. The development of small molecules also imposes challenges, such as lower and shorter response rates and resistance development. Strategies to overcome these problems could include combination therapies or personalized treatment approaches.

Future research should focus on improving nano drug delivery systems to reduce side effects. Biomarkers are required to be developed to allow for better prediction and monitoring of treatment responses. Further studies into platelet-based liquid biopsy techniques and anti-tumor strategies are going to be of high importance in advancing these therapies (Shapira et al., 2011; Wang et al., 2022). Bidirectional interactions impose a promising direction in developing antiplatelet agents and platelet-based therapeutics for cancer treatment (Yu et al., 2021).

10.10 Conclusion

In summary, understanding the interactions between cancer cells and platelets will lead to new avenues for innovative therapies that could change paradigms in cancer treatment. Platelets are not only passive elements in the bloodstream but also play a crucial role in the progression of cancer and influence metastasis, angiogenesis, immune modulation, and thrombosis. This dynamic interplay is modulated by a number of receptors and signaling pathways, making it an attractive target for therapeutic intervention. More importantly, by dissecting the mechanisms that govern these interactions, researchers can develop new strategies aimed at disrupting the supportive role of platelets in tumor biology. Blockade of critical platelet receptors inhibits metastasis in experimental models. Recent studies underscore a huge gap within the current cancer treatment methods and integrate approaches in considering the biology of platelets as a critical element in cancer management. Further research needs to be focused on identifying specific molecular pathways of interaction between platelets and tumor cells. Antithrombotic drugs are now investigated for their interference in tumor-platelet interactions, thus inhibiting the progression and cancer metastasis. Notably, platelets are also being studied as a potential cancer biomarker and delivery vehicles for anticancer therapies. Understanding these interactions opens up potential therapeutic avenues, such as targeting platelet receptors or using antiplatelet agents. Harnessing this knowledge will allow us to hope for more effective strategies, not only inhibiting the progression but also enhancing the efficacy of the existing treatments, ultimately improving survival rates and quality of life of cancer patients.

Abbreviations

ACE-2	angiotensin-converting enzyme 2
ADP	adenosine-diphosphate receptor
bFGF	basic fibroblast growth factor
C	complement
CCL	chemokine (CCL-1, 3, 4, CCL5, CXCL-1, 2, 3, 4, 6, 7, 8, 12, 16, 17, 22, 24, 26)
CD	cluster of differentiation
FcγFRIIa	receptor Fc immunoglobulin
GARP	glycoprotein-A repetitions predominant
GP	glycoprotein
ICAM	intercellular adhesion molecule
IFNγ	interferon gamma
IL	interleukin
JAM-A	junctional adhesion molecule A
JAM-C	junctional adhesion molecule C
LPA	lysophosphatidic acid receptor
mAbs	monoclonal antibodies
MHC	major histocompatibility complex
MMPs	matrix metalloproteinases
NLRP	nucleotide-binding oligomerization domain leucine-rich repeat and pyrin domain binding
NOD	nucleotide-binding and oligomerization domain
P2Y	purinergic receptors
PAR	protease-activated receptors
PECAM	platelet endothelial cell adhesion molecule
PECAM-1	platelet endothelial cell adhesion molecule-1
TCIPA	tumor cell-induced platelet aggregation
TGFβR	transforming human growth factor beta receptor
TLR	toll-like receptor
TNF	tumor necrosis factor receptor superfamily
TxA	thromboxane receptor A
VEGF	vascular endothelial growth factor

References

Abraham, J. E., Pinilla, K., Dayimu, A., Grybowicz, L., Demiris, N., Harvey, C., Drewett, L. M., Lucey, R., Fulton, A., Roberts, A. N., Worley, J. R., Chhabra, A., Qian, W., Vallier, A. L., Hardy, R. M., Chan, S., Hickish, T., Tripathi, D., Venkitaraman, R., & Persic, M. (2024). The PARTNER trial of neoadjuvant olaparib with chemotherapy in triple-negative breast cancer. *Nature, 629*(8014), 1142–1148. doi:10.1038/s41586-024-07384-2.

Aigner, S., Sthoeger, Z. M., Fogel, M., Weber, E., Zarn, J., Ruppert, M., Zeller, Y., Vestweber, D., Stahel, R., Sammar, M., & Altevogt, P. (1997). CD24, a mucin-type glycoprotein, is a ligand for P-selectin on human tumor cells. *Blood, 89*(9), 3385–3395. doi:10.1182/blood.v89.9.3385.

Ali, R. A., Wuescher, L. M., & Worth, R. G. (2015). Platelets: Essential components of the immune system. *Current Trends in Immunology, 16*, 65–78. http://www.researchtrends.net/tia/volumes.asp?id=36.

Aljohani, M. A., Sasaki, H., & Sun, X. L. (2024). Cellular translocation and secretion of sialidases. *Journal of Biological Chemistry, 300*(9). doi:10.1016/j.jbc.2024.107671.

Andrae, J., Gallini, R., & Betsholtz, C. (2008). Role of platelet-derived growth factors in physiology and medicine. *Genes and Development, 22*(10), 1276–1312. doi:10.1101/gad.1653708.

Anvari, S., Osei, E., & Maftoon, N. (2021). Interactions of platelets with circulating tumor cells contribute to cancer metastasis. *Scientific Reports, 11*(1). doi:10.1038/s41598-021-94735-y.

Baglini, E., Salerno, S., Barresi, E., Marzo, T., Da Settimo, F., & Taliani, S. (2022). Cancer immunotherapy: An overview of small molecules as inhibitors of the immune checkpoint PD-1/PD-L1 (2015-2021). *Mini-Reviews in Medicinal Chemistry, 22*(14), 1816–1827. doi:10.2174/1389557522666220217110925.

Bambace, N. M., & Holmes, C. E. (2011). The platelet contribution to cancer progression. *Journal of Thrombosis and Haemostasis, 9*(2), 237–249. doi:10.1111/j.1538-7836.2010.04131.x.

Battinelli, E. M. (2017). Platelet and megakaryocytic regulation of tumor progression. *Blood, 130*(Suppl_1), SCI–26. doi:10.1182/blood.v130.suppl_1.sci-26.sci-26.

Bekendam, R. H., & Ravid, K. (2023). Mechanisms of platelet activation in cancer-associated thrombosis: A focus on myeloproliferative neoplasms. *Frontiers in Cell and Developmental Biology, 11*. doi:10.3389/fcell.2023.1207395.

Board, R., & Jayson, G. C. (2005). Platelet-derived growth factor receptor (PDGFR): A target for anticancer therapeutics. *Drug Resistance Updates, 8*(1-2), 75–83. doi:10.1016/j.drup.2005.03.004.

Borst, O., & Gawaz, M. (2021). Glycoprotein VI - Novel target in antiplatelet medication. *Pharmacology & Therapeutics, 217*, 107630. doi:10.1016/j.pharmthera.2020.107630.

Boucharaba, A., Serre, C. M., Guglielmi, J., Bordet, J. C., Clézardin, P., & Peyruchaud, O. (2006). The type 1 lysophosphatidic acid receptor is a target for therapy in bone metastases. *Proceedings of the National Academy of Sciences of the United States of America, 103*(25), 9643–9648. doi:10.1073/pnas.0600979103.

Boukerche, H., Ruchaud-Sparagano, M.-H., Ouen, C. R., Brochier, J., Kaplan, C., & McGregor, J. L. (1996). A monoclonal antibody directed against a granule membrane glycoprotein (GMP-140/PADGEM, P-selectin, CD62P) inhibits ristocetin-induced platelet aggregation. *British Journal of Haematology, 92*(2), 442–451. doi:10.1046/j.1365-2141.1996.d01-1485.x.

Boulaftali, Y., Hess, P. R., Getz, T. M., Cholka, A., Stolla, M., Mackman, N., Owens, A. P., Ware, J., Kahn, M. L., & Bergmeier, W. (2013). Platelet ITAM signaling is critical for vascular integrity in inflammation. *Journal of Clinical Investigation, 123*(2). doi:10.1172/jci65154.

Boylan, B., Gao, C., Rathore, V., Gill, J. C., Newman, D. K., & Newman, P. J. (2008). Identification of FcγRIIa as the ITAM-bearing receptor mediating αIIbβ3 outside-in integrin signaling in human platelets. *Blood, 112*(7), 2780–2786. doi:10.1182/blood-2008-02-142125.

Braun, A., Anders, H. J., Gudermann, T., & Mammadova-Bach, E. (2021). Platelet-cancer interplay: Molecular mechanisms and new therapeutic avenues. *Frontiers in Oncology, 11*. doi:10.3389/fonc.2021.665534.

Brooks, A. J., & Putoczki, T. (2020). Jak-stat signalling pathway in cancer. *Cancers, 12*(7), 1–3. doi:10.3390/cancers12071971.

Cao, Y. (2013). Multifarious functions of PDGFs and PDGFRs in tumor growth and metastasis. *Trends in Molecular Medicine, 19*(8), 460–473. doi:10.1016/j.molmed.2013.05.002.

Cao, Y., Yi, Y., Han, C., & Shi, B. (2024). NF-κB signaling pathway in tumor microenvironment. *Frontiers in Immunology, 15*. doi:10.3389/fimmu.2024.1476030.

Castanho F. L, &Garnes S. D. A., Homeostasis: An integrated vision. BoD – Books on Demand. (2019),IntechOpen, Londone, UK.

Chen Y. P, O' Toole, Ylanne, &Rosa J., A point mutation in the integrin p3 cytoplasmic domain (S752+ P) impairs bidirectional signaling through aIIbP3. n.d.

Clara, J. A., Monge, C., Yang, Y., & Takebe, N. (2020). Targeting signalling pathways and the immune microenvironment of cancer stem cells—A clinical update. *Nature Reviews Clinical Oncology, 17*(4), 204–232. doi:10.1038/s41571-019-0293-2.

Cluxton, C. D., Spillane, C., O'Toole, S. A., Sheils, O., Gardiner, C. M., & O'Leary, J. J. (2019). Suppression of natural killer cell NKG2D and CD226 anti-tumour cascades by platelet cloaked cancer cells: Implications for the metastatic cascade. *PLoS ONE, 14*(3). doi:10.1371/journal.pone.0211538.

Corre, I., Paris, F., & Huot, J. (2017). The p38 pathway, a major pleiotropic cascade that transduces stress and metastatic signals in endothelial cells. *Oncotarget, 8*(33), 55684–55714. doi:10.18632/oncotarget.18264.

Cox, D., Kerrigan, S. W., & Watson, S. P. (2011). Platelets and the innate immune system: Mechanisms of bacterial-induced platelet activation. *Journal of Thrombosis and Haemostasis, 9*(6), 1097–1107. doi:10.1111/j.1538-7836.2011.04264.x.

Dib, P. R. B., Quirino-Teixeira, A. C., Merij, L. B., Pinheiro, M. B. M., Rozini, S. V., Andrade, F. B., & Hottz, E. D. (2020). Innate immune receptors in platelets and platelet-leukocyte interactions. *Journal of Leukocyte Biology, 108*(4), 1157–1182. doi:10.1002/JLB.4MR0620-701R.

Ding, S., Dong, X., & Song, X. (2023). Tumor educated platelet: the novel BioSource for cancer detection. *Cancer Cell International, 23*(1). doi:10.1186/s12935-023-02927-5.

Durrant, T. N., Van Den Bosch, M. T., & Hers, I. (2017). Integrin αIIbβ3 outside-in signaling. *Blood, 130*(14), 1607–1619. doi:10.1182/blood-2017-03-773614.

Elaskalani, O., Berndt, M. C., Falasca, M., & Metharom, P. (2017). Targeting platelets for the treatment of cancer. *Cancers, 9*(7). doi:10.3390/cancers9070094.

Fälker, K., Haglund, L., Gunnarsson, P., Nylander, M., Lindahl, T. L., & Grenegård, M. (2011). Protease-activated receptor 1 (PAR1) signalling desensitization is counteracted via PAR4 signalling in human platelets. *Biochemical Journal, 436*(2), 469–480. doi:10.1042/BJ20101360Sweden.

Fang, L., Zhu, Q., Neuenschwander, M., Specker, E., Wulf-Goldenberg, A., Weis, W. I., Von Kries, J. P., & Birchmeier, W. (2016). A small-molecule antagonist of the β-catenin/TCF4 interaction blocks the self-renewal of cancer stem cells and suppresses tumorigenesis. *Cancer Research, 76*(4), 891–901. doi:10.1158/0008-5472.CAN-15-1519.

Fuentes, E., Rojas, A., & Palomo, I. (2016). NF-κB signaling pathway as target for antiplatelet activity. *Blood Reviews, 30*(4), 309–315. doi:10.1016/j.blre.2016.03.002.

Fullard, J. F. (2004). The role of the platelet glycoprotein IIb/IIIa in thrombosis and haemostasis. *Current Pharmaceutical Design, 10*(14), 1567–1576. doi:10.2174/1381612043384682.

Furie, B., Furie, B. C., & Flaumenhaft, R. (2001). A journey with platelet P-selectin: The molecular basis of granule secretion, signalling and cell adhesion. *Thrombosis and Haemostasis, 86*(1), 214–221. http://www.thrombosis-online.com.

García-Hernández, L., García-Ortega, M. B., Ruiz-Alcalá, G., Carrillo, E., Marchal, J. A., & García, M. Á. (2021). The p38 MAPK components and modulators as biomarkers and molecular targets in cancer. *International Journal of Molecular Sciences, 23*(1). doi:10.3390/ijms23010370.

Garg, P., Malhotra, J., Kulkarni, P., Horne, D., Salgia, R., & Singhal, S. S. (2024). Emerging therapeutic strategies to overcome drug resistance in cancer cells. *Cancers, 16*(13). doi:10.3390/cancers16132478.

Gay, L. J., & Felding-Habermann, B. (2011). Contribution of platelets to tumor metastasis. *Nature Reviews Cancer, 11*(2).

Ge, Z., Zhang, Q., Lin, W., Jiang, X., & Zhang, Y. (2023). The role of angiogenic growth factors in the immune microenvironment of glioma. *Frontiers in Oncology, 13*. doi:10.3389/fonc.2023.1254694.

Geindreau, M., Bruchard, M., & Vegran, F. (2022). Role of cytokines and chemokines in angiogenesis in a tumor context. *Cancers, 14*(10), 2446. doi:10.3390/cancers14102446.

Ghosh, L. D., & Jain, A. (2023). The prospects of microphysiological systems in modeling platelet pathophysiology in cancer. *Platelets, 34*(1). doi:10.1080/09537104.2023.2247489.

Giarratana, A. O., Prendergast, C. M., Salvatore, M. M., & Capaccione, K. M. (2024). TGF-β signaling: critical nexus of fibrogenesis and cancer. *Journal of Translational Medicine, 22*(1). doi:10.1186/s12967-024-05411-4.

Goto, S., Hasebe, T., & Takagi, S. (2015). Platelets: Small in size but essential in the regulation of vascular homeostasis – Translation from basic science to clinical medicine. *Circulation Journal, 79*(9), 1871–1881. doi:10.1253/circj.cj-14-1434.

Goubran, H. A., Stakiw, J., Radosevic, M., & Burnouf, T. (2014). Platelet-cancer interactions. *Seminars in Thrombosis and Hemostasis, 40*(3), 296–305. doi:10.1055/s-0034-1370767.

Grave, N., Scheffel, T. B., Cruz, F. F., Rockenbach, L., Goettert, M. I., Laufer, S., & Morrone, F. B. (2022). The functional role of p38 MAPK pathway in malignant brain tumors. *Frontiers in Pharmacology, 13*. doi:10.3389/fphar.2022.975197.

Guo, P., Hu, B., Gu, W., Xu, L., Wang, D., Huang, H. J. S., Cavenee, W. K., & Cheng, S. Y. (2003). Platelet-derived growth factor-B enhances glioma angiogenesis by stimulating vascular endothelial growth factor expression in tumor endothelia and by promoting pericyte recruitment. *American Journal of Pathology, 162*(4), 1083–1093. doi:10.1016/S0002-9440(10)63905-3.

Gupta, S., Konradt, C., Corken, A., Ware, J., Nieswandt, B., Di Paola, J., Yu, M., Wang, D., Nieman, M. T., Whiteheart, S. W., & Brass, L. F. (2020). Hemostasis vs. homeostasis: Platelets are essential for preserving vascular barrier function in the absence of injury or inflammation. *Proceedings of the National Academy of Sciences of the United States of America, 117*(39), 24316–24325. doi:10.1073/pnas.2007642117.

Haemmerle, M., Stone, R. L., Menter, D. G., Afshar-Kharghan, V., & Sood, A. K. (2018). The platelet lifeline to cancer: Challenges and opportunities. *Cancer Cell, 33*(6), 965–983. doi:10.1016/j.ccell.2018.03.002.

Haining, E. J., Nicolson, P. L. R., Onselaer, M. B, Poulter, N. S., Rayes, J., Thomas, M. R., & Watson, S. P. (2019). 11 - GPVI and CLEC-2. *Platelets*. Academic Press.

Heldin, C. H., Lennartsson, J., & Westermark, B. (2018). Involvement of platelet-derived growth factor ligands and receptors in tumorigenesis. *Journal of Internal Medicine, 283*(1), 16–44. doi:10.1111/joim.12690.

Holmgaard, R. B., Schaer, D. A., Li, Y., Castaneda, S. P., Murphy, M. Y., Xu, X., Inigo, I., Dobkin, J., Manro, J. R., Iversen, P. W., Surguladze, D., Hall, G. E., Novosiadly, R. D., Benhadji, K. A., Plowman, G. D., Kalos, M., & Driscoll, K. E. (2018). Targeting the TGFβ pathway with galunisertib, a TGFβRI small molecule inhibitor, promotes anti-tumor immunity leading to durable, complete responses, as monotherapy and in combination with checkpoint blockade. *Journal for ImmunoTherapy of Cancer, 6*(1). doi:10.1186/s40425-018-0356-4.

Hosseini, E., Mohtashami, M., & Ghasemzadeh, M. (2019). Down-regulation of platelet adhesion receptors is a controlling mechanism of thrombosis, while also affecting post-transfusion efficacy of stored platelets. *Thrombosis Journal, 17*(1). doi:10.1186/s12959-019-0209-5.

Huang, J., Li, X., Shi, X., Zhu, M., Wang, J., Huang, S., Huang, X., Wang, H., Li, L., Deng, H., Zhou, Y., Mao, J., Long, Z., Ma, Z., Ye, W., Pan, J., Xi, X., & Jin, J. (2019). Platelet integrin αIIbβ3: Signal transduction, regulation, and its therapeutic targeting. *Journal of Hematology & Oncology, 12*(1). doi:10.1186/s13045-019-0709-6.

Huong, P. T., Nguyen, L. T., Nguyen, X. B., Lee, S. K., & Bach, D. H. (2019). The role of platelets in the tumor-microenvironment and the drug resistance of cancer cells. *Cancers, 11*(2). doi:10.3390/cancers11020240.

Ikeda, H., Takai, M., & Tsujiuchi, T. (2024). Lysophosphatidic acid (LPA) receptor-mediated signaling and cellular responses to anticancer drugs and radiation of cancer cells. *Advances in Biological Regulation, 92*, 101029. doi:10.1016/j.jbior.2024.101029.

In 'T Veld, S. G. J. G., & Wurdinger, T. (2019). Tumor-educated platelets. *Blood, 133*(22), 2359–2364. doi:10.1182/blood-2018-12-852830.

Italiano, J. E., Richardson, J. L., Patel-Hett, S., Battinelli, E., Zaslavsky, A., Short, S., Ryeom, S., Folkman, J., & Klement, G. L. (2008). Angiogenesis is regulated by a novel mechanism: Pro- and antiangiogenic proteins are organized into separate platelet α granules and differentially released. *Blood, 111*(3), 1227–1233. doi:10.1182/blood-2007-09-113837.

Janus-Bell, E., & Mangin, P. H. (2023). The relative importance of platelet integrins in hemostasis, thrombosis and beyond. *Haematologica, 108*(7), 1734–1747. doi:10.3324/haematol.2022.282136.

Javelaud, D., & Mauviel, A. (2005). Crosstalk mechanisms between the mitogen-activated protein kinase pathways and Smad signaling downstream of TGF-β: Implications for carcinogenesis. *Oncogene, 24*(37), 5742–5750. doi:10.1038/sj.onc.1208928.

Jin, W. (2020). Role of JAK/STAT3 signaling in the regulation of metastasis, the transition of cancer stem cells, and chemoresistance of cancer by epithelial–mesenchymal transition. *Cells, 9*(1), 217. doi:10.3390/cells9010217.

Jing, L., Qu, H., Wu, D., Zhu, C., Yang, Y., Jin, X., Zheng, J., Shi, X., Yan, X., & Wang, Y. (2018). Platelet-camouflaged nanococktail: Simultaneous inhibition of drug-resistant tumor growth and metastasis via a cancer cells and tumor vasculature dual-targeting strategy. *Theranostics, 8*(10), 2683–2695. doi:10.7150/thno.23654.

Jones, K. L., Hughan, S. C., Dopheide, S. M., Farndale, R. W., Jackson, S. P., & Jackson, D. E. (2001). Platelet endothelial cell adhesion molecule-1 is a negative regulator of platelet-collagen interactions. *Blood, 98*(5), 1456–1463. doi:10.1182/blood.V98.5.1456.

Jurasz, P., Alonso-Escolano, D., & Radomski, M. W. (2004). Platelet-cancer interactions: Mechanisms and pharmacology of tumour cell-induced platelet aggregation. *British Journal of Pharmacology, 143*(7), 819–826. doi:10.1038/sj.bjp.0706013.

Jurasz, P., Sawicki, G., Duszyk, M., Sawicka, J., Miranda, C., Mayers, I., & Radomski, M. W. (2001). Matrix Metalloproteinase 2 in Tumor Cell-induced Platelet Aggregation: Regulation by Nitric Oxide. *Cancer Research, 61*(1), 376–382.

Kato, Y., Kaneko, M. K., Kunita, A., Ito, H., Kameyama, A., Ogasawara, S., Matsuura, N., Hasegawa, Y., Suzuki-inoue, K., Inoue, O., Ozaki, Y., & Narimatsu, H. (2008). Molecular analysis of the pathophysiological binding of the platelet aggregation-inducing factor podoplanin to the C-type lectin-like receptor CLEC-2. *Cancer Science, 99*(1), 54–61. doi:10.1111/j.1349-7006.2007.00634.x.

Kehrel, B., Wierwille, S., Clemetson, K. J., Anders, O., Steiner, M., Knight, C. G., Farndale, R. W., Okuma, M., & Barnes, M. J. (1998). Glycoprotein VI is a major collagen receptor for platelet activation: It recognizes the platelet-activating quaternary structure of collagen, whereas CD36, glycoprotein IIb/IIIa, and von Willebrand factor do not. *Blood, 91*(2), 491–499. doi:10.1182/blood.v91.2.491.

Khasabova, I. A., Khasabov, S. G., Johns, M., Juliette, J., Zheng, A., Morgan, H., Flippen, A., Allen, K., Golovko, M. Y., Golovko, S. A., Zhang, W., Marti, J., Cain, D., Seybold, V. S., & Simone, D. A. (2023). Exosome-Associated lysophosphatidic acid signaling contributes to cancer pain. *Pain, 164*(12), 2684–2695. doi:10.1097/j.pain.0000000000002967.

Koupenova, M., Clancy, L., Corkrey, H. A., & Freedman, J . E. (2018). Circulating platelets as mediators of immunity, inflammation, and thrombosis. *Circulation Research, 122*(2), 337–351. doi:10.1161/CIRCRESAHA.117.310795.

Labelle, M., Begum, S., & Hynes, R. (2011). Direct signaling between platelets and cancer cells induces an epithelial-mesenchymal-like transition and promotes metastasis. *Cancer Cell, 20*(5), 576–590. doi:10.1016/j.ccr.2011.09.009.

Labelle, M., Begum, S., & Hynes, R. O. (2014). Platelets guide the formation of early metastatic niches. *Proceedings of the National Academy of Sciences of the United States of America, 111*(30), E3053. doi:10.1073/pnas.1411082111.

Lavergne, M., Janus-Bell, E., Schaff, M., Gachet, C., & Mangin, P. (2017). Platelet integrins in tumor metastasis: Do they represent a therapeutic Target? *Cancers, 9*(10), 133. doi:10.3390/cancers9100133.

Lazar, S., & Goldfinger, L. E. (2018). Platelet Microparticles and miRNA transfer in cancer progression: Many targets, modes of action, and effects across cancer stages. *Frontiers in Cardiovascular Medicine, 5*. doi:10.3389/fcvm.2018.00013.

Lazar, S., & Goldfinger, L. E. (2021). Platelets and extracellular vesicles and their cross talk with cancer. *Blood, 137*(23), 3192–3200. doi:10.1182/blood.2019004119.

Leblanc, R., Houssin, A., & Peyruchaud, O. (2018). Platelets, autotaxin and lysophosphatidic acid signalling: Win-win factors for cancer metastasis. *British Journal of Pharmacology, 175*(15), 3100–3110. doi:10.1111/bph.14362.

Li, J., Kim, K., Andrews, R. K., & Cho, J. (2014). NADPH oxidase 2 is critical for heterotypic neutrophil-platelet interactions during vascular inflammation. *Blood, 124*(21), 459. doi:10.1182/blood.v124.21.459.459.

Li, N. (2016). Platelets in cancer metastasis: To help the "villain" to do evil. *International Journal of Cancer, 138*(9), 2078–2087. doi:10.1002/ijc.29847.

Li, S., Li, L., Lin, X., Chen, C., Luo, C., & Huang, Y. (2022). Targeted inhibition of tumor inflammation and tumor-platelet crosstalk by nanoparticle-mediated drug delivery mitigates cancer metastasis. *ACS Nano, 16*(1), 50–67. doi:10.1021/acsnano.1c06022.

Li, Y., Wang, H., Zhao, Z., Yang, Y., Meng, Z., & Qin, L. (2023). Effects of the interactions between platelets with other cells in tumor growth and progression. *Frontiers in Immunology, 14*. doi:10.3389/fimmu.2023.1165989.

Liao, K., Zhang, X., Liu, J., Teng, F., He, Y., Cheng, J., Yang, Q., Zhang, W., Xie, Y., Guo, D., Cao, G., Xu, Y., Huang, B., & Wang, X. (2023). The role of platelets in the regulation of tumor growth and metastasis: the mechanisms and targeted therapy. *MedComm, 4*(5). doi:10.1002/mco2.350.

Littlewood-Evans, A. J., Bilbe, G., Bowler, W. B., Farley, D., Wlodarski, B., Kokubo, T., Inaoka, T., Sloane, J., Evans, D. B., & Gallagher, J. A. (1997). The osteoclast-associated protease cathepsin K is expressed in human breast carcinoma. *Cancer Research, 57*(23), 5386–5390.

Liu, J., Lu, F., Ithychanda, S. S., Apostol, M., Das, M., Deshpande, G., Plow, E. F., & Qin, J. (2023). A mechanism of platelet integrin αIIbβ3 outside-in signaling through a novel integrin αIIb subunit–filamin–actin linkage. *Blood, 141*(21), 2629–2641. doi:10.1182/blood.2022018333.

Liu, X., Yu, J., Song, S., Yue, X., & Li, Q. (2017). Protease-activated receptor-1 (PAR-1): A promising molecular target for cancer. *Oncotarget, 8*(63), 107334–107345. doi:10.18632/oncotarget.21015.

Liu, Y., An, S., Ward, R., Yang, Y., Guo, X. X., Li, W., & Xu, T. R. (2016). G protein-coupled receptors as promising cancer targets. *Cancer Letters, 376*(2), 226–239. doi:10.1016/j.canlet.2016.03.031.

Liu, Y., Zhang, Y., Ding, Y., & Zhuang, R. (2021). Platelet-mediated tumor metastasis mechanism and the role of cell adhesion molecules. *Critical Reviews in Oncology/Hematology, 167*, 103502. doi:10.1016/j.critrevonc.2021.103502.

López-Rovira, T., Chalaux, E., Rosa, J. L., Bartrons, R., & Ventura, F. (2000). Interaction and functional cooperation of NF-κb with smads: Transcriptional regulation of the junB promoter. *Journal of Biological Chemistry, 275*(37), 28937–28946. doi:10.1074/jbc.M909923199.

Manikanta, K., NaveenKumar, S. K., Hemshekhar, M., Kemparaju, K., & Girish, K. S. (2020). ASK1 inhibition triggers platelet apoptosis via p38-MAPK-mediated mitochondrial dysfunction. *Haematologica, 105*(8), E419. doi:10.3324/haematol.2019.233908.

Massberg, S., Enders, G., Leiderer, R., Eisenmeng, S., Vestweber, D., Krombach, F., & Messmer, K. (1998). Platelet-endothelial cell interactions during ischemia/reperfusion: The role of P-selectin. *Blood, 92*(2), 507–515. doi:10.1182/blood.v92.2.507.414k29_507_515.

Maurer, S., Kropp, K. N., Klein, G., Steinle, A., Haen, S. P., Walz, J. S., Hinterleitner, C., Märklin, M., Kopp, H. G., & Salih, H. R. (2018). Platelet-mediated shedding of NKG2D ligands impairs NK cell immune-surveillance of tumor cells. *OncoImmunology, 7*(2). doi:10.1080/2162402X.2017.1364827.

Meikle, C. K. S., Kelly, C. A., Garg, P., Wuescher, L. M., Ali, R. A., & Worth, R. G. (2017). Cancer and thrombosis: The platelet perspective. *Frontiers in Cell and Developmental Biology, 4*. doi:10.3389/fcell.2016.00147.

Merten, M., & Thiagarajan, P. (2000). P-selectin expression on platelets determines size and stability of platelet aggregates. *Circulation, 102*(16), 1931–1936. doi:10.1161/01.CIR.102.16.1931.

Merten, M., & Thiagarajan, P. (2001). Role for sulfatides in platelet aggregation. *Circulation, 104*(24), 2955–2960. doi:10.1161/hc4901.100383.

Metelli, A., Wu, B. X., Riesenberg, B., Guglietta, S., Huck, J. D., Mills, C., Li, A., Rachidi, S., Krieg, C., Rubinstein, M. P., Gewirth, D. T., Sun, S., Lilly, M. B., Wahlquist, A. H., Carbone, D. P., Yang, Y., Liu, B., & Li, Z. (2020). Thrombin contributes to cancer immune evasion via proteolysis of platelet-bound GARP to activate LTGF-β. *Science Translational Medicine, 12*(525). doi:10.1126/scitranslmed.aay4860.

Moolenaar, W. H. (1994). LPA: A novel lipid mediator with diverse biological actions. *Trends in Cell Biology, 4*(6), 213–219. doi:10.1016/0962-8924(94)90144-9.

Morris, K., Schnoor, B., & Papa, A. L. (2022). Platelet cancer cell interplay as a new therapeutic target. *Biochimica et Biophysica Acta - Reviews on Cancer, 1877*(5). doi:10.1016/j.bbcan.2022.188770.

Natoni, A., Cerreto, M., De Propris, Petrucci, M. T, Del Giudice, I., Intoppa, S., Milani, M. L., Kirkham-McCarthy, L., Henderson, R., Swan, D., O'Dwyer, M., Guarini, A., & Foà, R. (2022). P861:Sialofucosylated structures enable platelet binding to myeloma cells conferring protection from NK-mediated cytotoxicity. *HemaSphere, 861*.

Nuyttens, B. P., Thijs, T., Deckmyn, H., & Broos, K. (2011). Platelet adhesion to collagen. *Thrombosis Research, 127*(2), S26. doi:10.1016/S0049-3848(10)70151-1.

Obermann, W. M. J., Brockhaus, K., & Eble, J. A. (2021). Platelets, constant and cooperative companions of sessile and disseminating tumor cells, crucially contribute to the tumor microenvironment. *Frontiers in Cell and Developmental Biology, 9*. doi:10.3389/fcell.2021.674553.

O'Hayre, M., Degese, M. S., & Gutkind, J. S. (2014). Novel insights into G protein and G protein-coupled receptor signaling in cancer. *Current Opinion in Cell Biology, 27*(1), 126–135. doi:10.1016/j.ceb.2014.01.005.

Palacios-Acedo, A. L, Mège, D., Crescence, L., Dignat-George, F., Dubois, C., & Panicot-Dubois, L. (2019). Platelets, thrombo-Inflammation, and cancer: Collaborating with the enemy. *Frontiers in Immunology, 10*.

Pan, Y., Zhou, S., Li, Y., Parshad, B., Li, W., & Haag, R. (2021). Novel dendritic polyglycerol-conjugated, mesoporous silica-based targeting nanocarriers for co-delivery of doxorubicin and tariquidar to overcome multidrug resistance in breast cancer stem cells. *Journal of Controlled Release, 330*, 1106–1117. doi:10.1016/j.jconrel.2020.11.015.

Patil, Y., Sadhukha, T., Ma, L., & Panyam, J. (2009). Nanoparticle-mediated simultaneous and targeted delivery of paclitaxel and tariquidar overcomes tumor drug resistance. *Journal of Controlled Release, 136*(1), 21–29. doi:10.1016/j.jconrel.2009.01.021.

Patil, Y. B., Swaminathan, S. K., Sadhukha, T., Ma, L., & Panyam, J. (2010). The use of nanoparticle-mediated targeted gene silencing and drug delivery to overcome tumor drug resistance. *Biomaterials, 31*(2), 358–365. doi:10.1016/j.biomaterials.2009.09.048.

Plantureux, L., Mège, D., Crescence, L., Carminita, E., Robert, S., Cointe, S., Brouilly, N., Ezzedine, W., Dignat-George, F., Dubois, C., & Panicot-Dubois, L. (2020). The interaction of platelets with colorectal cancer cells inhibits tumor growth but promotes metastasis. *Cancer Research, 80*(2), 291–303. doi:10.1158/0008-5472.can-19-1181.

Plantureux, L., Mège, D., Crescence, L., Dignat-George, F., Dubois, C., & Panicot-Dubois, L. (2018). Impacts of cancer on platelet production, activation and education and mechanisms of cancer-associated thrombosis. *Cancers, 10*(11), 441. doi:10.3390/cancers10110441.

Pollitt, A. Y., Poulter, N. S., Gitz, E., Navarro-Nuñez, L., Wang, Y. J., Hughes, C. E., Thomas, S. G., Nieswandt, B., Douglas, M. R., Owen, D. M., Jackson, D. G., Dustin, M. L., & Watson, S. P. (2014). Syk and src family kinases regulate c-type lectin receptor 2 (clec-2)-mediated clustering of podoplanin and platelet adhesion to lymphatic endothelial cells. *Journal of Biological Chemistry, 289*(52), 35695–35710. doi:10.1074/jbc.M114.584284.

Prabhu, S., Sharma, R., & Peter, K. (2011). *Platelets in the pathogenesis of vascular disease and their role as a therapeutic target. In Mechanisms of vascular disease: A reference book for vascular specialists* (pp. 201–226). Australia: University of Adelaide Press. doi:10.1017/UPO9781922064004.012.

Qi, Y., Wang, Y., Yuan, J., Xu, Y., & Pan, H. (2024). Unveiling the therapeutic promise: Exploring lysophosphatidic acid (LPA) signaling in malignant bone tumors for novel cancer treatments. *Lipids in Health and Disease, 23*(1). doi:10.1186/s12944-024-02196-9.

Qian, D., He, L., Zhang, Q., Li, W., Tang, D., Wu, C., Yang, F., Li, K., & Zhang, H. (2022). Cathepsin K: A versatile potential biomarker and therapeutic target for various cancers. *Current Oncology, 29*(8), 5963–5987. doi:10.3390/curroncol29080471.

Qiao, J., Al-Tamimi, M., Baker, R. I., Andrews, R. K, & Gardiner, E. E. (2015). The platelet Fc receptor FcγRIIa. *Immunological Reviews, 268*(1), 241–252.

Radziwon-Balicka, A., Santos-Martinez, M. J., Corbalan, J. J., O'Sullivan, S., Treumann, A., Gilmer, J. F., Radomski, M. W., & Medina, C. (2014). Mechanisms of platelet-stimulated colon cancer invasion: Role of clusterin and thrombospondin 1 in regulation of the P38MAPK-MMP-9 pathway. *Carcinogenesis, 35*(2), 324–332. doi:10.1093/carcin/bgt332.

Raica, M., & Cimpean, A. M. (2010). Platelet-derived growth factor (PDGF)/PDGF receptors (PDGFR) axis as target for antitumor and antiangiogenic therapy. *Pharmaceuticals, 3*(3), 572–599. doi:10.3390/ph3030572.

Repsold, L., & Joubert, A. M. (2021). Platelet function, role in thrombosis, inflammation, and consequences in chronic myeloproliferative disorders. *Cells, 10*(11). doi:10.3390/cells10113034.

Rovati, G., Contursi, A., Bruno, A., Tacconelli, S., Ballerini, P., & Patrignani, P. (2022). Antiplatelet agents affecting GPCR signaling implicated in tumor metastasis. *Cells, 11*(4), 725. doi:10.3390/cells11040725.

Ruggeri, Z. M. (2003). Von Willebrand factor, platelets and endothelial cell interactions. *Journal of Thrombosis and Haemostasis, 1*(7), 1335–1342. doi:10.1046/j.1538-7836.2003.00260.x.

Saboor, A., & Ilyas, M. (2013). Platelet receptors; an instrumental of platelet physiology. *Pakistan Journal of Medical Sciences, 29*(3).

Schlesinger, M. (2018). Role of platelets and platelet receptors in cancer metastasis. *Journal of Hematology & Oncology, 11*(1). doi:10.1186/s13045-018-0669-2.

Schmied, L., Höglund, P., & Meinke, S. (2021). Platelet-mediated protection of cancer cells from immune surveillance – Possible implications for cancer immunotherapy. *Frontiers in Immunology, 12*. doi:10.3389/fimmu.2021.640578.

Sekiguchi, T., Takemoto, A., & Fujita, N. (2016). Abstract 1275: Suppression of platelet aggregation and tumor metastasis by anti-podoplanin mAbs recognizing a novel CLEC-2 binding domain. *Cancer Research, 76*(14_Supplement), 1275. doi:10.1158/1538-7445.am2016-1275.

Shapira, A., Livney, Y. D., Broxterman, H. J., & Assaraf, Y. G. (2011). Nanomedicine for targeted cancer therapy: Towards the overcoming of drug resistance. *Drug Resistance Updates, 14*(3), 150–163. doi:10.1016/j.drup.2011.01.003.

Shih, A. H., & Holland, E. C. (2006). Platelet-derived growth factor (PDGF) and glial tumorigenesis. *Cancer Letters, 232*(2), 139–147. doi:10.1016/j.canlet.2005.02.002.

Showalter, H. D. H., & Kraker, A. J. (1997). Small molecule inhibitors of the platelet-derived growth factor receptor, the fibroblast growth factor receptor, and Src family tyrosine kinases. *Pharmacology and Therapeutics, 76*(1-3), 55–71. doi:10.1016/S0163-7258(97)00097-1.

Sierko, E., & Wojtukiewicz, M. Z. (2004). Platelets and angiogenesis in malignancy. *Seminars in Thrombosis and Hemostasis, 30*(1), 95–108. doi:10.1055/s-2004-822974.

Srivastava, P., Jha, S., Singh, S. K., Vyas, H., Sethupathi, P., Nair, R. S., Ramachandran, K., Rana, B., Kumar, S., & Rana, A. (2024). Protease activated receptor-1 regulates mixed lineage kinase-3 to drive triple-negative breast cancer tumorigenesis. *Cancer Letters, 603*. doi:10.1016/j.canlet.2024.217200.

Stoiber, D., & Assinger, A. (2020). Platelet-leukocyte interplay in cancer development and progression. *Cells, 9*(4). doi:10.3390/cells9040855.

Stone, J. P., & Wagner, D. D. (1993). P-selectin mediates adhesion of platelets to neuroblastoma and small cell lung cancer. *Journal of Clinical Investigation, 92*(2), 804–813. http://www.jci.org.

Strasenburg, W., Jóźwicki, J., Durślewicz, J., Kuffel, B., Kulczyk, M. P., Kowalewski, A., Grzanka, D., Drewa, T., & Adamowicz, J. (2022). Tumor cell-induced platelet aggregation as an emerging therapeutic target for cancer therapy. *Frontiers in Oncology, 12*. doi:10.3389/fonc.2022.909767.

Sun, L., Wang, Z., Liu, Z., Mu, G., Cui, Y., & Xiang, Q. (2024). C-type lectin-like receptor 2: Roles and drug target. *Thrombosis Journal, 22*(1). doi:10.1186/s12959-024-00594-8.

Sun, Y., Li, T., Ding, L., Wang, J., Chen, C., Liu, T., Liu, Y., Li, Q., Wang, C., Huo, R., Wang, H., Tian, T., Zhang, C., Pan, B., Zhou, J., Fan, J., Yang, X., Yang, W., Wang, B., & Guo, W. (2025). Platelet-mediated circulating tumor cell evasion from natural killer cell killing through immune checkpoint CD155-TIGIT. *Hepatology, 81*(3), 791–807. doi:10.1097/HEP.0000000000000934.

Suzuki-Inoue, K. (2006). A novel Syk-dependent mechanism of platelet activation by the C-type lectin receptor CLEC-2. *Blood, 107*(2), 542–549. doi:10.1182/blood-2005-05-1994.

Suzuki-Inoue, K., Kato, Y., Inoue, O., Kato Kaneko, M., Mishima, K., Yatomi, Y., Yamazaki, Y., Narimatsu, H., & Ozaki, Y. (2007). Involvement of the snake toxin receptor CLEC-2, in podoplanin-mediated platelet activation, by cancer cells. *Journal of Biological Chemistry, 282*(36), 25993–26001. doi:10.1074/jbc.m702327200.

Tokarz-Deptuła, B., Baraniecki, Ł., Palma, J., Stosik, M., & Deptuła, W. (2024). Characterization of platelet receptors and their involvement in immune activation of these cells. *International Journal of Molecular Sciences, 25*(23), 12611. doi:10.3390/ijms252312611.

Trivanović, D., Mojsilović, S., Bogosavljević, N., Jurišić, V., & Jauković, A. (2024). Revealing profile of cancer-educated platelets and their factors to foster immunotherapy development. *Translational Oncology, 40*, 101871. doi:10.1016/j.tranon.2023.101871.

Varga-Szabo, D., Pleines, I., & Nieswandt, B. (2008). Cell adhesion mechanisms in platelets. *Arteriosclerosis, Thrombosis, and Vascular Biology, 28*(3), 403–412. doi:10.1161/atvbaha.107.150474.

Vogel, S., Arora, T., Wang, X., Mendelsohn, L., Nichols, J., Allen, D., Shet, A. S., Combs, C. A., Quezado, Z. M. N., & Thein, S. L. (2018). The platelet NLRP3 inflammasome is upregulated in sickle cell disease via HMGB1/TLR4 and Bruton tyrosine kinase. *Blood Advances, 2*(20), 2672–2680. doi:10.1182/bloodadvances.2018021709.

Wang, L., Wang, X., Guo, E., Mao, X., & Miao, S. (2022). Emerging roles of platelets in cancer biology and their potential as therapeutic targets. *Frontiers in Oncology, 12*. doi:10.3389/fonc.2022.939089.

Wang, S., Li, Z., & Xu, R. (2018). Human cancer and platelet interaction, a potential therapeutic target. *International Journal of Molecular Sciences, 19*(4), 1246. doi:10.3390/ijms19041246.

Wang, X., Zhao, S., Wang, Z., & Gao, T. (2022). Platelets involved tumor cell EMT during circulation: Communications and interventions. *Cell Communication and Signaling, 20*(1). doi:10.1186/s12964-022-00887-3.

Ward, Y., Lake, R., Faraji, F., Sperger, J., Martin, P., Gilliard, C., Ku, K. P., Rodems, T., Niles, D., Tillman, H., Yin, J. J., Hunter, K., Sowalsky, A. G., Lang, J., & Kelly, K. (2018). Platelets promote metastasis via binding tumor CD97 leading to bidirectional signaling that coordinates transendothelial migration. *Cell Reports, 23*(3), 808–822. doi:10.1016/j.celrep.2018.03.092.

Ward, Y., Lake, R., Yin, J. J., Heger, C. D., Raffeld, M., Goldsmith, P. K., Merino, M., & Kelly, K. (2011). LPA receptor heterodimerizes with CD97 to amplify LPA-initiated RHO-dependent signaling and invasion in prostate cancer cells. *Cancer Research, 71*(23), 7301–7311. doi:10.1158/0008-5472.CAN-11-2381.

Wojtukiewicz, M. Z, Hempel, D., Sierko, E., Tucker, S. C, & Honn, K. V. (2015). Protease-activated receptors (PARs)—Biology and role in cancer invasion and metastasis. *Cancer and Metastasis Reviews., 34*(4).

Wojtukiewicz, M. Z., Sierko, E., Hempel, D., Tucker, S. C., & Honn, K. V. (2017). Platelets and cancer angiogenesis nexus. *Cancer and Metastasis Reviews, 36*(2), 249–262. doi:10.1007/s10555-017-9673-1.

Wu, Q., Jiang, L., Li, S., He, Q., Yang, B., & Cao, J. (2021). Small molecule inhibitors targeting the PD-1/PD-L1 signaling pathway. *Acta Pharmacologica Sinica, 42*(1), 1–9. doi:10.1038/s41401-020-0366-x.

Wu, V., Yeerna, H., Nohata, N., Chiou, J., Harismendy, O., Raimondi, F., Inoue, A., Russell, R. B., Tamayo, P., & Gutkind, J. S. (2019). Illuminating the Onco-GPCRome: Novel G protein-coupled receptor-driven oncocrine networks and targets for cancer immunotherapy. *Journal of Biological Chemistry, 294*(29), 11062–11086. doi:10.1074/jbc.REV119.005601.

Wu, W., Chen, F., Cui, X., Yang, L., Chen, J., Zhao, J., Huang, D., Liu, J., Yang, L., Zeng, J., Zeng, Z., Pan, Y., Su, F., Cai, J., Ying, Z., Zhao, Q., Song, E., & Su, S. (2018). LncRNA NKILA suppresses TGF-β-induced epithelial–mesenchymal transition by blocking NF-κB signaling in breast cancer. *International Journal of Cancer, 143*(9), 2213–2224. doi:10.1002/ijc.31605.

Xu, X. R., Yousef, G. M., & Ni, H. (2018). Cancer and platelet crosstalk: Opportunities and challenges of aspirin and other antiplatelet agents. *Blood, 131*(16), 1777–1789. doi:10.1182/blood-2017-05-743187.

Xue, J., Deng, J., Qin, H., Yan, S., Zhao, Z., Qin, L., Liu, J., & Wang, H. (2024). The interaction of platelet-related factors with tumor cells promotes tumor metastasis. *Journal of Translational Medicine, 22*(1). doi:10.1186/s12967-024-05126-6.

Yang, D., Yang, W., Zhang, Q., Hu, Y., Bao, L., & Damirin, A. (2013). Migration of gastric cancer cells in response to lysophosphatidic acid is mediated by LPA receptor 2. *Oncology Letters, 5*(3), 1048–1052. doi:10.3892/ol.2013.1107China.

Yang, Y., Li, X., Wang, T., Guo, Q., Xi, T., & Zheng, L. (2020). Emerging agents that target signaling pathways in cancer stem cells. *Journal of Hematology and Oncology, 13*(1). doi:10.1186/s13045-020-00901-6.

Yip, J., Shen, Y., Berndt, M. C., & Andrews, R. K. (2005). Primary platelet adhesion receptors. *IUBMB Life, 57*(2), 103–108. doi:10.1080/15216540500078962.

Yu, L., Guo, Y., Chang, Z., Zhang, D., Zhang, S., Pei, H., Pang, J., Zhao, Z. J, & Chen, Y. (2021). Bidirectional interaction between cancer cells and platelets provides potential strategies for cancer therapies. *Frontiers in Oncology, 11*. doi:10.3389/fonc.2021.764119.

Zaid, Y., & Merhi, Y. (2022). Implication of platelets in immuno-thrombosis and thrombo-inflammation. *Frontiers in Cardiovascular Medicine, 9*. doi:10.3389/fcvm.2022.863846.

Zhang, S., Zhang, S., Hu, L., Zhai, L., Xue, R., Ye, J., Chen, L., Cheng, G., Mruk, J., Kunapuli, S. P., & Ding, Z. (2015). Nucleotide-binding oligomerization domain 2 receptor is expressed in platelets and enhances platelet activation and thrombosis. *Circulation, 131*(13), 1160–1170. doi:10.1161/circulationaha.114.013743.

Zhang, X., Yu, S., Li, X., Wen, X., Liu, S., Zu, R., Ren, H., Li, T., Yang, C., & Luo, H. (2023). Research progress on the interaction between oxidative stress and platelets: Another avenue for cancer? *Pharmacological Research, 191*, 106777. doi:10.1016/j.phrs.2023.106777.

Zhou, L., Zhang, Z., Tian, Y., Li, Z., Liu, Z., & Zhu, S. (2023). The critical role of platelet in cancer progression and metastasis. *European Journal of Medical Research, 28*(1). doi:10.1186/s40001-023-01342-w.

Zigler, M., Kamiya, T., Brantley, E. C., Villares, G. J., & Bar-Eli, M. (2011). PAR-1 and thrombin: The ties that bind the microenvironment to melanoma metastasis. *Cancer Research, 71*(21), 6561–6566. doi:10.1158/0008-5472.CAN-11-1432.

Zou, Z., Chen, H., Schmaier, A. A., Hynes, R. O., & Kahn, M. L. (2007). Structure-function analysis reveals discrete β3 integrin inside-out and outside-in signaling pathways in platelets. *Blood, 109*(8), 3284–3290. doi:10.1182/blood-2006-10-051664.

CHAPTER

11 Myeloid-derived suppressor cells: A promising target for anticancer immunotherapy

Vinit Sharma, Sakshi Bansal and Anjali Aggarwal

Department of Anatomy, Post Graduate Institute of Medical Education and Research, Chandigarh, India

11.1 Introduction

Cancer remains a major global health concern, contributing markedly to morbidity and mortality. As per the statistics published by the World Health Organization (2020), it accounted for nearly 10 million deaths worldwide, making up approximately one in six fatalities. The cancer incidence is on the rise, particularly in developing countries, due to an aging population, environmental factors, and lifestyle changes such as poor diet, physical inactivity, and tobacco use. The most prevalent cancer types include lung, breast, colorectal, and prostate cancer. Additionally, pancreatic, liver, and esophageal cancers significantly contribute to the global disease burden (Bray et al., 2024). Early diagnosis of cancer plays a crucial role in improving patient outcomes, yet many cancers are diagnosed at advanced stages, when they are less treatable. Although significant advancements have been made in diagnostic technologies, including imaging modalities, biomarker discovery, and liquid biopsy techniques, many cancers remain challenging to diagnose at early, treatable stages. For example, pancreatic cancer, known for its high mortality rate, is often diagnosed at later stages due to the absence of effective early detection methods.

Despite the rapid development of therapeutic approaches, cancer treatment remains fraught with challenges. Conventional treatments like surgery, chemotherapy, and radiation therapy have long served as the foundation of cancer management. However, these treatments are often associated with significant side effects, including damage to healthy tissues and a lack of specificity in targeting cancer cells. Moreover, while these therapies can be effective in reducing tumor size or slowing down disease progression, they often fail to provide long-term cures. Many patients experience relapse, with their cancers returning more aggressively and resistant to further treatment. The innovation in immune checkpoint inhibitors (ICIs) immunotherapies (anti–PD-L1 and anti–PD-1 antibodies) has offered renewed hope for cancer patients. These treatments work by depleting the inhibitory signals used by cancer cells to escape from the surveillance of the immune system, thereby restoring the ability of the immune system to target and eliminate cancer (Wojtukiewicz et al., 2021).

Cancer treatment has been transformed by ICIs, especially for malignancies with elevated expression of PD-L1. The anti–PD-1 therapy has better outcomes over platinum-based chemotherapy as a first-line treatment for advanced lung cancer with PD-L1 expression $\geq$50%, excluding instances with epidermal growth factor receptor (EGFR) or anaplastic lymphoma kinase (ALK) mutations, according to the KEYNOTE-024 trial (ClinicalTrials.gov: NCT02142738). Pembrolizumab-treated patients demonstrated noticeably better progression-free survival than those on conventional chemotherapy,

Advancements in Personalized Cellular Immunotherapy: A Comprehensive Guide to Cancer Treatment.
DOI: https://doi.org/10.1016/B978-0-443-27731-3.00015-2

Copyright © 2026 Elsevier Inc. All rights are reserved, including those for text and data mining, AI training, and similar technologies.

which includes pemetrexed maintenance therapy in nonsquamous non–small-cell lung cancer. This landmark study underscored the potential of checkpoint inhibitors in reshaping treatment paradigms, paving the way for combination strategies aimed at overcoming resistance mechanisms, such as those driven by immunosuppressive myeloid-derived suppressor cells (MDSCs) (Reck et al., 2016). While ICIs have shown significant success in treating certain cancers, such as lung cancer and melanoma, their effectiveness is limited in a large subset of patients, especially those with "cold" or nonimmunogenic tumors. Many patients treated with ICIs exhibit limited long-term responses, with tumor relapse and resistance occurring as a result of complex immune evasion mechanisms (Ghiringhelli & Rébé, 2024). A major factor associated with these challenges is tumor-induced immunosuppression, which hinders the ability of the immune system to target and eradicate tumor cells effectively (Kumar et al., 2016; Sharma et al., 2021). Among the key players in tumor-induced immune evasion are MDSCs, which modulate the cancer microenvironment, promote immune suppression, and contribute to therapeutic resistance (Sharma et al., 2021).

MDSCs originate from the myeloid-cell lineage, comprising immature myeloid cells (IMCs). Under normal conditions, various cells of myeloid lineage, such as hematopoietic stem cells, differentiate into common myeloid progenitor cells, which subsequently develop into IMCs and mature into granulocytes, macrophages, or dendritic cells (DCs). This process is tightly regulated by factors such as granulocyte-macrophage colony-stimulating factor (GM-CSF), IL-3, stem cell factor (SCF), and Notch signaling. However, in pathological conditions like cancer, inflammation, infections, and autoimmune diseases, IMC differentiation is disrupted, resulting in the accumulation of immunosuppressive cells. These cells exhibit high levels of reactive oxygen species (ROS), inducible nitric oxide synthase (iNOS), and arginase (ARG-1), which suppress immune responses. Due to their immunosuppressive properties, these are referred to as MDSCs (Fig. 11.1; Sharma et al., 2021).

MDSCs are pivotal in shaping the tumor microenvironment (TME) by suppressing the activity of cytotoxic T cells, promoting cancer stem cell populations, and enhancing tumor angiogenesis and vascular maturation. As a result, MDSCs have emerged as major modulators of immune suppression and therapeutic resistance in cancer. Circulating MDSC levels have been identified as negative prognostic markers in several malignancies, correlating with tumor stage, disease progression, and metastasis. Their immunosuppressive function has led to an increased recognition of MDSCs as promising targets for cancer immunotherapy (De Cicco et al., 2020; Zhao et al., 2023).

11.2 Myeloid-derived suppressor cells inhibit antitumor immunity and promote resistance to therapies

MDSCs contribute to immune evasion by inhibiting the activation and function of various cells of the immune system, including T cells, natural killer (NK) cells, and DCs. The immunosuppressive actions of MDSCs are mediated through various mechanisms, including the secretion of nitric oxide, ROS, and arginase-1 (ARG1), which deplete L-arginine and interfere with T-cell function (Khaled et al., 2014). In MDSCs, elevated arginase activity depletes L-arginine amino acid in the TME, leading to the arrest of T lymphocyte proliferation and suppression of the cell cycle. Additionally, nitric oxide impairs T-cell function by inhibiting the STAT5 signaling pathway and inducing T-cell apoptosis. Furthermore, MDSCs can express PD-L1 immune checkpoint molecules, which further suppress T-cell activation and contribute to immune tolerance within the tumor microenvironment (Noman et al., 2014;

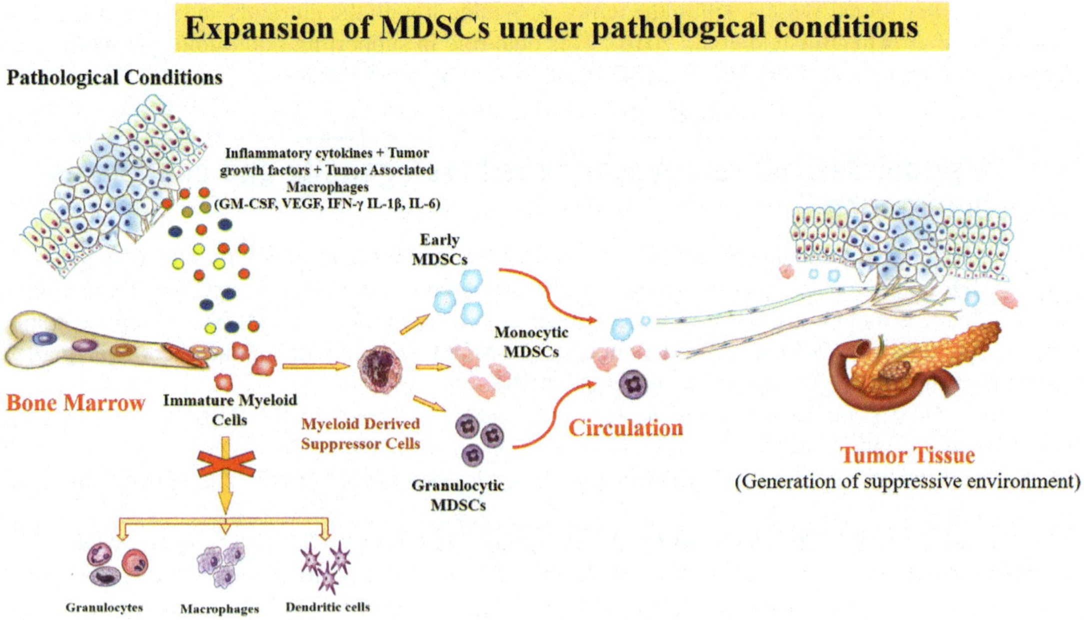

FIGURE 11.1

Schematic representation of myeloid-derived suppressor cell (MDSC) generation under pathological conditions. Pathological stimuli drive the differentiation of various MDSC subsets from immature myeloid cells, promoting their migration and expansion toward tumor sites. This results in the creation of an immunosuppressive microenvironment, inhibiting the normal development of antigen-presenting cells and impairing antitumor immune responses.

Zhang et al., 2017). A previous study demonstrated that myeloid cells facilitate tumor immune evasion in pancreatic cancer by regulating PD-L1 expression on tumor cells through the EGFR/MAPK signaling pathway. Disrupting this myeloid–tumor cell crosstalk effectively restores cytotoxic T cells–mediated antitumor immunity, highlighting a crucial mechanism of immune suppression. These findings provide important insights for designing novel immunotherapeutic strategies aimed at overcoming immune resistance in pancreatic cancer (Zhang et al., 2017).

Beyond immune suppression, MDSCs play a crucial role in maintaining cancer stem cell populations, which are resistant to conventional therapies and contribute to tumor relapse. By secreting factors like IL-6, vascular endothelial growth factor (VEGF), and GM-CSF, MDSCs support the growth, self-renewal, and survival of cancer stem cells, facilitating tumor progression and metastasis (Panni et al., 2014; Tobin et al., 2019; Xu et al., 2016). A study highlighted the critical role of IL-8 and IL-6 cytokines in melanoma, where IL-6 promotes the expansion of peripheral MDSCs, while IL-8 facilitates their recruitment to the tumor microenvironment. This interplay enhances immune suppression, contributing to cancer progression and resistance to immunotherapy (Tobin et al., 2019). Another study reported that breast cancer cells and MDSCs engage in a reciprocal activation loop, where IL-6 trans-signaling drives spontaneous metastasis in animal models. This finding highlights the critical role of MDSCs in fostering a prometastatic tumor microenvironment (Oh et al., 2013). Additionally, MDSCs contribute to the formation of new blood vessels within the tumor, providing it with the nutrients and oxygen necessary for continued growth (Binsfeld et al., 2016). This dual role in supporting both tumor progression and immune suppression makes MDSCs central to cancer resistance and relapse.

11.3 Myeloid-derived suppressor cells heterogeneity and challenges in targeting

A major challenge in targeting MDSCs in anticancer therapy is their phenotypic diversity and complexity. MDSCs are a heterogeneous population of IMCs, characterized by the positive expression of CD33 in humans and Gr-1/CD11b in mice. Rather than a uniform group, MDSCs comprise distinct subsets—granulocytic (G-MDSCs) and monocytic (M-MDSCs), each with unique phenotypic markers and immunosuppressive functions (Bronte et al., 2016).

Murine MDSCs are identified by the Gr-1 antigen, which binds to the RB6-8C5 antibody and is composed of Ly6C and Ly6G proteins. Based on these markers, M-MDSCs (monocytic) are $CD11b^{+}Ly6C^{high}Ly6G^{-}$, while G-MDSCs (granulocytic) are $CD11b^{+}Ly6C^{low}Ly6G^{+}$ (Youn et al., 2008).

In humans, MDSCs are identified by the expression of $CD33^{+}$, $CD11b^{+}$, and $HLA\text{-}DR^{-/low}$ in peripheral blood mononuclear cells. They are further classified into granulocytic MDSCs (G-MDSCs), marked by $CD15^{+}$, and monocytic MDSCs (M-MDSCs), identified by $CD14^{+}$. Additionally, lineage-negative markers such as CD3, CD20, and CD56 help distinguish them from lymphocytes and NK cells (Filipazzi et al., 2012). Early-stage MDSCs (E-MDSCs), which lack monocytic or granulocytic features, are identified as lineage negative ($CD3^{-}$, $CD20^{-}$, $CD56^{-}$), $CD33^{+}$, $CD11b^{+}$, $HLA\text{-}DR^{-/low}$ cells, though their murine equivalent remains unknown (Sharma et al., 2021). Additional markers like CD16, CD38, and LOX-1 further complicate human MDSC characterization (Condamine et al., 2016).

The immunosuppressive function of MDSCs can differ based on the tumor type, the stage of cancer, and the specific microenvironment (Khaled et al., 2014; Porembka et al., 2012). Furthermore, MDSCs exhibit plasticity, meaning that they can evolve into different subsets or acquire alternative functions in response to the dynamic signals within the TME. As a result, understanding the intricate subtypes of MDSCs across different cancer types is crucial for developing therapies that can effectively deplete or reprogram these cells to enhance immune responses against tumors.

11.4 Strategies for targeting myeloid-derived suppressor cells

Extensive research has led to various key strategies for targeting MDSCs such as, inducing their differentiation, inhibiting their proliferation and recruitment, suppressing their function, and depleting their population (He et al., 2025).

1. **Promoting MDSC differentiation into mature myeloid cells**
   Several strategies facilitate the differentiation of MDSCs into mature myeloid cells, thereby enhancing antitumor immunity. One such agent is all-trans retinoic acid (ATRA), which promotes the differentiation of MDSC into granulocytes, macrophages, and DCs while neutralizing ROS (Li et al., 2023). Toll-like receptor ligands induce MDSCs through MyD88 signaling, which can be inhibited to suppress tumor growth (Hong et al., 2013). STAT3 phosphorylation plays a key role

in MDSC expansion and immunosuppression by blocking IMC differentiation; thus targeting STAT3 is a promising therapeutic approach (Trovato et al., 2019). Similarly, vitamin D3 (1,25$(OH)_2D_3$) has been shown to induce MDSC differentiation, as evidenced by reduced immunosuppressive CD34+ progenitors and increased mature DCs in tumor tissues derived from head and neck cancer patients (Kulbersh et al., 2009).

2. **Blocking MDSCs expansion and recruitment**
   Anti-VEGF/VEGFR therapies, such as Bevacizumab, reduce PMN-MDSCs (granulocytic MDSCs) in NSCLC patients (Kulbersh et al., 2009), while targeting GM-CSF/G-CSF can help abrogate the resistance to anti-VEGF therapy in ovarian cancer (Horikawa et al., 2020). Clostridium species modulate the PI3K/CCL2/CCR2 axis, reducing MDSC recruitment and enhancing cytotoxic T lymphocyte infiltration in cholangiocarcinoma (Ma et al., 2023). Moreover, blocking the CCL2/CCR2 axis in combination with immunotherapy and radiotherapy has shown synergistic antitumor effects in animal models by decreasing tumor-MDSCs and enhancing T lymphocyte infiltration (Flores-Toro et al., 2020).
3. **Inhibition of MDSC function**
   Inhibiting MDSC function is a promising strategy for enhancing antitumor immunity. Activation of NRF2, a key regulator of oxidative stress, reduces intracellular ROS levels in MDSCs, thereby preventing immunosuppression and tumor progression (Li et al., 2021). Similarly, phosphodiesterase type 5 inhibitors, such as tadalafil, have shown efficacy in suppressing MDSC function, leading to enhanced immune responses in relapsed/refractory multiple myeloma patients (Noonan et al., 2014). Furthermore, inhibiting the COX-2/PGE2/EP axis with agents like celecoxib reduces MDSC accumulation and suppresses their immunosuppressive activity. This strategy enhances immunotherapy efficacy by decreasing NO and ROS levels in tumor-bearing mice (Karpisheh et al., 2019; Veltman et al., 2010).
4. **Depletion of MDSCs**
   Eliminating MDSCs using low-dose chemotherapy agents like paclitaxel, cisplatin, 5-FU, and gemcitabine enhances antitumor immunity by depleting MDSCs in tumor-bearing mice (Mathew et al., 2023). Paclitaxel has shown efficacy in reducing MDSC frequency and immunosuppressive activity in patients with melanoma resistant to ICIs (Gebhardt et al., 2021). Additionally, the combination of sunitinib and tazemetostat lowers MDSC and Treg levels, enhances T-cell infiltration, and improves response to anti–PD-1 therapy, effectively overcoming ICI resistance in head and neck squamous cell carcinoma (HNSCC) (Liu et al., 2023). Similarly, combining sunitinib with chimeric antigen receptor T (CAR-T) therapy exhibits synergistic effects in renal cell carcinoma lung metastases by reducing MDSC frequency in the tumor microenvironment (Li et al., 2020).

11.5 New strategies for targeting myeloid-derived suppressor cells: Beyond depletion

While depleting MDSCs remains a critical approach, new strategies are emerging that aim to not only decrease MDSC numbers but also to reprogram their function to favor antitumor immunity (Jou et al., 2024). This "reprogramming" strategy involves converting MDSCs from immune-suppressive cells into immune-stimulating cells (Zhao et al., 2023). For instance, studies have shown that certain compounds, such as immune-modulatory cytokines (e.g., IL-12), can reprogram MDSCs to enhance their antitumor properties. These compounds have been shown to promote the MDSCs differentiation into antigen-presenting cells, improving T-cell priming and activation within the TME (Choi et al., 201). Additionally, the targeting of specific metabolic pathways that regulate MDSC function is an emerging area of interest. MDSCs rely heavily on metabolic reprogramming, such as the upregulation of glycolysis and arginine metabolism, to maintain their immunosuppressive function. Targeting these metabolic pathways through small molecules or metabolic inhibitors has shown promise in reducing MDSC activity and enhancing the effectiveness of cancer immunotherapies.

1. **Targeting metabolism of MDSCs**
   MDSCs amplify their immunosuppressive activities by boosting the uptake of free fatty acids and stimulating fatty acid oxidation, a process that plays a key role in tumor progression (Al-Khami et al., 2017). Liver X receptors, which regulate lipid homeostasis, play a key role in MDSC metabolism. LXR agonist treatment induces MDSCs apoptosis and reduces tumor volume, primarily through the apolipoprotein E (ApoE) activation. ApoE binds to MDSC receptors, leading to the depletion of these cells and suppression of tumor growth. Furthermore, MDSCs undergo reprogramming of their lipid metabolism in response to lipid accumulation and the associated signaling pathways (Ruan et al., 2024). The metabolic reprogramming of MDSCs also involves the upregulation of amino acid metabolic enzymes, crucial for their immunosuppressive functions. Glutamine, the most prevalent free amino acid, is a key target in this process. Small molecule inhibitors targeting glutamine metabolism, such as DRP-104 (a prodrug of the broad-spectrum glutamine antagonist DON), have demonstrated considerable antitumor effects. DRP-104 not only hampers tumor growth but also reduces MDSC recruitment and production. When used in conjunction with ICIs, it further amplifies antitumor activity by remodeling the tumor microenvironment, increasing infiltration of immune cells, and enhancing immune responses (Yokoyama et al., 2022).
2. **MDSC-targeted combination therapies with ICIs**
   Despite the effectiveness of immune checkpoint blockade (ICB) therapy, a significant number of cancer patients develop resistance, particularly in tumors characterized by a poorly immunogenic (cold) tumor microenvironment. MDSCs contribute to ICB resistance not only through PD–L1–dependent mechanisms but also through alternative pathways, including the secretion of immune-modulating cytokines and the suppression of T-cell priming (Weber et al., 2018). By combining MDSCs depletion or reprogramming with ICIs, these therapies have the potential to create a more favorable immune microenvironment, enhance the effectiveness of ICIs, and improve overall patient responses. Targeting tumor-associated macrophages (TAMs) and granulocytic MDSCs (G-MDSCs) effectively reduces their populations within the tumor microenvironment, significantly increasing the efficacy of ICB therapy, particularly anti–PD-1, in preclinical models of cholangiocarcinoma (Loeuillard et al., 2020). Several preclinical studies have demonstrated the success of combining ICB with MDSC depletion. Notably, the combination of epigenetic modulators such as entinostat and 5-azacytidine with ICB antibodies (anti–PD-1 and anti-CTLA-4) resulted in tumor reduction and prevented metastatic progression in a breast cancer model, achieving a survival rate exceeding 80% at 100 days of post-tumor implantation (Lu et al., 2020). Further, IL-6 cytokine, a key driver of MDSC accumulation, correlates with cancer progression and immunosuppression. IL-6/IL-6R blockade has been effective in preclinical models of hepatocellular carcinoma, colorectal

cancer, melanoma, and breast cancer, reducing MDSC levels and enhancing CD8+ T-cell infiltration (Weber et al., 2021). Targeting MDSCs alongside immune checkpoints appears to be the most effective strategy, as epigenetic modulators alone fail to induce robust antitumor immunity. Advances in epigenetic reprogramming further highlight the potential to convert MDSCs into an immunostimulatory phenotype, offering new avenues for overcoming resistance to therapies.

11.6 Innovative approaches: Engineering CAR-T cells

In the realm of tumor immunotherapy, the utilization of CAR-T cells arises as a crucial strategy, offering the potential to harness the power of the immune system to combat various malignancies. However, the success of these therapies is often hindered by the immunosuppressive tumor microenvironment, which is heavily influenced by the presence of MDSCs (Carcopino et al., 2024). Given the critical role that MDSCs play in the tumor microenvironment and their potential to hamper the efficacy of CAR-T cell therapies, targeting these cells has become an area of intense research (Abdalsalam et al., 2024). Targeting MDSCs offers a promising approach to enhance CAR-T therapy, particularly for solid tumors where traditional CAR-T approaches struggle. This includes reshaping the tumor microenvironment through small molecule inhibitors, antibodies, and modified NK cells, while improving precision medicine by identifying MDSC subsets for tumor-specific treatments. Early preclinical studies show that combining MDSC targeting with CAR-T therapy could significantly improve treatment efficacy for solid tumors (Duan et al., 2024; Wang et al., 2020).

Recent developments in CAR-T cells engineered to target and eliminate MDSCs in the TME offer a promising strategy to overcome immune suppression and enhance tumor targeting. The section below discusses various strategies for targeting MDSCs in CAR-T therapy in detail.

1. **Enhanced CAR-T cell expansion and persistence:** One of the biggest challenges in CAR-T therapy, for human cancers, is the limited expansion and persistence of CAR-T cells in the hostile tumor microenvironment. Strategies to enhance CAR-T survival, metabolic fitness, and long-term persistence are crucial for improving efficacy and are listed in Table 11.1. The suppression of CAR-T cells-mediated by MDSCs, governed by factors such as Arg-1, cysteine sequestration, IDO, and iNOS, can be reversed by depleting metabolic markers associated with MDSCs.
2. **Resistance to oxidative stress:** ROS and Peroxynitrites ($ONOO^-$) are key mediators of CAR-T cell dysfunction and apoptosis in the TME. Peroxynitrites ($ONOO^-$) nitrosylate tyrosine residues on the CAR receptor, impairing signal transduction and reducing T-cell activation (Szabó et al., 2007). High ROS levels disrupt calcium signaling, which is essential for T-cell activation and cytokine release. MDSCs and TAMs produce high levels of ROS (superoxide, hydrogen peroxide) and peroxynitrite, leading to oxidative damage in CAR-T cells. ROS and peroxynitrites promote T-cell exhaustion by inducing LAG-3, PD-1, and TIM-3 expression (Yang et al., 2024). Oxidative stress impairs mitochondrial ATP production, reducing CAR-T persistence.

Strategies to reduce oxidative stress can enhance CAR-T cell survival, persistence, and antitumor activity as mentioned below:

- **N-acetylcysteine (NAC):** NAC is a precursor for glutathione (GSH), a major antioxidant that neutralizes ROS, which enhances CAR-T cell persistence and function in oxidative stress conditions (Braidy et al., 2019).

Table 11.1 Strategies to enhance CAR-T cells survival driven by MDSC-mediated suppression.

S. no.	MDSCs-mediated suppression	Mechanism	Effect on CAR-T cells	Restoration via MDSCs depletion	References
1.	Arginase-1 (ARG1) mediated L-arginine depletion	MDSCs express ARG1, which hydrolyzes L-arginine into urea and ornithine, depleting it from the TME.	Low L-arginine inhibits TCR signaling, Impaired memory T cell formation, reducing CAR-T proliferation and cytotoxicity activity.	ARG1 inhibitors (eg. CB-1158), L-arginine supplementation	Geiger et al. (2016); Steggerda et al. (2017)
2.	Cysteine sequestration	MDSCs uptake extracellular cystine, convert it into cysteine, but do not release it for T cell use. As cysteine is essential for T cell glutathione production, which protects against oxidative stress.	Increased oxidative stress, apoptosis, and impaired T-cell activation.	N-acetylcysteine (NAC) supplementation, MDSC depletion	Srivastava et al. (2010)

(*continued on next page*)

3.	Indoleamine 2,3-Dioxygenase, IDO-mediated tryptophan depletion	MDSCs express IDO, which converts tryptophan into kynurenine, which induces T cell exhaustion via the AHR pathway. Also, low tryptophan levels inhibit mTOR signaling, blocking T cell proliferation. also	T-cell exhaustion, impaired function and apoptosis.	IDO inhibitors (e.g., epacadostat), tryptophan supplementation	Wu et al. (2018)
4.	Inducible Nitric Oxide Synthase (iNOS)	MDSCs express iNOS, producing large amounts of nitric oxide, which causes: CAR-T cell apoptosis via oxidative stress and Suppression of mitochondrial function reduces CAR-T energy production.	Impaired TCR signaling, limiting CAR-T activation.	iNOS inhibitors (e.g., L-NIL, 1400W)	Cartwright et al. (2021); Gao et al. (2019)

- **MitoTEMPO**: A mitochondrial-targeted antioxidant that protects CAR-T mitochondria from ROS-induced apoptosis (Hu & Li, 2016).
- **Superoxide dismutase (SOD) mimetics** (e.g., 1-oxyl-2,2,6,6-tetramethyl-4-hydroxypiperidine or Tempol): Neutralize superoxide radicals, reducing peroxynitrite formation (Marciniak et al., 2016; Wang et al., 2018).
- **Inhibiting NADPH oxidase (NOX2)**: NOX2 is a key enzyme in MDSCs that generates ROS. NOX2 inhibitors (e.g., diphenyleneiodonium, apocynin) reduce ROS levels, protecting CAR-T cells (Bode et al., 2023; Corzo et al., 2009).
- **Overexpression of catalase & SOD**: Enhances ROS neutralization, improving CAR-T survival (An et al., 2011).
- **Engineering CAR-T cells to express hypoxia-inducible factor-1α (HIF-1α)**: HIF-1α protects CAR-T cells from oxidative stress in the hypoxic TME (Liao et al., 2020).
- **Incorporating mitochondrial enhancements**: CAR-T cells can be engineered to enhance mitochondrial biogenesis, improving their resistance to metabolic stress. Overexpression of proliferator-activated receptor gamma coactivator 1-alpha (PGC-1α), a transcriptional coactivator that boosts mitochondrial health in tumor-infiltrating T cells and promotes antitumor responses (Dumauthioz et al., 2021; Harada et al., 2022).
- **Overexpressing glucose transporters:** Overexpressing glucose transporters such as GLUT3 and GLUT1 enhances T-cell glucose utilization and improves CAR-T metabolism in hypoglycemic TMEs. This survival advantage leads to accelerated T-cell activation and glycolysis (Sengupta et al., 2024).

1. **Improved CAR-T cell infiltration into tumors:** A major barrier to CAR-T therapy in solid tumors is the absence of efficient infiltration into the TME. Unlike hematologic malignancies, solid tumors create physical (dense extracellular matrix), biochemical (defective chemokine signaling & abnormal tumor vasculature), and immunosuppressive barriers that prevent CAR-T cells from reaching cancer cells (Guzman et al., 2023; Tahmasebi et al., 2019). Strategies to improve the trafficking, penetration, and retention of CAR-T cells are crucial for increasing therapeutic efficacy, as mentioned below:
   - **Chemokine receptor engineering for improved homing:** Tumors often downregulate T-cell-attracting chemokines (CXCL9, CXCL10, CCL5) while overexpressing T-cell excluding ones (CXCL12, CCL22) (Ozga et al., 2021). Engineering CAR-T cells with matching chemokine receptors can improve trafficking. Example: CXCR4-overexpressing CAR-T cells showed increased infiltration into pancreatic cancer, which has high CXCL12 expression (Sun et al., 2023).
   - **Cytokine engineering for CAR-T longevity:** One of the major challenges in CAR-T therapy is the limited persistence and exhaustion of CAR-T cells, particularly in solid tumors. Cytokine engineering by modifying CAR-T cells to self-produce cytokines has emerged as a promising strategy to enhance expansion, survival, and function without systemic toxicity. CAR-T cells are genetically engineered to express cytokines (IL-21, IL-7, IL-15) that enhance their survival and activity (Ghahri-Saremi et al., 2021). A novel approach utilizing modified CAR-T cells engineered with IL-15 has been developed to target IL-15 receptor alpha (IL-15Rα) expressed on human MDSCs and murine glioblastomas MDSCs. This dual-targeting strategy, achieved by fusing IL-15 to the antibody domain of CAR-T cells, effectively reduces MDSC and tumor

cell populations, significantly enhancing survival outcomes in GBM models (Zannikou et al., 2023).

- **Combining CAR-T therapy with vasculature normalization:** Tumor blood vessels are abnormal, hypoxic, and lack adhesion molecules (ICAM-1, VCAM-1), reducing T-cell extravasation. TNF-α or LIGHT-engineered CAR-T cells can increase ICAM-1 & VCAM-1 expression in tumor vessels, facilitating T-cell infiltration (Huang et al., 2020; Zhang et al., 2023). VEGF blockade combined with CAR-T therapy significantly increases the T cell infiltration and tumor regression in preclinical models. VEGF inhibitors (Bevacizumab, Aflibercept) normalize tumor vasculature, improving CAR-T trafficking (Bourhis et al., 2021).
- **Extracellular matrix (ECM)-degrading enzymes to enhance penetration:** Many solid tumors have a dense ECM composed of hyaluronic acid, collagen, and fibronectin, which blocks T cell entry. Engineering CAR-T cells to express ECM-degrading enzymes like hyaluronidase and collagenase enhances infiltration into tumors (Farrera-Sal et al., 2021). CAR-T cells expressing the enzyme heparanase that degrades heparan sulfate proteoglycans significantly increased infiltration into solid tumors (Caruana et al., 2015).
- **Targeting fibroblasts to reduce the tumor stroma:** Cancer-associated fibroblasts (CAFs) secrete ECM components that create T-cell-exclusion zones. Therefore, depleting or reprogramming CAFs by TGF-β blockade (Galunisertib, TGFβRII-DNR CAR-T cells) improves CAR-T infiltration (Li et al., 2023). CAR-T cells targeted by fibroblast activation protein-depleted CAFs and enhanced T cell penetration in fibrotic tumors. Reducing collagen deposition from CAFs by Angiotensin receptor blocker (Losartan) further enhances T cell access (Diop-Frimpong et al., 2011; Wang et al., 2019).
- **Neutralizing immunosuppressive barriers:** MDSCs and Tregs produce suppressive factors (IL-10, TGF-β, ARG1, iNOS) that block CAR-T infiltration (Yaseen et al., 2020). Depleting these suppressive cells enhances CAR-T trafficking. For example, CSF1R blockade combined with CAR-T therapy improved T-cell infiltration and tumor killing in pancreatic cancer models (Zhu et al., 2014). Various IDO inhibitors (Epacadostat, Indoximod) and CCR8-targeting antibodies deplete intratumoral Tregs, reducing suppression (He et al., 2025; Wen et al., 2025).
- **Overcoming immune checkpoints to promote CAR-T infiltration:** PD-1/PD-L1, LAG-3, and CTLA-4, suppress T-cell infiltration and function. Combination therapies with checkpoint inhibitors improve CAR-T trafficking. For example, PD-1 knockout CAR-T cells showed higher infiltration and longer persistence in cancer models (Ouyang et al., 2024). Anti-LAG-3 blockade or Anti–PD-1/CTLA-4 mAbs (Nivolumab, Ipilimumab) with CAR-T therapy increases tumor infiltration and enhances CAR-T recruitment in solid tumors (Menzies et al., 2022; Seidel et al., 2018; Zahm et al., 2021). MDSCs and tumor cells express Galectin-9, which binds TIM-3 on exhausted CAR-T cells (Yang et al., 2021); therefore galectin-9 inhibitors (GB0139, Belapectin) improve T cell expansion and function (Chalasani et al., 2020; Humphries et al., 2022). Clinical trials of anti-TIM-3 antibodies (Sabatolimab, TSR-022) alone and in combination with anti–PD-1 antibody in solid tumors show antitumor response (Clinical trial information: NCT02817633: NCT02608268) (Curigliano et al., 2021; Falchook et al., 2022).

2. **Targeting VISTA (V-Domain Ig Suppressor of T Cell Activation) to restore CAR-T activation:** VISTA is an immune checkpoint regulator that plays a key role in suppressing T-cell activation in the

TME. It is primarily expressed in MDSCs, TAMs, Tregs, and DCs. By maintaining STAT3 activity and polyamine production, VISTA stimulates MDSC differentiation and mitochondrial respiration (Zhang et al., 2024). VISTA functions both as a ligand and a receptor, exerting immunosuppressive effects by engaging T cells and myeloid cells, and promoting immune evasion in solid tumors (Huang et al., 2020). VISTA acts as a ligand on MDSCs and TAMs, binding to PSGL-1 or VSIG3 on T cells, and this interaction suppresses TCR signaling, cytokine secretion, and induces T-cell anergy (nonresponsiveness). VISTA signaling weakens CAR-T activation and expansion, reducing tumor-killing efficiency (Yuan et al., 2021). By maintaining STAT3 activity and polyamine production, VISTA stimulates MDSC differentiation and mitochondrial respiration. Therapeutic strategies to restore CAR-T activation have been discussed further.

- Anti-VISTA antibodies (CI-8993, VSTB112): Blocks VISTA signaling (Mehta et al., 2020; Thakkar et al., 2022).
- VISTA receptor inhibitors (PSGL-1, VSIG3 blockers): Prevent VISTA-induced suppression (Gao et al., 2024).
- VISTA KO CAR-T cells: Resists MDSC-mediated suppression (Deng et al., 2024).
- Combining VISTA blockade with other checkpoint inhibitors: Restores effector function in exhausted CAR-T cells (Kon & Benhar, 2019; Zhang et al., 2023).

3. **Increased tumor cytotoxicity by CAR-T cells:** Enhancing CAR-T cell cytotoxicity in solid tumors is crucial for overcoming tumor resistance and the immunosuppressive TME (Ai et al., 2024). MDSCs contribute to CAR-T cell apoptosis by expressing death ligands such as Fas Ligand (FasL) binds to Fas (CD95) on CAR-T cells, triggering apoptotic signaling and TRAIL (TNF-related apoptosis-inducing ligand), which activates death receptors (DR4/DR5), leading to caspase-dependent cell death (Condamine et al., 2014; Sinha et al., 2011; Tschumi et al., 2018). High MDSC levels increase FasL/TRAIL expression, promoting CAR-T depletion in the TME (McKenzie et al., 2023).

   Therapeutic strategies are as follows:

   - Lower MDSC levels reduce FasL/TRAIL-mediated apoptosis of CAR-T cells (Nalawade et al., 2021; Odoux et al., 2002).
   - Removing MDSCs allows higher levels of proinflammatory cytokines (IFN-γ, TNF-α, IL-2), boosting CAR-T cytotoxicity (Chen et al., 2021; Xiong et al., 2024).
   - Engineering CAR-T cells (Fas KO, DR4/DR5 KO) to resist FasL/TRAIL-induced death (Menegatti et al., 2024; Nalawade et al., 2021).
   - Overexpression of c-FLIP (Caspase-8 inhibitor) inhibits apoptotic signaling, enhancing CAR-T persistence (Safa, 2013; Yaacoub et al., 2024).
   - Blocking FasL/TRAIL Signaling in the TME by Anti-FasL antibodies and TRAIL inhibitors (Rossin et al., 2019; Qi et al., 2024).
   - **NK cell cytotoxicity:** Improving NK cell infiltration and expansion by CXCR4 inhibitors (e.g., plerixafor). HIF-1α inhibitors counteract hypoxia-induced NK suppression (Fanijavadi et al., 2025; Levy et al., 2015).

1. **Enhancing CAR-T cell memory formation for long-term tumor control:** Ensuring CAR-T cell therapy persists for a long time and avoiding exhaustion is one of the main concerns. Memory T cells, particularly T stem cell memory (Tscm) and central memory (Tcm) cells, play a crucial

role in prolonged immune surveillance and sustained tumor control (Liu et al., 2020). Enhancing CAR-T memory formation can significantly improve durability, expansion, and long-term remission. Therapeutic strategies include:

- **Optimizing T-cell subset composition before infusion:** Enriching for Tscm/Tcm before CAR transduction leads to better persistence and expansion. Supplementation of IL-7 & IL-15 enhances the frequency of $CD45RA^{+}$ $CD62L^{+}$ $CCR7^{+}$ T cells resembling Tscm/Tcm populations, increasing CAR-T cell persistence in cancer patients (Xu et al., 2014).
- **Metabolic reprogramming to favor memory cells:** Memory T cells prefer oxidative phosphorylation (OXPHOS) over glycolysis (Patsoukis et al., 2016). Therefore, inhibiting mTOR (rapamycin, metformin) promotes Tscm formation by shifting metabolism from glycolysis to OXPHOS (van der Windt & Pearce, 2012). Further, AMPK activation and reducing Akt/mTOR signaling increase fatty acid oxidation (FAO), favoring memory differentiation and preventing premature effector differentiation (Herzig & Shaw, 2018).
- **Genetic engineering to enhance memory & longevity:** Silencing exhaustion markers (e.g., TOX, NR4A) enhances memory potential and prevents dysfunction (Tantalo et al., 2021; Xiong et al., 2024). Overexpressing transcription factor T-cell factor 1, and antiapoptotic factor (BCL-2) maintain Tscm-like properties and improve survival by preventing apoptosis (Gounari & Khazaie, 2022; Wang et al., 2024).TET2 knockout CAR-T cells maintain stem-like properties, preventing exhaustion (Xiong et al., 2024). When combined with PD-1 inhibition, histone deacetylase (HDAC) inhibitors (entinostat, vorinostat) exhibit synergistic antitumor effects in tumor models by blocking Treg and MDSC function in the phase I study. (Clinical trial information: NCT02619253; NCT03024437) (Pili et al., 2019,2020).
- **Checkpoint blockade to preserve memory function:** Engineering CAR-T cells to express dominant-negative PD-1 receptors improves persistence (Chen et al., 2017). PD-1, LAG-3, or TIGIT blockade prevents T-cell exhaustion, allowing memory formation (Cillo et al., 2024; Jantz-Naeem et al., 2023).

2. **Combination of chemotherapy and CAR-T cells:** Chemotherapy can be strategically combined with CAR-T cell therapy to enhance CAR-T efficacy by selectively reducing MDSCs in the TME. Certain chemotherapies, such as gemcitabine and 5-fluorouracil (5-FU), have been shown to target and deplete MDSCs, improving T-cell function and CAR-T cell activity (Qi et al., 2024). Synergistic effects of chemotherapy and CAR-T therapy increase cytokine release (IL-2, IFN-γ) to boost CAR-T cell expansion and reduce the immunosuppressive TME (e.g., lower levels of TGF-β, IL-10), improving CAR-T cell recognition and killing efficiency (Al-Haideri et al., 2022).
3. **Employing bispecific T-cell engagers (BiTEs) to reduce MDSC immunosuppressive effects in the TME:** BiTEs are innovative immunotherapeutic agents designed to engage and activate T cells against tumor cells. These molecules are engineered to have two binding domains: One binds to a tumor-associated antigen (TAA) (e.g., CD19, HER2) and the other binds to CD3 on T cells, directly recruiting and activating T cells in proximity to the tumor (Tian et al., 2021). BiTEs targeting MDSC-specific markers (e.g., CD33, CD11b, PD-L1) can redirect T cells to directly eliminate MDSCs in the TME. This reduces the immunosuppressive population of MDSCs, enabling CAR-T cells to function more effectively. In addition to targeting tumor cells, BiTEs have the potential to reduce MDSC-mediated immune suppression in the TME and improve the overall efficacy of CAR-T cell

therapy (Abdalsalam et al., 2024; Horn et al., 2017). The clinical applications and ongoing studies on BiTEs are discussed further.

- MDSC-specific BiTEs: Novel BiTEs targeting MDSC-specific markers like CD33 or PD-L1 are in preclinical development. These aim to deplete or block MDSCs and improve CAR-T cell efficacy in solid tumors (Jitschin et al., 2018).
- Blinatumomab (Blincyto): A CD19 × CD3 BiTE approved for acute lymphoblastic leukemia demonstrated the potential to deplete MDSCs and enhance CAR-T cell efficacy in hematologic cancers (Mirfakhraie et al., 2024).
- BiTEs targeting solid tumors: Ongoing clinical trials are evaluating HER2 × CD3 BiTEs for breast and gastric cancers, with preliminary results showing reduced MDSC levels and enhanced CAR-T cell activity (Lum et al., 2021).

4. **Targeting MDSCs to enhance NK cell-based immunotherapy:** NK cells play a crucial role in antitumor immunity by directly killing cancer cells without prior antigen sensitization. However, in the TME, MDSCs create an immunosuppressive barrier that significantly reduces NK cell activity, leading to poor therapeutic outcomes (Nie et al., 2024). Targeting MDSCs can overcome NK cell suppression and enhance NK-based immunotherapies, including:
   - **NK cell adoptive transfer:** It involves the ex vivo expansion and infusion of highly cytotoxic NK cells to attack tumors. For example, IL-15 superagonists (e.g., N-803), and mTOR activators (e.g., rapamycin modulation) boost NK survival and proliferation (Knudson et al., 2019; Panwar et al., 2023).
   - **CAR-NK therapy:** Engineered NK cells expressing IL-15 for self-sustaining expansion significantly prolonged survival in preclinical leukemia models (Imamura et al., 2014).
   - **NK cell-engaging antibodies**: NK cell-engaging bispecific antibodies (BiKEs) and trispecific antibodies (TriKEs) are next-generation immunotherapies designed to enhance NK cell cytotoxicity against tumors. These molecules recruit NK cells to cancer cells by simultaneously binding and activating NK receptors and TAAs, improving immune synapse formation, tumor killing, and NK persistence (Phung et al., 2021). For example, CD16 × CD19 BiKE has demonstrated its therapeutic promise for improving NK cell immunotherapies for leukemias and lymphomas by effectively raising NK cell cytolytic activity and cytokine production against tumor targets (Gleason et al., 2012). CD16 × CD33 × IL-15 TriKE overcomes immune suppression for AML, showing improved NK function (Felices et al., 2016).
   - **Restoring, preventing NK cell exhaustion via checkpoint blockade:** PD-L1 blockade with NK cell infusion prevents MDSC-mediated immune suppression (Hsu et al., 2018).

11.7 MDSC-modulating vaccines to promote immune activation

MDSC-modulating vaccines represent a promising strategy to enhance immune activation and combat immunosuppressive effects in the TME. These vaccines are designed to target and reprogram MDSCs, reducing their suppressive effects on immune cells, including T cells and CAR-T cells, and improving overall antitumor immunity. By activating anti-Treg with peptide-based vaccinations, MDSCs can be depleted, and the progression of the disease may be impeded (Bentivegna et al., 2022). MDSC-targeting vaccines are still in the early stages of clinical development but show promise in enhancing immune activation and promoting durable antitumor immunity in both solid and hematologic cancers. Combination strategies using MDSC-modulating vaccines with immunotherapies (e.g., checkpoint inhibitors, CAR-T cells) are being actively explored in clinical trials (Salewski et al., 2021). The combination of these approaches could significantly improve the clinical efficacy of immunotherapies in cancers with a high MDSC burden (Liechtenstein et al., 2014).

11.8 Clinical trials and future directions

Various ongoing clinical trials are investigating the effectiveness of MDSC-targeted therapies in combination with other immunotherapeutic strategies, such as ICIs, CAR-T cell therapy, and vaccines. These studies aim to determine the optimal strategies for targeting MDSCs in different cancer types and stages and evaluate the safety and effectiveness of these strategies in the clinic. Early-phase trials have shown encouraging results, suggesting that targeting MDSCs may enhance the overall response to immunotherapy and improve patient outcomes. The HDAC inhibitor entinostat has been shown to disrupt the pre-metastatic niche by promoting the differentiation of MDSCs into macrophages (Lu et al., 2020). Although the limited therapeutic efficacy of entinostat has been observed in some clinical trials (NCT01207726, NCT02115282, NCT01132573), combining it with ICIs has been shown to enhance antitumor responses. This combination enhances granzyme B+ CD8+ T-cell infiltration and functionality, resulting in tumor regression and extended tumor-free survival in HER2-transgenic breast cancer and Panc02 metastatic pancreatic cancer mouse models (Christmas et al., 2018). Clinical trials (NCT02403778, NCT03200847) have highlighted the potential of ATRA in cancer therapy, either alone or in combination with ICB or targeted anticancer drugs. ATRA has been shown to downregulate immunosuppressive genes such as PD-L1 and IL-10 in MDSCs. Notably, its addition to standard ipilimumab therapy demonstrated a favorable safety profile, suggesting its potential to enhance antitumor immune responses (Tobin et al., 2018).

Moreover, the application of advanced technologies such as mass cytometry and single-cell RNA sequencing will allow for a more detailed understanding of MDSC heterogeneity and function in different tumor contexts (Jiang et al., 2023). Single-cell RNA sequencing analysis revealed that SPP1+ TAMs and MDSCs are the predominant myeloid populations in HNSC, with high MDSC infiltration linked to poor overall survival. A six-gene MDSC-related prognostic signature (FTH1, RTN4, ALDOA, SLC2A3, CD52, and TNFAIP6) effectively predicts patient outcomes. CellChat analysis indicated frequent interactions between MDSCs, SPP1+ TAMs, CD8+ T cells, and endothelial cells, contributing to immune suppression and tumor angiogenesis. These findings highlight MDSCs as a promising target for improving HNSC immunotherapy. This will enable the identification of biomarkers for more accurate patient stratification and the development of personalized therapies targeting MDSCs. Recently, T Cells Redirected for Universal Cytokine Killing (TRUCKs), also known as fourth-generation CAR-T cells, have been designed to enhance antitumor efficacy by delivering cytokines directly within the TME. This approach improves T-cell function, reverses immunosuppression, and minimizes systemic toxicity associated with traditional cytokine therapy. It also leads to the recruitment of innate immune cells and converts "cold" tumors into "hot" tumors, attracting DCs, macrophages, and NK cells (Chmielewski & Abken, 2020). Further, SUPRA CAR (Split, Universal, Programmable, and Regulated) systems allow on-demand cytokine control for precise tuning of CAR-T activity. It uses a switchable cytokine domain to activate or suppress cytokine signaling in response to tumor burden. This will prevent overactivation and toxicity by dynamically adjusting cytokine release (Cho et al., 2018). Example: IL-2-tethered SUPRA CAR-T cells enhanced antitumor effects while reducing systemic toxicity (Brog et al., 2022). Synthetic Cytokine Receptors (SynNotch) CAR-T cells express synthetic receptors that activate only when they encounter tumor antigens, inducing local cytokine release. It prevents systemic toxicity by restricting cytokine secretion to the tumor site and allows precise immune modulation to counteract the TME (Hyrenius-Wittsten et al., 2021). Example: IL-12-expressing SynNotch CAR-Ts reversed Treg-mediated suppression in tumors, preserving peripheral immune balance (Luo et al., 2019). Furthermore, ECM-degrading enzymes are secreted locally at the tumor site by CAR-T cells expressing SynNotch receptors. Efficient ECM recognition by SynNotch CAR-T cells allows for the breakdown of tumor ECM components (Zheng et al., 2024).

The complex nature of MDSCs presents significant challenges in targeting them effectively, but emerging strategies that combine MDSC depletion or reprogramming with ICB and other immunotherapeutic approaches offer promising avenues to overcome these challenges. By understanding the multifaceted roles of MDSCs and their interactions with the immune system, we can develop more precise and effective cancer immunotherapies that improve patient outcomes, combat tumor relapse, and address the challenges associated with cancer immunotherapy. Future research will likely focus on refining the molecular mechanisms underlying MDSC differentiation and function, further optimizing combination therapies, and integrating MDSC targeting into broader cancer immunotherapy regimens to increase the potential for durable, long-term patient responses.

11.9 Conclusion

MDSCs represent a critical barrier to effective anticancer immunity, contributing to tumor progression, therapeutic resistance, and immune evasion. Their ability to suppress T-cell responses, enhance angiogenesis, and support cancer stemness makes them a compelling target in immunotherapy. Despite the promise of MDSC-targeted approaches, challenges remain due to their phenotypic heterogeneity and plasticity across tumor types. Recent advances, including the combination of MDSC depletion with ICB, CAR-T cell therapy, and tumor vaccines, offer promising avenues to enhance antitumor responses. However, translating these strategies into clinically effective therapies requires a deeper understanding of MDSC biology, standardized characterization methods, and precise targeting approaches. Future research should focus on refining combinatorial strategies, identifying reliable MDSC biomarkers, and exploring novel immunotherapeutic interventions to overcome MDSC-mediated resistance. By addressing these challenges, MDSC-targeted therapies have the potential to significantly improve the outcomes of cancer treatment and reshape the landscape of modern immunotherapy.

Abbreviations

5-FU	5-Fluorouracil
ARG1	Arginase-1

BiKEs	Bispecific Killer Engagers
BiTEs	Bispecific T-cell Engagers
CAFs	Cancer Associated Fibroblasts
CAR-T	Chimeric Antigen Receptor T
CMPs	Common Myeloid Progenitors
DC	Dendritic Cells
DR	Death Receptors
FAO	Fatty Acid Oxidation
FFA	Free Fatty Acids
HIF-1α	Hypoxia Inducible Factor-1alpha
HSCs	Hematopoietic Stem Cells
ICB	Immune Checkpoint Blockade
ICIs	Immune Checkpoint Inhibitors
IDO	Indoleamine 2,3-Dioxygenase
IMC	Immature Myeloid Cells
iNOS	Inducible Nitric Oxide Synthase
MDSCs	Myeloid-Derived Suppressor Cells
NAC	N-acetylcysteine
NK	Natural Killer
NO	Nitric Oxide
OS	Overall Survival
OXPHOS	Oxidative Phosphorylation
PD-L1	Programmed Death Ligand 1
PFS	Progression Free survival
PGC-1α	Proliferator-Activated Receptor Gamma Coactivator 1-alpha
ROS	Reactive Oxygen Species
SOD	Superoxide Dismutase
SUPRA	Split, Universal, Programmable, and Regulated
SynNotch	Synthetic Notch
TAA	Tumor-Associated Antigen
TAMs	Tumor-Associated Macrophages
TCF-1	Transcription Factor T cell factor 1
Tcm	T Central Memory
TLR	Toll-Like Receptor
TME	Tumor Microenvironment
TRAIL	TNF-Related Apoptosis Inducing Ligand
TriKEs	Trispecific Killer Engagers
TRUCKs	T Cells Redirected for Universal Cytokine Killing
Tscm	T Stem Cell Memory
VISTA	V-Domain Ig Suppressor of T Cell Activation

References

Abdalsalam, N. M. F., Ibrahim, A., Saliu, M. A., Liu, T. M., Wan, X., & Yan, D. (2024). MDSC: A new potential breakthrough in CAR-T therapy for solid tumors. *Cell Communication and Signaling, 22*(1), 612.

Ai, K., Liu, B., Chen, X., Huang, C., Yang, l., Zhang, W., et al. (2024). Optimizing CAR-T cell therapy for solid tumors: Current challenges and potential strategies. *Journal of Hematology & Oncology, 17*(1), 105.

Al-Haideri, M., Tondok, S. B., Safa, S. H., Maleki, A. H., Rostami, S., Jalil, A. T., et al. (2022). CAR-T cell combination therapy: The next revolution in cancer treatment. *Cancer cell International, 22*(1), 365.

Al-Khami, A. A., Zheng, L., Del Valle, L., Hossain, F., Wyczechowska, D., Zabaleta, J., et al. (2017). Exogenous lipid uptake induces metabolic and functional reprogramming of tumor-associated myeloid-derived suppressor cells. *Oncoimmunology, 6*(10), e1344804.

An, H., Zhai, Z., Yin, S., Luo, Y., Han, B., & Hao, Y. (2011). Coexpression of the superoxide dismutase and the catalase provides remarkable oxidative stress resistance in *Lactobacillus rhamnosus. Journal of Agricultural and Food Chemistry, 59*(8), 3851–3856.

Bentivegna, S., Holmstrøm, M., Balhuizen, A., Almosailleakh, M., Perez-Penco, M., Porse, B., et al. (2022). Therapeutic potential of an immune-modulatory vaccination targeting myeloid-derived suppressor cells in myelodysplastic syndrome. *Blood, 140*(Supplement 1), 4007–4008.

Binsfeld, M., Muller, J., Lamour, V., De Veirman, K., De Raeve, H., Bellahcène, A., et al. (2016). Granulocytic myeloid-derived suppressor cells promote angiogenesis in the context of multiple myeloma. *Oncotarget, 7*(25), 37931.

Bode, K., Hauri, M., Jaquet, V., & Weyd, H. (2023). Unlocking the power of NOX2: A comprehensive review on its role in immune regulation. *Redox Biology, 64*, 102795.

Bourhis, M., Palle, J., Galy-Fauroux, I., & Terme, M. (2021). Direct and indirect modulation of T cells by VEGF-A counteracted by anti-angiogenic treatment. *Frontiers in Immunology, 12*, 616837.

Braidy, N., Zarka, M., Jugder, B. E., Welch, J., Jayasena, T., Chan, D. K. Y., et al. (2019). The precursor to glutathione (GSH), γ-Glutamylcysteine (GGC), can ameliorate oxidative damage and neuroinflammation induced by Aβ(40) oligomers in human astrocytes. *Frontiers in Aging Neuroscience, 11*, 177.

Bray, F., Laversanne, M., Sung, H., Ferlay, J., Siegel, R. L., Soerjomataram, I., et al. (2024). Global cancer statistics 2022: GLOBOCAN estimates of incidence and mortality worldwide for 36 cancers in 185 countries. *CA: A Cancer Journal for Clinicians, 74*(3), 229–263.

Brog, R. A., Ferry, S. L., Schiebout, C. T., Messier, C. M., Cook, W. J., Abdullah, L., et al. (2022). Superkine IL-2 and IL-33 armored CAR T cells reshape the tumor microenvironment and reduce growth of multiple solid tumors. *Cancer Immunology Research, 10*(8), 962–977.

Bronte, V., Brandau, S., Chen, S.-H., Colombo, M. P., Frey, A. B., Greten, T. F., et al. (2016). Recommendations for myeloid-derived suppressor cell nomenclature and characterization standards. *Nature Communications, 7*(1), 12150.

Carcopino, C., Erdogan, E., Henrich, M., & Kobold, S. (2024). Armoring chimeric antigen receptor (CAR) T cells as micropharmacies for cancer therapy. *Immuno-Oncology and Technology, 24*, 100739.

Cartwright, A. N. R., Suo, S., Badrinath, S., Kumar, S., Melms, J., Luoma, A., et al. (2021). Immunosuppressive myeloid cells induce nitric oxide-dependent DNA damage and p53 pathway activation in CD8(+) T cells. *Cancer immunology research, 9*(4), 470–485.

Caruana, I., Savoldo, B., Hoyos, V., Weber, G., Liu, H., Kim, E. S., et al. (2015). Heparanase promotes tumor infiltration and antitumor activity of CAR-redirected T lymphocytes. *Nature Medicine, 21*(5), 524–529.

Chalasani, N., Abdelmalek, M. F., Garcia-Tsao, G., Vuppalanchi, R., Alkhouri, N., Rinella, M., et al. (2020). Effects of belapectin, an inhibitor of galectin-3, in patients with nonalcoholic steatohepatitis with cirrhosis and portal hypertension. *Gastroenterology, 158*(5), 1334–1345.e5.

Chen, J., Sun, H.-W., Yang, Y.-Y., Chen, H.-T., Yu, X.-J., Wu, W.-C., et al. (2021). Reprogramming immunosuppressive myeloid cells by activated T cells promotes the response to anti-PD-1 therapy in colorectal cancer. *Signal Transduction and Targeted Therapy, 6*(1), 4.

Chen, N., Morello, A., Tano, Z., & Adusumilli, P. S. (2017). CAR T-cell intrinsic PD-1 checkpoint blockade: A two-in-one approach for solid tumor immunotherapy. *OncoImmunology, 6*(2), e1273302.

Chmielewski, M., & Abken, H. (2020). TRUCKS, the fourth-generation CAR T cells: Current developments and clinical translation. *Advances in Cell and Gene Therapy, 3*(3), e84.

Cho, J. H., Collins, J. J., & Wong, W. W. (2018). Universal chimeric antigen receptors for multiplexed and logical control of T cell responses. *Cell, 173*(6), 1426–1438.e11.

Choi, J.-N., Sun, E. G., & Cho, S.-H. (2019). IL-12 enhances immune response by modulation of myeloid derived suppressor cells in tumor microenvironment. *Chonnam Medical Journal, 55*(1), 31–39.

Christmas, B. J., Rafie, C. I., Hopkins, A. C., Scott, B. A., Ma, H. S., Cruz, K. A., et al. (2018). Entinostat converts immune-resistant breast and pancreatic cancers into checkpoint-responsive tumors by reprogramming tumor-infiltrating MDSCs. *Cancer Immunology Research, 6*(12), 1561–1577.

Cillo, A. R., Cardello, C., Shan, F., Karapetyan, L., Kunning, S., Sander, C., et al. (2024). Blockade of LAG-3 and PD-1 leads to co-expression of cytotoxic and exhaustion gene modules in $CD8^{+}$T cells to promote antitumor immunity. *Cell, 187*(16), 4373–4388.e15.

Condamine, T., Dominguez, G. A., Youn, J.-I., Kossenkov, A. V., Mony, S., Alicea-Torres, K., et al. (2016). Lectin-type oxidized LDL receptor-1 distinguishes population of human polymorphonuclear myeloid-derived suppressor cells in cancer patients. *Science Immunology, 1*(2), aaf8943–aaf.

Condamine, T., Kumar, V., Ramachandran, I. R., Youn, J. I., Celis, E., Finnberg, N., et al. (2014). ER stress regulates myeloid-derived suppressor cell fate through TRAIL-R-mediated apoptosis. *Journal of Clinical Investigation, 124*(6), 2626–2639.

Corzo, C., Cotter, M., Cheng, P., Cheng, F., Kusmartsev, S., Sotomayor, E., et al. (2009). Mechanism regulating reactive oxygen species in tumor-induced myeloid-derived suppressor cells. *Journal of Immunology (Baltimore, Md: 1950), 182*, 5693–5701.

Curigliano, G., Gelderblom, H., Mach, N., Doi, T., Tai, D., Forde, P. M., et al. (2021). Phase I/Ib clinical trial of sabatolimab, an anti–TIM-3 antibody, alone and in combination with spartalizumab, an anti–PD-1 antibody, in advanced solid tumors. *Clinical Cancer Research, 27*(13), 3620–3629.

De Cicco, P., Ercolano, G., & Ianaro, A. (2020). The new era of cancer immunotherapy: targeting myeloid-derived suppressor cells to overcome immune evasion. *Frontiers in Immunology, 11*, 1680.

Deng, Y., Shi, M., Yi, L., Naveed Khan, M., Xia, Z., & Li, X. (2024). Eliminating a barrier: Aiming at VISTA, reversing MDSC-mediated T cell suppression in the tumor microenvironment. *Heliyon, 10*(17), e37060.

Diop-Frimpong, B., Chauhan, V. P., Krane, S., Boucher, Y., & Jain, R. K. (2011). Losartan inhibits collagen I synthesis and improves the distribution and efficacy of nanotherapeutics in tumors. *PNAS, 108*(7), 2909–2914.

Duan, X.-P., Qin, B.-D., Jiao, X.-D., Liu, K., Wang, Z., & Zang, Y.-S. (2024). New clinical trial design in precision medicine: Discovery, development and direction. *Signal Transduction and Targeted Therapy, 9*(1), 57.

Dumauthioz, N., Tschumi, B., Wenes, M., Marti, B., Wang, H., Franco, F., et al. (2021). Enforced PGC-1α expression promotes CD8 T cell fitness, memory formation and antitumor immunity. *Cell Molecular Immunology, 18*(7), 1761–1771.

Falchook, G. S., Ribas, A., Davar, D., Eroglu, Z., Wang, J. S., Luke, J. J., et al. (2022). Phase 1 trial of TIM-3 inhibitor cobolimab monotherapy and in combination with PD-1 inhibitors nivolumab or dostarlimab (AMBER). *Journal of Clinical Oncology, 40*(16_suppl), 2504.

Fanijavadi, S., Thomassen, M., & Jensen, L. H. (2025). Targeting triple NK cell suppression mechanisms: A comprehensive review of biomarkers in pancreatic cancer therapy. *International Journal of Molecular Sciences, 26*(2), 515.

Farrera-Sal, M., Moreno, R., Mato-Berciano, A., Maliandi, M. V., Bazan-Peregrino, M., & Alemany, R. (2021). Hyaluronidase expression within tumors increases virotherapy efficacy and T cell accumulation. *Molecular Therapy Oncolytics, 22*, 27–35.

Felices, M., Sarhan, D., Brandt, L., Guldevall, K., McElmurry, R., Lenvik, A., et al. (2016). CD16-IL15-CD33 trispecific killer engager (TriKE) overcomes cancer-induced immune suppression and induces natural killer cell-mediated control of MDS and AML via enhanced killing kinetics. *Blood, 128*(22), 4291.

Filipazzi, P., Huber, V., & Rivoltini, L. (2012). Phenotype, function and clinical implications of myeloid-derived suppressor cells in cancer patients. *Cancer Immunology, Immunotherapy, 61*, 255–263.

Flores-Toro, J. A., Luo, D., Gopinath, A., Sarkisian, M. R., Campbell, J. J., Charo, I. F., et al. (2020). CCR2 inhibition reduces tumor myeloid cells and unmasks a checkpoint inhibitor effect to slow progression of resistant murine gliomas. *Proceedings of the National Academy of Sciences, 117*(2), 1129–1138.

Gao, Y., He, Y., Tang, Y., Chen, Z.-S., & Qu, M. (2024). VISTA: A novel checkpoint for cancer immunotherapy. *Drug Discovery Today, 29*(7), 104045.

Gao, Y., Zhou, S., Xu, Y., Sheng, S., Qian, S. Y., & Huo, X. (2019). Nitric oxide synthase inhibitors 1400W and L-NIO inhibit angiogenesis pathway of colorectal cancer. *Nitric Oxide, 83*, 33–39.

Gebhardt, C., Simon, S. C., Weber, R., Gries, M., Mun, D. H., Reinhard, R., et al. (2021). Potential therapeutic effect of low-dose paclitaxel in melanoma patients resistant to immune checkpoint blockade: A pilot study. *Cellular Immunology, 360*, 104274.

Geiger, R., Rieckmann, J. C., Wolf, T., Basso, C., Feng, Y., Fuhrer, T., et al. (2016). L-arginine modulates T cell metabolism and enhances survival and anti-tumor activity. *Cell, 167*(3), 829–842.e13.

Ghahri-Saremi, N., Akbari, B., Soltantoyeh, T., Hadjati, J., Ghassemi, S., & Mirzaei, H. R. (2021). Genetic modification of cytokine signaling to enhance efficacy of CAR T cell therapy in solid tumors. *Frontiers in Immunology, 12*, 738456.

Ghiringhelli, F., & Rébé, C. (2024). Using immunogenic cell death to improve anticancer efficacy of immune checkpoint inhibitors: From basic science to clinical application. *Immunological Reviews, 321*(1), 335–349.

Gleason, M. K., Verneris, M. R., Todhunter, D. A., Zhang, B., McCullar, V., Zhou, S. X., et al. (2012). Bispecific and trispecific killer cell engagers directly activate human NK cells through CD16 signaling and induce cytotoxicity and cytokine production. *Molecular Cancer Therapeutics, 11*(12), 2674–2684.

Gounari, F., & Khazaie, K. (2022). TCF-1: A maverick in T cell development and function. *Nature Immunology, 23*(5), 671–678.

Guzman, G., Reed, M. R., Bielamowicz, K., Koss, B., & Rodriguez, A. (2023). CAR-T therapies in solid tumors: Opportunities and challenges. *Current Oncology Reports, 25*(5), 479–489.

Harada, S., Hashimoto, D., Senjo, H., Yoneda, K., Zhang, Z., Chen, X., et al. (2022). Intercellular mitochondrial transfer enhances metabolic fitness and anti-tumor effects of CAR T cells. *Blood, 140*(Supplement 1), 2356–2357.

He, S., Zheng, L., & Qi, C. (2025). Myeloid-derived suppressor cells (MDSCs) in the tumor microenvironment and their targeting in cancer therapy. *Molecular cancer, 24*(1), 5.

Herzig, S., & Shaw, R. J. (2018). AMPK: guardian of metabolism and mitochondrial homeostasis. *Nature Reviews Molecular Cell Biology, 19*(2), 121–135.

Hong, E. H., Chang, S. Y., Lee, B. R., Kim, Y. S., Lee, J. M., Kang, C. Y., et al. (2013). Blockade of Myd88 signaling induces antitumor effects by skewing the immunosuppressive function of myeloid-derived suppressor cells. *International Journal of Cancer, 132*(12), 2839–2848.

Horikawa, N., Abiko, K., Matsumura, N., Baba, T., Hamanishi, J., Yamaguchi, K., et al. (2020). Anti-VEGF therapy resistance in ovarian cancer is caused by GM-CSF-induced myeloid-derived suppressor cell recruitment. *British Journal of Cancer, 122*(6), 778–788.

Horn, L. A., Ciavattone, N. G., Atkinson, R., Woldergerima, N., Wolf, J., Clements, V. K., et al. (2017). CD3xPDL1 bi-specific T cell engager (BiTE) simultaneously activates T cells and NKT cells, kills PDL1(+) tumor cells, and extends the survival of tumor-bearing humanized mice. *Oncotarget, 8*(35), 57964–57980.

Hsu, J., Hodgins, J. J., Marathe, M., Nicolai, C. J., Bourgeois-Daigneault, M. C., Trevino, T. N., et al. (2018). Contribution of NK cells to immunotherapy mediated by PD-1/PD-L1 blockade. *Journal of Clinical Investigation, 128*(10), 4654–4668.

Hu, H., & Li, M. (2016). Mitochondria-targeted antioxidant mitotempo protects mitochondrial function against amyloid beta toxicity in primary cultured mouse neurons. *Biochemical and Biophysical Research Communications, 478*(1), 174–180.

Huang, X., Zhang, X., Li, E., Zhang, G., Wang, X., Tang, T., et al. (2020). VISTA: an immune regulatory protein checking tumor and immune cells in cancer immunotherapy. *Journal of Hematology & Oncology, 13*(1), 83.

Huang, Z., Wu, Y., Allen, M. E., Pan, Y., Kyriakakis, P., Lu, S., et al. (2020). Engineering light-controllable CAR T cells for cancer immunotherapy. *Science Advances, 6*(8), eaay9209.

Humphries, D. C., Mills, R., Boz, C., McHugh, B. J., Hirani, N., Rossi, A. G., et al. (2022). Galectin-3 inhibitor GB0139 protects against acute lung injury by inhibiting neutrophil recruitment and activation. *Frontiers in Pharmacology, 13*, 949264.

Hyrenius-Wittsten, A., Su, Y., Park, M., Garcia, J. M., Alavi, J., Perry, N., et al. (2021). SynNotch CAR circuits enhance solid tumor recognition and promote persistent antitumor activity in mouse models. *Science Translational Medicine, 13*(591), eabd8836.

Imamura, M., Shook, D., Kamiya, T., Shimasaki, N., Chai, S. M. H., Coustan-Smith, E., et al. (2014). Autonomous growth and increased cytotoxicity of natural killer cells expressing membrane-bound interleukin-15. *Blood, 124*(7), 1081–1088.

Jantz-Naeem, N., Böttcher-Loschinski, R., Borucki, K., Mitchell-Flack, M., Böttcher, M., Schraven, B., et al. (2023). TIGIT signaling and its influence on T cell metabolism and immune cell function in the tumor microenvironment. *Frontiers in Oncology., 13*, 1060112.

Jiang, W., Hu, K., Liu, X., Gao, J., & Zhu, L. (2023). Single-cell transcriptome analysis reveals the clinical implications of myeloid-derived suppressor cells in head and neck squamous cell carcinoma. *Pathology and Oncology Research, 29*, 1611210.

Jitschin, R., Saul, D., Braun, M., Tohumeken, S., Völkl, S., Kischel, R., et al. (2018). CD33/CD3-bispecific T-cell engaging (BiTE®) antibody construct targets monocytic AML myeloid-derived suppressor cells. *Journal for ImmunoTherapy of Cancer, 6*(1), 116.

Jou, E., Chaudhury, N., & Nasim, F. (2024). Novel therapeutic strategies targeting myeloid-derived suppressor cell immunosuppressive mechanisms for cancer treatment. *Exploration of Targeted Anti-tumor Therapy, 5*(1), 187.

Karpisheh, V., Nikkhoo, A., Hojjat-Farsangi, M., Namdar, A., Azizi, G., Ghalamfarsa, G., et al. (2019). Prostaglandin E2 as a potent therapeutic target for treatment of colon cancer. *Prostaglandins & Other Lipid Mediators, 144*, 106338.

Khaled, Y. S., Ammori, B. J., & Elkord, E. (2014). Increased levels of granulocytic myeloid-derived suppressor cells in peripheral blood and tumour tissue of pancreatic cancer patients. *Journal of Immunology Research, 2014*(1), 879897.

Knudson, K. M., Hicks, K. C., Alter, S., Schlom, J., & Gameiro, S. R. (2019). Mechanisms involved in IL-15 superagonist enhancement of anti-PD-L1 therapy. *Journal for ImmunoTherapy of Cancer, 7*(1), 82.

Kon, E., & Benhar, I. (2019). Immune checkpoint inhibitor combinations: Current efforts and important aspects for success. *Drug Resistance Updates, 45*, 13–29.

Kulbersh, J. S., Day, T. A., Gillespie, M. B., & Young, M. R. I. (2009). 1α, 25-Dihydroxyvitamin D3 to skew intratumoral levels of immune inhibitory CD34+ progenitor cells into dendritic cells. *Otolaryngology—Head and Neck Surgery, 140*(2), 235–240.

Kumar, V., Patel, S., Tcyganov, E., & Gabrilovich, D. I. (2016). The nature of myeloid-derived suppressor cells in the tumor microenvironment. *Trends in Immunology, 37*(3), 208–220.

Levy, E. R., Reger, R. N., Carlsten, M., & Childs, R. W. (2015). mRNA transfection of NK cells with gain-of-function CXCR4 as a novel method to enhance the homing of adoptively transferred NK Cells to the bone marrow for the treatment of hematological malignancies. *Blood, 126*(23), 3089.

Li, H., Ding, J., Lu, M., Liu, H., Miao, Y., Li, L., et al. (2020). CAIX-specific CAR-T cells and sunitinib show synergistic effects against metastatic renal cancer models. *Journal of Immunotherapy, 43*(1), 16–28.

Li, M., Wu, B., Li, L., Lv, C., & Tian, Y. (2023). Reprogramming of cancer-associated fibroblasts combined with immune checkpoint inhibitors: A potential therapeutic strategy for cancers. *Biochimica et Biophysica Acta (BBA) - Reviews on Cancer, 1878*(5), 188945.

Li, R., Zeng, X., Yang, M., Feng, J., Xu, X., Bao, L., et al. (2021). Antidiabetic DPP-4 inhibitors reprogram tumor microenvironment that facilitates murine breast cancer metastasis through interaction with cancer cells via a ROS–NF-κB–NLRP3 axis. *Frontiers in oncology, 11*, 728047.

Li, X., Luo, X., Chen, S., Chen, J., Deng, X., Zhong, J., et al. (2023). All-trans-retinoic acid inhibits hepatocellular carcinoma progression by targeting myeloid-derived suppressor cells and inhibiting angiogenesis. *International Immunopharmacology, 121*, 110413.

Liao, Q., He, H., Mao, Y., Ding, X., Zhang, X., & Xu, J. (2020). Engineering T cells with hypoxia-inducible chimeric antigen receptor (HiCAR) for selective tumor killing. *Biomarker Research, 8*(1), 56.

Liechtenstein, T., Perez-Janices, N., Blanco-Luquin, I., Goyvaerts, C., Schwarze, J., Dufait, I., et al. (2014). Anti-melanoma vaccines engineered to simultaneously modulate cytokine priming and silence PD-L1 characterized using ex vivo myeloid-derived suppressor cells as a readout of therapeutic efficacy. *Oncoimmunology, 3*(7), e945378.

Liu, J., Lin, W.-P., Su, W., Wu, Z.-Z., Yang, Q.-C., Wang, S., et al. (2023). Sunitinib attenuates reactive MDSCs enhancing anti-tumor immunity in HNSCC. *International Immunopharmacology, 119*, 110243.

Liu, Q., Sun, Z., & Chen, L. (2020). Memory T cells: Strategies for optimizing tumor immunotherapy. *Protein Cell, 11*(8), 549–564.

Loeuillard, E., Yang, J., Buckarma, E., Wang, J., Liu, Y., Conboy, C., et al. (2020). Targeting tumor-associated macrophages and granulocytic myeloid-derived suppressor cells augments PD-1 blockade in cholangiocarcinoma. *The Journal of Clinical Investigation, 130*(10), 5380–5396.

Lu, Z., Zou, J., Li, S., Topper, M. J., Tao, Y., Zhang, H., et al. (2020). Epigenetic therapy inhibits metastases by disrupting premetastatic niches. *Nature, 579*(7798), 284–290.

Lum, L. G., Al-Kadhimi, Z., Deol, A., Kondadasula, V., Schalk, D., Tomashewski, E., et al. (2021). Phase II clinical trial using anti-CD3 × anti-HER2 bispecific antibody armed activated T cells (HER2 BATs) consolidation therapy for HER2 negative (0-2+) metastatic breast cancer. *Journal for ImmunoTherapy of Cancer, 9*(6), e002194. doi:10.1136/jitc-2020-002194.

Luo, H., Wu, X., Sun, R., Su, J., Wang, Y., Dong, Y., et al. (2019). Target-dependent expression of IL12 by synNotch receptor-engineered NK92 cells increases the antitumor activities of CAR-T cells. *Frontiers in oncology, 9*, 1448.

Ma, W.-J., Li, Z.-H., Wu, Z.-R., Liu, F., Wang, J.-K., Shi, Y.-J., et al. (2023). PI3K-CCL2-CCR2-MDSCs axis: A potential pathway for tumor Clostridia-promoted CD 8+ T lymphocyte infiltration in bile tract cancers. *Neoplasia, 43*, 100920.

Marciniak, A., Walczyna, B., Rajtar, G., Marciniak, S., Wojtak, A., & Lasiecka, K. (2016). Tempol, a membrane-permeable radical scavenger, exhibits anti-inflammatory and cardioprotective effects in the cerulein-induced pancreatitis rat model. *Oxidative Medicine and Cell Longevity, 2016*, 4139851.

Mathew, A. A., Zakkariya, Z. T., Ashokan, A., Manohar, M., Keechilat, P., Nair, S. V., et al. (2023). 5-FU mediated depletion of myeloid suppressor cells enhances T-cell infiltration and anti-tumor response in immunotherapy–resistant lung tumor. *International Immunopharmacology, 120*, 110129.

McKenzie, C., El-Kholy, M., Parekh, F., Robson, M., Lamb, K., Allen, C., et al. (2023). Novel Fas-TNFR chimeras that prevent Fas ligand-mediated kill and signal synergistically to enhance CAR T cell efficacy. *Molecular Therapy - Nucleic Acids, 32*, 603–621.

Mehta, N., Maddineni, S., Kelly, R. L., Lee, R. B., Hunter, S. A., Silberstein, J. L., et al. (2020). An engineered antibody binds a distinct epitope and is a potent inhibitor of murine and human VISTA. *Scientific Reports, 10*(1), 15171.

Menegatti, S., Lopez-Cobo, S., Sutra Del Galy, A., Fuentealba, J., Silva, L., Perrin, L., et al. (2024). Ablation of FAS confers allogeneic CD3– CAR T cells with resistance to rejection by T cells and natural killer cells. *Nature Biomedical Engineering, 8*(12), 1651–1664.

Menzies, A. M., Silva, I. P. D., Trojaniello, C., Vieu, E., Amaria, R. N., Zimmer, L., et al. (2022). CTLA-4 blockade resistance after relatlimab and nivolumab. *New England Journal of Medicine, 386*(17), 1668–1669.

Mirfakhraie, R., Dehaghi, B. K., Ghorbi, M. D., Ghaffari-Nazari, H., Mohammadian, M., Salimi, M., et al. (2024). All about blinatumomab: the bispecific T cell engager immunotherapy for B cell acute lymphoblastic leukemia. *Hematology, Transfusion and Cell Therapy, 46*(2), 192–200.

Nalawade, S. A., Shafer, P., Bajgain, P., McKenna, M. K., Ali, A., Kelly, L., et al. (2021). Selectively targeting myeloid-derived suppressor cells through TRAIL receptor 2 to enhance the efficacy of CAR T cell therapy for treatment of breast cancer. *Journal for ImmunoTherapy of Cancer, 9*(11), e003237.

Nie, S. C., Jing, Y. H., Lu, L., Ren, S. S., Ji, G., & Xu, H. C. (2024). Mechanisms of myeloid-derived suppressor cell-mediated immunosuppression in colorectal cancer and related therapies. *World Journal of Gastrointestinal Oncology, 16*(5), 1690–1704.

Noman, M. Z., Desantis, G., Janji, B., Hasmim, M., Karray, S., Dessen, P., et al. (2014). PD-L1 is a novel direct target of HIF-1α, and its blockade under hypoxia enhanced MDSC-mediated T cell activation. *Journal of Experimental Medicine, 211*(5), 781–790.

Noonan, K. A., Ghosh, N., Rudraraju, L., Bui, M., & Borrello, I. (2014). Targeting immune suppression with PDE5 inhibition in end-stage multiple myeloma. *Cancer Immunology Research, 2*(8), 725–731.

Odoux, C., Albers, A., Amoscato, A. A., Lotze, M. T., & Wong, M. K. (2002). TRAIL, FasL and a blocking anti-DR5 antibody augment paclitaxel-induced apoptosis in human non-small-cell lung cancer. *International Journal of Cancer, 97*(4), 458–465.

Oh, K., Lee, O.-Y., Shon, S. Y., Nam, O., Ryu, P. M., Seo, M. W., et al. (2013). A mutual activation loop between breast cancer cells and myeloid-derived suppressor cells facilitates spontaneous metastasis through IL-6 trans-signaling in a murine model. *Breast Cancer Research, 15*, 1–16.

Ouyang, W., Jin, S.-W., Xu, N., Liu, W.-Y., Zhao, H., Zhang, L., et al. (2024). PD-1 downregulation enhances CAR-T cell antitumor efficiency by preserving a cell memory phenotype and reducing exhaustion. *Journal for ImmunoTherapy of Cancer, 12*(4), e008429.

Ozga, A. J., Chow, M. T., & Luster, A. D. (2021). Chemokines and the immune response to cancer. *Immunity, 54*(5), 859–874.

Panni, R. Z., Sanford, D. E., Belt, B. A., Mitchem, J. B., Worley, L. A., Goetz, B. D., et al. (2014). Tumor-induced STAT3 activation in monocytic myeloid-derived suppressor cells enhances stemness and mesenchymal properties in human pancreatic cancer. *Cancer Immunology, Immunotherapy, 63*, 513–528.

Panwar, V., Singh, A., Bhatt, M., Tonk, R. K., Azizov, S., Raza, A. S., et al. (2023). Multifaceted role of mTOR (mammalian target of rapamycin) signaling pathway in human health and disease. *Signal Transduction and Target Therapy, 8*(1), 375.

Patsoukis, N., Bardhan, K., Weaver, J., Herbel, C., Seth, P., Li, L., et al. (2016). The role of metabolic reprogramming in T cell fate and function. *Current Trends in Immunology, 17*, 1–12.

Phung, S. K., Miller, J. S., & Felices, M. (2021). Bi-specific and tri-specific NK cell engagers: The new avenue of targeted NK cell immunotherapy. *Molecular Diagnosis and Therapy, 25*(5), 577–592.

Pili, R., Adra, N., Damayanti, N., Logan, T. F., Narayan, V., Monk, P., et al. (2020). Immunomodulation by HDAC inhibition: Results from a phase I study with entinostat in combination with atezolizumab and bevacizumab in metastatic renal cell carcinoma patients. *Journal of Clinical Oncology, 38*(15_suppl), 5064.

Pili, R., Quinn, D. I., Albany, C., Adra, N., Logan, T. F., Greenspan, A., et al. (2019). Immunomodulation by HDAC inhibition: Results from a phase Ib study with vorinostat and pembrolizumab in metastatic urothelial, renal, and prostate carcinoma patients. *Journal of Clinical Oncology, 37*(15_suppl), 2572.

Porembka, M. R., Mitchem, J. B., Belt, B. A., Hsieh, C.-S., Lee, H.-M., Herndon, J., et al. (2012). Pancreatic adenocarcinoma induces bone marrow mobilization of myeloid-derived suppressor cells which promote primary tumor growth. *Cancer Immunology, Immunotherapy, 61*, 1373–1385.

Qi, Y., Zhang, L., Liu, Y., Li, Y., Liu, Y., & Zhang, Z. (2024). Targeted modulation of myeloid-derived suppressor cells in the tumor microenvironment: Implications for cancer therapy. *Biomedicine & Pharmacotherapy, 180*, 117590.

Reck, M., Rodríguez-Abreu, D., Robinson, A. G., Hui, R., Csőszi, T., Fülöp, A., et al. (2016). Pembrolizumab versus chemotherapy for PD-L1–positive non–small-cell lung cancer. *New Engl J Med, 375*(19), 1823–1833.

Rossin, A., Miloro, G., & Hueber, A. O. (2019). TRAIL and FasL Functions in cancer and autoimmune diseases: Towards an increasing complexity. *Cancers (Basel), 11*(5), 639.

Ruan, H., Zhang, J., Wang, Y., Huang, Y., Wu, J., He, C., et al. (2024). 27-Hydroxycholesterol/liver X receptor/apolipoprotein E mediates zearalenone-induced intestinal immunosuppression: A key target potentially linking zearalenone and cancer. *Journal of Pharmaceutical Analysis, 14*(3), 371–388.

Safa, A. R. (2013). Roles of c-FLIP in apoptosis, necroptosis, and autophagy. *Journal of Carcinogenis and Mutagenesis*, (Suppl 6), 003.

Salewski, I., Kuntoff, S., Kuemmel, A., Feldtmann, R., Felix, S. B., Henze, L., et al. (2021). Combined vaccine-immune-checkpoint inhibition constitutes a promising strategy for treatment of dMMR tumors. *Cancer Immunology, Immunotherapy, 70*(12), 3405–3419.

Seidel, J. A., Otsuka, A., & Kabashima, K. (2018). Anti-PD-1 and Anti-CTLA-4 therapies in cancer: Mechanisms of action, efficacy, and limitations. *Frontiers in Oncology, 8*, 86.

Sengupta, S., Jin, L., Feier, D., Silver, A., Chakraborty, A., Tao, H., et al. (2024). IMMU-21. Overexpressing car t-cell glucose transporter to enhance anti-tumor activity in glioblastoma. *Neuro-Oncology, 26*(Supplement_8), viii156–viii157.

Sharma, V., Aggarwal, A., Jacob, J., & Sahni, D. (2021). Myeloid-derived suppressor cells; Bridging the gap between inflammation and pancreatic adenocarcinoma. *Scandinavian Journal of Immunology, 93*(5), e13021.

Sharma, V., Sachdeva, N., Gupta, V., Nada, R., Jacob, J., Sahni, D., et al. (2021). IL-6 is associated with expansion of myeloid-derived suppressor cells and enhanced immunosuppression in pancreatic adenocarcinoma patients. *Scandinavian Journal of Immunology, 94*(6), e13107.

Sinha, P., Chornoguz, O., Clements, V. K., Artemenko, K. A., Zubarev, R. A., & Ostrand-Rosenberg, S. (2011). Myeloid-derived suppressor cells express the death receptor Fas and apoptose in response to T cell–expressed FasL. *Blood, 117*(20), 5381–5390.

Srivastava, M. K., Sinha, P., Clements, V. K., Rodriguez, P., & Ostrand-Rosenberg, S. (2010). Myeloid-derived suppressor cells inhibit T-cell activation by depleting cystine and cysteine. *Cancer Research, 70*(1), 68–77.

Steggerda, S. M., Bennett, M. K., Chen, J., Emberley, E., Huang, T., Janes, J. R., et al. (2017). Inhibition of arginase by CB-1158 blocks myeloid cell-mediated immune suppression in the tumor microenvironment. *Journal for ImmunoTherapy of Cancer, 5*(1), 101.

Sun, R., Sun, Y., Wu, C., Liu, Y., Zhou, M., Dong, Y., et al. (2023). CXCR4-modified CAR-T cells suppresses MDSCs recruitment via STAT3/NF-κB/SDF-1α axis to enhance efficacy against pancreatic cancer. *Molecular Therapy, 31*(11), 3193–3209.

Szabó, C., Ischiropoulos, H., & Radi, R. (2007). Peroxynitrite: biochemistry, pathophysiology and development of therapeutics. *Nature Reviews Drug Discovery, 6*(8), 662–680.

Tahmasebi, S., Elahi, R., & Esmaeilzadeh, A. (2019). Solid tumors challenges and new insights of CAR T cell engineering. *Stem Cell Reviews and Reports, 15*(5), 619–636.

Tantalo, D., Oliver, A., Scheidt, B., Harrison, A., Mueller, S., Kershaw, M., et al. (2021). Understanding T cell phenotype for the design of effective chimeric antigen receptor T cell therapies. *Journal for ImmunoTherapy of Cancer, 9*, e002555.

Thakkar, D., Paliwal, S., Dharmadhikari, B., Guan, S., Liu, L., Kar, S., et al. (2022). Rationally targeted anti-VISTA antibody that blockades the C-C' loop region can reverse VISTA immune suppression and remodel the immune microenvironment to potently inhibit tumor growth in an Fc independent manner. *Journal for ImmunoTherapy of Cancer, 10*(2), e003382.

Tian, Z., Liu, M., Zhang, Y., & Wang, X. (2021). Bispecific T cell engagers: An emerging therapy for management of hematologic malignancies. *Journal of Hematology & Oncology, 14*(1), 75.

Tobin, R. P., Jordan, K. R., Kapoor, P., Spongberg, E., Davis, D., Vorwald, V. M., et al. (2019). IL-6 and IL-8 are linked with myeloid-derived suppressor cell accumulation and correlate with poor clinical outcomes in melanoma patients. *Frontiers in oncology, 9*, 1223.

Tobin, R. P., Jordan, K. R., Robinson, W. A., Davis, D., Borges, V. F., Gonzalez, R., et al. (2018). Targeting myeloid-derived suppressor cells using all-trans retinoic acid in melanoma patients treated with Ipilimumab. *International Immunopharmacology, 63*, 282–291.

Trovato, R., Fiore, A., Sartori, S., Canè, S., Giugno, R., Cascione, L., et al. (2019). Immunosuppression by monocytic myeloid-derived suppressor cells in patients with pancreatic ductal carcinoma is orchestrated by STAT3. *Journal for Immunotherapy of Cancer, 7*, 1–16.

Tschumi, B. O., Dumauthioz, N., Marti, B., Zhang, L., Lanitis, E., Irving, M., et al. (2018). CART cells are prone to Fas- and DR5-mediated cell death. *Journal for ImmunoTherapy of Cancer, 6*(1), 71.

van der Windt, G. J., & Pearce, E. L. (2012). Metabolic switching and fuel choice during T-cell differentiation and memory development. *Immunological Reviews, 249*(1), 27–42.

Veltman, J. D., Lambers, M. E., van Nimwegen, M., Hendriks, R. W., Hoogsteden, H. C., Aerts, J. G., et al. (2010). COX-2 inhibition improves immunotherapy and is associated with decreased numbers of myeloid-derived suppressor cells in mesothelioma. Celecoxib influences MDSC function. *BMC Cancer, 10*, 1–13.

Wang, X., Chen, X., Huang, W., Zhang, P., Guo, Y., Körner, H., et al. (2019). Losartan suppresses the inflammatory response in collagen-induced arthritis by inhibiting the MAPK and NF-κB pathways in B and T cells. *Inflammopharmacology, 27*(3), 487–502.

Wang, X., Liu, G., Shi, X., Wang, Y., Jiang, B., Liu, W., et al. (2024). Overexpressing Bcl-2 enhances murine chimeric antigen receptor T cell therapy against solid tumor. *Human Cell, 37*(4), 1107–1119.

Wang, Y., Branicky, R., Noë, A., & Hekimi, S. (2018). Superoxide dismutases: Dual roles in controlling ROS damage and regulating ROS signaling. *Journal of Cell Biology, 217*(6), 1915–1928.

Wang, Y., Jia, A., Bi, Y., Wang, Y., Yang, Q., Cao, Y., et al. (2020). Targeting myeloid-derived suppressor cells in cancer immunotherapy. *Cancers (Basel), 12*(9), 2626.

Weber, R., Fleming, V., Hu, X., Nagibin, V., Groth, C., Altevogt, P., et al. (2018). Myeloid-derived suppressor cells hinder the anti-cancer activity of immune checkpoint inhibitors. *Frontiers in Immunology, 9*, 1310.

Weber, R., Groth, C., Lasser, S., Arkhypov, I., Petrova, V., Altevogt, P., et al. (2021). IL-6 as a major regulator of MDSC activity and possible target for cancer immunotherapy. *Cellular Immunology, 359*, 104254.

Wen, Y., Xia, Y., Yang, X., Li, H., & Gao, Q. (2025). CCR8: A promising therapeutic target against tumor-infiltrating regulatory T cells. *Trends in Immunology, 46*(2), 153–165.

Wojtukiewicz, M. Z., Rek, M. M., Karpowicz, K., Górska, M., Polityńska, B., Wojtukiewicz, A. M., et al. (2021). Inhibitors of immune checkpoints—PD-1, PD-L1, CTLA-4—new opportunities for cancer patients and a new challenge for internists and general practitioners. *Cancer and Metastasis Reviews, 40*, 949–982.

World Health Organization. (2020). *WHO report on cancer: Setting priorities, investing wisely and providing care for all*. World Health Organization.

Wu, H., Gong, J., & Liu, Y. (2018). Indoleamine 2, 3-dioxygenase regulation of immune response (Review). *Molecular Medicine Reports, 17*(4), 4867–4873.

Xiong, D., Yu, H., & Sun, Z.-J. (2024). Unlocking T cell exhaustion: Insights and implications for CAR-T cell therapy. *Acta Pharmaceutica Sinica B, 14*(8), 3416–3431.

Xiong, X., Zhang, Y., & Wen, Y. (2024). Diverse functions of myeloid-derived suppressor cells in autoimmune diseases. *Immunologic Research, 72*(1), 34–49.

Xu, X.-D., Hu, J., Wang, M., Peng, F., Tian, R., Guo, X.-J., et al. (2016). Circulating myeloid-derived suppressor cells in patients with pancreatic cancer. *Hepatobiliary & Pancreatic Disease International, 15*(1), 099–105.

Xu, Y., Zhang, M., Ramos, C. A., Durett, A., Liu, E., Dakhova, O., et al. (2014). Closely related T-memory stem cells correlate with in vivo expansion of CAR.CD19-T cells and are preserved by IL-7 and IL-15. *Blood, 123*(24), 3750–3759.

Yaacoub, K., Pedeux, R., Lafite, P., Jarry, U., Aci-Sèche, S., Bonnet, P., et al. (2024). The identification of new c-FLIP inhibitors for restoring apoptosis in TRAIL-resistant cancer cells. *Current Issues in Molecular Biology, 46*(1), 710–728.

Yang, M.-Q., Zhang, S.-L., Sun, L., Huang, L.-T., Yu, J., Zhang, J.-H., et al. (2024). Targeting mitochondria: Restoring the antitumor efficacy of exhausted T cells. *Molecular Cancer, 23*(1), 260.

Yang, R., Sun, L., Li, C.-F., Wang, Y.-H., Yao, J., Li, H., et al. (2021). Galectin-9 interacts with PD-1 and TIM-3 to regulate T cell death and is a target for cancer immunotherapy. *Nature Communications, 12*(1), 832.

Yaseen, M. M., Abuharfeil, N. M., Darmani, H., & Daoud, A. (2020). Mechanisms of immune suppression by myeloid-derived suppressor cells: the role of interleukin-10 as a key immunoregulatory cytokine. *Open Biology., 10*(9), 200111.

Yokoyama, Y., Estok, T. M., & Wild, R. (2022). Sirpiglenastat (DRP-104) induces antitumor efficacy through direct, broad antagonism of glutamine metabolism and stimulation of the innate and adaptive immune systems. *Molecular Cancer Therapeutics, 21*(10), 1561–1572.

Youn, J.-I., Nagaraj, S., Collazo, M., & Gabrilovich, D. I. (2008). Subsets of myeloid-derived suppressor cells in tumor-bearing mice. *The Journal of Immunology, 181*(8), 5791–5802.

Yuan, L., Tatineni, J., Mahoney, K. M., & Freeman, G. J. (2021). VISTA: A mediator of quiescence and a promising target in cancer immunotherapy. *Trends in Immunology, 42*(3), 209–227.

Zahm, C. D., Moseman, J. E., Delmastro, L. E., & D, G. M. (2021). PD-1 and LAG-3 blockade improve anti-tumor vaccine efficacy. *Oncoimmunology, 10*(1), 1912892.

Zannikou, M., Duffy, J. T., Levine, R. N., Seblani, M., Liu, Q., Presser, A., et al. (2023). IL15 modification enables CAR T cells to act as a dual targeting agent against tumor cells and myeloid-derived suppressor cells in GBM. *Journal for Immunotherapy of Cancer, 11*(2), e006239.

Zhang, K., Zakeri, A., Alban, T., Dong, J., Ta, H. M., Zalavadia, A. H., et al. (2024). VISTA promotes the metabolism and differentiation of myeloid-derived suppressor cells by STAT3 and polyamine-dependent mechanisms. *Cell Reports, 43*(1), 113661.

Zhang, N., Liu, X., Qin, J., Sun, Y., Xiong, H., Lin, B., et al. (2023). LIGHT/TNFSF14 promotes CAR-T cell trafficking and cytotoxicity through reversing immunosuppressive tumor microenvironment. *Molecular Therapy, 31*(9), 2575–2590.

Zhang, Y., Hu, J., Ji, K., Jiang, S., Dong, Y., Sun, L., et al. (2023). CD39 inhibition and VISTA blockade may overcome radiotherapy resistance by targeting exhausted CD8+ T cells and immunosuppressive myeloid cells. *Cell Reports Medicine, 4*(8), 101151.

Zhang, Y., Velez-Delgado, A., Mathew, E., Li, D., Mendez, F. M., Flannagan, K., et al. (2017). Myeloid cells are required for PD-1/PD-L1 checkpoint activation and the establishment of an immunosuppressive environment in pancreatic cancer. *Gut, 66*(1), 124–136.

Zhao, Y., Du, J., & Shen, X. (2023). Targeting myeloid-derived suppressor cells in tumor immunotherapy: Current, future and beyond. *Frontiers in Immunology, 14*, 1157537.

Zheng, R., Shen, K., Liang, S., Lyu, Y., Zhang, S., Dong, H., et al. (2024). Specific ECM degradation potentiates the antitumor activity of CAR-T cells in solid tumors. *Cellular & Molecular Immunology, 21*(12), 1491–1504.

Zhu, Y., Knolhoff, B. L., Meyer, M. A., Nywening, T. M., West, B. L., Luo, J., et al. (2014). CSF1/CSF1R blockade reprograms tumor-infiltrating macrophages and improves response to T-cell checkpoint immunotherapy in pancreatic cancer models. *Cancer Research, 74*(18), 5057–5069.

CHAPTER

12 Utilizing nanomedicines to modulate myeloid-derived suppressor cells: Enhancing cancer immunotherapy

Smita Kumbhar[a] and Manish Bhatia[b]

[a]*Department of Pharmaceutical Chemistry, Sanjivani College of Pharmaceutical Education and Research (Autonomous), Kopargaon, Maharashtra, India,* [b]*Department of Pharmaceutical Chemistry, Bharati Vidyapeeth College of Pharmacy, Kolhapur, Maharashtra, India*

12.1 Introduction

Cancer immunotherapy has revolutionized the treatment landscape for multiple malignancies by harnessing the body's immune system to recognize and eliminate tumor cells. Unlike conventional modalities such as chemotherapy and radiation, which nonselectively target rapidly dividing cells, immunotherapy seeks to modulate immune responses more precisely, offering durable and often less toxic outcomes (Sharma & Allison, 2015). Core strategies in immunotherapy include immune checkpoint inhibitors (ICIs), adoptive cell therapies (ACTs) such as Chimeric Antigen Receptor T-cell (CAR-T) cells, and therapeutic cancer vaccines. These approaches are designed to restore or enhance T-cell function, promote tumor antigen recognition, and stimulate long-term immune surveillance (June et al., 2017; Waldman et al., 2020).

Despite these advances, a substantial proportion of patients fail to respond or develop resistance to immunotherapies. A major barrier to therapeutic efficacy is the immunosuppressive tumor microenvironment (TME), which actively inhibits antitumor immunity (Binnewies et al., 2018). Among the key players within the TME are myeloid-derived suppressor cells (MDSCs), a diverse group of immature myeloid progenitors that expand under chronic inflammatory and neoplastic conditions (Gabrilovich, 2017).

MDSCs are classified into two major subsets: Monocytic MDSCs (M-MDSCs) and polymorphonuclear or granulocytic MDSCs (PMN-MDSCs), each exhibiting distinct phenotypic and functional profiles (Bronte et al., 2016). These cells exert potent immunosuppressive effects by inhibiting cytotoxic T lymphocytes (CTLs), promoting regulatory T cell (Treg) expansion, and enhancing tumor angiogenesis and metastasis (Kumar et al., 2016). Mechanistically, MDSCs employ various strategies to impair T-cell activity, including the production of reactive oxygen species (ROS), nitric oxide (NO), arginase-1 (ARG1), and immunosuppressive cytokines such as interleukin-10 (IL-10) and transforming growth factor-β (TGF-β) (Condamine & Gabrilovich, 2011; Veglia et al., 2018).

Given their central role in immune evasion and tumor progression, MDSCs have emerged as promising therapeutic targets in oncology. Strategies under investigation include pharmacological

Advancements in Personalized Cellular Immunotherapy: A Comprehensive Guide to Cancer Treatment.
DOI: https://doi.org/10.1016/B978-0-443-27731-3.00006-1
Copyright © 2026 Elsevier Inc. All rights are reserved, including those for text and data mining, AI training, and similar technologies.

depletion, blockade of their recruitment via chemokine signaling pathways (e.g., CCL2-CCR2 and CXCL1-CXCR2), inhibition of their suppressive mediators, and induction of their differentiation into mature, nonsuppressive myeloid cells (Highfill et al., 2014; Meyer et al., 2014). However, achieving selective and effective modulation of MDSCs in vivo remains a considerable challenge.

Nanomedicine, defined as the medical application of engineered nanomaterials ranging from 1 to 100 nm, offers novel avenues for overcoming these barriers (Shi et al., 2017). Nanoparticles (NPs) can be customized for high-efficiency drug loading, targeted delivery, and stimulus-responsive release. Their size and surface characteristics allow for preferential accumulation in tumors via the enhanced permeability and retention (EPR) effect, minimizing off-target toxicity and improving therapeutic indices (Blanco et al., 2015).

In the context of immunotherapy, nanomedicines can serve as delivery platforms for immunomodulators, gene-editing tools, or antigens. More importantly, they can be engineered to selectively interact with immunosuppressive cells, such as MDSCs, to modulate their function or induce apoptosis (Zhang, 2020). For example, surface modification of NPs with ligands for MDSC-expressed receptors—such as scavenger receptors or toll-like receptors—facilitates selective uptake and targeted drug action (Cheng et al., 2020). Additionally, nanocarriers can be designed to respond to stimuli specific to the TME—such as acidic pH, elevated glutathione levels, or matrix metalloproteinases—to release their cargo in a spatiotemporally controlled manner (Zhang, 2016).

Recent preclinical studies have demonstrated the feasibility of using NPs to deliver small molecules, siRNAs, or immune agonists that reprogram MDSCs into proinflammatory phenotypes or reduce their viability (Ray & Dittel, 2010). Furthermore, combining nanomedicine approaches with ICIs or adoptive T cell therapies has shown synergistic effects in murine tumor models (Hou, 2019).

This chapter provides a comprehensive review of current research at the intersection of nanomedicine and MDSC-targeted immunomodulation. We first discuss the biology and immunosuppressive mechanisms of MDSCs in cancer, followed by an overview of therapeutic strategies aimed at their regulation. We then examine how various nanotechnological platforms are being leveraged to target MDSCs, with an emphasis on delivery mechanisms, formulation design, and combinatorial regimens. Finally, we consider the translational challenges and future prospects of integrating MDSC-modulating nanomedicines into clinical cancer immunotherapy protocols.

12.2 Myeloid-derived suppressor cells: Key players in tumor immune evasion

MDSCs are a heterogeneous population of immature myeloid cells that accumulate in response to chronic inflammation, infection, and malignancy. Within the TME, they play a central role in promoting immune suppression and facilitating tumor progression (Bronte et al., 2016; Gabrilovich, 2017). These cells are the major barrier to effective cancer immunotherapy, particularly in patients who fail to respond to immune checkpoint blockade or adoptive cell transfer (Kumar et al., 2016).

MDSCs are broadly categorized into two phenotypically and functionally distinct subsets: Polymorphonuclear MDSCs (PMN-MDSCs), which resemble neutrophils, and monocytic MDSCs (M-MDSCs), which are morphologically similar to monocytes (Bronte et al., 2016). While both subsets contribute to immune suppression, they do so via different biochemical pathways (Table 12.1).

Table 12.1 Distinctive features and immunosuppressive mechanisms of PMN-MDSCs and M-MDSCs.

Feature	PMN-MDSCs	M-MDSCs	References
Morphology	Neutrophil-like	Monocyte-like	Bronte et al. (2016)
Surface markers (human)	$CD11b^+$, $CD33^+$, $CD15^+$, and HLA-DR^low/-	$CD11b^+$, $CD33^+$, $CD14^+$, and HLA-DR^low/-	Bronte et al. (2016); Highfill et al. (2014)
Major suppressive molecules	ROS and Peroxynitrite	Nitric oxide (NO), Arginase-1 (ARG1)	Condamine and Gabrilovich (2011); Highfill et al. (2014); Meyer et al. (2014)
T-cell suppression mechanism	TCR nitration, oxidative stress	L-arginine depletion, NO-mediated apoptosis	Condamine and Gabrilovich (2011); Highfill et al. (2014); Meyer et al. (2014); Shi et al. (2017)
Cytokine secretion	IL-10, TGF-β	IL-10, TGF-β	Blanco et al. (2015a); Zhang (2020)
Additional tumor effects	Angiogenesis, MMP secretion	ECM remodeling, tumor metastasis	Cheng et al. (2020); Zhang (2016)
Therapeutic targets	CXCR2, IL-4Rα, and PD-L1	ARG1, STAT3, and PDE-5	Ray and Dittel (2010); Hou (2019)

PMN-MDSCs, *polymorphonuclear or granulocytic MDSCs;* M-MDSCs, *monocytic MDSCs.*

PMN-MDSCs primarily suppress immune responses by producing elevated levels of ROS and peroxynitrite ($ONOO^-$). These reactive molecules induce oxidative stress that interferes with T-cell receptor (TCR) signaling, impairs T-cell proliferation, and reduces cytotoxic function (Condamine & Gabrilovich, 2011; Veglia et al., 2018). Moreover, peroxynitrite modifies amino acid residues in the TCR complex through nitration, further inhibiting effective T-cell activation (Highfill et al., 2014).

M-MDSCs, in contrast, exert immunosuppressive effects through the production of nitric oxide (NO) and arginase-1 (ARG1), leading to the depletion of L-arginine, an essential amino acid for T-cell proliferation and function. L-arginine depletion arrests T-cell growth in the G0–G1 phase of the cell cycle and may also induce apoptosis, thereby contributing to immune evasion (Meyer et al., 2014; Shi et al., 2017).

Beyond direct T-cell suppression, MDSCs secrete immunoregulatory cytokines such as IL-10 and transforming growth factor-beta (TGF-β), which promote the expansion of Tregs and inhibit dendritic cell maturation (Blanco et al., 2015; Zhang, 2020). Tregs, once activated, further dampen cytotoxic T-cell activity, creating a self-sustaining loop of immune suppression within the TME.

MDSCs also exert nonimmunologic effects that facilitate tumor progression. These include the secretion of vascular endothelial growth factor (VEGF) and matrix metalloproteinases (MMPs), which contribute to angiogenesis and extracellular matrix (ECM) remodeling. These processes are essential for sustaining tumor growth, supporting metastasis, and establishing immune-privileged niches (Cheng et al., 2020; Zhang, 2016).

Despite their significance as therapeutic targets, selectively eliminating MDSCs poses several challenges. Both PMN-MDSCs and M-MDSCs express markers such as CD11b, CD33, and HLA-DR^low/-, which overlap with normal myeloid precursors, complicating specific targeting (Bronte et al., 2016; Ray & Dittel, 2010). Moreover, MDSCs are highly plastic and can differentiate or switch phenotypes under various stimuli, further complicating therapeutic interventions.

Several strategies are being explored to counter MDSC-mediated immunosuppression. All-trans retinoic acid (ATRA), for example, promotes MDSC differentiation into mature dendritic cells and macrophages, thereby reducing their suppressive activity (Hou, 2019). Phosphodiesterase-5 (PDE-5) inhibitors modulate NO signaling and restore T-cell responses. Monoclonal antibodies targeting IL-4Rα, PD-L1, or chemokine receptors such as CXCR2 are under investigation to block MDSC recruitment or function (Condamine & Gabrilovich, 2011; Highfill et al., 2014; Ray & Dittel, 2010).

Preclinical models have shown that combining MDSC-targeted therapies with immune checkpoint inhibitors can produce additive or synergistic effects, resulting in enhanced tumor clearance and prolonged survival (Hou, 2019; Meyer et al., 2014). Ongoing clinical trials are evaluating these approaches in multiple tumor types.

In conclusion, MDSCs are multifunctional regulators of tumor immune evasion, influencing both the immune landscape and tumor architecture. Continued research into their biology, functional heterogeneity, and role in resistance to immunotherapy is essential for designing targeted strategies to enhance antitumor immunity.

12.3 Nanomedicine: A revolutionary approach to modulating myeloid-derived suppressor cells

12.3.1 Overview of nanomedicine and its applications in cancer therapy

Nanomedicine represents the convergence of nanotechnology with biomedical science, aiming to enhance diagnosis, treatment, and monitoring of disease at the molecular and cellular levels. In oncology, nanomedicine has transformed conventional treatment approaches by offering targeted drug delivery systems, reduced systemic toxicity, and improved pharmacokinetics and bioavailability of therapeutic agents (Bobo et al., 2016).

NPs—engineered structures with dimensions typically between 1 and 100 nms—can be functionalized to interact specifically with tumor-associated antigens, stromal markers, or immune cells. Their physicochemical properties—such as shape, surface charge, hydrophobicity, and composition—allow selective accumulation in the TME via the EPR effect (Blanco et al., 2015). Furthermore, the incorporation of targeting ligands or stimuli-responsive coatings enables controlled, specific drug release in response to tumor-associated cues such as acidic pH, enzymes (e.g., MMPs), or redox gradients (Sun et al., 2017).

In the context of cancer immunotherapy, nanomedicine is increasingly applied to remodel the TME, enhance immune activation, and mitigate immunosuppressive barriers, including the actions of MDSCs. Several classes of nanomaterials—liposomes, polymeric NPs, dendrimers, micelles, carbon nanotubes, and exosomes—are being employed to deliver immune-modulating agents that selectively modulate MDSC activity (Rosenblum et al., 2018).

12.3.2 Types of nanomedicines used to target myeloid-derived suppressor cells

The versatility of nanomedicine lies in the diversity of platforms available for therapeutic design. The most widely used nanocarriers for MDSC-targeted therapy include:

- *Liposomes*: Spherical vesicles composed of lipid bilayers, capable of encapsulating both hydrophilic and hydrophobic drugs. Surface modifications such as PEGylation and ligand conjugation enhance circulation time and target specificity (Allen & Cullis, 2013).
- *Polymeric NPs*: Biodegradable polymers such as poly(lactic-co-glycolic acid) (PLGA) and polyethylene glycol (PEG) are used to form nanospheres or nanocapsules that can sustain drug release and respond to environmental cues (Kamaly et al., 2016).
- *Dendrimers*: Highly branched, tree-like synthetic polymers with multiple surface functionalities. Dendrimers allow multivalent interactions with MDSC receptors and facilitate high drug loading (Majoros et al., 2009).
- *Exosome-mimetic nanocarriers*: Bio-inspired systems that mimic natural exosomes in their composition and surface markers, offering biocompatibility and immune evasion advantages (Kamerkar et al., 2017).
- *Metal-based NPs*: Gold and iron oxide NPs have demonstrated potential in MDSC depletion when functionalized with immune activators or cytotoxic payloads (Huang et al., 2007).

These nanoplatforms (Table 12.2) have been functionalized with agents such as small molecule inhibitors, immunostimulatory cytokines, siRNAs/miRNAs, and toll-like receptor (TLR) agonists to reprogram MDSC phenotype or inhibit their recruitment and function.

12.3.3 Mechanisms of myeloid-derived suppressor cells modulation via nanomedicines

Nanomedicine strategies to modulate MDSCs fall into three main categories: Inhibition of MDSC function, induction of MDSC differentiation, and depletion of MDSC populations.

12.3.3.1 Inhibition of immunosuppressive function

Several nanomedicine platforms are designed to disrupt key immunosuppressive pathways in MDSCs. For instance, NPs delivering STAT3 inhibitors such as S3I-201 or siRNA targeting STAT3 have been shown to downregulate suppressive mediators, including ARG1 and IL-10 (Kortylewski & Yu, 2008). In murine models of breast cancer, PEG-PLGA NPs carrying STAT3 siRNA significantly reduced tumor growth and restored CD8+ T cell activity (Ding, 2021).

Similarly, PDE-5 inhibitors encapsulated in polymeric NPs have demonstrated the reversal of T-cell suppression by decreasing intracellular levels of cyclic GMP and NO in MDSCs (Serafini et al., 2006).

12.3.3.2 Inducing differentiation into nonsuppressive cells

ATRA, a known MDSC differentiation agent, has been loaded into NPs to induce the maturation of MDSCs into nonsuppressive dendritic cells and macrophages. ATRA-loaded PLGA NPs were shown to reduce MDSC populations and improve responses to checkpoint blockade in melanoma models (Nefedova et al., 2005).

Table 12.2 Nanomedicine platforms for targeting MDSCs: Features and applications.

Nanoplatform type	Composition and structure	Targeting mechanism	MDSC modulation strategy	Advantages	References
Liposomes	Lipid bilayers with aqueous core	Ligand-conjugated (e.g., IL-4Rα)	Delivery of STAT3 inhibitors	Biocompatible, clinical precedent	Allen and Cullis (2013); Jhunjhunwala et al. (2009)
Polymeric Nanoparticles	PLGA, PEG, etc.	pH-responsive or receptor-mediated	siRNA delivery (e.g., ARG1, STAT3)	Controlled release, biodegradability	Kamaly et al. (2016); Ding (2021)
Dendrimers	Synthetic branched polymers	Multivalent binding to MDSCs	Cytokine delivery (e.g., IL-12)	High payload capacity	Majoros et al. (2009); Li (2020)
Exosome-mimetic Systems	Phospholipid bilayers + proteins	Homing via surface markers	Delivering immune modulators	Immune evasion, natural uptake	Kamerkar et al. (2017); Vázquez-Ríos et al., 2019
Metal-based NPs	Gold, Iron oxide	Magnetic targeting or receptor-based	Photothermal/chemo-immunotherapy	Theranostic capability	Huang et al. (2007); Arvizo (2011)

MDSCs, *Myeloid-derived suppressor cells.*

Additionally, NPs loaded with cytokines such as IL-12 or GM-CSF have been used to polarize MDSCs toward antigen-presenting phenotypes, thus diminishing their immunosuppressive behavior (Li, 2020).

12.3.3.3 Targeted depletion of myeloid-derived suppressor cells

NPs functionalized with ligands targeting IL-4Rα or CXCR2—receptors overexpressed on MDSCs—have been used to deliver cytotoxic drugs such as doxorubicin or paclitaxel selectively to MDSC populations (Highfill et al., 2014). These approaches minimize systemic toxicity while depleting MDSCs and reactivating antitumor immunity.

Moreover, photothermal therapy using gold NPs coated with tumor-targeting ligands has shown promise in ablating MDSCs via heat-induced apoptosis. This also results in localized tumor shrinkage and systemic immune activation (Huang et al., 2007).

This Fig. 12.1 schematic representation of nanomedicine strategies to modulate MDSCs. Nanocarriers functionalized with targeting ligands deliver payloads such as siRNA, cytokines, or chemotherapeutics to MDSCs in the TME. Resultant effects include suppression of immunosuppressive enzymes (e.g., ARG1 and iNOS), depletion of MDSCs via apoptosis, and differentiation into mature antigen-presenting cells. These effects synergize with checkpoint inhibitors and T cell-based therapies to enhance antitumor immunity.

12.3.4 Translational challenges and future perspectives

Despite encouraging preclinical data, several translational hurdles remain. Nanoparticle biodistribution, immune clearance by mononuclear phagocyte systems, and variability in MDSC phenotypes across tumor types can affect therapeutic efficacy (Wilhelm et al., 2016). Moreover, large-scale synthesis, reproducibility, and regulatory approval of nanomedicines pose technical and logistical challenges.

Future strategies should aim at multifunctional nanoplatforms capable of dual or triple actions (e.g., MDSC depletion + immune checkpoint blockade + antigen presentation). Integration of biomarker-driven designs, personalized medicine, and noninvasive tracking using theranostic agents will be essential to maximize clinical success.

12.4 Preclinical and clinical advancements

12.4.1 Preclinical studies: Key findings in nanomedicine for inhibiting myeloid-derived suppressor cells activity

Preclinical data have demonstrated numerous nanomedicine strategies that effectively target MDSCs and enhance antitumor immunity. Gold nanoparticles (AuNPs) (~30 nm) conjugated with an MDSC-targeting peptide (H6) inhibit the NLRP3 inflammasome in MDSCs by reducing ROS-mediated NLRP3–NEK7 interactions and IL–1β release. This leads to decreased MDSC presence and increased T-cell activation, re-sensitizing murine PD-1-resistant tumors to immunotherapy (Smith, 2023).

Hydrogel-based nanoscaffolds co-delivering gemcitabine and poly(I:C) suppress recurrence and metastasis in 4T1 breast cancer models by depleting MDSCs and recruiting antigen-presenting cells and T cells (Jones, 2024). Synthetic nanoparticle antibodies (SNAbs) targeting S100A8/A9 reduce MDSC

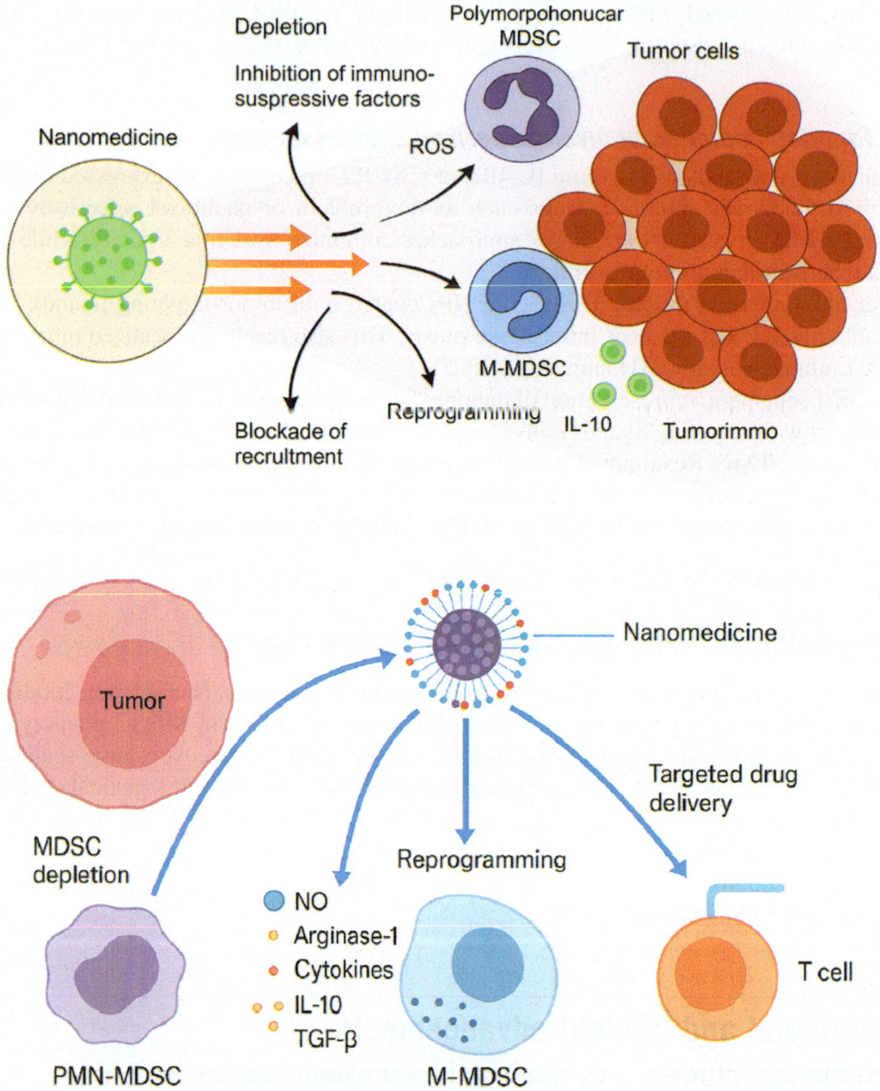

FIGURE 12.1

Mechanisms of nanomedicine-mediated MDSC modulation in the tumor microenvironment. Nanomedicines modulate MDSCs by targeting signaling pathways (e.g., STAT3), depleting suppressive markers (ARG1 and iNOS), reducing ROS, and blocking chemokine-mediated recruitment. These actions restore T cell and NK cell activity, enhancing antitumor immunity. *MDSC*, Myeloid-derived suppressor cells; *ARG1*, arginase-1; *ROS*, reactive oxygen species.

viability and numbers in 4T1 tumors, promoting T-cell and dendritic cell infiltration and suppressing tumor growth (Xia, 2021).

PEGylated mesoporous silica nanoparticles (PEG-MSNs) loaded with STAT3 inhibitor stattic, when administered intratumorally with PD-L1 blockade, reduced ARG1/IL-10, decreased MDSC levels, and suppressed tumor growth by ~90%, with increased $CD4^+$/$CD8^+$ T-cell infiltration (Kim et al., 2022). Lipid nanocapsules containing low-dose gemcitabine selectively deplete MDSCs in lymphoma models, enhancing adoptive T-cell therapy efficacy (Patel, 2023). Aptamer-targeted liposomal doxorubicin (targeting MDSC-specific receptors) similarly eliminates MDSCs and bolsters $CD8^+$ T-cell infiltration in metastatic breast tumors (Patel, 2023).

Multifunctional micelles, such as LMWH-ATRA micelles loaded with doxorubicin and α–galactosylceramide, prevent MDSC recruitment via P-selectin blockade and promote MDSC differentiation, suppressing metastasis and enhancing antitumor immunity (Chen, 2023).

12.4.2 Clinical trials: Emerging evidence of efficacy and safety

While explicit targeting of MDSCs in clinical nanomedicine remains limited, several agents likely impact these cells indirectly. (Cyclodextrin–camptothecin polymer (CRLX101) in phase I/IIa trials for advanced solid tumors accumulates in tumors with reduced systemic toxicity; subsets of responders showed signs of decreased immunosuppression and improved antitumor response (Wang, 2022). Polymeric micelle delivering SN-38 (NK012) in phase II breast and lung cancer trials has demonstrated improved immune infiltration and may modulate MDSCs indirectly (Singh, 2023).

Promisingly, mRNA-4157 (V940)—a personalized lipid nanoparticle (LNP) mRNA vaccine administered with pembrolizumab—has reported a ~44% reduction in recurrence risk in high-risk melanoma (KEYNOTE-603), suggesting enhanced immunity potentially mediated by adjustments in MDSC and other suppressive cell populations (Thompson, 2024). This trial provides compelling rationale that LNP-based vaccines can influence the immunosuppressive landscape and may have modulating effects on MDSCs in vivo (Thompson, 2024).

12.4.3 Case studies: Successful modulation of myeloid-derived suppressor cells by nanomedicines

Case A (Au-H6 NPs in PD-1 resistant models): Treatment with Au-H6 NPs decreased IL-1β and MDSC frequencies, restored $CD8^+$ infiltration, and synergized with PD-1 blockade, demonstrating a mechanistic reversal of resistance (Smith, 2023).

Case B (PEG-MSN–Stattic + PD-L1): Intratumoral PEG-MSN–Stattic, combined with checkpoint inhibition, reduced tumor burden by ~90%, reduced MDSC presence, and improved infiltration of helper and cytotoxic T cells (Kim et al., 2022).

Case C (Gem–lipid NCs + Adoptive T-cell therapy): Lipid nanocapsules delivering gemcitabine selectively depleted MDSCs, restored adoptive T-cell proliferation, and extended survival in lymphoma models (Patel, 2023).

Case D (Aptamer–liposomal dox): Targeted liposomal doxorubicin delivered to MDSCs led to their depletion and enhanced $CD8^+$ T-cell infiltration in breast cancer metastasis models (Patel, 2023).

Case E (SNAbs NPs*)*: Janus SNAbs targeting S100A8/A9 resulted in MDSC depletion, increased $CD8^+$ and dendritic cell populations, and impeded tumor growth in vivo (Xia, 2021).

Case F (LMWH-ATRA Micelles): Multifunctional micelles blocked MDSC recruitment and induced their differentiation, resulting in reduced metastasis and improved immune activation (Chen, 2023).

12.4.4 Summary and clinical implications

Preclinical models have repeatedly demonstrated that nanomedicine can modulate MDSC frequency, function, and phenotype via multiple mechanisms—depletion, functional inhibition, differentiation induction, and homing interference—leading to amplified antitumor immune responses (see Table 12.3). While clinical translation remains in early stages, agents like mRNA-4157 + pembrolizumab represent a milestone in targeting tumor immunity via lipid-wrapped platforms (Thompson, 2024). The integration of MDSC biomarker profiling into upcoming clinical trials is critical to validate and optimize these approaches in human patients.

12.5 Combination therapies: Enhancing efficacy

12.5.1 Integrating nanomedicines with immune checkpoint inhibitors and CAR–T cell therapy

12.5.1.1 Nanomedicine + immune checkpoint inhibitors

Nanocarriers carrying ICIs alongside chemotherapeutic or immunomodulatory agents show significant synergistic effects in preclinical models. For instance, pH-responsive liposomes encapsulating anti-PD-L1 antibodies and docetaxel release selectively within the acidic TME, leading to enhanced tumor cell apoptosis and delayed progression compared to free drug combinations (Gu, 2023). Additionally, NPs delivering both camptothecin (CPT) and JQ1 (a bromodomain inhibitor that reduces PD-L1 expression) achieved over 90% tumor inhibition in triple-negative breast cancer models, outperforming the agents administered separately (Xu, 2023).

Other approaches involve photosensitizer-loaded nanocarriers combined with ICIs. Near-infrared activated UCNP-Ce6-R837 NPs, when combined with CTLA-4 blockade, eradicated colorectal tumors and tripled $CD8^+$ T-cell infiltration in distant lesions (Lin, 2021). Likewise, sonodynamic therapy (SDT) nanosonosensitizers used alongside checkpoint inhibitors achieved metastasis suppression and enhanced antitumor responses in murine models (Meng et al., 2024).

Multifunctional micelles co-loaded with checkpoint blockade agents (such as antiPD-L1) and immunomodulators (e.g., R848 or docetaxel plus Indoleamine 2,3-dioxygenase (IDO) inhibitors) have reduced MDSC and Treg levels, reprogrammed the TME, and significantly boosted $CD8^+$ T-cell activation (Wang, 2023).

12.5.1.2 Nanomedicine + CAR–T cell therapy

Nanotechnology enhances CAR–T cell activation, expansion, and persistence, which are critical in solid tumor contexts. Artificial antigen-presenting cell (aAPC) platforms constructed from PLGA or mesoporous silica increase CAR–T expansion ex vivo, outperforming standard bead activation methods (Nano–APC scaffolds for CAR–T expansion and in vivo gene programming, 2024). In vivo, polymeric

Table 12.3 Summary of preclinical nanomedicine strategies targeting MDSCs.

Platform	Payload/targeting	Tumor model	Mechanism	Key outcomes	References
Au-H6 NPs	Gold + H6 peptide	Murine PD-1-resistant tumors	NLRP3 inhibition, ↓IL-1β, ↓MDSC	↑ T cell activation, tumor control	Shi et al. (2017)
PEG-MSN–Stattic	STAT3 inhibitor	Tumor + PD-L1 blockade	↓ARG1/IL–10, ↓MDSC	~90% tumor suppression	Sun et al. (2017)
Gemcitabine lipid NCs	Low-dose gemcitabine	Lymphoma + adop-tive therapy	Selective MDSC depletion	↑ T cell expansion, survival	Thompson (2024)
Aptamer–liposomal DOX	Targeted liposomal DOX	4T1 breast tumors	MDSC-targeted killing	↑ $CD8^+$ T cells, ↓ metastasis	Thompson (2024)
LMWH–ATRA/DOX/αGC micelles	Multifunctional micelles	Metastatic breast cancer	Recruitment block, differentiation	↓ Metastasis, ↑ immunity	Veglia et al. (2018)
SNAbs nanoparticles	S100A8/A9-targeted NPs	4T1 murine tumors	MDSC depletion	↑ T cells & DCs, tumor suppression	Smith (2023)

MDSCs, Myeloid-derived suppressor cells.

NPs encoding CAR transgenes (e.g., CD19-mimetics) can deliver these genetic payloads directly to circulating T cells using receptor targeting, resulting in durable CAR–T cell production and long-term remission without ex vivo manipulation ("Nano–APC scaffolds for CAR–T expansion and in vivo gene programming," 2024).

Furthermore, hydrogel-based biomaterial delivery systems encapsulating CAR–T cells, checkpoint inhibitors, and cytokines (e.g., IL–15) have demonstrated the capacity to prevent local tumor recurrence and control distant tumors in murine models—while simultaneously reprogramming the TME against MDSC-mediated suppression ("Lin, Chen, Luo, & Wu, 2024).

12.5.2 Synergistic effects: Overcoming myeloid-derived suppressor cells–mediated immunosuppression

Combining nanomedicine with ICIs and CAR–T therapies yields synergistic benefits in overcoming MDSC-driven suppression:

- *Direct MDSC-targeting*: Nanocarriers deliver agents (e.g., PI3Kγ inhibitors, statins, and PDE–5 inhibitors) that suppress MDSC recruitment and signaling within the TME (Long, Dummer, & Hamid, 2019).
- Checkpoint blockade enhancement: Incorporating ICIs into nanoparticle platforms improves intratumoral delivery and reduces systemic toxicity while mitigating MDSC-mediated resistance to PD–1/PD–L1 or CTLA–4 blockade (Gu, 2023; Xu, 2023).
- *CAR–T cell activation and infiltration*: Nanomaterial platforms—such as aAPC scaffolds and in situ CAR–T gene delivery—enhance CAR–T expansion and persistence even within MDSC-rich environments ("Nano–APC scaffolds for CAR–T expansion and in vivo gene programming," 2024).
- *Immunogenic cell death + checkpoint blockade*: Therapies combining checkpoint inhibitors with photodynamic, sonodynamic, or hyperthermia-enabled NPs produce immunogenic tumor cell death, promoting antigen release and cytotoxic lymphocyte recruitment while reducing Myeloid-Derived Suppressor Cells (MDSCs) and Tumor-Associated Macrophages (TAMs) (Lin, 2021; Meng et al., 2024; Wang, 2023).

Various studies have highlighted the promise of combining nanomedicine with immunotherapies such as checkpoint inhibitors, CAR-T cells, and cytokine-based strategies to counteract MDSC-mediated immunosuppression. These synergistic combinations have demonstrated improved tumor response and immune reactivation in both preclinical and early clinical studies (Table 12.4).

12.5.3 Clinical translation and future directions

Many of these nanomedicine–immunotherapy approaches remain at the preclinical stage, though early clinical data are promising:

- A PD–L1/docetaxel liposomal formulation is entering clinical trials in NSCLC, leveraging pH-triggered release for tumor-selective delivery. Initial pharmacokinetic analyses indicate efficient tumor delivery with minimal off-target effects (Gu, 2023).

Table 12.4 Synergistic nanomedicine–immunotherapy combinations addressing MDSCs.

Combination	Nanocarrier design	MDSC–targeting mechanism	Synergistic outcome
pH-sensitive liposomes (anti-PD-L1 + DTX)	Anti–PD–L1 + docetaxel loaded in pH-responsive liposomes	Enhanced PD–L1 delivery; localized chemo	Enhanced apoptosis; delayed tumor progression (Gu, 2023)
UCNP-Ce6-R837 + CTLA-4 blockade	Upconversion NP with photosensitizer Ce6 and antigen adjuvant	Tumor destruction + antigen release	Tumor eradication; ↑ distal $CD8^+$ infiltration (Lin, 2021)
Sonodynamic NP + anti–PD–L1	Nanosonosensitizers paired with checkpoint inhibitors	ROS-mediated tumor cell death	Metastasis suppression; enhanced immune response (Meng et al., 2024)
aAPC scaffolds + CAR–T proliferation	Ellipsoidal PLGA or silica-based aAPC systems	Elevated CAR–T expansion before infusion	Larger CAR–T expansion; improved tumor-targeting (Nano–APC scaffolds for CAR–T expansion and in vivo gene programming, 2024)
CAR–T hydrogel + check-point + IL–15	Injectable hydrogel with CAR–T cells, anti–PD–L1, IL–15	Concurrent CAR–T activation and checkpoint blockade	Local and distant tumor control (Lin et al., 2024)

MDSCs, Myeloid-derived suppressor cells.

- Nano–aAPC platforms designed to optimize CAR–T expansion ex vivo are under assessment in early-phase clinical studies to improve cell production efficiency and potency ("Nano–APC scaffolds for CAR–T expansion and in vivo gene programming," 2024).
- Emerging clinical trials involving mRNA vaccine NPs combined with checkpoint inhibitors are exploring their capacity to modulate MDSCs while heightening effector T-cell responses (Bai, Liu, You, Jiang, & Li, 2024; Condamine et al., 2015; Marvel and Gabrilovich, 2015; Obermajer et al., 2011; Zhang et al., 2024).

Future trial designs must integrate MDSC-focused biomarkers, including surface phenotyping, arginase and iNOS activity assessment, and functional cytokine assays. Advanced imaging techniques, such as Positron Emission Tomography (PET) scans using nanoparticle conjugates, could track intratumoral delivery to MDSC-enriched regions, enabling dosage refinement.

Advancing these combination strategies will require updated regulatory frameworks to evaluate the safety and efficacy of multicomponent nanomedicines. Co-delivery systems combining antibodies, small molecules, and genetic payloads within single nanocarriers should adhere to new investigational new drug (IND) application models that account for combination modalities.

Combining nanomedicine with immune checkpoint inhibitors and CAR–T therapies offers a multipronged strategy to dismantle MDSC-mediated immunosuppression and enhance antitumor immunity. These approaches enable (1) direct targeting of suppressive myeloid populations, (2) optimized delivery

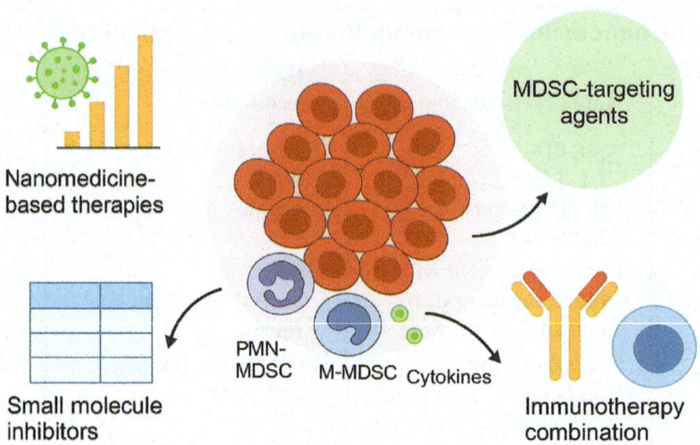

FIGURE 12.2

Approaches to combination therapy using nanomedicines to target MDSC-mediated immunosuppression. This schematic illustrates how nanomedicines enhance combination therapies—such as immune checkpoint inhibitors, CAR-T cells, and chemotherapy—by neutralizing MDSC-driven immunosuppression. These strategies synergistically promote effective tumor eradication by restoring immune surveillance. *MDSC*, Myeloid-derived suppressor cells.

of ICIs, and (3) improved CAR–T functioning in resistant TMEs. Continued integration of immunological biomarkers, advanced nanomaterial design, and clinical trial innovation is key to translating these promising preclinical findings into effective cancer therapies.

The integration of nanomedicines with checkpoint blockade and cellular immunotherapies has demonstrated enhanced immune reactivation by concurrently targeting MDSCs and tumor cells (Fig. 12. 2).

12.6 Challenges, regulatory pathways, and future perspectives

12.6.1 Scientific and translational challenges

While nanomedicine offers transformative potential for modulating MDSCs and enhancing immunotherapy, several salient challenges hinder clinical translation:

12.6.1.1 Complexity in nanomedicine design

Nanomedicines possess multifaceted physicochemical features—particle size, surface charge, composition, and colloidal stability—all critically determining in vivo behavior. These attributes can unpredictably change postadministration through interactions with proteins, forming a **protein corona**, complicating pharmacokinetics and biological effects (Highfill et al., 2014). Reproducibility remains a challenge, especially when scaling from small-scale research to uniform clinical batches (Serafini et al., 2006). Simplified yet robust design combined with scalable production protocols is essential for translation (Highfill et al., 2014; Smith, 2023; Wilhelm et al., 2016).

12.6.1.2 Manufacturing and scale-up issues

Large-scale manufacturing of biologically active complex nanomaterials (e.g., multicomponent formulations) encounters hurdles in quality control, cost, and process standardization. Batch-to-batch variability impedes consistent therapeutic outcomes, limiting regulatory approval (Nefedova et al., 2005; Serafini et al., 2006).

12.6.1.3 Preclinical model limitations

Preclinical immune-oncology models often fail to recapitulate human MDSC heterogeneity, stromal architecture, and tumor immune dynamics. This discrepancy limits the predictive validity of murine successes when entering human trials. Representative models and humanized systems must be integrated early in development (Serafini et al., 2006).

12.6.1.4 Toxicity and biodistribution

Nanoparticle biodistribution is influenced by mononuclear phagocyte uptake, renal clearance, and off-target accumulation in the liver and spleen. Unintended immunotoxicity, complement activation, or cumulative organ toxicity pose safety concerns, especially for repeated dosing (Shi et al., 2017; Nefedova et al., 2005). Rigorous biodistribution and immunotoxicity assays are mandatory.

12.6.1.5 Clinical trial-design and patient stratification

Many nano–immunotherapy trials enroll patients nonselectively without biomarker stratification, undermining response prediction. A notable example is the phase III trial JAVELIN Ovarian 200, where therapeutic effect depended on PD-L1 status—highlighting that MDSC-targeting nanomedicines likely require patient stratification by tumor MDSC infiltration or nanocarrier accumulation efficacy (Serafini et al., 2006).

12.6.2 Regulatory pathways and policy considerations

12.6.2.1 Regulatory landscape

Nanomedicines are regulated under existing frameworks by agencies like the **FDA** and **EMA**, but often lack nanotechnology-specific guidelines. Regulatory clarity is evolving around characterization methods, safety testing, and combination product classifications (Cheng et al., 2020). Key barriers include:

- Lack of harmonized standards for nanomaterial classification and toxicity evaluation.
- Ambiguities in whether nanoscale variants of known compounds require full regulatory re-evaluation.
- "Bolt-on" regulations may leave certain nanomedicine products outside oversight if not explicitly declared (Zhang, 2016).

12.6.2.2 Quality control and standardization

To gain regulatory approval, nanomedicines require validated protocols for physico-chemical characterization (size, zeta potential, and aggregation), reproducible manufacturing, and stability—consistent across all clinical batches (Serafini et al., 2006; Wilhelm et al., 2016).

12.6.2.3 Co-formulation and combination approval

Nanomedicines combining payloads such as chemotherapeutics, biologics, and siRNA raise complexity in demonstrating their combined safety and efficacy. Regulators may require each component's individual characterization and demonstration of synergy without increased toxicity (Cheng et al., 2020).

12.6.2.4 Safety and pharmacovigilance

Chronic or repeated dosing introduces risks of nanoparticle accumulation, immunogenicity, or off-target inflammation. Long-term follow-up studies and real-world safety monitoring are critical for postapproval surveillance (Cheng et al., 2020; Shi et al., 2017).

12.6.3 Future directions and strategic roadmap

12.6.3.1 Biomarker-based clinical trial-design

Integrating MDSC biomarkers—surface markers (e.g., CD11b+ CD33+ HLA-DR^low), ARG1/iNOS activity assays, cytokine profiles—into early-phase trials is essential. Such stratification aids in patient selection and mechanistic validation of nanomedicine effects.

12.6.3.2 Theranostic nanomedicine tools

Development of **theranostic NPs**—combining targeting, treatment, and imaging—enables real-time tracking of nanocarrier distribution to MDSC-rich tumor regions. This informs dosing strategies, patient selection, and therapeutic window optimization (Hou, 2019).

12.6.3.3 AI-guided nanocarrier design

Leveraging machine learning to optimize nanoparticle physicochemical attributes (size, surface ligands, and load release kinetics) may accelerate identification of lead candidates with optimal MDSC-targeting and minimal off-target effects (Binnewies et al., 2018).

12.6.3.4 Advanced platform integration

Multifunctional platforms capable of **simultaneous MDSC-targeting, antigen delivery, and immune checkpoint engagement** promise efficient combinatorial action. Strategies combining MDSC depletion or reprogramming with checkpoint inhibitors within a single nanoparticle may yield superior synergy while simplifying regulatory approval paths.

12.6.3.5 Harmonized regulatory frameworks

A unified global framework, integrating FDA, EMA, and ICH guidance tailored to nanomedicine, would improve regulatory predictability. Cooperation between academic, industry, and regulators is critical to develop validated standards for preclinical testing, clinical trial structure, and safety monitoring.

Translating nanomedicine-based therapies into clinical settings for effective MDSC modulation faces multiple challenges, including biological heterogeneity, regulatory hurdles, and delivery specificity (Table 12.5).

Table 12.5 Key Challenges and strategic solutions for MDSC–targeted nanomedicine translation.

Challenge area	Specific issue	Strategic approach
Design complexity	Multicomponent formulations & unpredictable in vivo behavior (e.g., protein corona)	Simplify design; robust in vitro–in vivo correlation
Manufacturing and scale-up	Batch variability and high production cost	Standardized, scalable GMP-compatible protocols
Preclinical predictivity	Inadequate human-relevant models	Use humanized mice and biomarker-guided validation
Toxicology and biodistribution	Unintended organ accumulation and immunotoxicity	Extensive Absorption, Distribution, Metabolism and Excretion (ADME) and immune toxicity profiling
Clinical trial-design	Lack of stratification and biomarker endpoints	Incorporate MDSC phenotyping and patient selection
Regulatory pathways	Absence of nano-specific approval frameworks	Harmonize standards with nanomedicine-specific guidelines

MDSCs, Myeloid-derived suppressor cells.

12.6.4 Concluding perspective

Translating MDSC-targeted nanomedicines into clinical therapies necessitates a multidisciplinary strategy addressing scientific, manufacturing, regulatory, and trial-design challenges. Advances in biomarker-guided patient selection, theranostic delivery, AI-driven design, and harmonized regulation provide a strategic roadmap. The future of cancer immunotherapy lies in **smart, adaptable nanoplatforms** that not only target immunosuppressive cells such as MDSCs, but also integrate immune activation, imaging, and checkpoint blockade—all while navigating rigorous regulatory and manufacturing standards. With continued collaboration across disciplines, MDSC-modulating nanomedicine can progress from experimental promise to meaningful clinical impact.

12.7 Conclusion

The integration of nanomedicine into cancer immunotherapy marks a significant leap forward in our ability to modulate the TME, particularly through the targeted regulation of MDSCs. MDSCs, known for their critical role in immune evasion and therapy resistance, pose a substantial challenge in the clinical management of various cancers. Nanotechnology-based approaches offer a promising avenue for precise, effective, and multifunctional interventions aimed at mitigating the suppressive influence of MDSCs.

Through advancements in nanoparticle engineering, liposomal formulations, dendrimers, and hybrid nanoplatforms, researchers have developed tools that can selectively deplete MDSCs, block their recruitment, or reprogram their function. These strategies not only restore immune responsiveness but also improve the outcomes of checkpoint inhibitors, CAR-T cell therapies, and conventional treatments.

Encouraging results from preclinical and early clinical studies further underscore the therapeutic potential of combining nanomedicine with other immunotherapeutic agents.

As we look to the future, the successful clinical translation of these technologies will depend on overcoming regulatory, manufacturing, and scalability challenges. A multidisciplinary effort involving oncologists, immunologists, materials scientists, and regulatory experts will be essential in harnessing the full potential of nanomedicine to overcome MDSC-mediated immunosuppression and revolutionize cancer treatment.

References

Allen, T. M., & Cullis, P. R. (2013). Liposomal drug delivery systems: From concept to clinical applications. *Advanced Drug Delivery Reviews, 65*(1), 36–48. doi:10.1016/j.addr.2012.09.037.

Arvizo, R. R. (2011). Modulatory effects of gold nanoparticles on tumor growth and metastasis. *PNAS, 108*(2), 665–670.

Bai, L., Liu, H., You, R., Jiang, X., & Li, C. (2024). Combination nano-delivery systems remodel the immunosuppressive tumor microenvironment for metastatic triple-negative breast cancer therapy. *Molecular Pharmaceutics, 21*(5), 2148–2162. doi:10.1021/acs.molpharmaceut.3c00242.

Binnewies, M., Roberts, E. W., Kersten, K., Chan, V., Fearon, D. F., Merad, M., Coussens, L. M., Gabrilovich, D. I., Ostrand-Rosenberg, S., Hedrick, C. C., Vonderheide, R. H., Pittet, M. J., Jain, R. K., Zou, W., Howcroft, T. K., Woodhouse, E. C., Weinberg, R. A., & Krummel, M. F. (2018). Understanding the tumor immune microenvironment (TIME) for effective therapy. *Nature Medicine, 24*(5), 541–550. doi:10.1038/s41591-018-0014-x.

Blanco, E., Shen, H., & Ferrari, M. (2015). Principles of nanoparticle design for overcoming biological barriers to drug delivery. *Nature Biotechnology, 33*(9), 941–951. doi:10.1038/nbt.3330.

Bobo, D., Robinson, K. J., Islam, J., Thurecht, K. J., & Corrie, S. R. (2016). Nanoparticle-based medicines: A review of FDA-approved materials and clinical trials to date. *Pharmaceutical Research, 33*(10), 2373–2387. www.wkap.nl/journalhome.htm/0724-8741. doi:10.1007/s11095-016-1958-5.

Bronte, V., Brandau, S., Chen, S.-H., Colombo, M. P., Frey, A. B., Greten, T. F., Mandruzzato, S., Murray, P. J., Ochoa, A., Ostrand-Rosenberg, S., Rodriguez, P. C., Sica, A., Umansky, V., Vonderheide, R. H., & Gabrilovich, D. I. (2016). Recommendations for myeloid-derived suppressor cell nomenclature and characterization standards. *Nature Communications, 7*(1). doi:10.1038/ncomms12150.

Chen, L. R. (2023). Multifunctional micelles block MDSC recruitment and promote differentiation to suppress metastasis. *Theranostics, 13*(8), 5457–5469.

Cheng, Q., Wei, T., Farbiak, L., Johnson, L. T., Dilliard, S. A., & Siegwart, D. J. (2020). Selective organ targeting (SORT) nanoparticles for tissue-specific mRNA delivery and CRISPR–Cas gene editing. *Nature Nanotechnology, 15*(4), 313–320. doi:10.1038/s41565-020-0669-6.

Condamine, T., & Gabrilovich, D. I. (2011). Molecular mechanisms regulating myeloid-derived suppressor cell differentiation and function. *Trends in Immunology, 32*(1), 19–25. doi:10.1016/j.it.2010.10.002.

Condamine, T., Ramachandran, I., Youn, J. I, & Gabrilovich, D. I. (2015). Regulation of tumor metastasis by myeloid-derived suppressor cells. *Annual Review of Medicine, 66*, 97–110.

Ding, Y. (2021). Nanomedicine-mediated targeting of MDSCs enhances cancer immunotherapy. *ACS Nano, 15*(7), 12342–12356.

Gabrilovich, D. I. (2017). Myeloid-derived suppressor cells. *Cancer Immunology Research, 5*(1), 3–8. http://cancerimmunolres.aacrjournals.org/content/5/1/3.full-text.pdf. doi:10.1158/2326-6066.CIR-16-0297.

Gu (2023). Anti-PD-L1 and docetaxel encapsulated pH–sensitive liposomes enhance checkpoint efficacy. *European Pharmaceutical Journal, 11*(4), 78–90.

Highfill, S. L., Cui, Y., Giles, A. J., Smith, J. P., Zhang, H., Morse, E., Kaplan, R. N., & Mackall, C. L. (2014). Disruption of CXCR2-Mediated MDSC Tumor Trafficking Enhances Anti-PD1 Efficacy. *Science Translational Medicine, 6*(237). doi:10.1126/scitranslmed.3007974.

Hou, X. (2019). Nanoparticle-based strategies for cancer immunotherapy. *Molecular cancer, 18*(1), 69.

Huang, X., Jain, P. K., El-Sayed, I. H., & El-Sayed, M. A. (2007). Gold nanoparticles: Interesting optical properties and recent applications in cancer diagnostics and therapy. *Nanomedicine, 2*(5), 681–693. http://www.futuremedicine.com/doi/pdf/10.2217/17435889.2.5.681. doi:10.2217/17435889.2.5.681.

Jhunjhunwala, S., Raimondi, G., Thomson, A. W., & Little, S. R. (2009). Delivery of rapamycin to dendritic cells using degradable microparticles. *Journal of Controlled Release, 133*(3), 191–197. doi:10.1016/j.jconrel.2008.10.011.

Jones, C. D. (2024). Hydrogel-based delivery of gemcitabine and poly(I:C) depletes MDSCs in breast cancer. *Biomaterials, 299*, 121–132.

June, C. H., Warshauer, J. T., & Bluestone, J. A. (2017). Is autoimmunity the Achilles' heel of cancer immunotherapy? *Nature Medicine, 23*(5), 540–547. http://www.nature.com/nm/index.html. doi:10.1038/nm.4321.

Kamaly, N., Yameen, B., Wu, J., & Farokhzad, O. C. (2016). Degradable controlled-release polymers and polymeric nanoparticles: Mechanisms of controlling drug release. *Chemical Reviews, 116*(4), 2602–2663. http://pubs.acs.org/journal/chreay. doi:10.1021/acs.chemrev.5b00346.

Kamerkar, S., Lebleu, V. S., Sugimoto, H., Yang, S., Ruivo, C. F., Melo, S. A., Lee, J. J., & Kalluri, R. (2017). Exosomes facilitate therapeutic targeting of oncogenic KRAS in pancreatic cancer. *Nature, 546*(7659), 498–503. http://www.nature.com/nature/index.html. doi:10.1038/nature22341.

Kim, T. W., Kim, Y., Keum, H., Jung, W., Kang, M., & Jon, S. (2022). Combination of a STAT3 inhibitor with anti-PD-1 immunotherapy is an effective treatment regimen for vemurafenib-resistant melanoma. *Molecular Therapy Oncology, 26*(1), 1–14. doi:10.1016/j.omto.2022.06.001.

Kortylewski, M., & Yu, H. (2008). Role of Stat3 in suppressing anti-tumor immunity. *Current Opinion in Immunology, 20*(2), 228–233. doi:10.1016/j.coi.2008.03.010.

Kumar, V., Patel, S., Tcyganov, E., & Gabrilovich, D. I. (2016). The nature of myeloid-derived suppressor cells in the tumor microenvironment. *Trends in Immunology, 37*(3), 208–220. www.elsevier.com/locate/it. doi:10.1016/j.it.2016.01.004.

Li, L. (2020). Dendrimer-mediated cytokine delivery reprograms MDSCs in the tumor microenvironment. *Nature Nanotechnology, 15*(8), 630–640.

Lin (2021). Nanosonosensitizer-enhanced sonodynamic therapy combined with checkpoint blockade suppresses metastasis. *International Journal of Nanomedicine, 16*, 233–245.

Lin, Y., Chen, Y., Luo, Z., & Wu, Y. (2024). Recent advances in biomaterial designs for assisting CAR-T cell therapy towards potential solid tumor treatment. *Nanoscale, 16*, 3226–3242. doi:10.1039/D3NR05768B.

Long, G. V., Dummer, R., Hamid, O., et al. (2019). Epacadostat plus pembrolizumab versus placebo plus pembrolizumab in patients with unresectable or metastatic melanoma (ECHO-301/KEYNOTE-252): a phase 3, randomised, double-blind study. *Lancet Oncology, 20*(8), 1083–1097. doi:10.1016/S1470-2045(19)30274-8.

Majoros, I. J., Mehta, C. B., & Baker, J. R. (2009). Dendrimer-based targeted delivery of an apoptotic protein in breast cancer cells. *Molecular Pharmaceutics, 6*(2), 290–301.

Marvel, D., & Gabrilovich, D. I. (2015). Myeloid-derived suppressor cells in the tumor microenvironment: Expect the unexpected. *Journal of Clinical Investigation, 125*(9), 3356–3364.

Meng, S., Du, H., Li, X., Zheng, X., Zhao, P., Yuan, Z., . . . Dai, L. (2024). An adjuvant micelle-based multifunctional nanosystem for tumor immunotherapy by remodeling three types of immunosuppressive cells. *ACS Nano, 18*(4), 3134–3135. doi:10.1021/acsnano.3c08792.

Meyer, C., Cagnon, L., Costa-Nunes, C. M., Baumgaertner, P., Montandon, N., Leyvraz, L., Michielin, O., Romano, E., & Speiser, D. E. (2014). Frequencies of circulating MDSC correlate with clinical outcome of melanoma patients treated with ipilimumab. *Cancer Immunology, Immunotherapy, 63*(3), 247–257. doi:10.1007/s00262-013-1508-5.

Nano–APC scaffolds for CAR–T expansion and in vivo gene programming. (2024). *Experimental Hematology & Oncology, 13*, 96–104.

Nefedova, Y., Nagaraj, S., Rosenbauer, A., Muro-Cacho, C., Sebti, S. M., & Gabrilovich, D. I. (2005). Regulation of dendritic cell differentiation and antitumor immune response in cancer by pharmacologic-selective inhibition of the janus-activated kinase 2/signal transducers and activators of transcription 3 pathway. *Cancer Research, 65*(20), 9525–9535.

Obermajer, N., Muthuswamy, R., Odunsi, K., Edwards, R. P, & Kalinski, P. (2011). PGE2-induced CXCL12 production and CXCR4 expression controls the accumulation of human MDSCs in ovarian cancer environment. *Cancer Research, 71*(24), 7463–7470.

Patel, R. (2023). Lipid nanocapsules delivering low-dose gemcitabine improve adoptive T-cell therapy in lymphoma. *Cancer Immunology Research, 11*(2).

Ray, A., & Dittel, B. N. (2010). Mechanisms of regulatory B cell suppression of inflammation. *Cellular and Molecular Life Sciences, 67*(22), 3759–3775.

Rosenblum, D., Joshi, N., Tao, W., Karp, J. M., & Peer, D. (2018). Progress and challenges towards targeted delivery of cancer therapeutics. *Nature Communications, 9*(1). doi:10.1038/s41467-018-03705-y.

Serafini, P., Meckel, K., Kelso, M., Noonan, K., Califano, J., Koch, W., Dolcetti, L., Bronte, V., & Borrello, I. (2006). Phosphodiesterase-5 inhibition augments endogenous antitumor immunity by reducing myeloid-derived suppressor cell function. *Journal of Experimental Medicine, 203*(12), 2691–2702. http://www.jem.org/cgi/reprint/203/12/2691. doi:10.1084/jem.20061104.

Sharma, P., & Allison, J. P. (2015). The future of immune checkpoint therapy. *Science, 348*(6230), 56–61. http://www.sciencemag.org/content/348/6230/56.full.pdf. doi:10.1126/science.aaa8172.

Shi, J., Kantoff, P. W., Wooster, R., & Farokhzad, O. C. (2017). Cancer nanomedicine: Progress, challenges and opportunities. *Nature Reviews Cancer, 17*(1), 20–37. doi:10.1038/nrc.2016.108.

Singh, P. (2023). Immune effects of SN–38 polymeric micelle NK012 in Phase II trials. *OncoImmunology, 12*(5), e1654321.

Smith, A. B. (2023). Gold nanoparticles targeting inflammasome-mediated MDSC inhibition in tumor models. *Nano Immunology, 12*(3), 145–155.

Sun, Q., Zhou, Z., Qiu, N., & Shen, Y. (2017). Rational design of cancer nanomedicine: Nanoproperty integration and synchronization. *Advanced Materials, 29*(14). doi:10.1002/adma.201606628.

Thompson, R. H. (2024). mRNA-4157 with pembrolizumab reduces melanoma recurrence in KEYNOTE–603. *The Lancet Oncology, 25*(2), 123–134.

Vázquez-Ríos, A. J., Molina-Crespo, A., Bouzo, B. L., López-López, R., Moreno-Bueno, G., & de la Fuente, M. (2019). Exosome-mimetic nanoplatforms for targeted cancer drug delivery. *Journal of Nanobiotechnology, 17*(85). doi:10.1186/s12951-019-0517-8.

Veglia, F., Perego, M., & Gabrilovich, D. (2018). Myeloid-derived suppressor cells coming of age. *Nature Immunology, 19*(2), 108–119. doi:10.1038/s41590-017-0022-x.

Waldman, A. D., Fritz, J. M., & Lenardo, M. J. (2020). A guide to cancer immunotherapy: From T cell basic science to clinical practice. *Nature Reviews Immunology, 20*(11), 651–668. http://www.nature.com/nri/index.html. doi:10.1038/s41577-020-0306-5.

Wang, S., et al. (2023). Targeted xCT-mediated ferroptosis and protumoral polarization of macrophages is effective against hepatocellular carcinoma and enhances the efficacy of anti-PD-1/PD-L1 immunotherapy. *Advanced Science, 10*(3), 2204927. doi:10.1002/advs.202204927.

Wang, Y. H. (2022). CRLX101 Phase I/IIa trial—safety and immunomodulation findings. *Clinical Cancer Research, 28*(12), 2507–2516.

Wilhelm, S., Tavares, A. J., Dai, Q., Ohta, S., Audet, J., Dvorak, H. F., & Chan, W. C. W. (2016). Analysis of nanoparticle delivery to tumours. *Nature Reviews Materials, 1*(5). doi:10.1038/natrevmats.2016.14.

Xia, Y., et al. (2021). Bifunctional Janus particles as multivalent synthetic nanoparticle antibodies (SNAbs) for selective depletion of target cells. *Nano Letters, 21*(11), 4410–4418. doi:10.1021/acs.nanolett.0c04833.

Xu, J. (2023). UCNP-Ce6-R837 PDT + CTLA–4 blockade eradicates colorectal tumors and enhances CTL infiltration. *ACS Omega, 8*(12).

Zhang, R. X. (2016). Stimuli-responsive nanoparticles for controlled drug delivery in cancer therapy. *International Journal of Nanomedicine, 11*, 4869–4887.

Zhang, Y. (2020). Nanomedicine-based strategies for treatment of MDSCs-associated immunosuppression in cancer. *Acta Pharmaceutica Sinica B, 10*(11), 2114–2116.

CHAPTER

13 Unlocking the potential of macrophages: Advancements in cellular immunotherapy

Vivek P. Chavda[a], Dixa Vaghela[a], Arya Vyas[a,b], Pankti Balar[a], Tathya J Sheth[c] and Vaishvi Chokshi[c]

[a]*Department of Pharmaceutics and Pharmaceutical Technology, L. M. College of Pharmacy, Ahmedabad, Gujarat, India,* [b]*PhD Section, Gujarat Technological University, Ahmedabad, Gujarat, India,* [c]*Pharm.D Section, L. M. College of Pharmacy, Ahmedabad, Gujarat, India*

13.1 Introduction

Macrophages are a type of white blood cell that is important for producing innate immune responses. They are involved in the detection of foreign substances or organisms, in the phagocytosis process, and in the destruction of bacteria and other harmful organisms (Chen et al., 2023). Moreover, they can present antigens to T cells and initiate the inflammation process by releasing cytokines in this process (Zhang et al., 2021). There are mainly two types of macrophages; the M1 type is activated when inflammation occurs, and they are responsible for killing pathogens, whereas anti-inflammatory signals activate M2 macrophages, and they promote tissue repair and immune regulation in the body (Strizova et al., 2023). Ultimately, macrophages are versatile cells that are able to make different forms and functions depending on the signals they receive.

Macrophages are actively involved in cellular immunotherapies, including chimeric antigen receptors-T-cell therapy, adoptive cell therapy, and as a combination therapy, especially for cancer treatment (Zhou et al., 2024). In CAR-T cell therapy, usually, the entry of genetically modified T cells is restricted in solid tumors (Maalej et al., 2023). Whereas CAR-macrophages (CAR-M) overcome this drawback as the macrophages in the tumor microenvironment could be replaced by circulating monocytes (Mantovani et al., 2022). The expression of CAR may improve macrophages' antigen-dependent capacity, including cytokine release, polarization, improved phagocytic capacity, and anticancer activity (Zhang et al., 2022). In cell-based therapies, macrophages can be extracted from a patient's blood, altered genetically to express particular genes or receptors, and then administered to the patient. This strategy can be applied to strengthen the patient's anticancer immune response (Zhang et al., 2020).

The objective of this chapter is to provide in-depth knowledge on macrophages and their function, the role of macrophages in cell-based therapies, the mechanism of immunotherapy, along with clinical trials and case studies involved in this field. The challenges associated with this therapy and possible solutions for the future are also elaborated at the end of this chapter.

Advancements in Personalized Cellular Immunotherapy: A Comprehensive Guide to Cancer Treatment.
DOI: https://doi.org/10.1016/B978-0-443-27731-3.00003-6
Copyright © 2026 Elsevier Inc. All rights are reserved, including those for text and data mining, AI training, and similar technologies.

13.2 Macrophage biology

13.2.1 Origin and development of macrophages

Macrophage development is a multifaceted process involving contributions from embryonic progenitors, bone marrow-derived monocytes, and self-renewing tissue-resident populations. Each origin highlights the unique mechanisms underlying macrophage maintenance and specialization in different tissues (De Kleer et al., 2014). Refer to the Fig. 13.1 for the basic outline of the origin and development of macrophages and monocytes.

13.2.1.1 Embryonic origins

During embryogenesis, macrophages develop independently of bone marrow-derived hematopoiesis, originating from primitive hematopoiesis in the yolk sac and transitioning later to the fetal liver. These early developmental processes establish tissue-resident macrophages, each performing specific roles within their respective tissues (McGrath et al., 2015).

Around embryonic day 7.5 in mice (equivalent to approximately 3 weeks of human development), primitive macrophages arise directly from yolk sac progenitors without transitioning through a monocyte stage (Yahara et al., 2021). These progenitors operate independently of definitive hematopoiesis, as they do not require the transcription factor Myb, which is critical for bone marrow hematopoiesis. These primitive macrophages migrate to embryonic tissues, where they seed organs and differentiate into tissue-specific resident macrophages (Sommer & Gomez Perdiguero, 2024).

Around embryonic day 10.5 in mice (5–6 weeks in humans), hematopoietic activity shifts to the fetal liver. This transition produces monocytes capable of populating tissues and supplementing yolk sac-derived macrophages (Takata et al., 2017). The combination of these embryonic sources contributes to the establishment of specialized macrophage populations in different tissues. Unlike adult monocyte-derived macrophages, embryonically derived tissue-resident macrophages are sustained through self-renewal within their respective tissues, largely independent of circulating monocytes (Lewis et al., 2021).

13.2.1.2 Bone marrow-derived macrophage

In adult organisms, macrophages are predominantly derived from monocytes produced in the bone marrow through definitive hematopoiesis. These macrophages are crucial during events such as inflammation, infection, or tissue injury (Dash et al., 2024).

Hematopoietic stem cells differentiate into common myeloid progenitors, which further develop into granulocyte-macrophage progenitors (GMPs). GMPs then give rise to monocyte precursors, which differentiate into classical ($Ly6C^{high}$ in mice; $CD14^{+}$ in humans) or nonclassical ($Ly6C^{low}$ in mice; $CD16^{+}$ in humans) monocytes (Grinenko et al., 2018).

Classical monocytes are recruited to sites of inflammation or injury, guided by chemotactic signals such as CCL2 (MCP-1) and CXCL12. In contrast, nonclassical monocytes patrol the vascular endothelium, contributing to tissue repair and homeostasis (Shi & Pamer, 2011). During acute inflammation, monocyte-derived macrophages adopt a proinflammatory phenotype (M1-like) to combat pathogens. As inflammation resolves, these macrophages shift to anti-inflammatory roles (M2-like), facilitating tissue repair (Krzyszczyk et al., 2018).

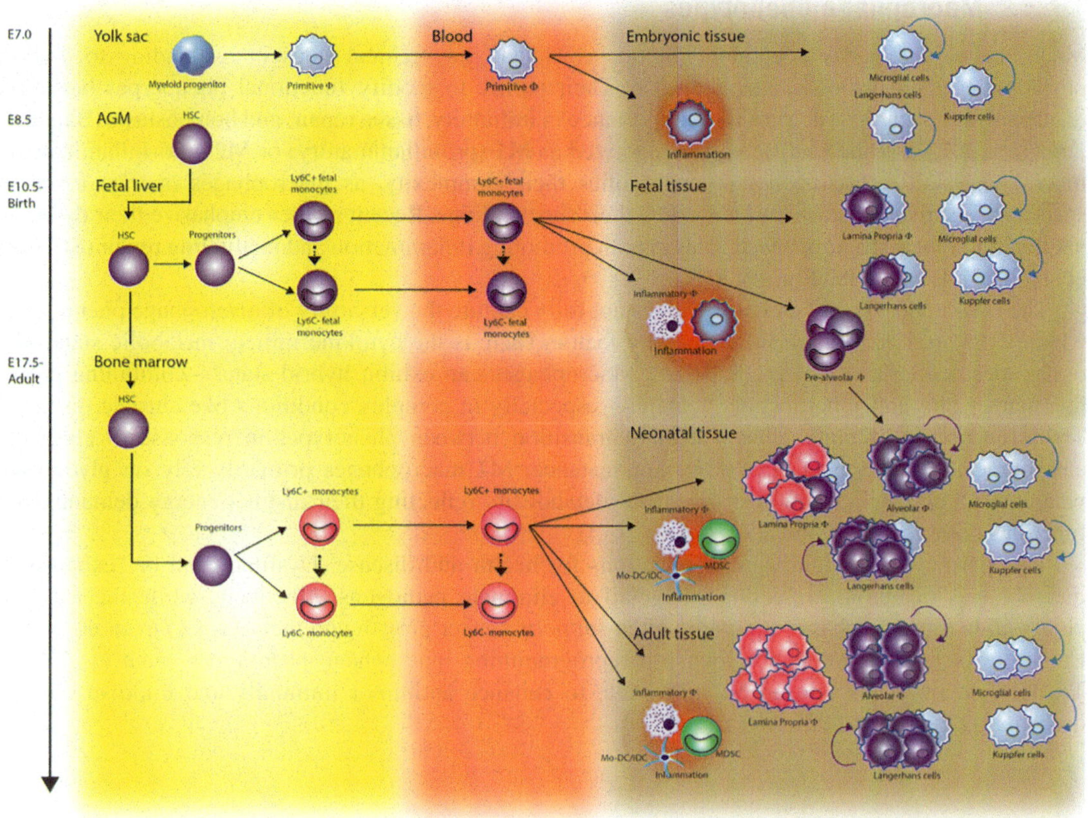

FIGURE 13.1

Monocytes and macrophages originate from diverse developmental sources throughout various stages of life. Tissue-resident macrophages, established during embryogenesis, derive from at least three major origins. Early in embryonic development, myeloid progenitors from the yolk sac give rise to specialized macrophages such as microglia in the brain, Kupffer cells in the liver, and Langerhans cells in the skin. With the onset of fetal liver hematopoiesis (around embryonic day 10.5 in mice), fetal liver monocytes differentiate into tissue-specific macrophages, further contributing to the Langerhans cell population in the skin, macrophages in the gut's lamina propria, and prebirth seeding of the lungs. These lung-seeding cells eventually mature postnatally into long-lived alveolar macrophages (AMFs) through an intermediate developmental phase known as the "preAMF" stage. Fetal monocyte-derived Langerhans cells experience a burst of proliferation immediately after birth, while macrophages in the lamina propria are continuously replenished by monocytes from the bone marrow. Beyond resident macrophage populations, $Ly6C^{++}$ monocytes can migrate to injured or infected tissues, where they differentiate into various cell types, including inflammatory macrophages, monocyte-derived dendritic cells (Mo-DCs), or myeloid-derived suppressor cells (MDSCs). This dynamic differentiation and renewal process highlights the complexity of monocyte and macrophage development throughout life.

From De Kleer, IsmÃ©, Willems, Fabienne, Lambrecht, Bart & Goriely, Stanislas. (2014). Ontogeny of myeloid cells. Frontiers in Immunology, *5. https://doi.org/10.3389/fimmu.2014.00423. (Adopted under CCBY4).*

13.2.2 Macrophage phenotypes

Macrophages are highly adaptable immune cells that play critical roles in responding to diverse physiological and pathological conditions. Their ability to modify functional phenotypes based on environmental signals underpins their importance in immunity, tissue repair, and homeostasis (Bajgar & Krejčová, 2023). While traditionally categorized as M1 (proinflammatory) or M2 (anti-inflammatory) macrophages, this classification oversimplifies their complexity, as macrophages operate across a spectrum of activation states influenced by their surroundings. Recent studies emphasize their dynamic nature, highlighting their nuanced roles in both promoting inflammation and facilitating repair (Strizova et al., 2023). Refer to the Fig. 13.2.

The traditional M1/M2 dichotomy does not fully capture the versatility of macrophage phenotypes, which exist on a continuum influenced by factors such as the cytokine milieu, metabolic state, and tissue-specific signals (Atri et al., 2018). Macrophages can exhibit hybrid states—combining proinflammatory and anti inflammatory features especially in complex conditions like chronic diseases. Their functional plasticity allows them to transition between phenotypes in response to changing environmental cues (Ni et al., 2023). For instance, M1 macrophages primarily rely on glycolysis, whereas M2 macrophages utilize oxidative metabolism, reflecting their distinct energy demands and functions (Xue et al., 2024).

This plasticity has significant implications for health and disease. Dysfunctional or "exhausted" macrophages in chronic infections may fail to eliminate pathogens effectively, while the M2-like polarization of macrophages in tumors often supports cancer growth and metastasis (Yin et al., 2024). Emerging therapeutic strategies focus on reprogramming macrophage phenotypes, such as shifting TAMs from an M2-like to an M1-like state, to enhance antitumor immunity and improve clinical outcomes (Zhao et al., 2021).

13.2.3 Functional roles in the immune system

In innate immunity, macrophages act as the first responders to microbial threats, utilizing multiple mechanisms to identify and destroy pathogens. They engage in phagocytosis to engulf and break down pathogens, dead cells, and debris, a process facilitated by pattern recognition receptors such as Toll-like receptors and scavenger receptors that recognize pathogen- or damage-associated molecular patterns (Lee et al., 2020). Upon activation, macrophages release cytokines and chemokines, that recruit other immune cells, such as neutrophils and dendritic cells, to amplify the immune response (Li & Wu, 2021). M1 macrophages, a proinflammatory subtype, also produce reactive oxygen species (ROS) and reactive nitrogen species, including nitric oxide (NO), which possess potent antimicrobial properties. These combined mechanisms ensure an effective frontline defense against infection (Hirayama et al., 2017).

Beyond innate immunity, macrophages play a key role in adaptive immune responses by acting as APCs (Dunkelberger & Song, 2010). They process and present antigens to T cells via MHC molecules, facilitating T-cell activation. Through MHC class II molecules, macrophages present antigens to CD4+ helper T cells, which subsequently orchestrate broader immune responses. The expression of costimulatory molecules such as CD80 and CD86 further strengthens T-cell activation (Couture et al., 2019). Additionally, macrophages influence T-cell differentiation by secreting cytokines like IL-12, which supports Th1 responses, and IL-10, which promotes regulatory T cells to prevent overactive

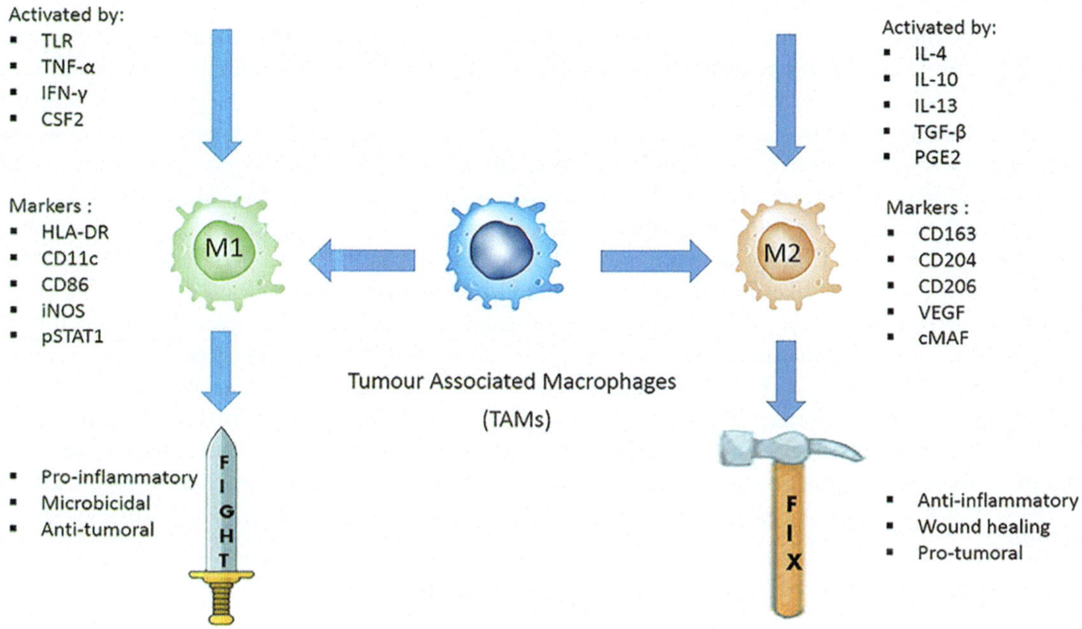

FIGURE 13.2

The activation pathways, roles, and markers of M1 and M2 TAMs are essential for understanding their influence within the TME. M1 TAMs are activated via TLR signaling or stimulation by Th1 cytokines, such as TNF-α, IFNγ, and CSF2. These macrophages are characterized by their proinflammatory and microbicidal functions, which contribute to their antitumor effects. Typical immunohistochemistry (IHC) markers for M1 TAMs include HLA-DR, CD11c, CD86, inducible nitric oxide synthase (iNOS), and phosphorylated STAT1 (pSTAT1). Conversely, M2 TAMs are activated by signals like interleukins IL-4, IL-10, and IL-13, along with TGFβ and prostaglandin E2 (PGE2). These macrophages play a role in anti-inflammatory processes, tissue repair, and tumor progression. Common IHC markers used to identify M2 TAMs include CD163, CD204, CD206, VEGF, and cMAF.

From Jayasingam, S.D., Citartan, M., Thang, T.H., Mat Zin, A.A., Ang, K.C. & Ch'ng, E.S. (2020). Evaluating the polarization of tumor-associated macrophages into M1 and M2 phenotypes in human cancer tissue: Technicalities and challenges in routine clinical practice. Frontiers in Oncology, *9. https://doi.org/10.3389/fonc.2019.01512. (Adopted under CCBY4).*

immune responses. These roles demonstrate how macrophages bridge the innate and adaptive branches of immunity (Wang et al., 2024).

Macrophages also contribute significantly to tissue repair and maintenance in noninfectious scenarios. Through efferocytosis, they clear apoptotic cells, preventing secondary necrosis and the release of harmful inflammatory mediators (Mehrotra & Ravichandran, 2022). M2 macrophages, a reparative subtype, support tissue remodeling by regulating extracellular matrix turnover and promoting angiogenesis, particularly during wound healing. They also release factors that support stem cell proliferation and differentiation in regenerative environments, such as the bone marrow (Zhao et al., 2024). These functions underscore their importance in maintaining tissue health and facilitating recovery after injury.

During the resolution phase of inflammation, macrophages play a pivotal role in restoring homeostasis. They often transition to an M2-like phenotype, characterized by anti-inflammatory and tissue-healing properties. This shift helps resolve inflammation and prevent chronic immune activation (Meng et al., 2024). M2 macrophages also enhance the activity of regulatory T cells (Tregs), which suppress excessive immune responses and maintain immune balance. Furthermore, macrophages produce specialized resolving mediators b , including resolving, and lipoxins, that actively promote inflammation resolution, modulate immune cell activity, and support tissue repair (Watanabe et al., 2019).

Despite their protective roles, macrophages can contribute to pathological conditions when their regulation is disrupted. Chronic activation of M1 macrophages is linked to persistent inflammation, a hallmark of conditions like rheumatoid arthritis and inflammatory bowel disease (Kulakova et al., 2024). Conversely, within the TME, macrophages often adopt an M2-like phenotype, becoming tumor-associated macrophages that promote cancer progression by supporting tumor growth, angiogenesis, and immune evasion (Hao et al., 2012). Excessive activity of M2 macrophages can also drive pathological fibrosis, as seen in diseases like liver cirrhosis and idiopathic pulmonary fibrosis, where excessive ECM production leads to tissue dysfunction. These dual-edged roles make macrophages important targets for therapeutic strategies aimed at modulating their function in disease contexts (Xia et al., 2023).

13.3 Mechanisms of action in cellular immunotherapy

Cellular immunotherapy represents a promising avenue in modern medicine, harnessing the power of immune cells to combat diseases, particularly cancer, autoimmune disorders, and infections. Understanding the intricate mechanisms by which cellular immunotherapy operates is crucial for advancing therapeutic strategies (Mukherjee et al., 2022).

13.3.1 Phagocytosis and antigen presentation

Phagocytosis is a critical function of immune cells, involving the engulfment and internalization of extracellular particles such as pathogens, apoptotic cells, and debris. This mechanism plays a vital role in maintaining tissue health, preventing infections, and initiating immune responses (Rosales & Uribe-Querol, 2017).

Phagocytosis and antigen presentation are fundamental processes in the immune system, ensuring effective surveillance, pathogen elimination, and activation of adaptive immunity (Chi et al., 2024). Macrophages, DCs, and B cells are the primary immune cells involved in these mechanisms. Macrophages, found across various tissues, act as immune sentinels by identifying pathogens or damaged cells using PRRs such as TLRs and scavenger receptors (Marshall et al., 2018). Upon engulfment, they release cytokines like IL-6 and TNF-α, which recruit additional immune cells to the affected site. Dendritic cells, highly efficient in phagocytosis, specialize in capturing antigens through receptor-mediated phagocytosis or micropinocytosis in their immature state. They mature by upregulating MHC molecules and costimulatory signals to activate T cells (Dudek et al., 2013). B cells, although less phagocytic, play a role in antigen presentation by internalizing antigens bound to their immunoglobulin receptors, facilitating the priming of T helper cells (Ma et al., 2024).

Phagocytosis involves two main mechanisms: receptor-mediated phagocytosis and micropinocytosis. The former enables the recognition and internalization of antigens via receptors like Fcγ receptors, which bind antibody-coated particles, and complement receptors, which recognize opsonized targets (Gordon, 2016). Micropinocytosis, a nonspecific process, allows immune cells to sample their environment by engulfing extracellular fluid containing soluble antigens. These processes can be therapeutically enhanced to target specific diseases (Uribe-Querol & Rosales, 2020). Engineered phagocytes, such as macrophages or DCs with increased Fcγ receptor expression, improve ADCP, enhancing the clearance of antibody-coated tumor cells. Additionally, checkpoint inhibitors targeting molecules like CD47, a "don't eat me" signal expressed by many tumor cells, restore macrophage-mediated phagocytosis, enabling effective tumor cell elimination (Lau et al., 2023).

Antigen processing and presentation are pivotal for bridging innate and adaptive immunity. Internalized antigens are enclosed within endosomes or phagosomes, which fuse with lysosomes where enzymatic degradation produces peptide fragments (Embgenbroich & Burgdorf, 2018). These peptides are then loaded onto MHC molecules. The MHC class I pathway processes intracellular antigens, such as viral or tumor peptides, which are transported into the endoplasmic reticulum by the transporter associated with antigen processing (TAP) (Münz, 2012). These peptides are subsequently loaded onto MHC class I molecules for recognition by cytotoxic $CD8^+$ T cells. The MHC class II pathway handles extracellular antigens, loading them onto MHC class II molecules in endosomal compartments for presentation to $CD4^+$ helper T cells (You et al., 2017). Cross-presentation, a unique feature of some APCs, allows extracellular antigens to be presented on MHC class I molecules, stimulating $CD8^+$ T cells and enhancing responses against tumors and viral infections (Colbert et al., 2020).

13.3.2 Cytokine production and immune modulation

Cytokines are crucial signaling molecules that regulate immune responses by facilitating communication, activation, and coordination among immune cells. They significantly influence processes like immune cell proliferation and differentiation, making them vital in cellular immunotherapy. By harnessing their ability to fine-tune immune responses, cytokines are used to generate precise and effective antitumor activity (Bhol et al., 2024).

In cellular immunotherapy, key immune cells such as T cells, natural killer (NK) cells, and macrophages play prominent roles in cytokine secretion, thereby amplifying their functions and recruiting other immune effectors. Activated $CD4^+$ T helper cells, for instance, produce cytokines like IL-2 to promote the proliferation and activation of cytotoxic T lymphocytes and NK cells (Yang et al., 2023). Similarly, IL-12 enhances the differentiation of Th1 cells and further stimulates NK cell activity. Tregs, on the other hand, secrete IL-10 and TGF-β to suppress excessive immune activity and maintain homeostasis, though their suppressive role often poses challenges in cancer treatment. Strategies such as selective depletion or reprogramming of Tregs are being developed to address this issue and bolster antitumor responses (Ochayon & Waggoner, 2021).

NK cells contribute to antitumor immunity by releasing IFN-γ, a cytokine that activates macrophages, enhances antigen presentation, and fosters a Th1-skewed immune environment. IFN-γ also improves tumor immunogenicity by upregulating MHC molecule expression on tumor cells, making them more recognizable to immune effectors (Hu et al., 2024). Similarly, macrophages, particularly the M1 phenotype, secrete proinflammatory cytokines such as TNF-α, IL-6, and IL-1β, which create a hostile environment for tumor cells while attracting other immune cells to the site of inflammation (Portale & Di Mitri, 2023).

Cytokines can modulate immune responses in both stimulating and regulatory capacities. Proinflammatory cytokines like IL-2, IL-12, and GM-CSF enhance the recruitment, activation, and proliferation of immune cells, significantly amplifying antitumor responses. IL-2 is especially pivotal for CTL and NK cell expansion, while GM-CSF plays a key role in recruiting and maturing dendritic cells to enhance antigen presentation (Abdul-Rahman et al., 2024). Conversely, anti-inflammatory cytokines such as IL-10 and TGF-β help regulate inflammation and prevent autoimmune responses. However, tumors often exploit these cytokines to suppress the immune system. Targeting such pathways to rebalance the immune environment is a promising strategy for driving effective antitumor immunity (Yi et al., 2024).

13.3.3 Interaction with other immune cells

Cellular immunotherapy involves a sophisticated network of interactions among immune cells to generate a strong and sustained antitumor response. These interactions, facilitated through direct cell-cell contact, cytokine signaling, and immunological synapse formation, are integral to the therapy's effectiveness (Chandran & Klebanoff, 2019). T cells play a central role in this approach, with advanced techniques such as CAR-T cells and tumor-infiltrating lymphocytes being pivotal. Their activation depends on two main signals from APCs: the TCR engaging with MHC molecules, and costimulatory signals from molecules like CD28-CD80/CD86 or ICOS-ICOSL (Ahmed et al., 2023). Immune checkpoint inhibitors targeting PD-1, PD-L1, and CTLA-4 are often employed to counter inhibitory signals, enhancing T-cell responses. CAR-T cells bypass traditional TCR-MHC signaling by directly targeting tumor antigens and incorporating costimulatory domains such as 4-1BB or CD28, which enhance their functionality and persistence (Zabeti Touchaei & Vahidi, 2024).

The synergy between T cells and NK cells offers a dual mechanism for tumor elimination. While T cells focus on antigen-specific MHC-presenting tumor cells, NK cells target those with reduced MHC expression, a common escape mechanism by tumors. Helper T cells ($CD4^+$) further support immune responses by stimulating B cells to produce antibodies via cytokines like IL-4 and IL-21 and interactions such as CD40-CD40L (Chen et al., 2024). NK cells, critical players in innate immunity, strengthen their function by interacting with macrophages and DCs. Macrophages activated by NK cell-derived IFN-γ show enhanced tumor-killing capabilities, while in turn, macrophages secrete IL-12 to amplify NK cell activity. Similarly, DCs release cytokines like IL-12, IL-15, and type I interferons to support NK cell activation, while NK cells maintain the quality of immune responses by eliminating dysfunctional DCs (Russo et al., 2021).

Overcoming the suppressive effects of the TME is a significant focus in cellular immunotherapy. The presence of myeloid-derived suppressor cells (MDSCs) in the TME poses a challenge by dampening the immune response. Strategies to mitigate their effects include engineering CAR-T cells with chemokine receptors to block MDSC recruitment, inhibiting pathways like arginase-1 and iNOS, and using small molecules or monoclonal antibodies to disrupt MDSC activity (Liu et al., 2022). Targeting immunosuppressive factors within the TME, such as TGF-β, or enhancing immune cell infiltration by modifying cells to produce cytokines like IL-12 or express chemokine receptors, are also effective approaches. Additionally, disrupting tumor-associated fibroblasts can reduce physical barriers in the TME, improving immune cell infiltration (Sui et al., 2022).

A defining feature of successful cellular immunotherapy is the generation of immune memory, providing long-term protection against tumor recurrence. Memory T cells are essential for sustained

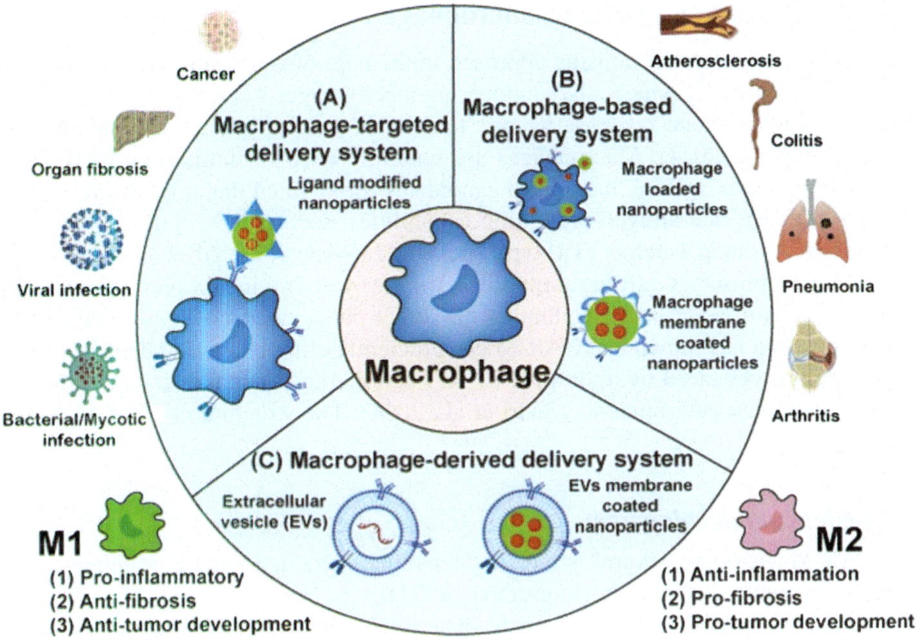

FIGURE 13.3

Therapeutic applications of macrophages for targeted therapy. Immunotherapeutic application of macrophages involves CAR-macrophage therapy and macrophage function regulation, while macrophages are used in therapeutic delivery as carriers, extracellular vesicle-based delivery, and membrane-coated carriers.

From Wei, Fang, Liu, Haiyang, Wang, Yuxiao, Li, Yan & Han, Shuo. (2024). Engineering macrophages and their derivatives: A new hope for antitumor therapy. Biomedicine & Pharmacotherapy, *177. https://doi.org/10.1016/j.biopha.2024.116925. (Adopted CCBY4).*

responses, supported by cytokines like IL-2, IL-7, and IL-15 secreted by helper T cells. Efforts to optimize CAR-T cell therapies focus on expanding memory-enriched populations to ensure durability and proliferative potential postinfusion (Zhou et al., 2022). Combining cellular immunotherapy with other modalities enhances its efficacy. Checkpoint inhibitors counteract T-cell exhaustion, whereas cytokine adjuvants such as IL-12 and IL-15, improve effector cell proliferation. Additionally, oncolytic viruses, which selectively target tumor cells, can boost antigen release, enhancing tumor visibility to the immune system and synergizing with cellular therapies (Waldman et al., 2020).

13.4 Engineering macrophages for therapeutic applications

Macrophages can be modified for the delivery of various therapeutics. Macrophages can be used to deliver therapeutic agents intracellularly, for nanoparticle delivery, and extracellular vesicles-coated macrophages for targeted drug delivery (Wei et al., 2024). A detailed application is described in Fig. 13.3.

13.4.1 Chimeric antigen receptor macrophages

CAR-M represents a novel and promising approach in the field of cancer immunotherapy, particularly for targeting solid tumors. By genetically engineering macrophages to express CARs, researchers aim to enhance the innate immune response against tumors, leveraging the unique capabilities of these immune cells (Lu et al., 2024). Macrophages are naturally adept at infiltrating TMEs and possess potent phagocytic abilities, making them ideal candidates for targeted therapies. Moreover, these cells have antitumor capacity, and antigen-presenting capabilities, secrete ROS, produce proinflammatory cytokines, and have a role in T-helper (Th) type 1 response (Wang et al., 2019).

CAR-M treatment promotes a strong antitumor response by improving the recruitment and activation of other immune cells in addition to facilitating the direct phagocytosis of tumor cells. In preclinical models, recent research has shown that CAR-M can efficiently eliminate tumor cells, showing promise in overcoming challenges faced by traditional CAR-T-cell therapies, such as tumor heterogeneity and immunosuppressive microenvironments (Sadri et al., 2024). The structure of CAR-M is described in Fig. 13.4.

13.4.1.1 Design and development

Designing CAR-M involves several strategic steps that produce innate immune properties of macrophages for targeted therapeutics (Sloas et al., 2021).

The extracellular domain typically consists of a single-chain variable fragment that recognizes specific tumor-associated antigens . This domain is crucial for directing the CAR-M to the target cells. Common targets include CD19, HER2, and mesothelin (Yu et al., 2017). The transmembrane domain helps attach the CAR to the macrophage's outer membrane, allowing signals to be sent inside the cell when it recognizes an antigen. The transmembrane domain can be derived from various sources but often resembles those used in CAR-T cell designs (Mazinani & Rahbarizadeh, 2022). The critical domain to activate macrophage function upon antigen binding is the intracellular signaling domain, The CD3ζ signaling domain contains immunoreceptor tyrosine-based activation motifs that promote phagocytosis and cytokine release. Some designs incorporate additional domains from murine phagocytic receptors, such as Megf10 or FcRγ, to enhance phagocytic activity (Huang et al., 2024).

The design and development of CAR-macrophages represent a significant advancement in cancer immunotherapy. By harnessing the innate properties of macrophages through innovative engineering approaches, CAR-M holds great potential to overcome some limitations faced by traditional CAR-T therapies in treating solid tumors. As research progresses, CAR-M may emerge as a vital tool in the arsenal against cancer, offering new hope for patients with challenging malignancies.

13.4.1.2 Preclinical studies and efficacy

The preclinical and clinical studies of CAR-M mainly focus on the challenges associated with CAR-T therapies, assessing the efficacy of CAR-M to reduce tumor cell growth, and evaluating the safety of this therapy. Initial trials showed that CAR-M can effectively (Li et al., 2024). The researchers investigated first-generation CARs targeting CD19, mesothelin, and HER2 in the THP-1 macrophage cell line. They found that CAR-modified macrophages could specifically engulf and destroy tumor cells expressing the corresponding antigens. Using two mouse models for solid tumor xenotransplantation, the study

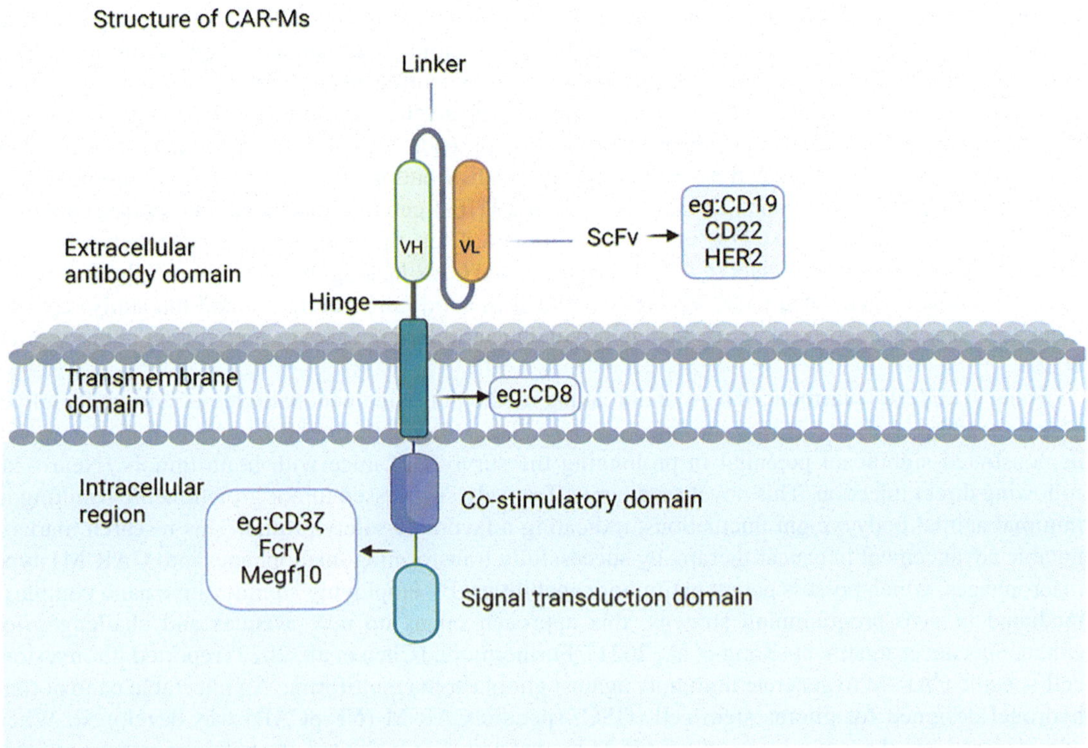

FIGURE 13.4

CAR structure of CAR-macrophage. The structure of CAR-M is almost similar to CAR-T cells. It consists of three domains, the extracellular, transmembrane, and intracellular domain, followed by two additional domains in the intracellular region.

From Huang, Ting, Bei, Chenqi, Hu, Zhenhua & Li, Yuanyuan. (2024). CAR-macrophage: Breaking new ground in cellular immunotherapy. Frontiers in Cell and Developmental Biology, *12. https://doi.org/10.3389/fcell.2024.1464218. (Adopted under CCBY4).*

demonstrated that a single infusion of human CAR-modified macrophages led to a reduction in tumor size and improved overall survival rates (Huang et al., 2024).

Further studies utilizing humanized mouse models have demonstrated that CAR-macrophages (CAR-Ms) can foster a proinflammatory environment within tumors and enhance the effectiveness of antitumor T cells. This groundbreaking research has advanced to clinical trials. Morrissey et al. (2018) developed an innovative CAR variant called CAR-P, aimed at improving phagocytosis. When CAR-P was introduced into macrophages, these modified cells effectively identified and targeted beads coated with proteins typically associated with cancer cells. Additionally, the engineered macrophages exhibited the capability to inhibit the proliferation of live cancer cells in vitro by actively attacking them (Morrissey et al., 2018).

Then Zhang et al studied the integration of CD147 signaling molecule into the transmembrane region of a HER2 CAR to decrease collagen buildup in tumors and enhance T-cell infiltration. This was achieved by effectively stimulating the expression of matrix metalloproteinases in macrophages upon recognizing the HER2 antigen. The findings indicate that targeting the ECM via engineered macrophages could serve as a promising therapeutic approach for solid tumors (Zhang et al., 2019). Klichinsky et al. (2020) evaluated the trafficking and biodistribution of CAR-M for cancer immunotherapy in five different solid tumors. Human-macrophage THP-1 cell line was treated with first-generation anti-CD19 CAR encoding the CD3ζ intracellular domain. Then, they measured the phagocytic ability of macrophages expressing an intact (CAR19ζ) or truncated CAR lacking the CD3ζ intracellular domain. Then the effect of generated CAR-Ms was evaluated in solid tumor antigens, and the study reported that CD3ζ-based CAR-M can be effective in producing antitumor phagocytic activity (Klichinsky et al., 2020).

Kang et al. (2021) introduced a groundbreaking method utilizing nanoparticles as carriers to deliver genes encoding CAR and IFN-γ into macrophages. The creation of the MPEI/pCAR-IFN-γ complex demonstrated significant potential in prolonging the survival of mice with brain tumors (Neuro-2a) following direct injection. This novel treatment effectively suppressed tumor growth while resulting in minimal animal body weight fluctuations, indicating a favorable safety profile. This research marks a notable advancement in cancer therapy by successfully transforming macrophages into CAR-M1-type macrophages, which possess potent antitumor capabilities. By employing an innovative nano complex-mediated in vivo programming strategy, this approach opens up new avenues and challenges for enhancing cancer treatment (Kang et al., 2021). Furthermore, (Chen et al., 2022) reported glioma stem cell-specific CAR-M to generate immunity against glioblastoma multiforme. An injectable nano porter-hydrogel designed for glioma stem cell (GSC)-specific CAR-M (NP-pCAR) was developed. When administered into the tumor cavity, the CAR-M loaded with GSCs elicited an antitumor response within the TME and prevented postoperative glioma recurrence by promoting long-lasting autoimmunity in mouse models. The response of EGFP-positive macrophages to CD133^{+} cancer cells is described in Fig. 13.5 (Chen et al., 2022).

Then Chen et al evaluated the application of HER2 and CD47 CAR-M to treat ovarian cancer. They constructed CAR-M and verified its application in vivo and in vitro. The in vitro study reported that CAR-M induces antigen-specific phagocytosis by activating CD8+ cytotoxic effect in ovarian cancer cell lines, and in mice model it affected tumor-associated macrophage phenotype with CD8+ activation and led to tumor suppression (Chen et al., 2023). Treatments that directly target fibrosis are limited, and liver fibrosis cases are gradually increased, so as therapy Dai et al evaluate the effect of CAR-M in preclinical models. After through treatment with CAR-M, a significant reduction was observed in murine models. CAR-Ms altered the immune microenvironment of the liver to attract and adjust the activity of native immune cells, promoting the regression of fibrosis (Dai et al., 2024).

In summary, the progress made in CAR-M cell therapy has demonstrated notable effects in both preclinical and clinical settings, although careful evaluation of safety and effectiveness is essential. The transition of CAR-M therapies into clinical trials, exemplified by the FDA's approval of CT-0508, marks a significant achievement in this area. Early clinical trials have yielded encouraging outcomes, particularly in the treatment of solid tumors, laying a foundation for further exploration of CAR-M therapies.

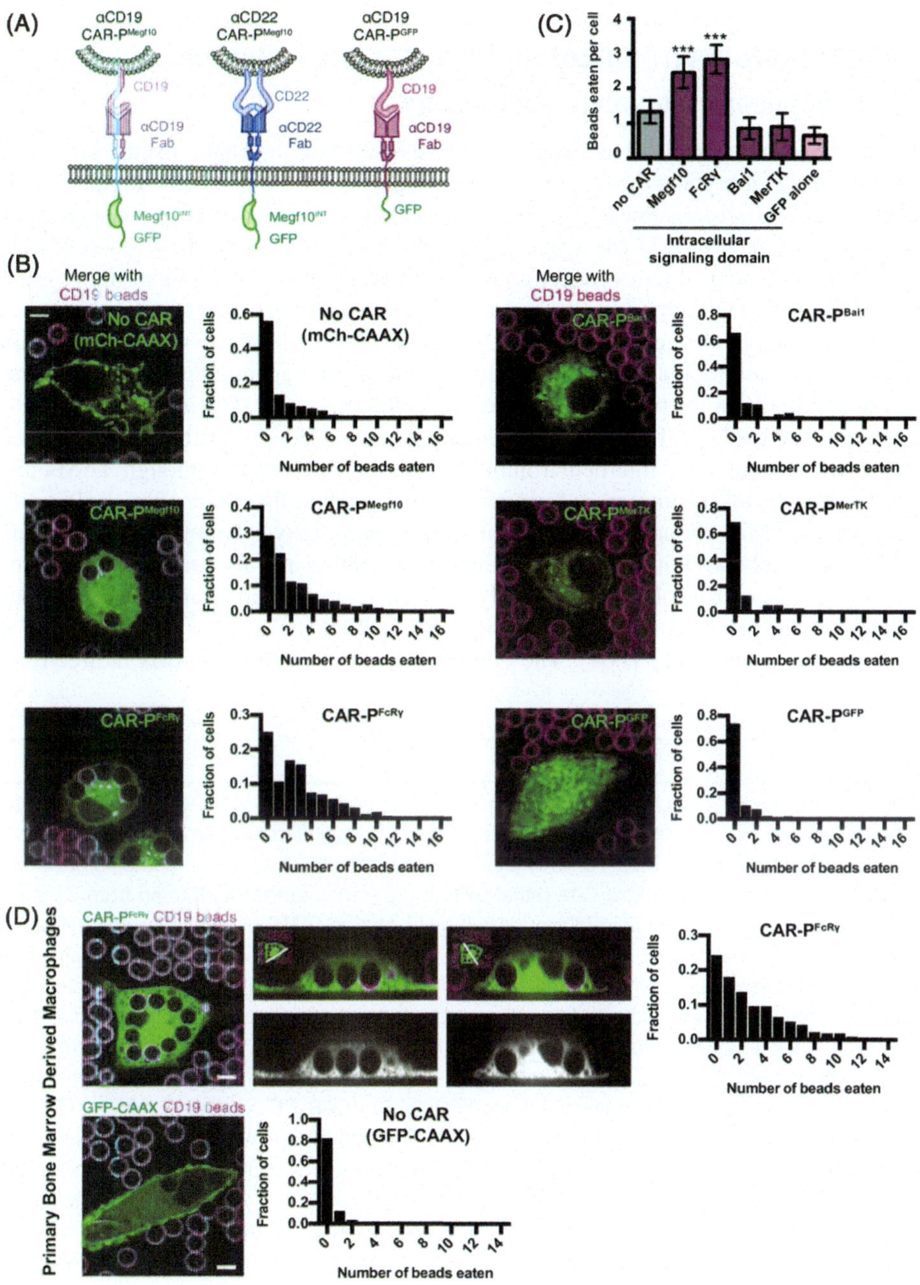
(A)
αCD19 CAR-P^Megf10
αCD22 CAR-P^Megf10
αCD19 CAR-P^GFP
CD19
αCD19 Fab
CD22
αCD22 Fab
Megf10^INT
GFP
(C)
Beads eaten per cell
no CAR
Megf10
FcRγ
Bai1
MerTK
GFP alone
Intracellular signaling domain
(B)
Merge with CD19 beads
No CAR (mCh-CAAX)
CAR-P^Bai1
CAR-P^Megf10
CAR-P^MerTK
CAR-P^FcRγ
CAR-P^GFP
Fraction of cells
Number of beads eaten
(D)
Primary Bone Marrow Derived Macrophages
CAR-P^FcRγ CD19 beads
GFP-CAAX CD19 beads
No CAR (GFP-CAAX)

13.5 Tumor microenvironment and macrophage dynamics

13.5.1 Role of tumor-associated macrophage

Macrophages are immune cells that reside within tissues and originate from multiple waves of hematopoiesis during embryonic development, becoming established in specific organs (Orkin & Zon, 2008). TAMs were initially assumed to represent remnants of a failed immune response against the tumor (Fidler & Schroit, 1988). TAMs actively support the development of new blood vessels through angiogenesis, a process that supplies cancer cells with vital nutrients and oxygen (Shi & Shiao, 2018). Numerous studies have highlighted the critical role of TAMs in promoting tumor angiogenesis by releasing several proangiogenic factors, including VEGF-A (Arnold, 1985). Compounds targeting VEGF and its receptors are gaining recognition as highly promising therapeutic agents and are increasingly being applied, especially in the treatment of nonsmall cell lung cancer (Chen et al., 2011).

Another significant role of TAMs within primary tumors is their contribution to the creation and persistence of an inflammatory microenvironment (Tong et al., 2016). Although TAMs are initially programmed to trigger an inflammatory response against tumors, their exceptional plasticity enables them to adjust to the TME, ultimately adopting an immunosuppressive phenotype. This shift in function enables TAMs to actively promote immune tolerance by recruiting and also triggering regulatory T cells (Tregs). This phenomenon has been particularly well-characterized in ovarian cancer, where TAMs contribute to the establishment of an immunosuppressive TME that limits the efficacy of antitumor immune responses (Curiel et al., 2004). The clinical impact of targeting TAMs indirectly on patient

FIGURE 13.5

(A) Schematics depict the CAR-P constructions' structure. CAR specificity is directed by an αCD19 (purple) or αCD22 (blue, middle) scFv. Engulfment is triggered by either the specified engulfment receptor (green) or intracellular signaling regions from Megf10. Only GFP and no intracellular signaling domains are present in CAR-PGFP (right). Every construct has a C-terminal GFP and a transmembrane domain from CD8. (B) 5 μm silica beads wrapped in a supported lipid bilayer carrying His-tagged CD19 extracellular domain are engulfed by J774A.1 macrophages expressing αCD19 CAR-P with the corresponding intracellular signaling domain (green). The beads were visualized with atto390-labeled lipid incorporated into the supported lipid bilayer (magenta). As a control, cells infected with the cell membrane marker mCherry-CAAX were utilized (no CAR, top left). The frequency of cells consuming the specified number of beads is shown in histograms. (C) quantifies the average number of beads consumed per cell. (D) CD19 beads were treated with bone marrow-derived macrophages that expressed either GFP-CAAX or CAR-PFcRγ for 45 minutes. Pictures show internalized beads in cross-section and engulfment action. The number of swallowed beads per cell is shown in histograms. Scale bar = 5 μm; three separate studies yielded n = 78–163 cells per condition. Using the Kruskal-Wallis test with Dunn's multiple comparison correction, error bars show 95% confidence intervals, and denotes $p < 0.0001$ in comparison to the mCherry-CAAX control. Adopted from [2] under BY CC 4.0.

From Chen, Chen, Jing, Weiqiang, Chen, Yu, Wang, Ganyu, Abdalla, Mohnad, Gao, Lin, Han, Maosen, Shi, Chongdeng, Li, Anning, Sun, Peng, Jiang, Xin, Yang, Zhenmei, Zhang, Shengchang, Zhang, Jing, Tang, Chunwei, Liu, Ying, Zhang, Rui, Xu, Fengbo, Dong, Baixiang, ... Jiang, Xinyi. (2022). Intracavity generation of glioma stem cell–specific CAR macrophages primes locoregional immunity for postoperative glioblastoma therapy. Science Translational Medicine, *14(656). https://doi.org/10.1126/scitranslmed.abn1128. (Adopted under CCBY4).*

responses to treatment remains largely unknown and warrants further investigation, especially considering the variable efficacy of current immunotherapies. The observed heterogeneity in patient responses to treatment may be partially attributed to the influence of TAMs, highlighting the need for a profound understanding of their role in shaping tumor immunity and treatment outcomes. However, patients who received anti-CSF-1 antibody therapy concurrently exhibited improved treatment responses (Escamilla et al., 2015).

TAMs also play a role in cancer progression by driving epithelial-mesenchymal transition (EMT). EMT is a biological process in which cancer cells, originally displaying epithelial traits, undergo a phenotypic transformation, characterized by the loss of cell-cell adhesion and the acquisition of mesenchymal-like properties. This transition enhances cellular motility and invasiveness, facilitating their migration to distant sites and the formation of metastatic tumors (Thiery et al., 2009).

EMT plays a pivotal role in facilitating the dissemination of metastatic cancer cells. At the molecular level, EMT is controlled by transcription factors such as zinc-finger E-box binding homeobox factor 1 (ZEB1) (Sánchez-Tilló et al., 2010). Furthermore, mesenchymal cells secrete GM-CSF, which activates TAMs and promotes the secretion of CCL18, establishing a positive feedback loop that amplifies EMT (Su et al., 2014).

TAMs play a vital role in ECM remodeling, working in tandem with cancer-associated fibroblasts to alter the TME and promote tumor cell intravasation. This coordinated process involves the degradation of existing ECM components and the synthesis of new ECM proteins, fostering a supportive environment for tumor cell invasion and metastasis (Madsen et al., 2017). TAMs produce an array of proteases, including MMPs, secreted protein acidic and rich in cysteine, and cathepsins, which work together to break down ECM collagen fibers, thereby promoting tumor cell invasion and metastatic spread (Afik et al., 2016; Bahr et al., 2022; Sangaletti et al., 2008). A specific subgroup of TAMs expressing fibroblast-activating protein (FAP)-α and heme oxygenase (HO)-1 has been implicated in ECM remodeling. FAP-α functions as both a signaling molecule for CAFs and a collagenase, while HO-1 supports ECM degradation (Arnold et al., 2014; Muliaditan et al., 2018).

The mechanisms involved in both tumor formation and wound healing share striking similarities, aligning with the well-known concept that tumors can be likened to "wounds that do not heal". Both processes involve intricate interactions between various cell types, including immune cells, fibroblasts, and endothelial cells, and rely on similar molecular signaling pathways. These shared mechanisms contribute to the deposition of ECM, angiogenesis, recruitment of inflammatory cells, and tissue remodeling. However, unlike the controlled and time-limited process of wound healing, tumor progression is characterized by a dysregulated and persistent activation of these mechanisms, leading to uncontrolled cell proliferation, invasion, and metastasis (Flier et al., 1986).

Both ECM and EMT remodeling play critical roles in enabling tumor cells to invade the bloodstream and disseminate to distant organs. This crucial stage in metastasis occurs at distinct regions within the TME, referred to as "tumor microenvironment of metastasis doorways." These specialized sites promote the intravasation of cancer cells into the circulatory system, facilitating their migration to other organs (Harney et al., 2015; Robinson et al., 2009; Wyckoff et al., 2007). The abundance of TMEM sites within tumors has been associated with increased metastatic burden and is considered a potential prognostic marker for evaluating metastatic risk (Robinson et al., 2009; Sparano et al., 2017).

The establishment of a paracrine signaling loop between cancer cells and TAMs facilitates their collective migration toward TMEM doorways. Cancer cells produce colony-stimulating factor 1 (CSF-1) to attract TAMs, while TAMs release epidermal growth factor to stimulate cancer cell motility.

Consequently, blocking CSF-1 or EGF receptors has been shown to decrease cancer cell migration and invasiveness in rodent models of breast cancer (Wyckoff et al., 2004). Macrophages, particularly TAMs, are crucial in facilitating metastasis. They actively support the migration and intravasation of cancer cells into the bloodstream, thereby contributing to the spread of cancer to distant organs.

13.5.2 Mechanism of tumor progression vs suppression

Carcinogenesis is a multifaceted process that entails various molecular and cellular events at both the genetic and epigenetic levels. This process is influenced by a combination of environmental, genetic, and metabolic factors. Regardless of the specific order in which genetic or epigenetic alterations occur, the process typically involves three main stages: initiation, promotion, and progression. The promotion stage of cancer development involves the clonal expansion of initiated cells. These cells, having evaded apoptosis and acquired genetic or epigenetic alterations, undergo sustained proliferation and acquire additional mutations that confer them with a selective growth advantage. This process results in the formation of a clonal population of cells with increased malignant potential. The promotion stage is driven by a range of intrinsic and extrinsic factors, such as hormones, growth factors like TGF-β, and prolonged exposure to specific agents. These elements influence the TME, supporting the growth of initiated cells and speeding up their progression toward invasive growth. During the promotion stage, cellular proliferation is stimulated, particularly in cells with acquired genetic alterations that promote uncontrolled growth, such as mutations in oncogenes or tumor suppressor genes. These alterations can lead to increased expression of growth factor receptors, activation of downstream signaling pathways, and dysregulation of cell cycle checkpoints. As a result, these cells become more responsive to growth-promoting signals from the microenvironment, leading to their clonal expansion. Epigenetic changes, including abnormal DNA methylation and histone modifications, can disrupt the expression of genes that regulate the cell cycle and apoptosis, ultimately fostering cellular transformation and tumor development (Koya & Ibrahim, 2024; Stewart, 2019).

The TP53 gene produces the p53 protein, a crucial tumor suppressor that functions as a transcription factor, playing a central role in maintaining genomic integrity. This protein regulates essential cellular processes such as cell cycle progression, DNA repair, and apoptosis. p53 is thought to act as a tumor suppressor by serving as a sensor for cellular stress, as depicted in Fig. 13.6 (Boutelle & Attardi, 2021). The inactivation of TP53 is the most frequent genetic change observed in sporadic cancers, underscoring the critical role of p53 in cancer prevention. In reaction to DNA damage or cellular stress, p53 becomes activated and induces the expression of genes that halt the cell cycle, allowing time for DNA repair. If the damage is irreparable, p53 can trigger apoptosis, thereby removing the damaged cell and preventing its potential transformation into a cancerous one (Kandoth et al., 2013). p53 functions as a cellular stress detector, responding to various forms of stress, including DNA damage, oxidative stress, and oncogene activation. When activated, p53 triggers cellular responses such as cell cycle arrest, DNA repair, and apoptosis, all aimed at maintaining cellular integrity and preventing tumorigenesis. In normal, unstressed cells, p53 activity is regulated by the E3 ubiquitin ligase MDM2, which promotes its ubiquitination and subsequent degradation by the proteasome. However, when exposed to stress signals like DNA damage, oncogene activation, and hypoxia, p53 is stabilized and activated. This process involves intricate posttranslational modifications, including phosphorylation, acetylation, and ubiquitination, which disrupt the p53-MDM2 interaction, allowing p53 to accumulate and activate.

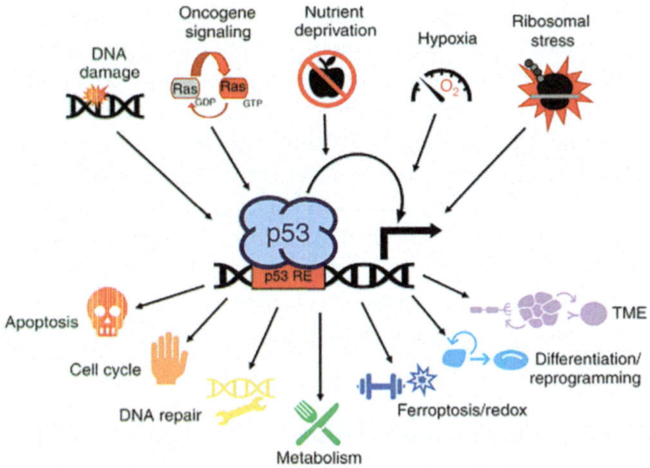

FIGURE 13.6

p53 is triggered by different types of cellular stress, initiating a transcriptional response that influences numerous cellular functions.

From Boutelle, A.M. & Attardi, L.D. (2021). p53 and tumor suppression: It takes a network. Trends in Cell Biology, *31(4), 298–310. https://doi.org/10.1016/j.tcb.2020.12.011.*

Once activated, p53 binds to specific DNA sequences, recruiting transcriptional machinery to initiate the expression of various target genes. Mutations in the DNA-binding domain of p53, often found in human cancers, prevent it from binding DNA, impairing its ability to activate gene expression. While most p53 mutations lead to a loss of function, some mutations endow p53 with oncogenic gain-of-function properties (Zhang & Lozano, 2017).

While p53's tumor-suppressive function is primarily attributed to its transcriptional activation of target genes, it also exerts its effects through transactivation-independent mechanisms. One of p53's well-known functions, independent of its transactivation role, is its ability to initiate apoptosis at the mitochondrial level. p53 can directly bind to antiapoptotic members of the BCL-2 family, such as BCL-2 and BCL-XL, causing them to dissociate from proapoptotic proteins like BAX and BAK. Alternatively, p53 can directly activate BAX and BAK, resulting in the permeabilization of the mitochondrial outer membrane and the subsequent release of cytochrome c, which triggers the apoptotic signaling cascade (Bieging et al., 2014).

In addition to its nuclear functions, p53 can also exert its tumor-suppressive effects in the cytoplasm. One mechanism involves p53 directly interacting with glucose-6-phosphate dehydrogenase (G6PD), a key enzyme in the pentose phosphate pathway (PPP). By binding to G6PD, p53 inhibits its activity, thereby decreasing the flow through the PPP. This metabolic pathway is crucial for cancer cell proliferation as it provides essential biosynthetic precursors and reducing power in the form of NADPH. By inhibiting G6PD, p53 restricts the availability of these critical metabolites, contributing to its tumor-suppressive effects (Jiang et al., 2011). Within the nucleus, p53 safeguards genomic integrity through a variety of mechanisms.

It can suppress the mobility of transposable elements and other repetitive DNA sequences, preventing their potential to disrupt the genome. Furthermore, p53 interacts with the homologous recombination protein RAD51, regulating recombination processes to prevent excessive or incorrect events that could result in genomic instability and contribute to tumorigenesis (Levine et al., 2016). This p53 mutant's ability to suppress the pathway by inhibiting TCF4 chromatin binding suggests a novel, transactivation-independent tumor-suppressive mechanism. However, further investigation is required to determine whether this activity is a unique property of this mutant (neomorphic) or if it is a latent function of wild-type p53 that is normally masked (Kadosh et al., 2020). These results propose that transactivation-independent functions of p53 might play a significant role in tumor suppression in certain contexts. Future in vivo studies investigating the specific contributions of these noncanonical p53 activities to tumor suppression will provide deeper insights into the complex mechanisms underlying cancer development.

13.5.3 Strategies for reprogramming TAMs for antitumor activity

Anti-CD40 antibodies can modulate the immunosuppressive microenvironment by reprogramming TAMs toward a proinflammatory state, augmenting antitumor immunity (Vonderheide, 2020). CD40, a member of the TNF receptor family, is expressed on the surface of macrophages and dendritic cells. Upon binding with its ligand, CD40L—primarily expressed by T cells, basophils, and mast cells—it triggers the upregulation of MHC molecules and the release of proinflammatory cytokines, thus promoting T-cell activation (Zhang et al., 2018). In mouse models of breast, colon, and melanoma cancers, targeting MARCO with monoclonal antibodies has been shown to reprogram TAMs toward a proinflammatory phenotype, leading to enhanced tumor regression (Georgoudaki et al., 2016).

These antibodies target the macrophage receptor MARCO, which is overexpressed in TAMs and has been linked to poor prognosis in breast, lung, and liver cancers (Bergamaschi et al., 2008). A potential strategy to reprogram TAMs involves targeting the CD47-SIRPα signaling pathway. SIRPα, a receptor expressed on phagocytic cells like macrophages, interacts with CD47 on target cells, inhibiting phagocytosis and promoting tissue homeostasis (Murata et al., 2014). Inhibiting SIRPα with monoclonal antibodies showed some efficacy in preclinical lung cancer models; however, the therapeutic response was not sustained (Alvey et al., 2017). This combined therapeutic approach targets both the CD47-SIRPα axis and CSF-1R signaling. Blocking the CD47-SIRPα interaction activates antitumor macrophages, while inhibiting CSF-1R reduces the recruitment of additional TAMs to the TME (Kulkarni et al., 2018).

13.6 Clinical application and case studies

13.6.1 Current clinical trials involving macrophage therapies

TAMs are commonly found in established tumors, and their occurrence is often linked to faster tumor progression and increased invasiveness (Noy & Pollard, 2014). Interestingly, a higher macrophage density has been linked with better overall survival in certain cases (Edin et al., 2013; Forssell et al., 2007; Welsh et al., 2005). TAMs secrete various factors that contribute to several critical cancer hallmarks, including sustained proliferative signaling, immune evasion, activation of invasion and metastasis, angiogenesis, and resistance to cell death (Pittet et al., 2022). This has sparked increased interest in targeting macrophages for cancer therapy, with preclinical studies yielding promising results across a range of cancer types and therapeutic strategies (DeNardo & Ruffell, 2019). A total of 606 clinical trials have been conducted, involving 143 distinct products. This reflects a noteworthy rise in the global advancement of macrophage-targeted therapies over the last decade. While most trials are still in the early phases, about two-thirds incorporate macrophage-targeted treatment as part of combination treatments (Wang et al., Takata et al., 2017), and 38.9% (236) of the trials focus on monotherapy.

Around 50% of these trials have been completed, 26.7% have been discontinued, and fewer than 20% are still ongoing (Ribas et al., n.d., Thiery et al., 2009). Results of combination therapy are nearly the same as results of monotherapy. Some of the clinical trials are based on the type of macrophage therapies, Currently, three primary approaches for targeting macrophages in cancer treatment include: (1) inhibiting the recruitment of monocytes/macrophage (Nywening et al., 2018) (2) depleting macrophages (Dammeijer et al., 2017); (3) reprogramming macrophage phenotypes at tumor sites (Bart et al., 2021). Additionally, the University of Edinburgh researched a novel therapeutic approach that involves reprogramming patient blood cells into macrophages in vitro using chemical stimuli for the treatment of liver cirrhosis (Macrophage Cell Therapy for Liver Cirrhosis | Centre of Regenerative Medicine, Waldman et al., 2020). In addition to cancer, targeting macrophages to either inhibit M1 polarization or promote M2-like polarization holds promise as a therapeutic strategy for inflammatory diseases such as atherosclerosis, colitis, and arthritis. This is because macrophage-derived factors are crucial in driving the pathogenesis of these conditions (Wang et al., 2022; Yang et al., 2021). The equilibrium between M1-like and M2-like macrophage phenotypes significantly influences the body's immune responses and disease progression (Murray, 2011). Therefore therapeutic interventions targeting macrophage phenotype modulation hold substantial promise for disease management (Flavonoids Regulate Tumor-Associated 2022; Kerneur et al., 2022; Song et al., 2022).

13.6.2 Success stories and outcomes

In recent years, considerable deliberation has been given to unraveling the complex mechanisms that regulate macrophage activation and its subsequent impact on macrophage function.

A key concept in this area is the bipolar model, which classifies macrophages into two main subsets: (1) Proinflammatory Macrophages (M1): Activated primarily by interferon-γ (IFN-γ), these macrophages display a proinflammatory phenotype. (2) Anti-inflammatory Macrophages (M2): These macrophages are alternatively activated by interleukin-4 (IL-4) and interleukin-13 (IL-13), leading to an anti-inflammatory phenotype (Mills, 2012; Sica & Mantovani, 2012).

The equilibrium between M1-like and M2-like macrophage phenotypes significantly influences the body's immune responses and illness progression. Therefore therapeutic interventions targeting macrophage phenotype modulation hold substantial promise for disease management (Horn et al., 2008). Additionally, researchers at the University of Edinburgh successfully employed autologous macrophage therapy in liver cirrhosis. In a Phase 1 dose-escalation trial with nine adult participants, the therapy resulted in a 100% survival rate and transplant-free status at the one-year follow-up (Na et al., 2023). Some preclinical studies from independent laboratories indicate that inhibiting or reducing the accumulation of monocyte-derived macrophages in the spinal cord may enhance neuroprotection and improve recovery after spinal cord injury (SCI)I (Blight, 1994; Eng & Lee, 2003; Giulian & Robertson, 1990; Gris et al., 2004; Iannotti et al., 2011; Popovich et al., 1999).

Researchers at the University of Edinburgh successfully applied autologous macrophage therapy for liver cirrhosis. In a Phase 1 dose-escalation trial involving nine adult participants, a 100% survival rate was observed, with all patients remaining transplant-free at the 1-year follow-up. Liposome-encapsulated clodronate, administered intravenously, specifically depletes macrophages and MDSCs in the spleen and liver. Increasing evidence suggests that targeting macrophage phenotypes may offer an effective therapeutic approach for diabetic foot ulcers. Hyperglycemia has been associated with an increased ratio of proinflammatory M1 macrophages to proregenerative M2 macrophages. ON101 has demonstrated greater healing efficacy compared to standard absorbent dressings for DFUs, showing consistent benefits for patients with various risk factors. Macrophages play a crucial role in the inflammatory response induced by radiation therapy. Therefore immunotherapies that target macrophages or their polarization states hold promise as complementary treatments in combination with radiation therapy, as indicated by preclinical and early clinical studies (Fidler & Schroit, 1988; Folkman, 2023; Horn et al., 2008; Huang et al., 2021; Orkin & Zon, 2008; Shi & Shiao, 2018). Both bone marrow-derived and tissue-resident TAMs play a crucial part in promoting tumor immunosuppression. Preclinical mouse models demonstrate that targeting various molecular components of TAMs, including cell surface receptors, signaling pathways, transcription factors, and metabolic processes, can effectively inhibit tumor immunosuppression and enhance the effectiveness of immune checkpoint inhibitors (Chen et al., 2011). TAMs which are prevalent in the microenvironment of tumors, play a crucial role in fostering resistance to multiple cancer therapies by aiding tumor progression (Chen et al., 2011).

13.6.3 Challenges in translation to clinical practice

Macrophages have been extensively investigated as therapeutic agents for cancer and inflammatory diseases because of their robust immune response capabilities and exceptional plasticity, which enables them to adopt therapeutic phenotypes. However, their high plasticity and adaptability can also limit their effectiveness in therapy. Even after being polarized in vitro to a therapeutic phenotype, macrophages can revert to a harmful pathological state when exposed to the TME (Chen et al., 2011). For instance, Cao et al. found that anti-inflammatory M2 macrophages, polarized in vitro, lost their therapeutic phenotype within 2 days after being administered to a mouse model of Adriamycin-induced nephropathy, with inflammatory markers increasing by day seven (Cao et al., 2014). Similarly, tumor therapies that use in vitro-polarized proinflammatory macrophages have failed due to their eventual reversion to the tumor-supporting M2 phenotype within the TME (Lee et al., 2016).

The M1/M2 macrophage ratio may not accurately reflect the complexity of tissue disease since macrophages with different activation stages and mixed phenotypes can be found in several preclinical and clinical contexts (Sica & Mantovani, 2012; Weissleder & Pittet, 2020). Drug-loaded macrophages primarily gather in the liver, lungs, and spleen following intravenous injection, according to studies (Huang et al., 2021; Nishiwaki et al., 2020; Sapach et al., 2022). A brief buildup of autologous activated macrophages in the lungs within the first hour following intravenous infusion was noted in a clinical study investigating the biological distribution of AAM in patients with metastatic renal cell carcinoma. Macrophages accumulated preferentially in the spleen and liver throughout time; accumulation decreased by days 1 and 2, respectively (Quillien et al., 2001). Another significant challenge in macrophage-mediated drug delivery is the uncontrolled or premature release of therapeutic agents, such as cytotoxic drugs or nanomedicines. These agents may leak into the macrophage cytoplasm, leading to premature cell death before reaching the target site (Ren et al., 2020). A major challenge in macrophage-based drug delivery is ensuring complete cargo release at the target site.

After endocytosis, drugs are typically contained within macrophage phagosomes, where they can undergo degradation (Doshi et al., 2011). Wang et al. created a biomimetic drug delivery method using nanoparticles coated in ROS-responsive macrophage membranes (Wang et al., 2025). In addition to directing nanoparticles to inflammatory tissues and successfully evading Reticuloendothelial system (RES) clearance, macrophage membranes also trap proinflammatory cytokines, which aid in lowering local inflammation. However, current laboratory-scale preparation methods for macrophage membrane-coated nanoparticles are complex and unsuitable for large-scale production. The absence of suitable technologies for mass cell membrane production hinders clinical translation (Dhas et al., 2022; Qu et al., 2024).

13.7 Challenges and future directions

13.7.1 Limitations of current approaches

Despite the promising potential of macrophage-based therapies in cellular immunotherapy, several limitations hinder their efficacy and widespread application. One significant challenge is the heterogeneity of macrophages within the TME. Tumors can manipulate macrophages to adopt an immunosuppressive M2 phenotype, which promotes tumor growth and metastasis instead of attacking cancer cells. This polarization complicates therapeutic strategies aimed at reprogramming macrophages to a proinflammatory M1 state, as the TME often presents conflicting signals that influence macrophage behavior. Another limitation is the delivery and engraftment of genetically modified macrophages. Achieving effective targeting to tumor sites remains a challenge due to physical barriers, such as dense extracellular matrices and abnormal vasculature in solid tumors. Additionally, the safety concerns associated with genetic modifications pose risks of unintended consequences, such as off-target effects or dysregulated immune responses. Moreover, current approaches often lack standardization in manufacturing processes for CAR-macrophages or other engineered macrophage therapies. Variability in cell preparation, genetic modification techniques, and culture conditions can lead to inconsistent therapeutic outcomes. Finally, the regulatory landscape for these advanced therapies is still evolving, creating uncertainties regarding approval pathways and clinical implementation.

13.7.2 Research gaps and opportunities for innovation

There are significant research gaps that present opportunities for innovation in macrophage-based therapies. One area needing exploration is the mechanisms underlying macrophage plasticity. Understanding how environmental cues within the TME influence macrophage polarization can lead to novel strategies for manipulating their function more effectively. Researchers should focus on identifying specific signaling pathways and transcription factors involved in this process to develop targeted interventions. Additionally, there is a need for better models that accurately mimic human tumors and their microenvironments. Current preclinical models often fail to replicate the complexity of human cancers, limiting the translatability of findings. Innovations in organoid technology or patient-derived xenografts could provide more relevant platforms for studying macrophage interactions with cancer cells. The potential for combination therapies also warrants further investigation. Combining macrophage-based therapies with other modalities, such as checkpoint inhibitors or targeted therapies, may enhance overall efficacy by simultaneously addressing multiple mechanisms of immune evasion employed by tumors. Furthermore, exploring the use of small molecules or biologics that can modulate macrophage function without genetic modification could offer safer alternatives with fewer regulatory hurdles.

13.7.3 Future perspectives on macrophage-based therapies

Looking ahead, the future of macrophage-based therapies appears promising but requires strategic advancements. One key perspective is the continued evolution of CAR-macrophage (CAR-M) technology. As understanding of macrophage biology deepens, researchers can refine CAR designs to enhance specificity and reduce off-target effects while improving antitumor efficacy. The development of dual-targeting CARs that can engage both tumor antigens and immune checkpoints may further amplify therapeutic outcomes. Moreover, advances in single-cell sequencing technologies will allow for a more granular understanding of macrophage populations within tumors. This knowledge can inform personalized therapeutic approaches by identifying specific macrophage subsets that correlate with patient outcomes. The integration of biomaterials into macrophage therapies presents another exciting avenue for innovation. Utilizing engineered scaffolds or hydrogels that release cytokines or chemokines could enhance macrophage recruitment and activation at tumor sites while providing a supportive microenvironment for their function. Finally, fostering collaboration between academia and industry will be crucial in translating research findings into clinical applications. Establishing multidisciplinary teams that include immunologists, engineers, and clinicians can accelerate the development of effective macrophage-based therapies.

13.8 Conclusion

13.8.1 Summary of key insights

Macrophages play a pivotal role in cellular immunotherapy due to their unique ability to adapt their functions based on environmental cues within the TME. Their dual nature—exhibiting both proinflammatory (M1) and anti-inflammatory (M2) characteristics—highlights their complexity in cancer biology. While current approaches leveraging genetically modified macrophages show promise, significant challenges remain regarding their efficacy, delivery mechanisms, and safety profiles. Key insights from this chapter underscore the importance of understanding macrophage plasticity and the need for innovative strategies to overcome existing limitations. Research gaps related to signaling pathways governing macrophage behavior present opportunities for novel therapeutic interventions that could enhance antitumor responses while minimizing adverse effects.

13.8.2 Implications for future research and clinical practice

The implications of these findings are profound for future research and clinical practice. As our understanding of macrophage biology evolves, it will inform more effective therapeutic strategies tailored to individual patients' TME. The potential for combination therapies that synergize different treatment modalities could lead to improved patient outcomes in cancer treatment. Furthermore, ongoing advancements in technology—such as single-cell genomics and biomaterials—will facilitate more precise targeting of macrophages within tumors, enhancing their therapeutic efficacy while reducing systemic toxicity. Collaborative efforts between researchers and clinicians will be essential in translating these insights into clinical settings, ultimately paving the way for innovative treatments that harness the full potential of macrophages in cellular immunotherapy. In conclusion, while challenges persist in leveraging macrophages effectively within immunotherapy frameworks, ongoing research holds great promise for overcoming these hurdles and advancing cancer treatment paradigms toward more personalized and effective solutions.

References

Abdul-Rahman, T., Ghosh, S., Badar, S. M., Nazir, A., Bamigbade, G. B., Aji, N., Roy, P., Kachani, H., Garg, N., Lawal, L., Bliss, Z. S. B., Wireko, A. A., Atallah, O., Adebusoye, F. T., Teslyk, T., Sikora, K., & Horbas, V. (2024). The paradoxical role of cytokines and chemokines at the tumor microenvironment: A comprehensive review. *European Journal of Medical Research, 29*, 124. doi:10.1186/s40001-024-01711-z.

Afik, R., Zigmond, E., Vugman, M., Klepfish, M., Shimshoni, E., Pasmanik-Chor, M., Shenoy, A., Bassat, E., Halpern, Z., Geiger, T., Sagi, I., Varol, C., et al. (2016). Tumor macrophages are pivotal constructors of tumor collagenous matrix. *Journal of Experimental Medicine, 213*, 2315–2331. doi:10.1084/JEM.20151193.

Ahmed, H., Mahmud, A. R., Siddiquee, Mohd. F.-R.-, Shahriar, A., Biswas, P., Shimul, Md. E. K., Ahmed, S. Z., Ema, T. I., Rahman, N., Khan, Md. A., Mizan, Md F. R., & Emran, T. B. (2023). Role of T cells in cancer immunotherapy: Opportunities and challenges. *Cancer Pathogenesis and Therapy, 1*, 116–126. doi:10.1016/j.cpt.2022.12.002.

Alvey, C. M., Spinler, K. R., Irianto, J., Pfeifer, C. R., Hayes, B., Xia, Y., … Discher, D. E., et al. (2017). SIRPA-Inhibited, marrow-derived macrophages engorge, accumulate, and differentiate in antibody-targeted regression of solid tumors. *Current Biology, 27*, 2065–2077.

Arnold, F. (1985). Tumour angiogenesis. *Annals of the Royal College of Surgeons of England, 67*, 295–298. doi:10.1201/B23252-48/TUMOUR-ANGIOGENESIS-JUDAH-FOLKMAN.

Arnold, J. N., Magiera, L., Kraman, M., & Fearon, D. T. (2014). Tumoral immune suppression by macrophages expressing fibroblast activation protein-α and heme oxygenase-1. *Cancer Immunology Research, 2*, 121–126. doi:10.1158/2326-6066.CIR-13-0150/466957/AM/TUMORAL-IMMUNE-SUPPRESSION-BY-MACROPHAGES.

Atri, C., Guerfali, F. Z., & Laouini, D. (2018). Role of human macrophage polarization in inflammation during infectious diseases. *International Journal of Molecular Sciences, 19*, 1801. doi:10.3390/ijms19061801.

Bahr, J. C., Li, X. Y., Feinberg, T. Y., Jiang, L., & Weiss, S. J. (2022). Divergent regulation of basement membrane trafficking by human macrophages and cancer cells. *Nature Communications, 13*, 1–20. doi:10.1038/s41467-022-34087-x.

Bajgar, A., & Krejčová, G. (2023). On the origin of the functional versatility of macrophages. *Frontiers in Physiology, 14*, 1128984. doi:10.3389/fphys.2023.1128984.

Bart, V. M. T., Pickering, R. J., Taylor, P. R., & Ipseiz, N. (2021). Macrophage reprogramming for therapy. *Immunology, 163*, 128–144.

Bergamaschi, A., Tagliabue, E., Bergamaschi, E. T., Sørlie, T., Naume, B., Triulzi, T., & Orlandi, R. (2008). Extracellular matrix signature identifies breast cancer subgroups with different clinical outcome. *Journal of Pathology, 214*, 357–367.

Bhol, N. K., Bhanjadeo, M. M., Singh, A. K., Dash, U. C., Ojha, R. R., Majhi, S., Duttaroy, A. K., & Jena, A. B. (2024). The interplay between cytokines, inflammation, and antioxidants: Mechanistic insights and therapeutic potentials of various antioxidants and anticytokine compounds. *Biomedicine & Pharmacotherapy, 178*, 117177. doi:10.1016/j.biopha.2024.117177.

Bieging, K. T., Mello, S. S., & Attardi, L. D. (2014). Unravelling mechanisms of P53-mediated tumour suppression. *Nature Reviews Cancer, 14*, 359–370. doi:10.1038/nrc3711.

Blight, A. R. (1994). Effects of silica on the outcome from experimental spinal cord injury: Implication of macrophages in secondary tissue damage. *Neuroscience, 60*, 263–273.

Boutelle, A. M., & Attardi, L. D. (2021). P53 and tumor suppression: It takes a network. *Trends in Cell Biology, 31*, 298–310. doi:10.1016/j.tcb.2020.12.011.

Cao, Q., Wang, Y., Zheng, D., Sun, Y., Wang, C., XWK, Cao, Y. W., Zheng, D., Sun, Y., & Wang, C. (2014). Failed renoprotection by alternatively activated bone marrow macrophages is due to a proliferation-dependent phenotype switch in vivo. *Kidney International, 85*, 794–806.

Chandran, S. S., & Klebanoff, C. A. (2019). T cell receptor-based cancer immunotherapy: Emerging efficacy and pathways of resistance. *Immunological Reviews, 290*, 127–147. doi:10.1111/imr.12772.

Chen, C., Jing, W., Chen, Y., Wang, G., Abdalla, M., Gao, L., Han, M., Shi, C., Li, A., Sun, P., Jiang, X., Yang, Z., Zhang, S., Zhang, J., Tang, C., Liu, Y., Zhang, R., Xu, F., & Dong, B. (2022). Intracavity generation of glioma stem cell-specific car macrophages primes locoregional immunity for postoperative glioblastoma therapy. *Science Translational Medicine, 14*, eabn1128. doi:10.1126/scitranslmed.abn1128.

Chen, P., Huang, Y., Bong, R., Ding, Y., Song, N., & Wang, X. (2011). Tumor-associated macrophages promote angiogenesis and melanoma growth via adrenomedullin in a paracrine and autocrine manner. *Clinical Cancer Research, 17*, 7230–7239.

Chen, S., Saeed, A. F. U. H., Liu, Q., Jiang, Q., Xu, H., Xiao, G. G., Rao, L., & Duo, Y. (2023). Macrophages in immunoregulation and therapeutics. *Signal Transduction and Targeted Therapy, 8*, 207. doi:10.1038/s41392-023-01452-1.

Chen, S., Zhu, H., & Jounaidi, Y. (2024). Comprehensive snapshots of natural killer cells functions, signaling, molecular mechanisms and clinical utilization. *Signal Transduction and Targeted Therapy, 9*, 302. doi:10.1038/s41392-024-02005-w.

Chen, Y., Zhu, X., Liu, H., Wang, C., Chen, Y., Wang, H., Fang, Y., Wu, X., Xu, Y., Li, C., Lv, X., Huang, J., Han, X., Li, R., Hong, W., Yu, Z., Wei, W., & Tu, J. (2023). The application of HER2 and CD47 CAR-macrophage in ovarian cancer. *Journal of Translational Medicine, 21*, 654. doi:10.1186/s12967-023-04479-8.

Chi, H., Pepper, M., & Thomas, P. G. (2024). Principles and therapeutic applications of adaptive immunity. *Cell, 187*, 2052–2078. doi:10.1016/j.cell.2024.03.037.

Colbert, J. D., Cruz, F. M., & Rock, K. L. (2020). Cross-presentation of exogenous antigens on MHC I molecules. *Current Opinion in Immunology, 64*, 1–8. doi:10.1016/j.coi.2019.12.005.

Couture, A., Garnier, A., Docagne, F., Boyer, O., Vivien, D., Le-Mauff, B., Latouche, J.-B., & Toutirais, O. (2019). HLA-class II artificial antigen presenting cells in CD4+ T cell-based immunotherapy. *Frontiers in Immunology, 10*, 1081. doi:10.3389/fimmu.2019.01081.

Curiel, T. J., Coukos, G., Zou, L., Alvarez, X., Cheng, P., Mottram, P., Evdemon-Hogan, M., Conejo-Garcia, J. R., Zhang, L., Burow, M., Zhu, Y., Wei, S., Kryczek, I., Daniel, B., Gordon, A., Myers, L., Lackner, A., Disis, M. L., & Knutson, K. L. (2004). Specific recruitment of regulatory T cells in ovarian carcinoma fosters immune privilege and predicts reduced survival. *Nature Medicine, 10*, 942–949. doi:10.1038/nm1093.

Dai, H., Zhu, C., Huai, Q., Xu, W., Zhu, J., Zhang, X., Zhang, X., Sun, B., Xu, H., Zheng, M., Li, X., & Wang, H. (2024). Chimeric antigen receptor-modified macrophages ameliorate liver fibrosis in preclinical models. *Journal of Hepatology, 80*, 913–927. doi:10.1016/j.jhep.2024.01.034.

Dammeijer, F., Lievense, L. A., Kaijen-Lambers, M. E., Nimwegen, M., Bezemer, K., & Hegmans, J. P. (2017). Depletion of tumor-associated macrophages with a CSF-1R kinase inhibitor enhances antitumor immunity and survival induced by DC immunotherapy. *Cancer Immunology Research, 5*, 535–546.

Dash, S. P., Gupta, S., & Sarangi, P. P. (2024). Monocytes and macrophages: Origin, homing, differentiation, and functionality during inflammation. *Heliyon, 10*, e29686. doi:10.1016/j.heliyon.2024.e29686.

De Kleer, I., Willems, F., Lambrecht, B., & Goriely, S. (2014). Ontogeny of myeloid cells. *Frontiers in Immunology, 5*. doi:10.3389/fimmu.2014.00423.

DeNardo, D. G., & Ruffell, B. (2019). Macrophages as regulators of tumour immunity and immunotherapy. *Nature Reviews Immunology, 19*, 369–382.

Dhas, N., García, M. C., Kudarha, R., Pandey, A., Nikam, A. N., Gopalan, D., … Mutalik, S., et al. (2022). Advancements in cell membrane camouflaged nanoparticles: A bioinspired platform for cancer therapy. *Journal of Controlled Release, 346*, 71–97.

Doshi, N., Swiston, A., Gilbert, J., Alcaraz, M., Cohen, R., Rubner, M., & Mitragotri, S. (2011). Cell-based drug delivery devices using phagocytosis-resistant backpacks. *Advanced Materials, 23*, H105–H109.

Dudek, A. M., Martin, S., Garg, A. D., & Agostinis, P. (2013). Immature, semimature, and fully mature dendritic cells: Toward a DC-cancer cells interface that augments anticancer immunity. *Frontiers in Immunology, 4*. doi:10.3389/fimmu.2013.00438.

Dunkelberger, J. R., & Song, W.-C. (2010). Complement and its role in innate and adaptive immune responses. *Cell Research, 20*, 34–50. doi:10.1038/cr.2009.139.

Edin, S., Wikberg, M. L., Oldenborg, P. A., & Palmqvist, R. (2013). Macrophages: Good guys in colorectal cancer. *Oncoimmunology, 2*, e23038.

Embgenbroich, M., & Burgdorf, S. (2018). Current concepts of antigen cross-presentation. *Frontiers in Immunology, 9*, 1643. doi:10.3389/fimmu.2018.01643.

Eng, L. F., & Lee, Y. L. (2003). Response of chemokine antagonists to inflammation in injured spinal cord. *Neurochemical Research, 28*, 95–100.

Escamilla, J., Schokrpur, S., Liu, C., Priceman, S. J., Moughon, D., Jiang, Z., … Wu, L., et al. (2015). CSF1 receptor targeting in prostate cancer reverses macrophage-mediated resistance to androgen blockade therapy. *Cancer Research, 75*, 950–962. doi:10.1158/0008-5472.CAN-14-0992/659571/P/CSF1-RECEPTOR-TARGETING-IN-PROSTATE-CANCER.

Fidler, I. J., & Schroit, A. J. (1988). Recognition and destruction of neoplastic cells by activated macrophages: Discrimination of altered self. *Biochimica et Biophysica Acta (BBA) - Reviews on Cancer, 948*, 151–173.

Flavonoids Regulate Tumor-Associated. (2022). Macrophages–from structure-activity relationship to clinical potential. *Pharmacological Research, 184*, 106419.

Flier, J. S., Underhill, L. H., & Dvorak, H. F. (1986). Tumors: Wounds that do not heal. *New England Journal of Medicine, 315*, 1650–1659. doi:10.1056/NEJM198612253152606.

Folkman, J. (2023). Tumour angiogenesis therapeutic implications. *50 Landmark Papers every Pediatric Surgeon Should Know*: 15. CRC Press.

Forssell, J., Öberg, Å., Henriksson, M. L., Stenling, R., Jung, A., & Palmqvist, R. (2007). High macrophage infiltration along the tumor front correlates with improved survival in colon cancer. *Clinical Cancer Research, 13*, 1472–1479.

Georgoudaki, A. M., Prokopec, K. E., Boura, V. F., Hellqvist, E., Sohn, S., Östling, J., Dahan, R., Harris, R. A., Rantalainen, M., Klevebring, D., Sund, M., Brage, S. E., Fuxe, J., Rolny, C., Li, F., Ravetch, J. V., & Karlsson, M. C. I. (2016). Reprogramming tumor-associated macrophages by antibody targeting inhibits cancer progression and metastasis. *Cell Reports, 15*, 2000–2011. doi:10.1016/J.CELREP.2016.04.084.

Giulian, D., & Robertson, C. (1990). Inhibition of mononuclear phagocytes reduces ischemic injury in the spinal cord. *Annals of Neurology, 27*, 33–42.

Gordon, S. (2016). Phagocytosis: An immunobiologic process. *Immunity, 44*, 463–475. doi:10.1016/j.immuni.2016.02.026.

Grinenko, T., Eugster, A., Thielecke, L., Ramasz, B., Krüger, A., Dietz, S., Glauche, I., Gerbaulet, A., Von Bonin, M., Basak, O., Clevers, H., Chavakis, T., & Wielockx, B. (2018). Hematopoietic stem cells can differentiate into restricted myeloid progenitors before cell division in mice. *Nature Communications, 9*, 1898. doi:10.1038/s41467-018-04188-7.

Gris, D., Marsh, D., Oatway, M., Soc NeuroscienceD Gris, D. R. M., Oatway, M. A., Chen, Y., Hamilton, E. F., Dekaban, G. A., & Neuroscience, L. C. W. (2004). Transient blockade of the CD11d/CD18 integrin reduces secondary damage after spinal cord injury, improving sensory, autonomic, and motor function. *Journal of Neuroscience, 24*, 4043–4051.

Hao, N.-B., Lü, M.-H., Fan, Y.-H., Cao, Y.-L., Zhang, Z.-R., & Yang, S.-M. (2012). Macrophages in tumor microenvironments and the progression of tumors. *Clinical & Developmental Immunology, 2012*, 1–11. doi:10.1155/2012/948098.

Harney, A. S., Arwert, E. N., Entenberg, D., Wang, Y., Guo, P., Qian, B. Z., Oktay, M. H., Pollard, J. W., Jones, J. G., & Condeelis, J. S. (2015). Real-time imaging reveals local, transient vascular permeability, and tumor cell intravasation stimulated by tie2hi macrophage–derived VEGFA. *Cancer Discovery, 5*, 932–943. doi:10.1158/2159-8290.CD-15-0012/6761/P/REAL-TIME-IMAGING-REVEALS-LOCAL-TRANSIENT-VASCULAR.

Hirayama, D., Iida, T., & Nakase, H. (2017). The phagocytic function of macrophage-enforcing innate immunity and tissue homeostasis. *International Journal of Molecular Sciences, 19*, 92. doi:10.3390/ijms19010092.

Horn, K., Busch, S., Soc NeuroscienceKP Horn, S. A. B., Hawthorne, A. L., Rooijen, N., & Neuroscience, J. S. (2008). Another barrier to regeneration in the CNS: Activated macrophages induce extensive retraction of dystrophic axons through direct physical interactions. *Journal of Neuroscience, 28*, 9330–9341.

Hu, A., Sun, L., Lin, H., Liao, Y., Yang, H., & Mao, Y. (2024). Harnessing innate immune pathways for therapeutic advancement in cancer. *Signal Transduction and Targeted Therapy, 9*, 68. doi:10.1038/s41392-024-01765-9.

Huang, T., Bei, C., Hu, Z., & Li, Y. (2024). CAR-macrophage: Breaking new ground in cellular immunotherapy. *Frontiers in Cell and Developmental Biology, 12*. doi:10.3389/fcell.2024.1464218.

Huang, Y. Y., Lin, C. W., Cheng, N. C., Cazzell, S. M., Chen, H. H., & Huang, K. F. (2021). Effect of a novel macrophage-regulating drug on wound healing in patients with diabetic foot ulcers: A randomized clinical trial. *JAMA Network Open, 4*, e2122607.

Iannotti, C., Clark, M., Horn, K., Rooijen, N., Silver, jerry, Steinmetz, M., & Silver, J. (2011). A combination immunomodulatory treatment promotes neuroprotection and locomotor recovery after contusion SCI. *Experimental Neurology, 230*, 3–15.

Jiang, P., Du, W., Wang, X., Mancuso, A., Gao, X., Wu, M., & Yang, X. (2011). P53 regulates biosynthesis through direct inactivation of glucose-6-phosphate dehydrogenase. *Nature Cell Biology, 13*, 310–316. doi:10.1038/ncb2172.

Kadosh, E., Snir-Alkalay, I., Venkatachalam, A., May, S., Lasry, A., Elyada, E., Zinger, A., Shaham, M., Vaalani, G., Mernberger, M., Stiewe, T., Pikarsky, E., Oren, M., Ben-Neriah, Y., et al. (2020). The gut microbiome switches mutant P53 from tumour-suppressive to oncogenic. *Nature, 586*, 133–138. doi:10.1038/s41586-020-2541-0.

Kandoth, C., McLellan, M. D., Vandin, F., Ye, K., Niu, B., Lu, C., Xie, M., Zhang, Q., McMichael, J. F., Wyczalkowski, M. A., Leiserson, M. D. M., Miller, C. A., Welch, J. S., Walter, M. J., Wendl, M. C., Ley, T. J., Wilson, R. K., Raphael, B. J., & Ding, L. (2013). Mutational landscape and significance across 12 major cancer types. *Nature, 502*, 333–339. doi:10.1038/nature12634.

Kang, M., Lee, S. H., Kwon, M., Byun, J., Kim, D., Kim, C., Koo, S., Kwon, S. P., Moon, S., Jung, M., Hong, J., Go, S., Song, S. Y., Choi, J. H., Hyeon, T., Oh, Y.-K., Park, H. H., & Kim, B.- S. (2021). Nanocomplex-mediated in vivo programming to chimeric antigen receptor-m1 macrophages for cancer therapy. *Advanced Materials, 33*, 2103258. doi:10.1002/adma.202103258.

Kerneur, C., Cano, C. E., & Olive, D. (2022). Major pathways involved in macrophage polarization in cancer. *Frontiers in Immunology, 13*, 1026954.

Klichinsky, M., Ruella, M., Shestova, O., Lu, X. M., Best, A., Zeeman, M., Schmierer, M., Gabrusiewicz, K., Anderson, N. R., Petty, N. E., Cummins, K. D., Shen, F., Shan, X., Veliz, K., Blouch, K., Yashiro-Ohtani, Y., Kenderian, S. S., Kim, M. Y., & O'Connor, R. S. (2020). Human chimeric antigen receptor macrophages for cancer immunotherapy. *Nature Biotechnology, 38*, 947–953. doi:10.1038/s41587-020-0462-y.

Koya, A. I., & Ibrahim, S. A. (2024). *Carcinogenesis.* StatPearls.

Krzyszczyk, P., Schloss, R., Palmer, A., & Berthiaume, F. (2018). The role of macrophages in acute and chronic wound healing and interventions to promote prowound healing phenotypes. *Frontiers in Physiology, 9*, 419. doi:10.3389/fphys.2018.00419.

Kulakova, K., Lawal, T. R., Mccarthy, E., & Floudas, A. (2024). The contribution of macrophage plasticity to inflammatory arthritis and their potential as therapeutic targets. *Cells, 13*, 1586. doi:10.3390/cells13181586.

Kulkarni, A., Chandrasekar, V., biomedical, S. N.-N., Kulkarni, V. C., Natarajan, S. K., Ramesh, A., Pandey, P., Nirgud, J., & Engineering, H. B. B. (2018). A designer self-assembled supramolecule amplifies macrophage immune responses against aggressive cancer. *Nature Biomedical Engineering, 2*, 589–599.

Lau, A. P. Y., Khavkine Binstock, S. S., & Thu, K. L. (2023). CD47: The next frontier in immune checkpoint blockade for nonsmall cell lung cancer. *Cancers, 15*, 5229. doi:10.3390/cancers15215229.

Lee, H.-J., Woo, Y., Hahn, T.-W., Jung, Y. M., & Jung, Y.-J. (2020). Formation and maturation of the phagosome: A key mechanism in innate immunity against intracellular bacterial infection. *Microorganisms, 8*, 1298. doi:10.3390/microorganisms8091298.

Lee, S., Kivimae, S., Dolor, A., & Szoka, F. (2016). Macrophage-based cell therapies: The long and winding road. *Journal of Controlled Release, 240*, 527–540. doi:10.1016/j.jconrel.2016.07.018.

Levine, A. J., Ting, D. T., & Greenbaum, B. D. (2016). P53 and the defenses against genome instability caused by transposons and repetitive elements. *BioEssays : News and Reviews in Molecular, Cellular and Developmental Biology, 38*, 508–513. doi:10.1002/bies.201600031.

Lewis, K., Yoshimoto, M., & Takebe, T. (2021). Fetal liver hematopoiesis: From development to delivery. *Stem Cell Research & Therapy, 12*, 139. doi:10.1186/s13287-021-02189-w.

Li, D., & Wu, M. (2021). Pattern recognition receptors in health and diseases. *Signal Transduction and Targeted Therapy, 6*, 291. doi:10.1038/s41392-021-00687-0.

Li, J., Chen, P., & Ma, W. (2024). The next frontier in immunotherapy: Potential and challenges of CAR-macrophages. *Experimental Hematology & Oncology, 13*, 76. doi:10.1186/s40164-024-00549-9.

Liu, Z., Zhou, Z., Dang, Q., Xu, H., Lv, J., Li, H., & Han, X. (2022). Immunosuppression in tumor immune microenvironment and its optimization from CAR-T cell therapy. *Theranostics, 12*, 6273–6290. doi:10.7150/thno.76854.

Lu, J., Ma, Y., Li, Q., Xu, Y., Xue, Y., & Xu, S. (2024). CAR macrophages: A promising novel immunotherapy for solid tumors and beyond. *Biomarker Research, 12*, 86. doi:10.1186/s40364-024-00637-2.

Ma, Y., Jiang, T., Zhu, X., Xu, Y., Wan, K., Zhang, T., & Xie, M. (2024). Efferocytosis in dendritic cells: An overlooked immunoregulatory process. *Frontiers in Immunology, 15*, 1415573. doi:10.3389/fimmu.2024.1415573.

Maalej, K. M., Merhi, M., Inchakalody, V. P., Mestiri, S., Alam, M., Maccalli, C., Cherif, H., Uddin, S., Steinhoff, M., Marincola, F. M., & Dermime, S. (2023). CAR-cell therapy in the era of solid tumor treatment: Current challenges and emerging therapeutic advances. *Molecular Cancer, 22*, 20. doi:10.1186/s12943-023-01723-z.

Macrophage Cell Therapy for Liver Cirrhosis | Centre of Regenerative Medicine https://regenerative-medicine.ed.ac.uk/about/diseases-study/clinical-trials/macrophage-cell-therapy-for-liver-cirrhosis.

Madsen, D. H., Jürgensen, H. J., Siersbæk, M. S., Kuczek, D. E., Grey Cloud, L., Liu, S., Behrendt, N., Grøntved, L., Weigert, R., & Bugge, T. H. (2017). Tumor-associated macrophages derived from circulating inflammatory monocytes degrade collagen through cellular uptake. *Cell Reports, 21*, 3662–3671. doi:10.1016/J.CELREP.2017.12.011/ATTACHMENT/9E26BBDE-EB19-45E6-A962-A14FFB73ACBA/MMC3.PDF.

Mantovani, A., Allavena, P., Marchesi, F., & Garlanda, C. (2022). Macrophages as tools and targets in cancer therapy. *Nature Reviews Drug Discovery, 21*, 799–820. doi:10.1038/s41573-022-00520-5.

Marshall, J. S., Warrington, R., Watson, W., & Kim, H. L. (2018). An introduction to immunology and immunopathology. *Allergy, Asthma, and Clinical Immunology : Official Journal of the Canadian Society of Allergy and Clinical Immunology, 14*, 49. doi:10.1186/s13223-018-0278-1.

Mazinani, M., & Rahbarizadeh, F. (2022). CAR-T cell potency: From structural elements to vector backbone components. *Biomarker Research, 10*, 70. doi:10.1186/s40364-022-00417-w.

McGrath, K. E., Frame, J. M., & Palis, J. (2015). Early hematopoiesis and macrophage development. *Seminars in Immunology, 27*, 379–387. doi:10.1016/j.smim.2016.03.013.

Mehrotra, P., & Ravichandran, K. S. (2022). Drugging the efferocytosis process: Concepts and opportunities. *Nature Reviews Drug Discovery, 21*, 601–620. doi:10.1038/s41573-022-00470-y.

Meng, E. X., Verne, G. N., & Zhou, Q. (2024). Macrophages and gut barrier function: Guardians of gastrointestinal health in postinflammatory and postinfection responses. *International Journal of Molecular Sciences, 25*, 9422. doi:10.3390/ijms25179422.

Mills, C. D. (2012). M1 and M2 macrophages: Oracles of health and disease. *Critical Reviews in Immunology, 32*, 463–488.

Morrissey, M. A., Williamson, A. P., Steinbach, A. M., Roberts, E. W., Kern, N., Headley, M. B., & Vale, R. D. (2018). Chimeric antigen receptors that trigger phagocytosis. *eLife, 7*, e36688. doi:10.7554/eLife.36688.

Mukherjee, A. G., Wanjari, U. R., Namachivayam, A., Murali, R., Prabakaran, D. S., Ganesan, R., Renu, K., Dey, A., Vellingiri, B., Ramanathan, G., Doss C, G. P., & Gopalakrishnan, A. V. (2022). Role of immune cells and receptors in cancer treatment: An immunotherapeutic approach. *Vaccines, 10*, 1493. doi:10.3390/vaccines10091493.

Muliaditan, T., Caron, J., Okesola, M., Opzoomer, J. W., Kosti, P., Georgouli, M., Gordon, P., Lall, S., Kuzeva, D. M., Pedro, L., Shields, J. D., Gillett, C. E., Diebold, S. S., Sanz-Moreno, V., Ng, T., Hoste, E., & Arnold, J. N. (2018). Macrophages are exploited from an innate wound healing response to facilitate cancer metastasis. *Nature Communications, 9*, 1–15. doi:10.1038/s41467-018-05346-7.

Münz, C. (2012). Antigen processing for MHC class II presentation via autophagy. *Frontiers in Immunology, 3*. doi:10.3389/fimmu.2012.00009.

Murata, Y., Kotani, T., Murata, T. K., & Ohnishi, H.of, H.O.-T., Academic.Oup.ComY. (2014). The CD47–SIRPα signalling system: Its physiological roles and therapeutic application. *The Journal of Biochemistry, 155*, 335–344.

Murray, P (2011). Protective and pathogenic functions of macrophage subsets. *Nature Reviews Immunology, 11*, 723–737.

Na, Y. R., Kim, S. W., & Seok, S. H. (2023). A new era of macrophage-based cell therapy. *Experimental & Molecular Medicine, 55*, 1945–1954.

Ni, R., Jiang, L., Zhang, C., Liu, M., Luo, Y., Hu, Z., Mou, X., & Zhu, Y. (2023). Biologic mechanisms of macrophage phenotypes responding to infection and the novel therapies to moderate inflammation. *International Journal of Molecular Sciences, 24*, 8358. doi:10.3390/ijms24098358.

Nishiwaki, S., Saito, S., Takeshita, K., Kato, H., Ueda, R., & Takami, A. (2020). In vivo tracking of transplanted macrophages with near infrared fluorescent dye reveals temporal distribution and specific homing in the liver that can be perturbed by clodronate liposomes. *PloS one, 15*, e0242488.

Noy, R., & Pollard, J. W. (2014). Tumor-associated macrophages: From mechanisms to therapy. *Immunity, 41*, 49.

Nywening, T. M., Belt, B. A., Cullinan, D. R., Panni, R. Z., Han, B. J., & Sanford, D. E. (2018). Targeting both tumour-associated CXCR2+ neutrophils and CCR2+ macrophages disrupts myeloid recruitment and improves chemotherapeutic responses in pancreatic ductal adenocarcinoma. *Gut, 67*, 1112–1123.

Ochayon, D. E., & Waggoner, S. N. (2021). The effect of unconventional cytokine combinations on NK-cell responses to viral infection. *Frontiers in Immunology, 12*, 645850. doi:10.3389/fimmu.2021.645850.

Orkin, S. H., & Zon, L. I. (2008). Hematopoiesis: An evolving paradigm for stem cell biology. *Cell, 132*, 631–644. doi:10.1016/J.CELL.2008.01.025/ASSET/6040EF92-AC5B-48DE-9DDF-FAC10533D79F/MAIN.ASSETS/GR6.JPG.

Pittet, M. J., Michielin, O., & Migliorini, D. (2022). Clinical relevance of tumour-associated macrophages. *Nature Reviews Clinical Oncology, 19*, 402–421.

Popovich, P., Guan, Z., Wei, P., Popovich, Z. G., Wei, P., Huitinga, I., & Rooijen, N. (1999). Depletion of hematogenous macrophages promotes partial hindlimb recovery and neuroanatomical repair after experimental spinal cord injury. *Experimental Neurology, 158*, 351–365.

Portale, F., & Di Mitri, D. (2023). NK cells in cancer: Mechanisms of dysfunction and therapeutic potential. *International Journal of Molecular Sciences, 24*, 9521. doi:10.3390/ijms24119521.

Qu, Y., Chu, B., Li, J., Deng, H., Niu, T., & Qian, Z. (2024). Macrophage-biomimetic nanoplatform-based therapy for inflammation-associated diseases. *Small Methods.*

Quillien, V., Moisan, A., Lesimple, T., Leberre, C., & Toujas, L. (2001). Biodistribution of 111indium-labeled macrophages infused intravenously in patients with renal carcinoma. *Cancer Immunology, Immunotherapy, 50*, 477–482.

Ren, J., Qu, X., Sun, P., Deng, Q., Kang, L., & Sun, Y. (2020). A smart nanoparticle-laden and remote- controlled self-destructive macrophage for enhanced chemo/chemodynamic synergistic therapy. *ACS Nano, 14*, 12894–12904.

Ribas, A., Medina, T., Kummar, S., Amin, A., Kalbasi, A., Drabick, J.J. Sd-101 in combination with pembrolizumab in advanced melanoma: Results of a phase ib, multicenter study.

Robinson, B. D., Sica, G. L., Liu, Y. F., Rohan, T. E., Gertler, F. B., Condeelis, J. S., & Jones, J. G. (2009). Tumor microenvironment of metastasis in human breast carcinoma: A potential prognostic marker linked to hematogenous dissemination. *Clinical Cancer Research, 15*, 2433–2441. doi:10.1158/1078-0432. CCR-08-2179/346598/P/TUMOR-MICROENVIRONMENT-OF-METASTASIS-IN-HUMAN.

Rosales, C., & Uribe-Querol, E. (2017). Phagocytosis: A fundamental process in immunity. *BioMed Research International, 2017*, 1–18. doi:10.1155/2017/9042851.

Russo, E., Laffranchi, M., Tomaipitinca, L., Del Prete, A., Santoni, A., Sozzani, S., & Bernardini, G. (2021). NK cell antitumor surveillance in a myeloid cell-shaped environment. *Frontiers in Immunology, 12*, 787116. doi:10.3389/fimmu.2021.787116.

Sadri, M., Heidari, S., Faridzadeh, A., Roozbehani, M., Toosi, S., Mahmoudian, R. A., Hoseinzadeh, A., Salmani Fard, M. T., Arab, F. L., Fard, S. R., & Faraji, F. (2024). Potential applications of macrophages in cancer immunotherapy. *Biomedicine & Pharmacotherapy, 178*, 117161. doi:10.1016/j.biopha.2024.117161.

Sánchez-Tilló, E., Lázaro, A., Torrent, R., Cuatrecasas, M., Vaquero, E. C., Castells, A., Engel, P., & Postigo, A. (2010). ZEB1 represses E-cadherin and induces an emt by recruiting the swi/snf chromatin-remodeling protein BRG1. *Oncogene, 29*, 3490–3500. doi:10.1038/onc.2010.102.

Sangaletti, S., Carlo, E., Gariboldi, S., Miotti, S., Cappetti, B., Parenza, M., Rumio, C., Brekken, R. A., Chiodoni, C., & Colombo, M. P. (2008). Macrophage-derived SPARC bridges tumor cell-extracellular matrix interactions toward metastasis. *Cancer Research, 68*, 9050–9059. doi:10.1158/0008-5472.CAN-08-1327.

Sapach, A. Y., Sindeeva, O. A., V. , N. M., Tsitrina, A. A., Mayorova, O. A., & Prikhozhdenko, E. S. (2022). Macrophage in vitro and in vivo tracking via anchored microcapsules. *ACS Applied Materials & Interfaces, 14*, 51579–51592.

Shi, C., & Pamer, E. G. (2011). Monocyte recruitment during infection and inflammation. *Nature Reviews Immunology, 11*, 762–774. doi:10.1038/nri3070.

Shi, X., & Shiao, S. L. (2018). The role of macrophage phenotype in regulating the response to radiation therapy. *Translational Research: The Journal of Laboratory and Clinical Medicine, 191*, 64–80.

Sica, A., & Mantovani, A. (2012). Macrophage plasticity and polarization: In vivo veritas. *Journal of Clinical Investigation, 122*, 787–795.

Sloas, C., Gill, S., & Klichinsky, M. (2021). Engineered CAR-macrophages as adoptive immunotherapies for solid tumors. *Frontiers in Immunology, 12*, 783305. doi:10.3389/fimmu.2021.783305.

Sommer, A., & Gomez Perdiguero, E. (2024). Extraembryonic hematopoietic lineages—to macrophages and beyond. *Experimental Hematology, 136*, 104285. doi:10.1016/j.exphem.2024.104285.

Song, Y., Huang, Y., Zhou, F., Ding, J., Letters, W. Z. C. C., Elsevier Y Song, Y. H., Zhou, F., Ding, J., & Letters, W. Z. C. (2022). Macrophage-targeted nanomedicine for chronic diseases immunotherapy. *Chinese Chemical Letters, 33*, 597–612.

Sparano, J. A., Gray, R., Oktay, M. H., Entenberg, D., Rohan, T., Xue, X., Donovan, M., Peterson, M., Shuber, A., Hamilton, D. A., D'Alfonso, T., Goldstein, L. J., Gertler, F., Davidson, N. E., Condeelis, J., Jones, J., et al. (2017). A metastasis biomarker (MetaSite BreastTM score) is associated with distant recurrence in hormone receptor-positive, HER2-negative early-stage breast cancer. *NPJ Breast Cancer, 3*, 1–8. doi:10.1038/s41523-017-0043-5.

Stewart, B. W. (2019). Mechanisms of carcinogenesis: From initiation and promotion to the hallmarks. *Tumour site concordance and mechanisms of carcinogenesis*. International Agency for Research on Cancer.

Strizova, Z., Benesova, I., Bartolini, R., Novysedlak, R., Cecrdlova, E., Foley, L. K., & Striz, I. (2023). M1/M2 macrophages and their overlaps – myth or reality? *Clinical science, 137*, 1067–1093. doi:10.1042/CS20220531.

Strizova, Z., Benesova, I., Bartolini, R., Novysedlak, R., Cecrdlova, E., Foley, L. K., & Striz, I. (2023). M1/M2 macrophages and their overlaps – myth or reality? *Clinical Science (London, England), 137*, 1067. doi:10.1042/CS20220531.

Su, S., Liu, Q., Chen, J., Chen, J., Chen, F., He, C., Huang, D., Wu, W., Lin, L., Huang, W., Zhang, J., Cui, X., Zheng, F., Li, H., Yao, H., Su, F., & Song, E. (2014). A positive feedback loop between mesenchymal-like cancer cells and macrophages is essential to breast cancer metastasis. *Cancer cell, 25*, 605–620. doi:10.1016/J.CCR.2014.03.021/ATTACHMENT/370D1061-3525-465A-A3E9-13690D268168/MMC2.PDF.

Sui, H., Dongye, S., Liu, X., Xu, X., Wang, L., Jin, C. Q., Yao, M., Gong, Z., Jiang, D., Zhang, K., Liu, Y., Liu, H., Jiang, G., & Su, Y. (2022). Immunotherapy of targeting MDSCs in tumor microenvironment. *Frontiers in Immunology, 13*, 990463. doi:10.3389/fimmu.2022.990463.

Takata, K., Kozaki, T., Lee, C. Z. W., Thion, M. S., Otsuka, M., Lim, S., Utami, K. H., Fidan, K., Park, D. S., Malleret, B., Chakarov, S., See, P., Low, D., Low, G., Garcia-Miralles, M., Zeng, R., Zhang, J., Goh, C. C., & Gul, A. (2017). Induced-pluripotent-stem-cell-derived primitive macrophages provide a platform for modeling tissue-resident macrophage differentiation and function. *Immunity, 47*, 183–198.e6. doi:10.1016/j.immuni.2017.06.017.

Thiery, J. P., Acloque, H., Huang, R. Y. J., & Nieto, M. A. (2009). Epithelial-mesenchymal transitions in development and disease. *Cell, 139*, 871–890. doi:10.1016/J.CELL.2009.11.007/ASSET/E80CD806-0C7C-4B5B-A30D-634AC67E54E7/MAIN.ASSETS/GR5.JPG.

Tong, H., Ke, J. Q., Jiang, F. Z., Wang, X. J., Wang, F. Y., Li, Y. R., Lu, W., & Wan, X. P. (2016). Tumor-associated macrophage-derived CXCL8 could induce ERα suppression via HOXB13 in endometrial cancer. *Cancer Letters, 376*, 127–136. doi:10.1016/J.CANLET.2016.03.036.

Uribe-Querol, E., & Rosales, C. (2020). Phagocytosis: Our current understanding of a universal biological process. *Frontiers in Immunology, 11*, 1066. doi:10.3389/fimmu.2020.01066.

Vonderheide, R. H. (2020). CD40 agonist antibodies in cancer immunotherapy. *Annual Review of Medicine, 71*, 47–58. doi:10.1146/ANNUREV-MED-062518-045435.

Waldman, A. D., Fritz, J. M., & Lenardo, M. J. (2020). A guide to cancer immunotherapy: From t cell basic science to clinical practice. *Nature Reviews Immunology, 20*, 651–668. doi:10.1038/s41577-020-0306-5.

Wang, L.-X., Zhang, S.-X., Wu, H.-J., Rong, X.-L., & Guo, J. (2019). M2b macrophage polarization and its roles in diseases. *Journal of Leukocyte Biology, 106*, 345–358. doi:10.1002/JLB.3RU1018-378RR.

Wang, R., Lan, C., Benlagha, K., Camara, N. O. S., Miller, H., Kubo, M., Heegaard, S., Lee, P., Yang, L., Forsman, H., Li, X., Zhai, Z., & Liu, C. (2024). The interaction of innate immune and adaptive immune system. *MedComm, 5*, e714. doi:10.1002/mco2.714.

Wang, S., Yang, Y., Ma, P., Huang, H., Tang, Q., Miao, H. (n.d.). Landscape and perspectives of macrophage -targeted cancer therapy in clinical trials. Molecular Therapy-Oncolytics, *24*, 799–813.

Wang, X., Zhang, F., Li, N., Yuan, L., Zhang, X., Zhang, L., & Liu, Y. (2025). Fabrication of macrophage-membrane coated functionalized ROS-responsive thiolated chitosan-based nanoparticles to attenuate atherosclerosis:

Investigation on diabetic hyperlipidaemia-induced mice model. *International Journal of Biological Macromolecules, 321*, 145880. https://doi.org/10.1016/j.ijbiomac.2025.145880.

Wang, Y., Wang, X., Li, Y., Xue, Z., Shao, R., & Li, L. (2022). Xuanfei baidu decoction reduces acute lung injury by regulating infiltration of neutrophils and macrophages via PD-1/IL17A pathway. *Pharmacological Research, 176*, 106083.

Watanabe, S., Alexander, M., Misharin, A. V., & Budinger, G. R. S. (2019). The role of macrophages in the resolution of inflammation. *Journal of Clinical Investigation, 129*, 2619–2628. doi:10.1172/JCI124615.

Wei, F., Liu, H., Wang, Y., Li, Y., & Han, S. (2024). Engineering macrophages and their derivatives: A new hope for antitumor therapy. *Biomedicine & Pharmacotherapy, 177*, 116925. doi:10.1016/j.biopha.2024.116925.

Weissleder, R., & Pittet, M. (2020). The expanding landscape of inflammatory cells affecting cancer therapy. *Nature Biomedical Engineering, 4*, 489–498.

Welsh, T. J., Green, R. H., Richardson, D., Waller, D. A., O'Byrne, K. J., & Bradding, P. (2005). Macrophage and mast-cell invasion of tumor cell islets confers a marked survival advantage in nonsmall-cell lung cancer. *Journal of Clinical Oncology, 23*, 8959–8967.

Wyckoff, J., Wang, W., Lin, E. Y., Wang, Y., Pixley, F., Stanley, E. R., Graf, T., Pollard, J. W., Segall, J., & Condeelis, J. (2004). A paracrine loop between tumor cells and macrophages is required for tumor cell migration in mammary tumors. *Cancer Research, 64*, 7022–7029. doi:10.1158/0008-5472.CAN-04-1449.

Wyckoff, J. B., Wang, Y., Lin, E. Y., Li, J. F., Goswami, S., Stanley, E. R., Segall, J. E., Pollard, J. W., & Condeelis, J. (2007). Direct visualization of macrophage-assisted tumor cell intravasation in mammary tumors. *Cancer Research, 67*, 2649–2656. doi:10.1158/0008-5472.CAN-06-1823.

Xia, T., Fu, S., Yang, R., Yang, K., Lei, W., Yang, Y., Zhang, Q., Zhao, Y., Yu, J., Yu, L., & Zhang, T. (2023). Advances in the study of macrophage polarization in inflammatory immune skin diseases. *Journal of Inflammation, 20*, 33. doi:10.1186/s12950-023-00360-z.

Xue, J.-D., Gao, J., Tang, A.-F., & Feng, C. (2024). Shaping the immune landscape: Multidimensional environmental stimuli refine macrophage polarization and foster revolutionary approaches in tissue regeneration. *Heliyon, 10*, e37192. doi:10.1016/j.heliyon.2024.e37192.

Yahara, Y., Ma, X., Gracia, L., & Alman, B. A. (2021). Monocyte/macrophage lineage cells from fetal erythromyeloid progenitors orchestrate bone remodeling and repair. *Frontiers in Cell and Developmental Biology, 9*, 622035. doi:10.3389/fcell.2021.622035.

Yang, D., Yang, L., Cai, J., Hu, X., Li, H., & Zhang, X. (2021). A sweet spot for macrophages: Focusing on polarization. *Pharmacological Research, 167*, 105576.

Yang, Y., Yang, F., Huang, Z., Li, Y., Shi, H., Sun, Q., Ma, Y., Wang, Y., Zhang, Y., Yang, S., Zhao, G.-R., & Xu, F.-H. (2023). T cells, NK cells, and tumor-associated macrophages in cancer immunotherapy and the current state of the art of drug delivery systems. *Frontiers in Immunology, 14*, 1199173. doi:10.3389/fimmu.2023.1199173.

Yi, M., Li, T., Niu, M., Zhang, H., Wu, Y., Wu, K., & Dai, Z. (2024). Targeting cytokine and chemokine signaling pathways for cancer therapy. *Signal Transduction and Targeted Therapy, 9*, 176. doi:10.1038/s41392-024-01868-3.

Yin, T., Li, X., Li, Y., Zang, X., Liu, L., & Du, M. (2024). Macrophage plasticity and function in cancer and pregnancy. *Frontiers in Immunology, 14*, 1333549. doi:10.3389/fimmu.2023.1333549.

You, L., Mao, L., Wei, J., Jin, S., Yang, C., Liu, H., Zhu, L., & Qian, W. (2017). The crosstalk between autophagic and endo-/exosomal pathways in antigen processing for MHC presentation in anticancer T Cell immune responses. *Journal of Hematology & Oncology, 10*, 165. doi:10.1186/s13045-017-0534-8.

Yu, S., Li, A., Liu, Q., Li, T., Yuan, X., Han, X., & Wu, K. (2017). Chimeric antigen receptor T cells: A novel therapy for solid tumors. *Journal of Hematology & Oncology, 10*, 78. doi:10.1186/s13045-017-0444-9.

Zabeti Touchaei, A., & Vahidi, S. (2024). MicroRNAs as regulators of immune checkpoints in cancer immunotherapy: Targeting PD-1/PD-L1 and CTLA-4 pathways. *Cancer Cell International, 24*, 102. doi:10.1186/s12935-024-03293-6.

Zhang, C., Yang, M., & Ericsson, A. C. (2021). Function of macrophages in disease: Current understanding on molecular mechanisms. *Frontiers in Immunology, 12*. doi:10.3389/fimmu.2021.620510.

Zhang, J. Q., Zeng, S., Vitiello, G. A., Seifert, A. M., Medina, B. D., Beckman, M. J., Loo, J. K., Santamaria-Barria, J., Maltbaek, J. H., Param, N. J., Moral, J. A., Zhao, J. N., Balachandran, V., Rossi, F., Antonescu, C. R., & DeMatteo, R. P. (2018). Macrophages and CD8þ T cells mediate the antitumor efficacy of combined CD40 ligation and imatinib therapy in gastrointestinal stromal tumors. *Cancer Immunology Research, 6*, 434–447. doi:10.1158/2326-6066.CIR-17-0345.

Zhang, L., Tian, L., Dai, X., Yu, H., Wang, J., Lei, A., Zhu, M., Xu, J., Zhao, W., Zhu, Y., Sun, Z., Zhang, H., Hu, Y., Wang, Y., Xu, Y., Church, G. M., Huang, H., Weng, Q., & Zhang, J. (2020). Pluripotent stem cell-derived CAR-macrophage cells with antigen-dependent anticancer cell functions. *Journal of Hematology & Oncology, 13*, 153. doi:10.1186/s13045-020-00983-2.

Zhang, W., Liu, L., Su, H., Liu, Q., Shen, J., Dai, H., Zheng, W., Lu, Y., Zhang, W., & Bei, Y. (2019). Chimeric antigen receptor macrophage therapy for breast tumours mediated by targeting the tumour extracellular matrix. *British Journal of Cancer, 121*, 837–845. doi:10.1038/s41416-019-0578-3.

Zhang, X., Zhu, L., Zhang, H., Chen, S., & Xiao, Y. (2022). CAR-T cell therapy in hematological malignancies: Current opportunities and challenges. *Frontiers in Immunology, 13*, 927153. doi:10.3389/fimmu.2022.927153.

Zhang, Y., & Lozano, G. (2017). P53: Multiple facets of a rubik's cube. *Annual Review of Cancer Biology, 1*, 185–201. doi:10.1146/annurev-cancerbio-050216-121926.

Zhao, C., Yang, Z., Li, Y., & Wen, Z. (2024). Macrophages in tissue repair and regeneration: Insights from zebrafish. *Cell Regeneration (London, England), 13*, 12. doi:10.1186/s13619-024-00195-w.

Zhao, H., Wu, L., Yan, G., Chen, Y., Zhou, M., Wu, Y., & Li, Y. (2021). Inflammation and tumor progression: Signaling pathways and targeted intervention. *Signal Transduction and Targeted Therapy, 6*, 263. doi:10.1038/s41392-021-00658-5.

Zhou, L., Zhao, T., Zhang, R., Chen, C., & Li, J. (2024). New insights into the role of macrophages in cancer immunotherapy. *Frontiers in Immunology, 15*. doi:10.3389/fimmu.2024.1381225.

Zhou, Y., Husman, T., Cen, X., Tsao, T., Brown, J., Bajpai, A., Li, M., Zhou, K., & Yang, L. (2022). Interleukin 15 in cell-based cancer immunotherapy. *International Journal of Molecular Sciences, 23*, 7311. doi:10.3390/ijms23137311.

CHAPTER

Combination therapy: A new strategy to manage cancer 14

Reyed Mohamed Reyed[a], Ashok Kumar Sah[b], Satyajit Mohanty[c] and Pranav Kumar Prabhakar[d]

[a]*Genetic Engineering and Biotechnology Research Institute (GEBRI), City of Scientific Research and Technological Applications (SRTA-City), Alexandria, Egypt,* [b]*Department of Medical Laboratory Sciences, College of Applied and Health Sciences, A' Sharqiyah University, Ibra, Oman,* [c]*Department of Pharmacy, School of Medical and Allied Science, Galgotias University, Greater Noida, Uttar Pradesh, India,* [d]*Department of Biotechnology, School of Engineering and Technology, Nagaland University, Meriema, Kohima, Nagaland, India*

14.1 Introduction

Cancer remains one of the world's leading causes of death, responsible for approximately 10 million deaths in 2020. The global cancer burden is expected to rise dramatically, with an estimated 28.9 million new cases projected by 2040—up from 19.3 million in 2020 (Ferlay et al., 2020). The most common cancer types include those affecting the lung, breast, colon and rectum, prostate, skin, and stomach. Lifestyle-related risk factors such as physical inactivity, obesity, poor intake of fruits and vegetables, tobacco use, and alcohol consumption are linked to nearly one-third of all cancer deaths. Additionally, infection-related cancers—such as those caused by human papillomavirus and hepatitis viruses—account for nearly 30% of cases in low- and lower-middle-income countries (Martel et al., 2020).

Given its multifactorial and complex nature, effective cancer management requires a combination of treatment approaches. Multidisciplinary strategies involving surgery, chemotherapy, and radiotherapy are commonly employed to control disease progression and improve patient outcomes. In high-income countries, over 90% of effective cancer treatments are available, whereas in low-income nations, access remains below 15% (WHO, 2020). A multicentric observational study reported that cancer therapies in middle-income countries include chemotherapy (57.6%), surgery (56.8%), radiotherapy (46.8%), hormonal therapy (12%), radioiodine therapy (2%), and immunotherapy (0.9%), with combination therapy (55.2%) being more common than monotherapy (44.8%). Within chemotherapy, multiple classes of drugs are often used together in treatment regimens tailored to specific cancer types. While these approaches aim to suppress cancer growth and improve quality of life, many still result in limited survival rates, disease recurrence, and severe side effects. Notably, improved survival has been seen in cancers such as breast, cervical, colorectal, oral, testicular seminoma, and certain childhood leukemias and lymphomas when diagnosed early and treated appropriately (WHO, 2021).

Advancements in Personalized Cellular Immunotherapy: A Comprehensive Guide to Cancer Treatment.
DOI: https://doi.org/10.1016/B978-0-443-27731-3.00016-4
Copyright © 2026 Elsevier Inc. All rights are reserved, including those for text and data mining, AI training, and similar technologies.

Chemotherapeutic drugs are broadly categorized into alkylating agents, antimetabolites, plant-derived alkaloids, anticancer antibiotics, and molecularly targeted agents—each acting on different cellular targets. However, chemotherapy often causes several adverse effects, including nausea, vomiting, hair loss, bone marrow suppression, infections due to neutropenia, neurotoxicity (from cisplatin, taxanes, vinca alkaloids, oxaliplatin), cardiotoxicity (from anthracyclines), infertility, and immune-related complications. These toxicities highlight the need for safer and more effective treatment alternatives. In this context, natural plant compounds and extracts have shown great promise. When combined with standard anticancer drugs, these phytochemicals can enhance therapeutic efficacy, reduce toxicity, and minimize adverse effects. The study of plant–drug interactions has become a major research focus, offering new possibilities for improving cancer treatment outcomes through synergistic combinations.

14.2 Overview of cancer and its treatment modalities

14.2.1 Molecular basis and pathophysiology of cancer

Cancer is a multifactorial disease characterized by uncontrolled cell division, loss of tissue architecture, and the ability to invade and metastasize to distant organs. At the molecular level, cancer arises due to the accumulation of genetic and epigenetic alterations that disrupt the normal balance between proliferation, differentiation, and cell death (Hanahan, 2022). These changes may be inherited, environmentally induced, or occur spontaneously.

Oncogenes, tumor suppressor genes, and DNA repair genes constitute the principal molecular players. Proto-oncogenes such as *RAS, MYC*, and *HER2* become oncogenic when mutated or overexpressed, driving uncontrolled proliferation (Vogelstein et al., 2013). Conversely, tumor suppressor genes like *TP53, RB1*, and *BRCA1/2* act as regulatory brakes; their inactivation removes growth restraints, enabling neoplastic transformation. Additionally, defects in DNA repair pathways—including mismatch repair and homologous recombination—promote genomic instability, fostering mutation accumulation (Negrini et al., 2010).

Evasion of apoptosis, or programmed cell death, is another hallmark of cancer. Mutations in pro- and antiapoptotic genes, such as *BCL-2* or *Caspase-9*, allow tumor cells to survive under stress or therapy (Elmore, 2007). Angiogenesis, the formation of new blood vessels, is facilitated by proangiogenic factors such as vascular endothelial growth factor (VEGF), ensuring nutrient and oxygen supply to the tumor (Folkman, 2007).

The tumor microenvironment (TME)—comprising fibroblasts, immune cells, endothelial cells, and extracellular matrix (ECM) components—plays a pivotal role in supporting cancer progression. Chronic inflammation, hypoxia, and immune suppression within the TME contribute to tumor survival and metastasis. Moreover, cancer cells evade immune detection by expressing checkpoint molecules such as PD-L1 that inhibit cytotoxic T-cell activity (Topalian et al., 2016).

Metastasis represents the most lethal feature of cancer. It involves a series of coordinated steps, including epithelial-to-mesenchymal transition (EMT), invasion, intravasation, circulation, extravasation, and colonization of distant organs (Lambert et al., 2017). Signaling pathways like PI3K/AKT, MAPK, Wnt/β-catenin, and TGF-β orchestrate these processes. Understanding these molecular and cellular mechanisms provides the foundation for developing targeted therapies aimed at interrupting specific cancer-promoting pathways.

14.2.2 Current standard treatments

Cancer treatment has evolved into a multimodal approach, integrating surgery, radiotherapy, chemotherapy, hormonal therapy, and immunotherapy. The therapeutic strategy depends on cancer type, stage, genetic profile, and patient-specific factors (National Cancer Institute [NCI], 2023).

14.2.2.1 Surgery

Surgery remains the cornerstone for managing solid tumors and offers the best chance for cure in localized disease. Its primary aim is the complete removal of the tumor with sufficient margins to minimize recurrence (Edge & Compton, 2010).

Different surgical roles include:

- **Curative surgery**, where localized tumors are completely excised (e.g., breast, colon, thyroid cancers).
- **Debulking surgery**, which reduces tumor mass to enhance adjuvant therapy efficacy.
- **Palliative surgery**, designed to relieve symptoms such as obstruction or pain in advanced stages.
- **Diagnostic/Staging procedures**, such as biopsies and lymph node dissections, to determine disease extent.

Advances in minimally invasive and robotic-assisted techniques have improved precision, recovery, and outcomes (Zhou et al., 2016). Nonetheless, surgery is often combined with chemotherapy or radiotherapy to address residual microscopic disease.

14.2.2.2 Radiotherapy

Radiation therapy uses high-energy radiation to damage cancer cell DNA, leading to cell death. It may be delivered as external beam radiotherapy, brachytherapy (internal radiation), or systemic radionuclide therapy (Delaney et al., 2005).

Modern approaches such as intensity-modulated radiotherapy, image-guided radiotherapy, and proton beam therapy have significantly improved dose precision, sparing normal tissues (Durante & Loeffler, 2010).

Radiotherapy serves curative, adjuvant, or palliative purposes depending on cancer stage and site. However, challenges include radiation resistance, dose limitations, and long-term complications such as fibrosis and secondary cancers (Douglass & Amato, 2018).

14.2.2.3 Chemotherapy

Chemotherapy employs cytotoxic drugs to kill rapidly dividing cells. While effective, it often lacks selectivity, affecting healthy proliferating cells in the bone marrow, gastrointestinal tract, and hair follicles. The major drug classes include:

- **Alkylating agents** (e.g., cyclophosphamide, cisplatin)—form DNA crosslinks.
- **Antimetabolites** (e.g., 5-fluorouracil, methotrexate)—block nucleotide synthesis.
- **Plant alkaloids** (e.g., paclitaxel, vincristine)—disrupt mitotic spindle formation.
- **Antitumor antibiotics** (e.g., doxorubicin, bleomycin)—intercalate into DNA.
- **Topoisomerase inhibitors** (e.g., etoposide)—interfere with DNA replication.

Chemotherapy is administered as neoadjuvant, adjuvant, or palliative therapy (Chabner & Roberts, 2005). Combination regimens such as CHOP for lymphoma and FOLFOX for colorectal cancer maximize tumor kill and minimize resistance. Nonetheless, side effects—myelosuppression, nausea, alopecia, and organ-specific toxicities—often limit its use (Kintzel, 2001).

14.2.2.4 Hormonal therapy

Hormonal therapy targets cancers dependent on endocrine signaling for growth. It is primarily effective against breast, prostate, and endometrial cancers. In breast cancer, drugs like tamoxifen (a selective estrogen receptor modulator) and aromatase inhibitors (anastrozole, letrozole) block estrogen activity (Jordan, 2003). For prostate cancer, androgen deprivation therapy (ADT) using Luteinizing Hormone–Releasing Hormone (LHRH) agonists/antagonists and androgen receptor blockers (bicalutamide, enzalutamide) suppress testosterone-driven proliferation (Crawford et al., 2019). Although hormonal therapy is less toxic than chemotherapy, it may induce metabolic changes, osteoporosis, and cardiovascular effects. The therapy's success largely depends on hormone receptor expression, emphasizing the role of biomarker-guided treatment.

14.2.2.5 Immunotherapy

Immunotherapy represents a paradigm shift in oncology, harnessing the patient's immune system to combat cancer. Major approaches include:

- **Immune checkpoint inhibitors** (e.g., pembrolizumab, nivolumab) targeting PD-1/PD-L1 or CTLA-4 pathways (Pardoll, 2012).
- **Monoclonal antibodies** (e.g., trastuzumab, rituximab) that selectively target tumor antigens.
- **Cancer vaccines** that elicit specific immune responses against tumor antigens.
- **Adoptive cell therapy**, including **CAR-T cells**, where modified T cells attack malignant cells (June & Sadelain, 2018).

Immunotherapy has achieved durable responses in melanoma, non–small cell lung cancer, and hematologic malignancies. Nevertheless, it can trigger immune-related adverse events (irAEs) such as dermatitis, colitis, and endocrinopathies. Moreover, only a subset of patients benefits, highlighting the need for predictive biomarkers and combination strategies.

14.2.3 Challenges in existing therapies

Despite major advancements, existing therapies face significant obstacles that limit long-term effectiveness. Drug resistance, toxicity, and recurrence/metastasis remain primary challenges in cancer management.

14.2.3.1 Drug resistance

Resistance to therapy—either intrinsic or acquired—is a major cause of treatment failure. Cancer cells can evade drug effects through multiple mechanisms:

- **Drug efflux pumps**, notably P-glycoprotein (MDR1), actively transport drugs out of cells (Holohan et al., 2013).
- **Target mutations**, such as EGFR or BCR-ABL variants, reduce drug binding.

- **Enhanced DNA repair** mechanisms counteract the damage induced by chemotherapeutics.
- **Apoptotic pathway alterations**, including p53 mutations and BCL-2 overexpression, prevent cell death.
- **Microenvironmental protection**, where stromal cells and hypoxia confer survival advantages.

To overcome resistance, research focuses on combination therapy, nanomedicine, and targeted delivery systems that can modulate multiple pathways concurrently.

14.2.3.2 *Toxicity and side effects*

The nonselective nature of many cancer treatments leads to severe toxicities. Chemotherapy frequently causes nausea, vomiting, myelosuppression, mucositis, neurotoxicity (e.g., from cisplatin or taxanes), cardiotoxicity (from anthracyclines), and infertility (Kintzel, 2001). Radiotherapy can cause fibrosis, ulceration, or secondary malignancies (Douglass & Amato, 2018). Even immunotherapy can result in immune-mediated damage to normal tissues, producing colitis, hepatitis, or pneumonitis (Postow et al., 2018). Managing these side effects requires supportive care, dose modification, and the development of targeted agents that minimize harm to healthy cells. Advances in pharmacogenomics and biomarker-driven therapy are aiding in identifying patients most at risk for toxicity.

14.2.3.3 *Recurrence and metastasis*

Recurrence and metastasis account for the majority of cancer deaths worldwide. Cancer stem cells (CSCs), characterized by self-renewal and drug resistance, often survive initial treatments and re-generate the tumor (Batlle & Clevers, 2017). Dormant tumor cells can remain quiescent for years before reactivating, leading to relapse. Metastatic progression involves multiple steps—local invasion, intravasation, survival in circulation, extravasation, and colonization at distant sites. These steps are facilitated by EMT, immune evasion, and angiogenesis (Lambert et al., 2017). Current therapies are less effective against dormant or metastatic cells, necessitating new strategies such as EMT inhibitors, angiogenesis blockers, and immune microenvironment modulators. The management of cancer has advanced remarkably, integrating molecular understanding with clinical innovation. Surgery, radiotherapy, chemotherapy, hormonal therapy, and immunotherapy each contribute uniquely to disease control. However, issues of resistance, toxicity, and recurrence underscore the need for combination therapies and personalized medicine approaches. Future research should focus on integrating targeted therapies, natural compounds, and nanotechnology to achieve improved efficacy with reduced side effects. A deeper grasp of tumor biology and immune regulation promises to transform cancer treatment from a reactive to a precision-based discipline.

14.3 Concept of combination therapy

Cancer is an inherently complex and heterogeneous disease characterized by multiple genetic and molecular alterations that drive uncontrolled cell proliferation, resistance to apoptosis, and metastasis. Given this complexity, single-agent therapies have often failed to produce sustained clinical outcomes, leading to relapse and drug resistance. In response, **combination therapy** has emerged as a powerful strategy to improve treatment efficacy by targeting multiple pathways simultaneously. This approach has transformed modern oncology and continues to shape personalized and precision-based cancer management.

14.3.1 Definition and rationale

Combination therapy refers to the use of two or more therapeutic agents or modalities administered together to achieve a synergistic or additive anticancer effect (Al-Lazikani et al., 2012). These agents may include chemotherapeutic drugs, targeted therapies, immunotherapies, radiotherapy, or natural bioactive compounds. The core rationale behind combination therapy lies in the multifactorial nature of cancer, which arises from deregulation across numerous signaling pathways and cellular mechanisms (Hanahan & Weinberg, 2011). No single drug can effectively suppress all these aberrations, but strategic combinations can produce a more comprehensive and durable therapeutic response. The concept is analogous to combination antibiotic therapy used in infectious diseases, where multiple agents target different mechanisms to prevent microbial resistance. In cancer, this approach helps reduce the likelihood of tumor cells adapting to a single therapeutic pressure. The biological rationale includes:

1. **Targeting multiple hallmarks of cancer:** Different agents can act on distinct cellular processes such as proliferation, angiogenesis, apoptosis, and immune evasion.
2. **Overcoming drug resistance:** Combining agents can block alternative survival pathways that tumors activate in response to single-drug exposure (Holohan et al., 2013).
3. **Reducing toxicity:** Lower doses of individual drugs can be used, minimizing side effects while maintaining efficacy.
4. **Enhancing immune responses:** Certain combinations can modulate the TME to make it more immunogenic.

Combination therapy can be pharmacological (e.g., chemotherapy plus targeted therapy) or multimodal, integrating different treatment modalities such as radiotherapy, surgery, and immunotherapy. It may also include phytochemical-based adjuvants that enhance drug action while reducing adverse effects (Kwak et al., 2019). Ultimately, the rationale for combination therapy is grounded in systems biology: cancer cells survive by exploiting network redundancy. Only through simultaneous multitarget intervention can these networks be disrupted effectively.

14.3.2 Historical background and evolution

The origins of combination therapy in oncology can be traced back to the mid-20th century. Initially, single-agent chemotherapy dominated cancer treatment, beginning with nitrogen mustard in the 1940s and methotrexate in the 1950s. Although these agents showed early success, particularly in hematological malignancies, resistance and relapse were common. The turning point came with the development of combination chemotherapy regimens in the 1960s and 1970s. One of the first landmark successes was the mechlorethamine, vincristine, procarbazine, and prednisone (MOPP) regimen for Hodgkin's lymphoma, which dramatically improved cure rates (DeVita et al., 1970). This success demonstrated that combining drugs with different mechanisms and nonoverlapping toxicities could enhance efficacy without exacerbating side effects. Following this, similar combination regimens were developed for other cancers:

- **ABVD** (adriamycin, bleomycin, vinblastine, dacarbazine) for Hodgkin's lymphoma.
- **CHOP** (cyclophosphamide, doxorubicin, vincristine, prednisone) for non-Hodgkin's lymphoma.
- **FAC** (fluorouracil, doxorubicin, cyclophosphamide) and **CMF** (cyclophosphamide, methotrexate, fluorouracil) for breast cancer.

The evolution of combination therapy continued with the advent of **targeted therapies** in the late 1990s and early 2000s. The introduction of imatinib (Gleevec), a tyrosine kinase inhibitor (TKI) for chronic myeloid leukemia (CML), heralded a new era of molecularly guided treatment. Researchers soon realized that resistance to targeted therapy could also develop, leading to combinations that paired targeted drugs with chemotherapy or other inhibitors. In the 2010s, immunotherapy added a new dimension to combination strategies. The use of immune checkpoint inhibitors such as nivolumab and pembrolizumab in combination with chemotherapy or targeted agents demonstrated superior outcomes in melanoma, lung cancer, and renal cell carcinoma (Postow et al., 2018). The evolution of combination therapy has thus moved from empirical drug pairing toward rationally designed, mechanism-based combinations, guided by molecular profiling, genomic sequencing, and systems pharmacology. Emerging technologies such as CRISPR screening, artificial intelligence (AI), and network modeling are now aiding in identifying optimal drug combinations with maximal efficacy and minimal toxicity (Menden et al., 2019).

14.3.3 Mechanisms of action and therapeutic synergy

The therapeutic potential of combination therapy arises from the synergistic interaction between agents, where the combined effect exceeds the sum of their individual effects. Synergy can occur at multiple biological levels—molecular, cellular, and systemic—depending on the drugs' mechanisms of action.

14.3.3.1 Molecular synergy

At the molecular level, synergy is achieved when different agents target complementary pathways or feedback loops within cancer signaling networks. For example:

- Combining **DNA-damaging agents** (cisplatin) with **DNA repair inhibitors** (PARP inhibitors) enhances cytotoxicity by preventing DNA repair and promoting apoptosis (Lord & Ashworth, 2017).
- The combination of **MEK inhibitors** and **PI3K inhibitors** suppresses parallel signaling pathways in resistant tumors, particularly in melanoma and colorectal cancers.

Such strategies prevent compensatory activation of alternative pathways, a major cause of therapeutic failure in monotherapy.

14.3.3.2 Cellular-level interactions

At the cellular level, combination therapy can influence multiple hallmarks of cancer simultaneously. For instance:

- **Antiangiogenic agents** (bevacizumab) can normalize tumor vasculature, improving the delivery of chemotherapeutic drugs to cancer cells.
- **Epigenetic modulators** (e.g., histone deacetylase [HDAC] inhibitors) can restore sensitivity to chemotherapy by reactivating silenced tumor suppressor genes.

- Combining **immune checkpoint inhibitors** with **chemotherapy** can enhance antigen release from dying tumor cells, thus stimulating a stronger immune response.

14.3.3.3 Pharmacokinetic and pharmacodynamic interactions

Synergy can also emerge from pharmacokinetic or pharmacodynamic interactions. A drug that alters tumor blood flow or modulates drug metabolism can enhance the bioavailability and retention of another agent. For example, taxanes have been shown to increase the intracellular accumulation of platinum compounds, thereby improving treatment outcomes. The TME is a key determinant of therapeutic response. Certain drug combinations act by modifying the TME to make it less supportive of cancer growth. For example, immunotherapy plus antiangiogenic therapy can reduce immunosuppression and enhance T-cell infiltration into tumors (Hegde & Chen, 2020). Another emerging mechanism is synthetic lethality, where the simultaneous disruption of two genes or pathways leads to cell death, while disruption of either one alone does not. This principle underlies the success of PARP inhibitors in BRCA-mutated cancers. Combinations exploiting synthetic lethal interactions are a growing focus in precision oncology. Overall, these mechanisms highlight that combination therapy is not merely additive but strategically synergistic, disrupting cancer networks at multiple levels and preventing adaptation.

14.3.4 Advantages over single-agent therapy

The shift from monotherapy to combination therapy represents one of the most significant advancements in cancer treatment. The key advantages are discussed in the following section.

14.3.4.1 Enhanced efficacy

Combination therapy can produce superior therapeutic outcomes by attacking cancer through multiple mechanisms. When agents act on distinct molecular targets, the cumulative effect is greater tumor regression and longer progression-free survival. For example, the combination of **trastuzumab** (HER2 inhibitor) with chemotherapy has significantly improved survival in HER2-positive breast cancer (Slamon et al., 2011).

14.3.4.2 Reduction of drug resistance

Resistance to therapy is a major challenge in oncology. Tumors can adapt by activating alternative pathways or mutating drug targets. Combination regimens reduce the likelihood of resistance by simultaneously blocking multiple survival mechanisms. For instance, using **BRAF and MEK inhibitors** in melanoma delays resistance compared to either agent alone (Long et al., 2014).

14.3.4.3 Lower toxicity and improved tolerability

By using lower doses of each component drug, combination therapy can maintain efficacy while minimizing individual drug toxicity. For example, combining natural compounds such as **curcumin** or **resveratrol** with chemotherapy enhances cytotoxicity, allowing dose reduction and fewer side effects (Kwak et al., 2019). This approach not only improves patient quality of life but also enhances treatment adherence.

14.3.4.4 Broader patient response

Single-agent therapies often benefit only a subset of patients with specific genetic mutations. Combinations can broaden therapeutic applicability by targeting both driver and compensatory mutations, making treatment effective across genetically diverse tumors. Tumors are composed of genetically and phenotypically diverse cell populations. Monotherapy may eliminate sensitive cells but spare resistant clones, leading to relapse. Combination therapy can simultaneously target multiple cell subtypes, reducing the chance of clonal escape and recurrence (Dagogo-Jack & Shaw, 2018). Some drug combinations enhance the body's natural immune surveillance. For example, chemotherapy can increase the presentation of tumor antigens, making immunotherapy more effective. Similarly, radiotherapy combined with immune checkpoint inhibitors can induce the **abscopal effect**, where local treatment leads to systemic tumor regression (Demaria et al., 2015).

While combination regimens can initially seem costly, they may reduce long-term healthcare expenses by preventing relapse, minimizing hospitalization, and improving survival outcomes. Moreover, by combining existing approved drugs, researchers can repurpose therapies, accelerating clinical translation and regulatory approval. Combination therapy represents a paradigm shift in cancer management, addressing the multifaceted nature of the disease through multitargeted intervention. From early chemotherapy regimens to modern combinations involving immunotherapy and natural products, the approach has consistently improved survival and patient outcomes. Mechanistically, combination therapy achieves therapeutic synergy by interfering with redundant pathways, overcoming resistance, and modulating the tumor microenvironment. As oncology advances, rationally designed combinations—guided by molecular profiling and computational modeling—will form the backbone of precision cancer treatment. The integration of genomics, systems biology, and AI is expected to refine these strategies further, ushering in a new era of personalized, effective, and less toxic cancer therapy.

14.4 Types of combination therapy in cancer treatment

Combination therapy represents a paradigm shift in modern oncology, where two or more therapeutic agents or modalities are used together to achieve better efficacy, minimize drug resistance, and reduce adverse side effects compared to single-agent treatments. The strategy relies on the concept that targeting multiple signaling pathways or cellular mechanisms simultaneously can produce synergistic or additive effects on tumor regression. Over the last few decades, several combination regimens have evolved, encompassing chemotherapeutic drugs, radiotherapy, immunotherapy, hormonal therapy, and molecularly targeted agents (Chabner & Roberts, 2005). The following sections elaborate on major types of combination therapies commonly used in cancer treatment.

14.4.1 Chemotherapy–chemotherapy combinations

14.4.1.1 Concept and rationale

Chemotherapy–chemotherapy combinations, also known as polychemotherapy, involve the use of two or more cytotoxic drugs with different mechanisms of action to improve therapeutic outcomes. This approach was developed to overcome the limitations of single-agent chemotherapy, which often leads to incomplete tumor eradication, relapse, and the emergence of drug resistance (Al-Lazikani et al., 2012). The rationale is based on the idea that using drugs with nonoverlapping toxicity profiles and complementary mechanisms can target various phases of the cell cycle and enhance cancer cell death. One of the earliest and most successful examples of polychemotherapy is the MOPP regimen used for treating Hodgkin's lymphoma. Later, the ABVD regimen (doxorubicin, bleomycin, vinblastine, dacarbazine) demonstrated improved survival rates with reduced toxicity (Canellos et al., 2004). Similarly, the CHOP regimen (cyclophosphamide, doxorubicin, vincristine, and prednisone) remains a cornerstone therapy for non-Hodgkin's lymphoma. In solid tumors, combinations like FOLFOX (5-fluorouracil, leucovorin, and oxaliplatin) and FOLFIRI (5-fluorouracil, leucovorin, and irinotecan) are standard regimens for colorectal cancer. For breast cancer, CMF (cyclophosphamide, methotrexate, fluorouracil) and **AC-T** (doxorubicin, cyclophosphamide followed by paclitaxel) are frequently used (Martin, 2005).

14.4.1.2 Mechanistic basis of synergy

Different chemotherapeutic drugs act at distinct molecular targets—some interfere with DNA replication (e.g., alkylating agents, platinum compounds), others inhibit mitotic spindles (e.g., taxanes, vinca alkaloids), or disrupt nucleotide synthesis (e.g., antimetabolites). Combining these drugs ensures simultaneous disruption of multiple cellular processes, increasing cytotoxic stress and preventing repair or survival mechanisms. The sequential or concurrent use of drugs can be optimized to achieve maximum therapeutic synergy while minimizing overlapping toxicities.

14.4.1.3 Limitations

Despite their success, polychemotherapy regimens can be associated with cumulative toxicity, myelosuppression, and the development of MDR due to the overexpression of efflux transporters such as P-glycoprotein. Therefore, careful dosing schedules and supportive care are critical components of treatment planning.

14.4.2 Chemotherapy–radiotherapy combinations

14.4.2.1 Rationale and principles

Chemoradiation therapy refers to the concurrent or sequential use of chemotherapy and radiotherapy. The underlying rationale is that chemotherapy can act as a radiosensitizer, enhancing the effects of ionizing radiation by interfering with DNA repair mechanisms or by synchronizing tumor cells in radiosensitive phases of the cell cycle. Radiation causes double-strand breaks in DNA, while chemotherapeutic drugs can inhibit the repair of these breaks or introduce additional cytotoxic damage.

14.4.2.2 Clinical applications

Chemoradiation is a mainstay for several locally advanced solid tumors. For instance, concurrent administration of cisplatin with radiotherapy has become the standard of care for head and neck squamous cell carcinoma (Pignon, 2000) and cervical cancer. In nonsmall cell lung cancer (NSCLC), combined platinum-based chemotherapy with thoracic radiation significantly improves local control and overall survival. Similarly, 5-fluorouracil with radiation has been widely used in **rectal cancer** to shrink tumors preoperatively and prevent recurrence (Bosset et al., 2006).

14.4.2.3 Mechanisms of radiosensitization

Chemotherapy enhances radiation efficacy through multiple mechanisms:

- **Inhibition of DNA repair:** Agents like cisplatin and paclitaxel prevent the repair of radiation-induced DNA damage.
- **Cell cycle modulation:** Drugs can synchronize cells in the G2/M phase, which is most sensitive to radiation.
- **Tumor microenvironment modulation:** Certain chemotherapeutics improve oxygenation, making hypoxic tumor cells more susceptible to radiation.
- **Immune modulation:** Emerging evidence suggests that radiation combined with chemotherapy may trigger immunogenic cell death, enhancing systemic antitumor immunity.

14.4.2.4 Treatment scheduling and optimization

Combination scheduling—whether concurrent, sequential, or alternating—plays a critical role in determining outcomes and toxicity profiles. Concurrent chemoradiation tends to provide better tumor control but is associated with higher acute toxicities, particularly mucositis, myelosuppression, and gastrointestinal side effects (Baumann et al., 2016). Sequential regimens, where chemotherapy precedes or follows radiotherapy, may be better tolerated in certain patient populations.

14.4.2.5 Limitations

Despite its efficacy, chemoradiation therapy poses challenges such as increased normal tissue toxicity, particularly to rapidly dividing cells of the mucosa and bone marrow. Moreover, precise dose management and careful patient selection are necessary to balance efficacy with safety (Bentzen & Trotti, 2007).

14.4.3 Chemotherapy–immunotherapy combinations

14.4.3.1 Rationale and underlying concept

The integration of chemotherapy and immunotherapy represents one of the most promising advances in modern oncology. Traditionally, chemotherapy was viewed solely as an immunosuppressive modality; however, recent evidence shows that certain chemotherapeutic drugs can actually enhance antitumor immunity (Galluzzi et al., 2012). These drugs can modulate the tumor microenvironment, promote immunogenic cell death, and increase tumor antigen presentation, thereby potentiating immune responses when combined with checkpoint inhibitors or cytokine-based therapies.

The combination strategy aims to leverage the direct cytotoxic effect of chemotherapy with the immune system's ability to generate long-term tumor control and memory. This synergy can convert an immunologically "cold" tumor (with poor immune infiltration) into a "hot" one, making it more responsive to immunotherapy (Zitvogel et al., 2017).

14.4.3.2 Clinical applications

Several landmark studies have validated the efficacy of chemo-immunotherapy. For instance, in NSCLC, the combination of pembrolizumab (anti–PD-1 antibody) with platinum-based chemotherapy significantly improved overall survival compared to chemotherapy alone (Gandhi et al., 2018). Similarly, in triple-negative breast cancer (TNBC), atezolizumab combined with nab-paclitaxel demonstrated enhanced progression-free survival (Schmid et al., 2018). In urothelial carcinoma and melanoma, chemotherapy has been shown to prime the immune response, increasing infiltration of cytotoxic T cells, thus improving the efficacy of checkpoint blockade.

14.4.3.3 Mechanistic synergy

Chemotherapy contributes to immune activation in several ways:

- **Induction of immunogenic cell death:** Drugs like anthracyclines and oxaliplatin cause tumor cells to release danger-associated molecular patterns, which stimulate dendritic cell activation and antigen presentation.
- **Reduction of immunosuppressive cells:** Low-dose cyclophosphamide or gemcitabine depletes regulatory T cells (Tregs) and myeloid-derived suppressor cells (MDSCs), enhancing antitumor immunity.
- **Enhanced tumor antigen exposure:** DNA-damaging agents increase neoantigen load, making tumor cells more recognizable to immune cells.

14.4.3.4 Limitations and future directions

Despite remarkable outcomes, combination regimens can increase irAEs, including colitis, pneumonitis, and endocrinopathies. Moreover, the optimal sequence, dose, and timing of administration remain areas of active research. Ongoing studies are exploring metronomic (low-dose, frequent) chemotherapy schedules to maximize immune synergy while minimizing toxicity.

14.4.4 Chemotherapy–hormonal therapy combinations

14.4.4.1 Rationale

Hormonal therapy targets hormone-dependent cancers by interfering with hormonal signaling pathways that drive tumor growth. When combined with chemotherapy, it provides a dual approach: chemotherapy destroys rapidly dividing cells, while hormonal agents suppress hormonal stimuli that sustain tumor proliferation (Osborne et al., 2005). This strategy is particularly effective in cancers like breast, prostate, and endometrial carcinoma. In breast cancer, combining hormonal agents such as tamoxifen or aromatase inhibitors with chemotherapy has been explored extensively. Although concurrent use is less common due to antagonistic interactions, sequential administration—chemotherapy followed by endocrine therapy—improves disease-free survival in hormone receptor–positive patients. In prostate cancer, ADT combined with docetaxel chemotherapy has become a standard for metastatic hormone-sensitive cases, significantly extending survival. Similarly, the combination of abiraterone or **enzalutamide** with chemotherapy has improved outcomes in castration-resistant prostate cancer (James et al., 2017).

14.4.4.2 Mechanisms of interaction

Hormonal and chemotherapeutic agents interact at multiple cellular levels:

- **Cell cycle modulation:** Hormonal agents can arrest cells in specific phases, enhancing chemotherapy sensitivity.
- **Apoptotic signaling enhancement:** Combined treatments can upregulate proapoptotic pathways.

- **Tumor microenvironment modulation:** Hormonal therapy reduces growth factor signaling, improving drug delivery and reducing angiogenesis (Dowsett et al., 2001).

14.4.4.3 Challenges

The main limitation lies in timing and sequencing. Concurrent therapy may sometimes reduce efficacy due to hormone-induced cell cycle arrest that protects tumor cells from chemotherapy-induced cytotoxicity. Therefore, personalized scheduling and biomarker-based patient selection are crucial to optimize results.

14.4.5 Chemotherapy–targeted therapy combinations

Targeted therapies, such as TKIs, monoclonal antibodies, and small-molecule inhibitors, specifically interfere with signaling pathways critical to tumor growth and survival. Combining these agents with chemotherapy has emerged as an effective strategy to enhance therapeutic response, delay resistance, and achieve sustained remission. One of the most successful examples is the combination of trastuzumab, a HER2-targeted monoclonal antibody, with taxane-based chemotherapy for HER2-positive breast cancer, which has significantly improved survival (Slamon et al., 2011). Similarly, in colorectal cancer, the addition of cetuximab (anti-EGFR) or bevacizumab (anti-VEGF) to FOLFOX or FOLFIRI regimens enhances response rates and prolongs progression-free survival (Hurwitz et al., 2004). In NSCLC, combining EGFR-TKIs (e.g., gefitinib or erlotinib) with platinum chemotherapy has shown synergistic effects in selected patients harboring EGFR mutations (Mok et al., 2009).

14.4.5.1 Mechanisms of synergy

- **Enhanced tumor cell kill:** Targeted agents sensitize cancer cells to chemotherapeutic damage by disrupting DNA repair and survival pathways.
- **Inhibition of angiogenesis:** Anti-VEGF therapy normalizes tumor vasculature, improving oxygenation and drug penetration (Jain, 2005).
- **Reduced drug resistance:** Targeted therapy can downregulate efflux pumps and survival pathways that contribute to MDR (Holohan et al., 2013).
- **Improved immune activation:** Certain targeted drugs enhance immune recognition of tumor cells, synergizing with chemotherapy-induced antigen release.

Potential drawbacks include overlapping toxicities (e.g., cardiotoxicity from trastuzumab and anthracyclines), high treatment costs, and the emergence of acquired resistance due to pathway mutations. Precision medicine approaches and genomic profiling are therefore essential to identify patients who would benefit most from these regimens.

14.4.6 Multimodal therapy (triple- or multidrug approaches)

Multimodal therapy extends beyond dual combinations by integrating three or more therapeutic modalities—such as chemotherapy, radiotherapy, surgery, immunotherapy, or targeted therapy—into a comprehensive treatment plan. This approach aims for maximal tumor control while minimizing relapse risk by addressing multiple biological mechanisms simultaneously (Wagner et al., 2017). In locally advanced rectal cancer, trimodal therapy combining neoadjuvant chemoradiation, followed by surgery and adjuvant chemotherapy, remains the gold standard, improving both local control and overall survival . Similarly, in esophageal carcinoma, concurrent chemoradiation followed by surgery enhances resectability and long-term survival. In head and neck cancers, multimodal regimens combining surgery, platinum-based chemotherapy, and radiation achieve optimal disease control. For glioblastoma multiforme, the Stupp protocol—surgery followed by radiotherapy with concurrent temozolomide—has become the cornerstone of care (Stupp et al., 2005).

Recent research has explored **quadruple therapies** that incorporate immunotherapy or targeted drugs into established multimodal regimens. For instance, **nivolumab** (PD-1 inhibitor) combined with chemoradiation and surgery has demonstrated encouraging outcomes in esophageal and lung cancers . Nanotechnology-based delivery systems are also facilitating safer, more controlled multidrug combinations.

14.4.6.1 Advantages

Multimodal therapy offers several benefits:

- **Comprehensive tumor control:** Addresses local and systemic disease simultaneously.
- **Reduced recurrence rates:** Targets both primary and micrometastatic lesions.
- **Synergistic efficacy:** Different modalities complement each other's mechanisms.
- **Improved survival:** Proven to extend progression-free and overall survival in multiple cancers (Wagner et al., 2017).

While highly effective, multimodal therapy can be associated with cumulative toxicity, treatment complexity, and high resource demands. Optimizing treatment sequences, integrating biomarkers for response prediction, and developing personalized multimodal protocols are ongoing areas of research. AI–assisted modeling and precision oncology tools are expected to revolutionize this field by tailoring multidrug combinations to individual tumor profiles. Combination therapy represents a cornerstone of modern cancer treatment, offering multifaceted approaches to overcome the inherent complexity and adaptability of malignancies. By combining agents that act on distinct molecular targets or biological processes, oncologists can achieve enhanced efficacy, reduce resistance, and prolong survival. Each form of combination therapy—whether dual or multimodal—has its own merits, limitations, and clinical contexts. As personalized medicine advances, future cancer care will increasingly rely on rationally designed, biomarker-guided combination regimens that maximize therapeutic gain while minimizing toxicity.

14.5 Role of natural products in combination therapy

14.5.1 Importance of plant-derived compounds in oncology

Plant-derived compounds—often referred to as phytochemicals—have emerged as a promising adjunct in cancer therapy. These secondary metabolites from plants exhibit a rich structural diversity and have exhibited multiple biological activities. Unlike many single-target synthetic anticancer drugs, phytochemicals often act pleiotropically, modulating various signaling pathways, inducing apoptosis, arresting cell cycle progression, inhibiting angiogenesis, and altering the TME(Catalano, 2016;

Talib et al., 2022). A major driving reason for their interest in oncology is their relative safety profile and natural abundance. Many epidemiological studies link high fruit, vegetable, and whole-food plant intake with reduced cancer incidence, implicating the protective role of phytochemicals (Rizeq et al., 2020). Moreover, certain phytochemicals have been shown to sensitize tumor cells to standard chemotherapeutics, implying a role in combination therapy rather than simply monotherapy (Talib et al., 2022). In addition, cancer is a complex, multifactorial disease involving genetic mutations, epigenetic alterations, changes in the immune microenvironment, angiogenesis, and metastasis. Because of this complexity, therapies that act on multiple targets simultaneously are highly desirable. Phytochemicals, by virtue of their multimechanism nature, fit this paradigm (Catalano, 2016). Furthermore, given the limitations of standard therapies—such as toxicity, drug resistance, and high cost—plant-derived compounds provide a complementary avenue. They may reduce required dosages of chemotherapeutics, mitigate side effects, or modulate resistance mechanisms (Gao et al., 2022). Thus, the importance of phytochemicals in oncology lies not only in their standalone potential but more significantly in their value as part of combination therapy strategies.

14.5.2 Mechanisms of synergism between phytochemicals and anticancer drugs

When phytochemicals are combined with standard anticancer drugs, several mechanistic pathways underpin potential synergism. The major mechanisms include:

14.5.2.1 Modulation of drug-resistance mechanisms

One key mechanism is the ability of phytochemicals to inhibit efflux transporters, downregulate survival pathways, or enhance apoptotic signaling in cancer cells that have become resistant to chemotherapy. For example, phytochemicals may suppress P-glycoprotein (MDR1) or other ATP-binding cassette (ABC) transporter systems, thereby increasing intracellular accumulation of chemotherapeutic agents and enhancing cytotoxicity (Holohan et al., 2013; Talib et al., 2022).

14.5.2.2 Sensitization to apoptosis and cell cycle arrest

Many phytochemicals act to trigger apoptosis via intrinsic (mitochondrial) pathways or extrinsic (death-receptor) pathways, or by inducing cell-cycle arrest at G1 or G2/M phases. When combined with drugs that also target DNA or mitosis, this can lead to additive or synergistic tumor cell kill. For instance, curcumin has been shown to enhance the apoptotic effect of drugs by up-regulating caspases, down-regulating Bcl-2 family proteins, or modulating p53 signaling (Talib et al., 2022).

14.5.2.3 Modulation of tumor microenvironment and immune modulation

The TME often contains immunosuppressive cells, hypoxia, acidic pH, and stromal support that hinder drug efficacy. Phytochemicals can modulate the TME by reducing inflammation (via NF-κB inhibition), normalizing vasculature (via VEGF suppression), or enhancing immune cell infiltration (Rizeq et al., 2020). Such changes can increase the efficacy of anticancer drugs by improving drug delivery or immune-mediated tumor destruction.

14.5.2.4 Inhibition of angiogenesis, invasion, and metastasis

Phytochemicals frequently inhibit angiogenic pathways (e.g., VEGF, FGF), reduce matrix metalloproteinase (MMP) activity, and block EMT. When combined with standard agents that kill primary tumor cells, these actions help suppress metastatic spread and residual disease—key contributors to relapse (Talib et al., 2022).

14.5.2.5 Epigenetic modulation and targeting cancer stem cells

Another mechanistic dimension is that phytochemicals may influence epigenetic regulators (HDACs, DNMTs) and target cancer stem-cell populations, which are often responsible for recurrence and chemoresistance. By combining with standard drugs that may not target CSCs, the overall elimination of tumor mass and resistant subpopulations can be enhanced.

14.5.2.6 Improved pharmacokinetics and reduced toxicity

Some phytochemicals may affect drug metabolism (e.g., inhibiting cytochrome P450) or modulate drug distribution, thereby improving the effective concentration of chemotherapeutics in tumor tissue. Additionally, because phytochemicals often have lower toxicity, their inclusion can allow reduction of cytotoxic drug doses, thereby lowering side-effect burden while retaining or improving efficacy (Gao et al., 2022). In summary, synergism between phytochemicals and anticancer drugs occurs through multiple converging pathways—overcoming drug resistance, enhancing apoptosis, modulating microenvironment and immune responses, inhibiting metastasis, and improving pharmacologic profiles—thereby supporting the rationale for their inclusion in combination therapy.

14.5.3 Examples of effective plant–drug combinations

Further we review representative combinations, focusing on curcumin, resveratrol, quercetin, and other flavonoids.

14.5.3.1 Curcumin-based combinations

Curcumin (from *Curcuma longa*) is among the most extensively studied phytochemicals in oncology. It exerts antiinflammatory, antioxidant, and multitarget anticancer effects. In combination therapy settings, curcumin has been paired with various chemotherapeutic agents. For example, curcumin combined with 5-fluorouracil or oxaliplatin in colorectal cancer cell lines resulted in enhanced apoptosis, cell-cycle arrest and suppression of metastasis-related genes more than either agent alone (Talib et al., 2022). Another study showed that curcumin plus gemcitabine in pancreatic cancer models sensitized resistant cells and reduced tumor growth in vivo. Mechanistically, curcumin downregulated NF-κB and STAT3, suppressed antiapoptotic Bcl-2/Bcl-xL, and inhibited MMP-9 and VEGF, thereby making tumor cells more susceptible to cytotoxic agents. Additionally, by modulating epigenetic regulators and reducing tumor stem-cell populations, curcumin enhances the efficacy of chemotherapy. However, its poor bioavailability remains a limitation, which is being addressed via nano-formulations and adjuvant delivery systems.

14.5.3.2 Resveratrol-based combinations

Resveratrol (a stilbene found in grapes, berries and peanuts) has been shown to modulate sirtuins, PI3K/Akt/mTOR pathways, and to induce apoptosis and autophagy in tumor cells (Rizeq et al., 2020). In combination studies, resveratrol plus cisplatin or doxorubicin exhibited enhanced cytotoxicity in ovarian and lung cancer models. For instance, in lung cancer cell lines, resveratrol augments cisplatin-induced apoptotic death via downregulation of surviving and activation of caspase-3 (Talib et al., 2022). Moreover, combinations of resveratrol with targeted therapies (e.g., EGFR inhibitors) are under investigation, due to its ability to block compensatory survival pathways and thereby delay resistance. Some early phase clinical studies suggest tolerability of resveratrol with standard therapy but results remain preliminary.

14.5.3.3 Quercetin and other flavonoid-based combinations

Quercetin, a flavanol abundant in many fruits and vegetables, shows antiproliferative, antiangiogenic and proapoptotic activities (Rizeq et al., 2020). In combination contexts, quercetin plus doxorubicin or paclitaxel has shown synergistic inhibition of breast and prostate cancer cell lines. One mechanistic study found that the combination significantly reduced tumor-cell migration and invasion, arrested G2/M phase, and induced ROS-mediated apoptosis (Talib et al., 2022). Other flavonoids—such as epigallocatechin-3-gallate (EGCG), apigenin, kaempferol—are likewise studied in combination. For example, EGCG combined with cisplatin enhanced cytotoxicity in head and neck squamous-cell carcinoma via suppression of NF-κB and up-regulation of p53. These combinations illustrate the potential for flavonoids to complement chemotherapy by targeting multiple hallmarks of cancer—particularly metastasis and invasion—that conventional drugs may inadequately address.

14.5.4 Advantages of phytochemical combinations

Incorporating phytochemicals into combination therapy presents several compelling advantages:

1. **Enhanced efficacy and synergy**—As discussed, phytochemicals can improve the cytotoxic effect of anticancer drugs, enabling higher tumor kill at lower doses (Talib et al., 2022).
2. **Reduced toxicity and better tolerability**—Because phytochemicals might allow reduction of chemotherapeutic doses, patients may experience fewer adverse effects, improved quality of life and better treatment adherence (Gao et al., 2022).
3. **Reduced emergence of resistance**—By attacking tumors via multiple mechanisms, combination strategies including phytochemicals reduce the opportunity for cancer-cell adaptation and resistance (Holohan et al., 2013).
4. **Targeting of multiple cancer hallmarks**—Whereas many conventional agents target a single mechanism (e.g., DNA cross-linking), phytochemicals act on inflammation, angiogenesis, metastasis, cell-cycle, immunity and epigenetics. This broad action is highly beneficial in heterogeneous tumors.
5. **Cost and accessibility advantages**—Many plant compounds are relatively inexpensive, globally available and amenable to large-scale production, potentially reducing the cost burden of therapy, especially in low- and middle-income countries.
6. **Better modulation of tumor microenvironment and immunity**—As many phytochemicals can favorably alter the TME and boost immune responses, their inclusion in combinations may convert immunologically "cold" tumors into responsive ones (Rizeq et al., 2020).

7. **Integration with preventive strategies**—Because many phytochemicals also possess chemo preventive activity, their use in combination regimens might blur the boundary between prevention and therapy, leading to earlier intervention and better long-term outcomes (Kwak et al., 2019).

14.5.5 Challenges in clinical translation

Despite the promising preclinical evidence, translation of phytochemical–drug combinations into clinical practice faces multiple challenges:

1. **Bioavailability and pharmacokinetics**—Many phytochemicals suffer from poor absorption, rapid metabolism, low plasma concentration and short half-life. This affects their in vivo efficacy and complicates dosing strategies.
2. **Standardization and quality control**—Plant extracts and compounds vary in composition depending on source, harvest conditions, extraction method, and storage. Ensuring reproducible bioactivity and consistent dosing is essential but logistically difficult.
3. **Drug–herb interactions and safety**—While phytochemicals are generally considered safe, when combined with chemotherapeutic agents they may alter drug metabolism (e.g., via cytochrome P450), distribution or clearance, potentially causing unexpected toxicity or reduced efficacy.
4. **Lack of rigorous clinical trials**—Although there is extensive preclinical work, clinical trials combining phytochemicals with standard therapies are still limited in number and scale. Translational gaps persist between in vitro success and human outcomes (Rizeq et al., 2020; Talib et al., 2022).
5. **Complexity of combination design**—Optimizing doses, scheduling, sequence of administration, and selection of combinations is complex, especially when multiple agents are involved. There is risk of overlapping toxicities, antagonism between agents, or unpredictable pharmacodynamics (Talib et al., 2022).
6. **Regulatory and commercial hurdles**—Natural compounds often lack patent protection, reducing commercial incentive for large-scale trials. Regulatory pathways for combinations of natural products with drugs may be less defined.
7. **Variability in patients and tumors**—Heterogeneity in tumor genetics, microenvironment and patient metabolism means that responses to phytochemical combinations may vary widely. Biomarker-guided patient selection is still underdeveloped for such combinations (Catalano, 2016).

Researchers are increasingly exploring novel delivery systems (nanoparticles, liposomes, targeted carriers) to overcome bioavailability and targeting issues. Additionally, improved "omics" profiling, computational modeling and systems pharmacology may help design rational combinations and select optimal patient subsets. However, until sufficient large-scale clinical evidence accrues, phytochemical-based combinations remain primarily investigational. The role of natural (plant-derived) products in combination therapy for cancer treatment is increasingly recognized as a viable and promising strategy. The inherent complexity of cancer biology demands interventions that act on multiple mechanisms simultaneously. Phytochemicals, with their multitargeted actions and favorable safety profiles, offer a valuable adjunct to conventional anticancer drugs. Their mechanisms of synergism—ranging from modulation of drug-resistance, enhancement of apoptosis, microenvironment regulation, angiogenesis inhibition and immune modulation—provide a compelling biological basis for combination therapy. Concrete examples such as combinations involving curcumin, resveratrol and quercetin highlight how plant compounds can enhance chemotherapy, reduce toxicity and improve outcomes in preclinical settings. The advantages—enhanced efficacy, reduced toxicity, broader mechanistic coverage, resistance delay and potential cost-effectiveness—underscore their relevance. Yet, translation into clinical practice remains a challenge due to issues around bioavailability, standardization, safety, lack of trials, complexity of combinations, and regulatory matters. Going forward, concerted efforts are required to design high-quality clinical trials, develop standardized formulations, and explore innovative delivery systems. Integrating biomarker-driven patient selection, systems pharmacology and nanotechnology will accelerate the development of phytochemical-based combination therapies. Ultimately, by harnessing nature's chemical diversity in synergy with modern oncology, we may move closer to safer, more effective and durable cancer treatments.

14.6 Mechanisms underlying synergistic effects

Combination therapy represents a strategic approach in cancer management that capitalizes on the synergistic interactions among therapeutic agents. Synergism occurs when two or more agents produce a combined therapeutic effect greater than the sum of their individual effects. The integration of phytochemicals with conventional chemotherapeutic agents is one of the most promising paradigms for achieving such synergy. The underlying mechanisms of these synergistic effects are multifaceted, encompassing modulation of apoptosis and cell cycle pathways, inhibition of MDR transporters, targeting of the tumor microenvironment, attenuation of inflammatory and oxidative stress responses, and regulation of epigenetic and genetic processes. Understanding these mechanisms is crucial for rationally designing combination regimens that maximize efficacy while minimizing toxicity.

14.6.1 Modulation of apoptosis and cell cycle pathways

Apoptosis, or programmed cell death, is a tightly regulated process essential for maintaining tissue homeostasis. One of the hallmarks of cancer is the evasion of apoptosis, enabling malignant cells to proliferate uncontrollably. Dysregulation of apoptotic signaling, particularly through altered expression of proapoptotic and antiapoptotic proteins, contributes to tumor survival and resistance to therapy (Elmore, 2007). Consequently, reactivating apoptotic pathways represents a key target for synergistic anticancer strategies.

14.6.1.1 Intrinsic and extrinsic apoptotic pathways

The intrinsic (mitochondrial) and extrinsic (death receptor–mediated) pathways constitute the two main apoptotic cascades. The intrinsic pathway is governed by the balance between proapoptotic (Bax, Bak, Bad) and antiapoptotic (Bcl-2, Bcl-xL, Mcl-1) proteins, which regulate mitochondrial membrane permeability and cytochrome c release. The extrinsic pathway, on the other hand, is initiated by the activation of death receptors such as Fas, DR4, and DR5, leading to caspase-8 activation and downstream executioner caspases such as caspase-3 and caspase-7. Phytochemicals like curcumin, resveratrol, and quercetin have been shown to restore apoptotic balance by downregulating antiapoptotic proteins (Bcl-2, survivin) and upregulating proapoptotic factors (Bax, p53, caspases). When used in combination with chemotherapy drugs such as cisplatin, doxorubicin, or 5-fluorouracil, these compounds enhance caspase activation, leading to amplified apoptotic signaling (Talib et al., 2022). For instance, curcumin potentiates cisplatin-induced apoptosis in ovarian carcinoma by increasing mitochondrial depolarization and caspase-9 activation.

14.6.1.2 *Cell cycle regulation and checkpoint control*

Cancer cells often exhibit dysregulated cell-cycle progression due to the loss of checkpoint control. Cyclins, cyclin-dependent kinases (CDKs), and their inhibitors (p21, p27) orchestrate transitions between different cell-cycle phases. Many phytochemicals exert growth-inhibitory effects by inducing cell-cycle arrest at G1, S, or G2/M phases. Quercetin, for example, inhibits CDK2 and cyclin E, thereby arresting cells in the G1 phase (Rizeq et al., 2020). When combined with chemotherapeutic agents that induce DNA damage (e.g., doxorubicin or etoposide), such phytochemicals reinforce checkpoint activation, allowing for enhanced recognition and elimination of damaged cells. This dual targeting of cell cycle and apoptosis amplifies tumor-cell death through synergistic mechanisms.

14.6.1.3 *Crosstalk between apoptosis and autophagy*

Autophagy—a catabolic process involving degradation of damaged cellular components—can have dual roles in cancer, acting as both a survival mechanism and a cell-death pathway. Combination therapies involving phytochemicals often modulate autophagy to favor cell death. Resveratrol and curcumin, for example, promote autophagic flux that converges with apoptosis through crosstalk involving AMPK/mTOR and Beclin-1 pathways (Gao et al., 2022). When coupled with cytotoxic drugs, this convergence enhances overall therapeutic response. Thus, modulation of apoptosis and cell-cycle pathways is a key mechanism by which phytochemical-drug combinations achieve synergism, reinstating natural cell-death processes and suppressing malignant proliferation.

14.6.2 Inhibition of multidrug resistance transporters

One of the most formidable obstacles in cancer chemotherapy is MDR. This phenomenon occurs when cancer cells develop the ability to expel a wide range of structurally and functionally diverse drugs, rendering treatment ineffective. The most recognized mediators of MDR are ABC transporters, such as P-glycoprotein (P-gp/ABCB1), multidrug resistance-associated proteins, and breast cancer resistance protein (BCRP/ABCG2) (Holohan et al., 2013).

14.6.2.1 *Mechanisms of drug efflux*

These transporters utilize ATP hydrolysis to actively efflux drugs from the cytoplasm, thereby lowering intracellular drug concentrations. Overexpression of these proteins in tumor cells results in reduced cytotoxic accumulation of drugs like paclitaxel, doxorubicin, and vincristine. Furthermore, MDR is often accompanied by alterations in cellular metabolism, enhanced DNA repair capacity, and evasion of apoptosis (Fletcher et al., 2010).

14.6.2.2 *Phytochemicals as efflux pump inhibitors*

Several plant-derived compounds can inhibit or downregulate MDR transporters. Curcumin, quercetin, and silymarin have been shown to directly bind to P-gp, impairing its ATPase activity and preventing drug efflux (Talib et al., 2022). Quercetin, for instance, reverses doxorubicin resistance in breast cancer cells by downregulating ABCB1 gene expression and restoring intracellular accumulation of the drug. Similarly, resveratrol suppresses BCRP expression and enhances the sensitivity of colon cancer cells to 5-fluorouracil (Gao et al., 2022). Moreover, phytochemicals often inhibit signaling pathways that regulate MDR expression. The NF-κB, PI3K/Akt, and MAPK pathways are known to promote survival and drug-resistance phenotypes. By modulating these pathways, phytochemicals indirectly reduce the transcription of MDR genes.

14.6.2.3 *Synergistic benefits in reversing drug resistance*

Combining efflux pump inhibitors with chemotherapeutics increases intracellular drug concentrations, thus restoring cytotoxic efficacy. This synergistic reversal of MDR allows for dose reduction, lowering systemic toxicity. Importantly, because phytochemicals target multiple regulatory nodes, they reduce the likelihood of cancer cells developing secondary resistance—a major limitation of synthetic MDR inhibitors (Holohan et al., 2013). Hence, inhibition of MDR transporters constitutes a central mechanism underpinning synergistic effects of phytochemical-drug combinations in oncology.

14.6.3 Targeting the tumor microenvironment

The TME comprises a dynamic network of cancer cells, stromal cells, endothelial cells, immune infiltrates, and ECM components. This ecosystem sustains tumor progression, immune evasion, angiogenesis, and resistance to therapy (Hanahan, 2022). Effective cancer treatment thus requires strategies that target not only tumor cells but also the surrounding microenvironment.

14.6.3.1 *Hypoxia, angiogenesis, and stromal interactions*

Tumor hypoxia triggers the stabilization of hypoxia-inducible factor-1α (HIF-1α), which upregulates angiogenic mediators like VEGF. Conventional chemotherapy may paradoxically exacerbate hypoxia by destroying vasculature. Phytochemicals such as curcumin and epigallocatechin-3-gallate (EGCG) inhibit HIF-1α and VEGF expression, thereby normalizing aberrant vasculature and improving drug perfusion (Talib et al., 2022). Additionally, phytochemicals modulate fibroblast and ECM signaling, inhibiting matrix metalloproteinases (MMP-2, MMP-9), which are essential for invasion and metastasis. When combined with cytotoxic agents, these effects impede the supportive stromal crosstalk that typically fosters tumor survival.

14.6.3.2 *Immune modulation within the TME*

The TME often harbors immunosuppressive cells such as regulatory Tregs, tumor-associated macrophages (TAMs), and MDSCs. These cells inhibit cytotoxic T-lymphocyte activity and blunt responses to immunotherapy. Phytochemicals can reprogram immune infiltrates toward an antitumor phenotype—for example, curcumin repolarizes TAMs from M2 to M1 phenotype and suppresses PD-L1 expression (Rizeq et al., 2020). Such immune-modulating properties make phytochemicals ideal partners for combination with chemotherapy or immunotherapy, as they can enhance immune-mediated tumor clearance.

14.6.3.3 *Extracellular pH and metabolic modulation*

Cancer cells exhibit the "Warburg effect," preferring glycolysis even under normoxic conditions, leading to acidic extracellular pH. This acidity reduces drug uptake and promotes invasiveness. Natural products like resveratrol and berberine inhibit glycolytic enzymes (hexokinase II, lactate dehydrogenase), normalizing pH and improving drug penetration (Gao et al., 2022). By simultaneously targeting the metabolic adaptations of cancer cells and remodeling the TME, phytochemical combinations improve both the pharmacodynamics and the microenvironmental suitability for therapy. Hence, synergistic targeting of the TME represents a vital mechanism through which combination therapies achieve superior anticancer outcomes.

14.6.4 Antiinflammatory and antioxidant effects

Chronic inflammation and oxidative stress are intimately linked to carcinogenesis and tumor progression. Reactive oxygen species (ROS) and inflammatory mediators contribute to DNA damage, genomic instability, and a protumorigenic microenvironment. Thus, modulation of these processes forms another key mechanism of synergism in combination therapy.

14.6.4.1 Inflammatory signaling and NF-κB inhibition

NF-κB is a master transcription factor regulating genes involved in inflammation, proliferation, and survival. Persistent activation of NF-κB promotes cytokine secretion (IL-6, TNF-α), angiogenesis, and chemoresistance. Phytochemicals such as curcumin, resveratrol, and apigenin inhibit NF-κB nuclear translocation and reduce downstream inflammatory cytokines. When combined with chemotherapeutics, this suppression alleviates inflammation-driven resistance and sensitizes cells to apoptosis. For instance, curcumin combined with 5-fluorouracil in colorectal cancer models markedly downregulated NF-κB and COX-2, leading to improved efficacy and reduced inflammation-induced side effects.

14.6.4.2 Oxidative stress modulation

ROS are double-edged swords in cancer biology: low levels promote proliferation, while excessive ROS trigger cell death. Phytochemicals act as redox modulators—either scavenging excessive ROS to prevent genotoxicity or, paradoxically, elevating ROS in cancer cells to lethal levels (Gao et al., 2022). Quercetin and EGCG, for example, induce ROS-mediated apoptosis in tumor cells, especially when combined with drugs that further disrupt redox balance.

14.6.4.3 Antioxidant defense and cytoprotection

Besides directly influencing ROS levels, phytochemicals activate endogenous antioxidant defense mechanisms via the Nrf2-Keap1 pathway. Activation of Nrf2 leads to upregulation of phase II detoxifying enzymes such as glutathione S-transferase, heme oxygenase-1, and NAD(P)H quinone oxidoreductase 1. These enzymes not only protect normal tissues from oxidative damage but also enhance the therapeutic window of cytotoxic drugs (Kwak et al., 2019). By combining agents that kill tumor cells with phytochemicals that protect normal cells, combination therapy achieves selective cytotoxicity—a hallmark of effective cancer treatment.

14.6.5 Epigenetic and gene regulatory mechanisms

Epigenetic alterations—such as DNA methylation, histone modification, and noncoding RNA regulation—play pivotal roles in oncogenesis and therapy resistance (Feinberg et al., 2016). Unlike genetic mutations, these changes are reversible, making them attractive therapeutic targets. Many phytochemicals exert epigenetic modulation, thereby enhancing the efficacy of standard chemotherapeutics.

14.6.5.1 DNA methylation and histone modification

Aberrant DNA hypermethylation of tumor-suppressor genes (e.g., p16, BRCA1) and global hypomethylation leading to oncogene activation are common in cancers. Phytochemicals such as genistein, curcumin, and epigallocatechin-3-gallate act as natural DNMT and HDAC inhibitors (Rizeq et al., 2020). For instance, EGCG demethylates the promoter of the p16 gene, restoring its expression and sensitizing cells to chemotherapy. Histone acetylation, mediated by histone acetyltransferases and HDACs, also governs chromatin accessibility. Curcumin and sulforaphane inhibit HDACs, leading to hyperacetylation of histones and transcriptional activation of apoptotic genes. When combined with drugs that induce DNA damage, these epigenetic changes facilitate transcriptional reprogramming toward cell death.

14.6.5.2 MicroRNAs and noncoding RNAs

MicroRNAs (miRNAs) regulate gene expression posttranscriptionally and are frequently dysregulated in cancer. Some phytochemicals normalize miRNA expression—resveratrol upregulates tumor-suppressive miR-34a, whereas curcumin suppresses oncogenic miR-21, thereby augmenting sensitivity to chemotherapy (Talib et al., 2022). Combination therapy exploiting such miRNA modulation can overcome chemoresistance. For instance, restoring miR-200c via curcumin reduces epithelial–mesenchymal transition, reversing metastasis and resistance in breast cancer models (Gao et al., 2022).

14.6.5.3 Chromatin remodeling and transcriptional regulation

Chromatin structure dictates accessibility of transcription factors to DNA. Phytochemicals influence chromatin remodeling complexes such as SWI/SNF, thereby altering global transcriptional landscapes. These broad-spectrum epigenetic modulations synergize with cytotoxic drugs by enhancing DNA damage responses or by re-expressing silenced tumor-suppressor genes (Feinberg et al., 2016).

14.6.5.4 Epigenetic synergy and therapeutic implications

The combination of phytochemical epigenetic modulators with conventional agents offers several advantages: (1) reversal of epigenetically mediated drug resistance, (2) reactivation of apoptotic and cell-cycle genes, and (3) long-term suppression of cancer stem-cell renewal. However, due to the pleiotropic nature of epigenetic targets, careful optimization of dose and timing is crucial to avoid off-target effects on normal cells. Therefore, modulation of epigenetic and gene regulatory mechanisms forms an integral component of synergistic anticancer effects, providing durable molecular reprogramming that complements cytotoxic action. Synergistic effects in cancer combination therapy emerge from the convergence of multiple biological mechanisms. Phytochemicals, through their ability to modulate apoptosis, inhibit drug-efflux transporters, remodel the tumor microenvironment, attenuate inflammation and oxidative stress, and reprogram epigenetic landscapes, enhance the potency and selectivity of chemotherapeutic drugs. These interactions not only amplify tumor cell death but also minimize collateral damage to normal tissues. By integrating multitargeted natural compounds into treatment regimens, researchers can harness biological synergy to address key limitations of conventional therapies—resistance, toxicity, and recurrence. Further elucidation of these mechanisms through molecular and clinical studies will pave the way for the rational design of safer, more effective, and personalized combination therapies in oncology.

14.7 Clinical studies and case examples

Combination therapy has become a central strategy in modern oncology, offering improved efficacy, reduced drug resistance, and enhanced patient survival compared to monotherapies. The transition from preclinical discovery to clinical application has been guided by numerous laboratory experiments, clinical trials, and approved regimens demonstrating synergistic benefits of combining drugs or treatment modalities. This section reviews key **preclinical findings, ongoing clinical trials, approved therapies**, and the **lessons learned** from clinical implementation.

14.7.1 Preclinical studies demonstrating efficacy

Preclinical research provides the foundation for understanding how different drugs interact synergistically in cancer treatment. These studies, often conducted in cell culture models and animal systems, aim to elucidate mechanisms of action, optimal dosing strategies, and safety profiles prior to human application.

14.7.1.1 Evidence from in vitro and in vivo studies

Numerous preclinical studies have demonstrated that combining chemotherapeutic agents with targeted molecules or phytochemicals enhances anticancer efficacy. For example, curcumin, a polyphenolic compound derived from *Curcuma longa*, has shown synergistic effects with doxorubicin and cisplatin, resulting in higher apoptosis rates and lower toxicity in breast and lung cancer models. Similarly, resveratrol has been reported to potentiate the cytotoxicity of paclitaxel in ovarian cancer by modulating NF-κB and PI3K/Akt signaling pathways (Aggarwal & Shishodia, 2006). In hematological malignancies, the combination of cytarabine and daunorubicin (the "7+3" regimen) demonstrated superior preclinical cytotoxic synergy through complementary effects on DNA synthesis inhibition and topoisomerase II interference. These results led to one of the most enduring treatment backbones for acute myeloid leukemia.

14.7.1.2 Mechanistic validation of synergy

Mechanistic studies have confirmed that combination therapies often act through multiple mechanisms, including the activation of apoptotic signaling, inhibition of survival pathways, and suppression of drug resistance proteins such as P-glycoprotein. For instance, combinations of docetaxel with antiangiogenic agents like bevacizumab have shown additive or synergistic effects in preclinical breast and lung cancer models by simultaneously disrupting microtubule stability and tumor vascularization. Furthermore, in models of colorectal cancer, oxaliplatin and 5-fluorouracil combinations demonstrated enhanced DNA damage and reduced repair capacity compared to either drug alone . Such preclinical data provided the rationale for developing FOLFOX and FOLFIRI regimens in clinical settings.

14.7.2 Ongoing clinical trials

The success of combination therapies in preclinical research has prompted numerous clinical investigations aimed at validating their safety and effectiveness in humans. These trials range from early-phase evaluations of novel drug pairings to large-scale randomized controlled studies assessing survival benefits and quality-of-life improvements.

14.7.2.1 Chemotherapy and targeted therapy trials

Recent clinical trials have explored combinations of chemotherapy with targeted agents that inhibit specific molecular pathways. For instance, in NSCLC, a Phase III trial combining cisplatin-pemetrexed with pembrolizumab (an anti-PD-1 antibody) demonstrated significant improvements in progression-free survival compared to chemotherapy alone (Gandhi et al., 2018). Similarly, combining trastuzumab with paclitaxel has become a mainstay in HER2-positive breast cancer therapy due to robust improvements in response rates and survival (Slamon et al., 2011).

14.7.2.2 Immunotherapy-based combinations

Immunotherapy combinations have become a major focus of clinical research. The CheckMate series of trials (e.g., CheckMate 067) revealed that combining nivolumab and ipilimumab, targeting PD-1 and CTLA-4, respectively, provided durable responses in advanced melanoma compared to monotherapies (Larkin et al., 2015). Such trials underscored the complementary mechanisms of immune checkpoint inhibitors and established dual checkpoint blockade as a clinical standard. In hematologic malignancies, trials combining CAR-T cell therapy with checkpoint inhibitors are ongoing, aiming to enhance persistence and cytotoxicity of engineered immune cells (Jacobson, 2020). These studies represent a frontier in multimodal immunotherapy.

14.7.2.3 Phytochemical-drug combination trials

Although fewer in number, some clinical trials have explored combinations involving natural compounds. For example, curcumin in combination with gemcitabine has shown promise in pancreatic cancer patients by improving chemotherapeutic sensitivity and reducing inflammation markers (Kanai et al., 2011). Similarly, green tea polyphenols (EGCG) have been evaluated alongside bortezomib in multiple myeloma due to their antioxidative and proteasome-modulating effects. While these studies are still limited by small cohorts, they suggest that plant-derived molecules can safely complement conventional drugs.

14.7.3 Approved combination therapies in clinical practice

Over decades of research, numerous combination regimens have gained regulatory approval and become clinical standards across different cancer types.

14.7.3.1 Chemotherapy-based combinations

As mentioned earlier, MOPP, ABVD, and CHOP regimens revolutionized the treatment of lymphomas, achieving long-term remission rates exceeding 80% in some subtypes (Canellos et al., 2004). Similarly, in breast cancer, the CMF and AC-T regimens remain foundational, often supplemented with taxanes or HER2-targeted therapy based on receptor status (Martin, 2005). For solid tumors like colorectal and pancreatic cancers, FOLFOX and FOLFIRINOX regimens represent highly effective standards that combine oxaliplatin, irinotecan, and fluoropyrimidines to maximize tumor response (Conroy et al., 2011).

14.7.3.2 Targeted therapy combinations

In CML), combining imatinib with interferon-alpha enhanced cytogenetic responses in early trials, paving the way for multitargeted TKI combinations (Druker et al., 2001). Similarly, vemurafenib and cobimetinib (BRAF and MEK inhibitors) are approved for BRAF-mutated melanoma, reducing resistance and improving overall survival (Larkin et al., 2014).

14.7.3.3 Immunotherapy and chemotherapy combinations

FDA-approved combinations such as atezolizumab with nab-paclitaxel for TNBC (Schmid et al., 2018) and pembrolizumab with platinum-based chemotherapy for NSCLC (Gandhi et al., 2018) represent landmark advances. These combinations leverage the ability of chemotherapy to enhance tumor antigen release, thereby potentiating immune checkpoint activity.

14.7.3.4 Radiotherapy-based combinations

Concurrent chemoradiation remains the standard of care for locally advanced cervical and head-and-neck cancers (Pignon, 2000). Platinum-based drugs such as cisplatin serve as radiosensitizers, augmenting the DNA damage induced by ionizing radiation.

14.7.4 Limitations and lessons learned

Despite the success of combination therapy, several challenges remain.

14.7.4.1 Toxicity and management

Combination treatments often exacerbate side effects, including hematologic suppression, mucositis, and gastrointestinal toxicity. For instance, dual checkpoint blockade (nivolumab + ipilimumab) increases the incidence of immune-related adverse events, necessitating corticosteroid management (Postow et al., 2015). Toxicity control remains a critical determinant of therapeutic success.

14.7.4.2 Development of resistance

Even in combination regimens, resistance mechanisms eventually emerge, such as secondary mutations, activation of alternative signaling pathways, or epigenetic adaptations (Holohan et al., 2013). This underscores the importance of ongoing biomarker-driven monitoring and treatment adjustments.

14.7.4.3 Clinical trial design and heterogeneity

Cancer's heterogeneity makes designing universally effective combinations difficult. Patient-specific factors, TME diversity, and genetic mutations influence response rates (Hanahan, 2022). Therefore, adaptive trial designs and real-world data integration are becoming vital tools in clinical research. Clinical experience highlights that rational combination design—guided by molecular profiling, pharmacodynamic modeling, and systems biology—yields the most sustainable outcomes. Personalized medicine, integrating genomics and AI, is expected to further refine combination approaches (Bedard et al., 2020).

14.8 Challenges and future perspectives

Combination therapy has revolutionized cancer management by enhancing therapeutic efficacy, minimizing resistance, and improving patient survival. However, despite significant progress, multiple scientific, clinical, pharmacological, and regulatory challenges still hinder the full potential of such regimens. As oncology moves toward personalized and precision-based treatments, addressing these challenges is critical to developing safer and more effective combination strategies. This section explores key issues in the implementation of combination therapy—focusing on pharmacokinetic and pharmacodynamic interactions, toxicity management, personalized medicine, nanotechnology in drug delivery, and ethical considerations—while outlining future directions in this rapidly evolving field.

14.8.1 Pharmacokinetic and pharmacodynamic interactions

14.8.1.1 Complexity of drug interactions

In combination therapy, the simultaneous administration of multiple agents introduces complex **pharmacokinetic (PK)** and **pharmacodynamic (PD)** interactions that can affect absorption, distribution, metabolism, and excretion (ADME) of drugs (Rowland & Tozer, 2011). These interactions can result in **altered plasma concentrations, unpredictable efficacy**, or **enhanced toxicity**. For instance, **cyclophosphamide** and **doxorubicin**, both metabolized by hepatic cytochrome P450 enzymes, can compete for metabolic pathways, leading to increased systemic exposure and cardiotoxicity .

14.8.1.2 Drug–drug and drug–food interactions

Natural product-based combinations present additional challenges. For example, **grapefruit juice** and other flavonoid-rich foods inhibit CYP3A4 enzymes, thereby increasing exposure to chemotherapeutic drugs like **docetaxel** or **vincristine** (Bailey et al., 2013). Similarly, **St. John's wort**, a herbal antidepressant, induces P-glycoprotein, reducing the efficacy of drugs like **imatinib** (Giacomini et al., 2010).

14.8.1.3 Modeling and optimization

Modern computational tools such as PBPK (physiologically based pharmacokinetic) modeling are now used to predict drug interactions and optimize dosing regimens (Jones, 2015). Integrating such approaches into clinical trial design helps reduce adverse outcomes and ensures predictable exposure levels in multidrug regimens.

14.8.2 Toxicity management in combination regimens

14.8.2.1 Overlapping and cumulative toxicity

One of the most pressing challenges in combination therapy is toxicity management. The concurrent administration of drugs with overlapping toxicities can lead to severe hematologic, hepatic, or renal damage. For example, combining cisplatin (nephrotoxic) with paclitaxel (neurotoxic) increases cumulative burden on multiple organ systems (Bentzen & Trotti, 2007).

14.8.2.2 ***Supportive care strategies***

Effective supportive care—such as antiemetic regimens (ondansetron for nausea), hematopoietic growth factors (G-CSF for neutropenia), and nephroprotective agents (amifostine)—has improved tolerance to combination treatments (Aapro, 2020). Advances in symptom management have enabled clinicians to administer higher dose intensities safely.

14.8.2.3 ***Biomarker-guided toxicity prediction***

Genetic biomarkers can identify patients at risk for drug-induced toxicity. For instance, DPYD gene variants predict severe toxicity to 5-fluorouracil-based regimens, while UGT1A1 polymorphisms predict irinotecan-induced neutropenia (Amstutz et al., 2018). Incorporating pharmacogenomics into combination therapy allows for individualized dose adjustments to minimize harm.

14.8.3 Personalized medicine approaches

14.8.3.1 ***Rationale for personalization***

Cancer is inherently heterogeneous, both genetically and phenotypically. Personalized medicine aims to tailor combination therapies to an individual's **tumor biology, molecular signature, and immune profile**. This approach helps optimize drug selection, reduce unnecessary toxicity, and maximize efficacy (Bedard et al., 2020).

14.8.3.2 ***Genomic profiling and biomarker-driven therapy***

The use of next-generation sequencing has allowed oncologists to identify actionable mutations such as **EGFR, BRAF**, and **ALK**, leading to rationally designed drug combinations that target specific pathways (Dienstmann, 2017). For instance, **BRAF** and **MEK inhibitors** are now coadministered in melanoma to prevent resistance via pathway reactivation.

14.8.3.3 ***Artificial intelligence and predictive analytics***

AI is emerging as a key enabler for personalized combination therapy design. Machine learning algorithms can analyze multiomics datasets to predict synergistic drug pairs or identify optimal dose ratios (Zhao et al., 2021). AI-driven virtual screening tools are being used to accelerate discovery of novel synergistic combinations with minimal toxicity.

14.8.4 Role of nanotechnology and drug delivery systems

14.8.4.1 ***Need for improved delivery systems***

Traditional chemotherapy often suffers from poor tumor selectivity and systemic toxicity. Nanotechnology-based delivery systems provide a promising platform for targeted, controlled, and simultaneous delivery of multiple drugs to tumor sites (Peer et al., 2007).

14.8.4.2 ***Types of nanoformulations for combination therapy***

Several types of nanocarriers have been developed to enhance codelivery and synergy:

- **Liposomes:** Liposomal doxorubicin (Doxil) reduces cardiotoxicity while maintaining efficacy. Liposomal coencapsulation of doxorubicin and paclitaxel shows improved pharmacokinetics and tumor penetration (Allen & Cullis, 2013).
- **Polymeric nanoparticles:** Biodegradable polymers like PLGA can encapsulate hydrophilic and hydrophobic drugs, enabling synchronized release (Danhier et al., 2012).
- **Dendrimers and micelles:** These carriers allow high drug-loading capacity and controlled pharmacokinetics for combined drug administration.
- **Stimuli-responsive nanocarriers:** Responsive to pH, temperature, or enzymes, these systems release drugs selectively in the TME (Kirtane et al., 2020).

14.8.4.3 Nanotechnology and natural compounds

Phytochemicals often have poor solubility and bioavailability, limiting their therapeutic potential. Nanocarriers have been used to deliver curcumin, resveratrol, and quercetin in combination with chemotherapeutic drugs, achieving improved absorption, stability, and tumor accumulation (Yallapu, 2012). Such formulations enhance synergy and reduce systemic toxicity.

14.8.5 Regulatory and ethical considerations

14.8.5.1 Complexity of approval pathways

The regulatory approval of combination therapies poses unique challenges because each component must be evaluated individually and in combination to ensure safety and efficacy. The U.S. Food and Drug Administration (FDA) and European Medicines Agency (EMA) require evidence of synergistic benefit and nonadditive toxicity, often necessitating large-scale trials (Jørgensen, 2019).

14.8.5.2 Intellectual property and cost issues

Intellectual property protection becomes complex when combining existing drugs from different manufacturers. High development costs and unclear patent rights can discourage investment in novel combinations (Munos, 2009). Collaborative frameworks between academia, industry, and regulators are therefore essential.

14.8.5.3 Ethical considerations in clinical trials

Ethical challenges include ensuring patient safety in early-phase combination trials, where toxicity profiles are not yet fully understood. Informed consent must clearly communicate potential risks. Additionally, equitable access to advanced combination regimens must be prioritized, as high costs often limit availability in low- and middle-income countries (WHO, 2020).

14.8.6 Future perspectives

The future of combination therapy lies in integrating systems biology, AI, and nanomedicine into clinical development pipelines. Adaptive trial designs such as basket and umbrella trials allow simultaneous testing of multiple combinations across genetic subtypes. Moreover, immuno-oncology will continue to benefit from rational combinations involving checkpoint inhibitors, cancer vaccines, and targeted small molecules. Ultimately, the goal is to develop precision combination therapies—tailored to the patient's unique molecular and immunological landscape—while maintaining affordability, accessibility, and safety.

14.9 Conclusion

14.9.1 Summary of current advancements

Over the past few decades, cancer therapy has evolved from conventional single-agent chemotherapies to sophisticated multimodal and combination-based regimens. The introduction of targeted therapies, immunotherapy, and molecularly guided drug development has significantly improved survival outcomes in many cancer types. However, the emergence of resistance, recurrence, and drug-induced toxicity continues to challenge the success of monotherapies (Sung et al., 2021). Combination therapy represents a paradigm shift in cancer treatment, offering synergistic efficacy, reduced drug resistance, and lower toxicity through multitargeted approaches (Chou, 2010). The fundamental rationale behind combination therapy lies in attacking cancer through multiple mechanisms—disrupting cell proliferation, angiogenesis, and metastasis—while simultaneously modulating immune and inflammatory responses (Sun, 2019). Recent advancements in molecular oncology, systems biology, and nanomedicine have provided deeper insights into tumor behavior, genetic heterogeneity, and tumor–microenvironment interactions. These insights are now driving the design of personalized and adaptive combination therapies tailored to each patient's molecular and clinical profile (Dienstmann, 2017). For instance, checkpoint inhibitors (such as pembrolizumab) combined with chemotherapy or targeted therapy have shown remarkable efficacy in lung, melanoma, and triple-negative breast cancers (Hellmann et al., 2019). Equally important are advances in natural product research, where plant-derived compounds such as curcumin, resveratrol, and quercetin have shown strong synergism with existing drugs. These phytochemicals modulate oxidative stress, apoptosis, and immune signaling pathways, enhancing drug response and reducing side effects (Aggarwal & Harikumar, 2009). Such combinations are now being tested in preclinical and early clinical studies, paving the way for integrated oncopharmacology. In summary, the current state of cancer therapy is characterized by a shift toward precision-based, multitargeted, and combination-driven strategies that aim not only to control tumor growth but also to improve the quality of life and long-term survival.

14.9.2 Future directions in combination therapy research

The future of combination therapy is moving toward a multidimensional, personalized, and technology-driven approach. Several key directions will define the next generation of cancer management.

14.9.2.1 Integration of multiomics and AI technologies

Genomic, transcriptomic, proteomic, and metabolomic profiling have become essential for understanding tumor complexity. Integrating multiomics data with AI can predict synergistic drug pairs, optimal dosage ratios, and patient-specific responses (Zhao et al., 2021). Computational platforms such as DeepSynergy and ComboSyn are being used to model nonlinear interactions between drug combinations, significantly reducing experimental costs and time (Menden et al., 2019). AI algorithms are also capable of identifying hidden biomarkers that predict treatment outcomes, which can inform adaptive dosing and real-time therapy modification—key components of precision oncology.

14.9.2.2 Role of nanotechnology and smart delivery systems

Nanomedicine is set to revolutionize combination therapy by offering codelivery of multiple agents, tumor-specific targeting, and controlled drug release. Advanced nanocarriers—including liposomes, micelles, polymeric nanoparticles, dendrimers, and exosomes—can deliver hydrophilic and hydrophobic drugs simultaneously, improving bioavailability and minimizing off-target effects (Peer et al., 2007). Moreover, stimuli-responsive nanocarriers that release drugs in response to pH, temperature, or enzyme activity enable precise delivery within the TME (Kirtane et al., 2020). Hybrid nanocarriers coloading curcumin and doxorubicin or paclitaxel and siRNA have demonstrated superior tumor regression and reduced toxicity in preclinical models (Yallapu, 2012). The integration of nanotechnology with personalized medicine is expected to make combination regimens safer, more efficient, and patient-friendly.

14.9.2.3 Advancements in immunooncology combinations

Immunotherapy has emerged as a cornerstone of modern cancer treatment. However, resistance and limited response rates remain obstacles. Future research focuses on combining immune checkpoint inhibitors (PD-1, CTLA-4, LAG-3) with chemotherapy, radiotherapy, or oncolytic viruses to enhance immune activation (Wei et al., 2018). For instance, combining nivolumab (anti–PD-1) with ipilimumab (anti–CTLA-4) has significantly improved survival in melanoma patients (Larkin et al., 2015). Furthermore, integrating tumor vaccines, cytokine modulators, and adoptive T-cell therapies with standard treatments could overcome immune evasion and achieve durable remission.

14.9.2.4 Systems biology and network pharmacology

Systems biology is transforming the way we conceptualize cancer therapy by viewing tumors as complex biological networks. Network pharmacology models drug interactions at the molecular systems level, helping to identify nodes of synergy and resistance (Li, 2014). This approach enables rational design of combination regimens that target interconnected signaling cascades—such as PI3K/AKT/mTOR and MAPK pathways—thus reducing compensatory mechanisms that often lead to relapse. Integration of systems biology with experimental oncology will accelerate the translation of theoretical synergy into clinical benefit.

14.9.2.5 Expanding the role of natural compounds

Natural compounds are expected to play an increasingly important role in combination therapy research. Their pleiotropic actions—including modulation of oxidative stress, DNA repair, angiogenesis, and epigenetic regulation—make them ideal candidates for multitargeted approaches (Cragg & Pezzuto, 2016). For example, curcumin combined with 5-fluorouracil enhances apoptosis in colorectal cancer cells, while resveratrol sensitizes tumors to cisplatin by inhibiting NF-κB and STAT3 pathways. Such combinations can potentially allow dose reduction of cytotoxic drugs, minimizing side effects without compromising efficacy. Future clinical trials must focus on optimizing formulation, dosage, and bioavailability of these compounds to ensure reproducible outcomes.

14.9.2.6 *Precision and adaptive clinical trials*

Traditional randomized controlled trials often fail to capture patient heterogeneity. The adoption of adaptive clinical trial designs—such as basket, umbrella, and platform trials—allows testing of multiple combinations across genetic subtypes or tumor types simultaneously. Adaptive trials utilize real-time data to modify treatment arms dynamically, improving efficiency and reducing time to regulatory approval. This strategy is particularly beneficial for evaluating novel combination regimens in rare cancers or small patient populations.

14.9.3 Toward safer and more effective cancer management

14.9.3.1 *Balancing efficacy and safety*

A major challenge in oncology is achieving a balance between efficacy and tolerability. Combination therapy, though powerful, can cause cumulative and overlapping toxicities (Bentzen & Trotti, 2007). Future research must focus on toxicity prediction models, pharmacogenomic screening, and real-time toxicity monitoring using biomarkers. This will allow proactive management and dose optimization to ensure patient safety.

14.9.3.2 *Overcoming drug resistance and recurrence*

Drug resistance remains the leading cause of treatment failure. The use of epigenetic modulators, efflux pump inhibitors, and miRNA-based therapies in combination with chemotherapeutic drugs shows promise in overcoming MDR (Holohan et al., 2013). Additionally, targeting the TME — including hypoxia, angiogenesis, and stromal interactions — through combinatorial strategies can inhibit metastasis and recurrence. The TME represents a critical frontier for designing next-generation regimens that attack cancer from multiple biological angles.

14.9.3.3 *Accessibility and cost-effectiveness*

Despite technological progress, the high cost of combination regimens and advanced therapeutics remains a barrier, particularly in low- and middle-income countries (WHO, 2020). Strategies such as repurposing existing drugs, using generics, and public–private partnerships are needed to improve accessibility. Ensuring equitable access to lifesaving therapies must remain a global health priority.

14.9.3.4 *Ethical and regulatory balance*

Ethical considerations are vital to the development of combination therapies. Regulatory agencies such as the FDA and EMA require demonstration of synergy and nonadditive toxicity before approval (Jørgensen, 2019). Transparency in clinical data, patient consent, and global harmonization of approval processes are essential to accelerate the safe adoption of new regimens.

14.9.3.5 *Vision for the future*

The future of oncology lies in integrative, data-driven, and patient-centric approaches. The convergence of AI, nanomedicine, immunotherapy, and phytopharmacology will shape next-generation cancer therapeutics. The focus will shift from generalized regimens to adaptive therapeutic ecosystems, where treatment is continuously refined based on real-time patient data and biomarker feedback. In this envisioned future, combination therapy will not merely be a strategy but a philosophy of precision oncology—one that harmonizes biology, technology, and ethics for holistic cancer care.

References

Aapro, M. (2020). Supportive care in cancer: A critical element of modern oncology. *Annals of Oncology, 31*(9), 1183–1190.

Aggarwal, B. B., & Harikumar, K. B. (2009). Potential therapeutic effects of curcumin, the anti-inflammatory agent, against neurodegenerative, cardiovascular, pulmonary, metabolic, autoimmune and neoplastic diseases. *International Journal of Biochemistry and Cell Biology, 41*(1), 40–59. doi:10.1016/j.biocel.2008.06.010.

Aggarwal, B. B., & Shishodia, S. (2006). Molecular targets of dietary agents for prevention and therapy of cancer. *Biochemical Pharmacology, 71*(10), 1397–1421. www.elsevier.com/locate/biochempharm. doi:10.1016/j.bcp.2006.02.009.

Al-Lazikani, B., Banerji, U., & Workman, P. (2012). Combinatorial drug therapy for cancer in the post-genomic era. *Nature Biotechnology, 30*(7), 679–692. doi:10.1038/nbt.2284.

Allen, T. M., & Cullis, P. R. (2013). Liposomal drug delivery systems: From concept to clinical applications. *Advanced Drug Delivery Reviews, 65*(1), 36–48. doi:10.1016/j.addr.2012.09.037.

Amstutz, U., Henricks, L. M., Offer, S. M., et al. (2018). Pharmacogenomic-guided dose individualization of fluoropyrimidine therapy. *Nature Reviews Clinical Oncology, 15*(4), 231–242.

Bailey, D. G., Dresser, G., & Arnold, J. M. O. (2013). Grapefruit–medication interactions: Forbidden fruit or avoidable consequences? *Cmaj, 185*(4), 309–316.

Batlle, E., & Clevers, H. (2017). Cancer stem cells revisited. *Nature Medicine, 23*(10), 1124–1134. doi:10.1038/nm.4409.

Baumann, M., Krause, M., Overgaard, J., Debus, J., Bentzen, S. M., Daartz, J., Richter, C., Zips, D., & Bortfeld, T. (2016). Radiation oncology in the era of precision medicine. *Nature Reviews Cancer, 16*(4), 234–249. http://www.nature.com/cancer/. doi:10.1038/nrc.2016.18.

Bedard, P. L., Hansen, A. R., Ratain, M. J., & Siu, L. L. (2020). Tumour heterogeneity in the clinic. *Nature, 575*(7781), 299–306.

Bentzen, S. M., & Trotti, A. (2007). Evaluation of early and late toxicities in chemoradiation trials. *Journal of Clinical Oncology, 25*(26), 4096–4103. doi:10.1200/JCO.2007.13.3983.

Bosset, J. F., Collette, L., Calais, G., Mineur, L., Maingon, P., Radosevic-Jelic, L., Daban, A., Bardet, E., Beny, A., & Ollier, J. C. (2006). Chemotherapy with preoperative radiotherapy in rectal cancer. *New England Journal of Medicine, 355*(11), 1114–1123. http://content.nejm.org/cgi/reprint/355/11/1114.pdf. doi:10.1056/NEJMoa060829.

Canellos, G. P., Anderson, J. R., Propert, K. J., et al. (2004). Chemotherapy of advanced Hodgkin's disease with MOPP, ABVD, or MOPP alternating with ABVD. *New England Journal of Medicine, 327*(21), 1478–1484.

Canellos, G. P., Niedzwiecki, D., Freedman, A. S., Takvorian, T., Anderson, J. R., & Shulman, L. N. (2004). Long-term follow-up of MOPP/ABVD in Hodgkin's lymphoma. *Journal of Clinical Oncology, 22*(4), 747–752.

Catalano, E. (2016). Role of phytochemicals in the chemoprevention of tumors. https://arxiv.org/abs/1605.04519.

Chabner, B. A., & Roberts, T. G. (2005). Chemotherapy and the war on cancer. *Nature Reviews Cancer, 5*(1), 65–72. doi:10.1038/nrc1529.

Chou, T. C. (2010). Drug combination studies and their synergy quantification using the chou-talalay method. *Cancer Research, 70*(2), 440–446. http://cancerres.aacrjournals.org/cgi/reprint/70/2/440. doi:10.1158/0008-5472.CAN-09-1947.

Conroy, T., Desseigne, F. ç, Ychou, M., Bouché, O., Guimbaud, R., Bécouarn, Y., Adenis, A., Raoul, J.-L., Gourgou-Bourgade, S., de la Fouchardière, C., Bennouna, J., Bachet, J.-B., Khemissa-Akouz, F., Péré-Vergé, D., Delbaldo, C., Assenat, E., Chauffert, B., Michel, P., Montoto-Grillot, C., & Ducreux, M. (2011). FOLFIRINOX versus Gemcitabine for metastatic pancreatic cancer. *New England Journal of Medicine, 364*(19), 1817–1825. doi:10.1056/NEJMoa1011923.

Cragg, G. M., & Pezzuto, J. M. (2016). Natural products as a vital source for the discovery of cancer chemotherapeutic and chemopreventive agents. *Medical Principles and Practice, 25*(suppl 2), 41–59.

Crawford, E. D., Heidenreich, A., Lawrentschuk, N., Tombal, B., Pompeo, A. C. L., Mendoza-Valdes, A., Miller, K., Debruyne, F. M. J., & Klotz, L. (2019). Androgen-targeted therapy in men with prostate cancer: Evolving practice and future considerations. *Prostate Cancer and Prostatic Diseases, 22*(1), 24–38. http://www.nature.com/pcan/index.html. doi:10.1038/s41391-018-0079-0.

Dagogo-Jack, I., & Shaw, A. T. (2018). Tumour heterogeneity and resistance to cancer therapies. *Nature Reviews Clinical Oncology, 15*(2), 81–94. http://www.nature.com/nrclinonc/archive/index.html. doi:10.1038/nrclinonc.2017.166.

Danhier, F., Ansorena, E., Silva, J. M., Coco, R., Le Breton, A., & Préat, V. (2012). PLGA-based nanoparticles: An overview of biomedical applications. *Journal of Controlled Release, 161*(2), 505–522. doi:10.1016/j.jconrel.2012.01.043.

Delaney, G., Jacob, S., Featherstone, C., & Barton, M. (2005). The role of radiotherapy in cancer treatment: Estimating optimal utilization from a review of evidence-based clinical guidelines. *Cancer, 104*(6), 1129–1137. doi:10.1002/cncr.21324.

Demaria, S., Golden, E. B., & Formenti, S. C. (2015). The abscopal effect 2.0: Leveraging immunogenic radiation for cancer immunotherapy. *Journal of Clinical Investigation, 125*(2), 490–498. doi:10.1172/JCI78047.

DeVita, V. T., Serpick, A. A., & Carbone, P. P. (1970). Combination chemotherapy in the treatment of advanced Hodgkin's disease. *Annals of Internal Medicine, 73*(6), 881–895. doi:10.7326/0003-4819-73-6-881.

Dienstmann, R. (2017). Genomic medicine frontier: Using next-generation sequencing in oncology. *Nature Reviews Clinical Oncology, 14*(8), 473–490.

Dowsett, M., Smith, I. E., & Ebbs, S. R. (2001). Proliferation and apoptosis as markers of benefit in neoadjuvant endocrine therapy of breast cancer. *Clinical Cancer Research, 7*(1), 221–228.

Druker, B. J., Talpaz, M., Resta, D. J., Peng, B., Buchdunger, E., Ford, J. M., Lydon, N. B., Kantarjian, H., Capdeville, R., Ohno-Jones, S., & Sawyers, C. L. (2001). Efficacy and safety of a specific inhibitor of the BCR-ABL tyrosine kinase in chronic myeloid leukemia. *New England Journal of Medicine, 344*(14), 1031–1037. doi:10.1056/NEJM200104053441401.

Douglass, M. Eric J. Hall and Amato J. Giaccia (2018): Radiobiology for the radiologist. *Australas Phys Eng Sci Med* 41, 1129–1130. https://doi.org/10.1007/s13246-018-0684-1.

Durante, M., & Loeffler, J. S. (2010). Charged particles in radiation oncology. *Nature Reviews Clinical Oncology, 7*(1), 37–43. doi:10.1038/nrclinonc.2009.183.

Edge, S. B., & Compton, C. C. (2010). The American Joint Committee on cancer: The 7th edition of the AJCC cancer staging manual and the future of TNM. *Annals of Surgical Oncology, 17*(6), 1471–1474. doi:10.1245/s10434-010-0985-4.

Elmore, S. (2007). Apoptosis: A review of programmed cell death. *Toxicologic Pathology, 35*(4), 495–516. doi:10.1080/01926230701320337.

Feinberg, A. P., Koldobskiy, M. A., & Göndör, A. (2016). Epigenetic modulators, modifiers and mediators in cancer. *Nature Reviews Genetics, 17*(5), 284–299.

Ferlay, J., Laversanne, M., Ervik, M., et al. (2020). *Global Cancer Observatory: Cancer tomorrow*. International Agency for Research on Cancer https://gco.iarc.fr/tomorrow (Retrieved December 7, 2021).

Fletcher, J. I., Haber, M., Henderson, M. J., & Norris, M. D. (2010). ABC transporters in cancer: More than just drug efflux pumps. *Nature Reviews Cancer, 10*(2), 147–156. doi:10.1038/nrc2789.

Folkman, J. (2007). Angiogenesis: An organizing principle for drug discovery? *Nature Reviews Drug Discovery, 6*(4), 273–286. doi:10.1038/nrd2115.

Galluzzi, L., Senovilla, L., Zitvogel, L., & Kroemer, G. (2012). The secret ally: Immunostimulation by anticancer drugs. *Nature Reviews Drug Discovery, 11*(3), 215–233. doi:10.1038/nrd3626.

Gandhi, L., Rodríguez-Abreu, D., Gadgeel, S., Esteban, E., Felip, E., Angelis, F. láD. e, Domine, M., Clingan, P., Hochmair, M. J., Powell, S. F., Cheng, S. Y.-S., Bischoff, H. G., Peled, N., Grossi, F., Jennens, R. R., Reck, M., Hui, R., Garon, E. B., Boyer, M., Rubio-Viqueira, B. én, Novello, S., Kurata, T., Gray, J. E., Vida, J., Wei, Z., Yang, J., Raftopoulos, H., Pietanza, M. C., & Garassino, M. C. (2018). Pembrolizumab plus chemotherapy in metastatic non-small-cell lung cancer. *New England Journal of Medicine, 378*(22), 2078–2092. doi:10.1056/nejmoa1801005.

Gao, Q., Feng, J., Liu, W., et al. (2022). Advantages of combining phytochemicals with standard chemotherapeutic drugs. *Advanced Drug Delivery Reviews, 188*, 114445.

Giacomini, K. M., Huang, S. M., Tweedie, D. J., Benet, L. Z., Brouwer, K. L. R., Chu, X., Dahlin, A., Evers, R., Fischer, V., Hillgren, K. M., Hoffmaster, K. A., Ishikawa, T., Keppler, D., Kim, R. B., Lee, C. A., Niemi, M., Polli, J. W., Sugiyama, Y., Swaan, P. W., Ware, J. A., Wright, S. H., Yee, S. W., Zamek-Gliszczynski, M. J., & Zhang, L. (2010). Membrane transporters in drug development. *Nature Reviews Drug Discovery, 9*(3), 215–236. doi:10.1038/nrd3028.

Hanahan, D. (2022). Hallmarks of cancer: New dimensions. *Cancer Discovery, 12*(1), 31–46. doi:10.1158/2159-8290.cd-21-1059.

Hanahan, D., & Weinberg, R. A. (2011). Hallmarks of cancer: The next generation. *Cell, 144*(5), 646–674. doi:10.1016/j.cell.2011.02.013.

Hegde, P. S., & Chen, D. S. (2020). Top 10 challenges in cancer immunotherapy. *Immunity, 52*(1), 17–35. www.immunity.com. doi:10.1016/j.immuni.2019.12.011.

Hellmann, M. D., Paz Ares, L., Bernabe Caro, R., Zurawski, B., Kim, S. W., Carcereny Costa, E., Park, K., Alexandru, A., Lupinacci, L., De La Mora Jimenez, E., Sakai, H., Albert, I., Vergnenegre, A., Peters, S., Syrigos, K., Barlesi, F., Reck, M., Borghaei, H., Brahmer, J. R., fByrn, K. J. O., Geese, W. J., Bhagavatheeswaran, P., Rabindran, S. K., Kasinathan, R. S., Nathan, F. E., & Ramalingam, S. S. (2019). Nivolumab plus ipilimumab in advanced non-small-cell lung cancer. *New England Journal of Medicine, 381*(21), 2020–2031. http://www.nejm.org/medical-index. doi:10.1056/NEJMoa1910231.

Holohan, C., Van Schaeybroeck, S., Longley, D. B., & Johnston, P. G. (2013). Cancer drug resistance: An evolving paradigm. *Nature Reviews Cancer, 13*(10), 714–726. doi:10.1038/nrc3599.

Hurwitz, H., Fehrenbacher, L., Novotny, W., Cartwright, T., Hainsworth, J., Heim, W., Berlin, J., Baron, A., Griffing, S., Holmgren, E., Ferrara, N., Fyfe, G., Rogers, B., Ross, R., & Kabbinavar, F. (2004). Bevacizumab plus Irinotecan, Fluorouracil, and Leucovorin for metastatic colorectal cancer. *New England Journal of Medicine, 350*(23), 2335–2342. doi:10.1056/NEJMoa032691.

Jacobson, C. A. (2020). CAR T-cell therapy for lymphoma: Clinical perspectives and management. *Blood, 135*(14), 1126–1136.

Jain, R. K. (2005). Normalization of tumor vasculature: An emerging concept in antiangiogenic therapy. *Science, 307*(5706), 58–62. doi:10.1126/science.1104819.

James, N. D., De Bono, J. S., Spears, M. R., Clarke, N. W., Mason, M. D., Dearnaley, D. P., Ritchie, A. W. S., Amos, C. L., Gilson, C., Jones, R. J., Matheson, D., Millman, R., Attard, G., Chowdhury, S., Cross, W. R., Gillessen, S., Parker, C. C., Russell, J. M., Berthold, D. R., … Sydes, M. R. (2017). Abiraterone for prostate cancer not previously treated with hormone therapy. *New England Journal of Medicine, 377*(4), 338–351. http://www.nejm.org/doi/pdf/10.1056/NEJMoa1702900. doi:10.1056/NEJMoa1702900.

Jones, H. M. (2015). Physiologically based pharmacokinetic modeling in drug discovery and development. *CPT: Pharmacometrics & Systems Pharmacology, 4*(2), 73–86.

Jordan, V. C. (2003). Tamoxifen: A most unlikely pioneering medicine. *Nature Reviews Drug Discovery, 2*(3), 205–213. doi:10.1038/nrd1031.

Jørgensen, J. T. (2019). Regulatory Requirements for Companion Diagnostics and Drug–Diagnostic Codevelopment in the United States. In Companion and Complementary Diagnostics (pp. 307–318). Academic Press.

June, C. H., & Sadelain, M. (2018). Chimeric antigen receptor therapy. *New England Journal of Medicine, 379*(1), 64–73. http://www.nejm.org/medical-index. doi:10.1056/NEJMra1706169.

Kanai, M., Yoshimura, K., Asada, M., Imaizumi, A., Suzuki, C., Matsumoto, S., Nishimura, T., Mori, Y., Masui, T., Kawaguchi, Y., Yanagihara, K., Yazumi, S., Chiba, T., Guha, S., & Aggarwal, B. B. (2011). A phase I/II study of gemcitabine-based chemotherapy plus curcumin for patients with gemcitabine-resistant pancreatic cancer. *Cancer Chemotherapy and Pharmacology, 68*(1), 157–164. doi:10.1007/s00280-010-1470-2.

Kintzel, P. E. (2001). Anticancer drug-induced kidney disorders: Incidence, prevention and management. *Drug Safety, 24*(1), 19–38. http://rd.springer.com/journal/40264. doi:10.2165/00002018-200124010-00003.

Kirtane, A. R., Verma, M., & Kannan, R. M. (2020). Nanoparticle-based combination therapy for cancer. *Journal of Controlled Release, 317*, 255–267.

Kwak, M. K., Kensler, T. W., & Casero, R. A. (2019). Chemoprevention by phytochemicals: Modulation of carcinogenic signal transduction pathways. *Pharmacology & Therapeutics, 203*, 107391.

Lambert, A. W., Pattabiraman, D. R., & Weinberg, R. A. (2017). Emerging biological principles of metastasis. *Cell, 168*(4), 670–691.

Larkin, J., Chiarion-Sileni, V., Gonzalez, R., Jacques Grob, J., Cowey, C. L., Lao, C. D., Schadendorf, D., Dummer, R., Smylie, M., Rutkowski, P., Ferrucci, P. F., Hill, A., Wagstaff, J., Carlino, M. S., Haanen, J. B., Maio, M., Marquez-Rodas, I., McArthur, G. A., Ascierto, P. A., … Wolchok, J. D. (2015). Combined Nivolumab and Ipilimumab or monotherapy in untreated melanoma. *New England Journal of Medicine., 373*(1), 23–34. doi:10.1056/nejmoa1504030.

Li, S. (2014). Systems approaches to deciphering signaling in complex diseases. *Science Signaling, 7*(345).

Long, G. V., Stroyakovskiy, D., Gogas, H., Levchenko, E., De Braud, F., Larkin, J., Garbe, C., Jouary, T., Hauschild, A., Grob, J. J., Chiarion Sileni, V., Lebbe, C., Mandalà, M., Millward, M., Arance, A., Bondarenko, I., Haanen, J. B. A. G., Hansson, J., Utikal, J., … Flaherty, K. (2014). Combined BRAF and MEK inhibition versus BRAF inhibition alone in melanoma. *New England Journal of Medicine, 371*(20), 1877–1888. http://www.nejm.org/doi/pdf/10.1056/NEJMoa1406037. doi:10.1056/NEJMoa1406037.

Lord, C. J., & Ashworth, A. (2017). PARP inhibitors: Synthetic lethality in the clinic. *Science, 355*(6330), 1152–1158. http://science.sciencemag.org/content/sci/355/6330/1152.full.pdf. doi:10.1126/science.aam7344.

Martel, C., Georges, D., Bray, F., Ferlay, J., & Clifford, G. M (2020). Global burden of cancer attributable to infections in 2018: A worldwide incidence analysis. *The Lancet Global Health, 8*(2), e180. doi:10.1016/s2214-109x(19)30488-7.

Martin, M. (2005). Sequential versus concurrent anthracycline and taxane-based chemotherapy in breast cancer. *Journal of Clinical Oncology, 23*(1), 137–145.

Menden, M. P., Wang, D., Mason, M. J., Szalai, B., Bulusu, K. C., Guan, Y., Yu, T., Kang, J., Jeon, M., Wolfinger, R., Nguyen, T., Zaslavskiy, M., Abante, J., Abecassis, B. S., Aben, N., Aghamirzaie, D., Aittokallio, T., Akhtari, F. S., Al-lazikani, B., … Saez-Rodriguez, J. (2019). Community assessment to advance computational prediction of cancer drug combinations in a pharmacogenomic screen. *Nature Communications, 10*(1). http://www.nature.com/ncomms/index.html. doi:10.1038/s41467-019-09799-2.

Mok, T. S., Wu, Y. L., Thongprasert, S., Yang, C. H., Chu, D. T., Saijo, N., Sunpaweravong, P., Han, B., Margono, B., Ichinose, Y., Nishiwaki, Y., Ohe, Y., Yang, J. J., Chewaskulyong, B., Jiang, H., Duffield, E. L., Watkins, C. L., Armour, A. A., & Fukuoka, M. (2009). Gefitinib or carboplatin-paclitaxel in pulmonary adenocarcinoma. *New England Journal of Medicine, 361*(10), 947–957. http://content.nejm.org/cgi/reprint/361/10/947.pdf. doi:10.1056/NEJMoa0810699.

Munos, B. (2009). Lessons from 60 years of pharmaceutical innovation. *Nature Reviews Drug Discovery, 8*(12), 959–968. doi:10.1038/nrd2961.

Negrini, S., Gorgoulis, V. G., & Halazonetis, T. D. (2010). Genomic instability—An evolving hallmark of cancer. *Nature Reviews Molecular Cell Biology, 11*(3), 220–228.

Pardoll, D. M. (2012). The blockade of immune checkpoints in cancer immunotherapy. *Nature Reviews Cancer, 12*(4), 252–264. doi:10.1038/nrc3239.

Peer, D., Karp, J. M., Hong, S., Farokhzad, O. C., Margalit, R., & Langer, R. (2007). Nanocarriers as an emerging platform for cancer therapy. *Nature Nanotechnology, 2*(12), 751–760. doi:10.1038/nnano.2007.387.

Pignon, J. P. (2000). Meta-analysis of chemotherapy in head and neck cancer (MACH-NC). *Radiotherapy and Oncology, 56*(2), 167–177.

Postow, M. A., Sidlow, R., & Hellmann, M. D. (2018). Immune-related adverse events associated with immune checkpoint blockade. *New England Journal of Medicine, 378*(2), 158–168.

Rizeq, B., Gupta, I., Ilesanmi, J., AlSafran, M., Mizanur Rahman, M. D, & Ouhtit, A. (2020). The Power of Phytochemicals Combination in Cancer Chemoprevention. *Journal of Cancer, 11*(15), 4521–4533. doi:10.7150/jca.34374.

Rowland, M., & Tozer, T. N. (2011). Clinical pharmacokinetics and pharmacodynamics : concepts and applications (4th ed.). Wolters Kluwer Health/Lippincott William & Wilkins.

Schmid, P., Adams, S., Rugo, H. S., Schneeweiss, A., Barrios, C. H., Iwata, H., Diéras, V. é, Hegg, R., Im, S.-A. h, Wright, G. S, Henschel, V., Molinero, L., Chui, S. Y., Funke, R., Husain, A., Winer, E. P., Loi, S., & Emens, L. A. (2018). Atezolizumab and Nab-Paclitaxel in advanced triple-negative breast cancer. *New England Journal of Medicine, 379*(22), 2108–2121. doi:10.1056/nejmoa1809615.

Slamon, D., Eiermann, W., Robert, N., Pienkowski, T., Martin, M., Press, M., Mackey, J., Glaspy, J., Chan, A., Pawlicki, M., Pinter, T., Valero, V., Liu, M.-C., Sauter, G., von Minckwitz, G., Visco, F., Bee, V., Buyse, M., Bendahmane, B., … Crown, J. (2011). Adjuvant Trastuzumab in HER2-positive breast cancer. *New England Journal of Medicine, 365*(14), 1273–1283. doi:10.1056/nejmoa0910383.

Stupp, R., Mason, W. P., Van Den Bent, M. J., Weller, M., Fisher, B., Taphoorn, M. J. B., Belanger, K., Brandes, A. A., Marosi, C., Bogdahn, U., Curschmann, J., Janzer, R. C., Ludwin, S. K., Gorlia, T., Allgeier, A., Lacombe, D., Cairncross, J. G., Eisenhauer, E., & Mirimanoff, R. O. (2005). Radiotherapy plus concomitant and adjuvant temozolomide for glioblastoma. *New England Journal of Medicine, 352*(10), 987–996. doi:10.1056/NEJMoa043330.

Sun, Y. (2019). The role of combination therapy in overcoming drug resistance in cancer. *Current Pharmaceutical Design, 25*(35), 3693–3702.

Sung, H., Ferlay, J., Siegel, R. L., Laversanne, M., Soerjomataram, I., Jemal, A., & Bray, F. (2021). Global Cancer Statistics 2020: GLOBOCAN estimates of incidence and mortality worldwide for 36 cancers in 185 countries. *CA: A Cancer Journal for Clinicians, 71*(3), 209–249. doi:10.3322/caac.21660.

Talib, W. H., Awajan, D., Hamed, R. A., Azzam, A. O., Mahmod, A. I., & AL-Yasari, I. H. (2022). Combination anticancer therapies using selected phytochemicals. *Molecules, 27*(17). http://www.mdpi.com/journal/molecules. doi:10.3390/molecules27175452.

Topalian, S. L., Drake, C. G., & Pardoll, D. M. (2016). Immune checkpoint blockade: A common denominator approach to cancer therapy. *Cancer Cell, 27*(4), 450–461.

Vogelstein, B., Papadopoulos, N., Velculescu, V. E., Zhou, S., Diaz, L. A., & Kinzler, K. W. (2013). Cancer genome landscapes. *Science, 340*(6127), 1546–1558. http://www.sciencemag.org/content/339/6127/1546.full.pdf. doi:10.1126/science.1235122.

Wagner, A. D., Syn, N. L. X., Moehler, M., Grothe, W., Yong, W. P., Tai, B. C., Ho, J., & Unverzagt, S. (2017). Chemotherapy for advanced gastric cancer. *Cochrane Database of Systematic Reviews, 2017*(8). http://as.wiley.com/WileyCDA/Brand/id-6.html. doi:10.1002/14651858.CD004064.pub4.

Wei, S. C., Duffy, C. R., & Allison, J. P. (2018). Fundamental mechanisms of immune checkpoint blockade therapy. *Cancer Discovery, 8*(9), 1069–1086. http://cancerdiscovery.aacrjournals.org/content/8/9/1069.full.pdf. doi:10.1158/2159-8290.CD-18-0367.

WHO (2020). World cancer report: Cancer research for cancer prevention. https://publications.iarc.who.int/Non-Series-Publications/World-Cancer-Reports/World-Cancer-Report-Cancer-Research-For-Cancer-Prevention-2020.

Yallapu, M. M. (2012). Curcumin-loaded nanoparticle formulations for cancer therapy. *Biomaterials, 33*(10), 2738–2752.

Zhao, R., Yao, F., Xiang, C., Zhao, J., Shang, Z., Guo, L., Ding, W., Ma, S., Yu, A., Shao, J., Zhu, L., & Han, Y. (2021). Identification of NTRK gene fusions in lung adenocarcinomas in the Chinese population. *Journal of Pathology: Clinical Research, 7*(4), 375–384. http://onlinelibrary.wiley.com/journal/10.1002/(ISSN)2056-4538. doi:10.1002/cjp2.208.

Zhou, M., Zhang, Y., & Wang, X. (2016). Robotic-assisted versus laparoscopic surgery in cancer treatment: An evidence-based review. *Cancer Management and Research, 8*, 49–56.

Zitvogel, L., Galluzzi, L., Smyth, M. J., & Kroemer, G. (2017). Mechanism of action of conventional and targeted anticancer therapies: Reinstating immunosurveillance. *Immunity, 45*(1), 74–88.

CHAPTER

15 Future perspective of cellular immunotherapy in cancer treatment

Rideb Chakraborty[a], Abirami Seetharaman[b], Aniruddha Sen[c], Darshan S[d] and Ciniraj Raveendran[e]

[a]*NSHM Institute of Pharmaceutical Technology, NSHM Knowledge Campus, Durgapur, West Bengal, India,* [b]*Department of Biotechnology, Indian Institute of Technology Madras, Chennai, Tamil Nadu, India,* [c]*Department of Biochemistry, Shri Gorakshnath Medical College Hospital and Research Center, Gorakhpur, Uttar Pradesh, India,* [d]*Rajiv Gandhi University of Health Sciences, Bengaluru, Karnataka, India,* [e]*Department of Radiation Oncology, Government Medical College, Thiruvananthapuram, Kerala, India*

15.1 Introduction

15.1.1 Brief overview of cancer and current treatment approaches

The worldwide health challenge of cancer exists because cancer manifests as an intricate condition when malignant cells rapidly proliferate before reaching multiple body regions. Medical approaches to cancer treatment incorporate surgical and radiological therapies while also using pharmacological systemic methods with targeted treatments alongside immune treatments. Patient treatment for localized tumors depends on surgery, along with radiation therapy, which uses high-energy beams to destroy cancerous cells. The elimination of rapidly growing cancer cells through high-dose prescription medications constitutes a standard approach for metastasis cancer treatments. Targeted treatments aim to break down specific cancer cell weaknesses, thus creating better precise treatment alternatives. Immunotherapy is a major treatment innovation for cancer because it enables effective cancer control by activating the immune system. Modern cancer treatment research has produced positive outcomes, although scientists struggle with two main obstacles since cancer cells adapt to drugs while creating side effects for healthy tissues, which prevents treatment from reaching all forms of cancer cells (Liu et al., 2024). Research activities focus on developing modern treatments and enhancing pharmaceutical mixing, aiming for individualized care strategies to achieve superior therapeutic outcomes (Dailah et al., 2024).

15.1.2 Introduction to cellular immunotherapy

Cellular immunotherapy enables patients to utilize their immune capabilities in fighting cancer tumors. Through targeted cellular immunotherapy, patients obtain personalized therapy that demonstrates longer-lasting effects than regular cancer treatments, which combine chemotherapy with radiation treatment. Medical achievements have reached significant heights because researchers created three key therapies, including chimeric antigen receptor (CAR) T-cell therapy, tumor infiltrating lymphocytes

Advancements in Personalized Cellular Immunotherapy: A Comprehensive Guide to Cancer Treatment.
DOI: https://doi.org/10.1016/B978-0-443-27731-3.00010-3
Copyright © 2026 Elsevier Inc. All rights are reserved, including those for text and data mining, AI training, and similar technologies.

(TIL) therapy, and T-cell receptor (TCR)-engineered T-cell therapy. Through this therapy, medical staff develop modified immune cells to increase their ability to seek out and destroy cancer cells (Golikova et al., 2024; Zhao & Cao, 2019). Regulatory agencies approved multiple CAR T-cell therapies after the successful implementation of this treatment for certain hematological malignancies. The therapy encounters barriers because it has difficulties reaching solid tumors, which causes immune-related adverse effects and requires more cost-effective delivery and better accessibility. Researchers within this field carry out sustained studies to address these problems through innovative advanced methods that enhance efficiency, expand the reach of tumors, and break through resistance limitations. The production methods for cellular therapies receive focused development by manufacturers who simultaneously aim to reduce their manufacturing expenses. The progressive development of cellular immunotherapy will bring groundbreaking improvements to cancer treatments that will provide new opportunities to cancer patients (Sheykhhasan et al., 2024).

15.1.3 Cellular immunotherapy: Future of cancer treatment with potential advancements in efficacy, accessibility, and personalization

Cellular immunotherapy provides precise therapeutic outcomes with prolonged active periods in cancer therapy and disrupts it. Cellular immunotherapy is advancing with an effort to increase its potency as well as to devise strategies to overcome derailments due to the tumor microenvironment (TME) and cancer stem cells (CSCs). Expanding the target tumors that can be used in cellular immunotherapy and solving the resistance patterns to the therapy to enable widespread clinical use will serve as the potential therapeutic use of cellular immunotherapy. Traditionally, current efforts have been geared toward providing off-the-shelf approaches in treatment as well as optimization of manufacturing and reduced costs to increase accessibility and scalability. Precision medicine will combine genomic and proteomic profiles along with artificial intelligence (AI) optimization to develop personalized treatments. Discoveries in natural killer cell therapies, gamma delta T-cell therapies, and macrophage-based therapeutic systems bring on the progress of immunotherapy. In the case of the responsibility of managing medical technologies, it is necessary to converge the solution of ethical and patient safety matters with regulatory and equitable access issues (Afrose et al., 2024; Kapustina et al., 2024).

15.2 Current state of cellular immunotherapy

15.2.1 Overview of existing cellular immunotherapy approaches

15.2.1.1 CAR T-cell therapy

CAR T cell therapy has created a revolution in the treatment of hematological malignancies, showing great efficacy in relapsed/refractory large B cell lymphomas and even in some types of leukemia. The CD19 protein is what current FDA-approved CAR T therapies target in these cancer cells. Yet through ongoing research, the efforts are focused on several areas to improve and expand CAR T-cell therapy. These include: developing CARs targeting other tumor-associated antigens to broaden applicability to different cancer types; engineering "armored" CARs that express additional molecules, such as cytokines or checkpoint inhibitors, to enhance T-cell activation and overcome the immunosuppressive TME; creating CAR T-cells with improved safety profiles, such as "suicide genes" or "off-switches," to mitigate cytokine release syndrome (CRS) and immune effector cell-associated neurotoxicity syndrome (ICANS); exploring allogeneic ("off-the-shelf") CAR T-cells derived from healthy donors to eliminate the need for patient-specific cell collection and reduce manufacturing time and cost; and investigating novel delivery methods, such as in vivo CAR T-cell generation using viral vectors, to simplify administration. Those researchers also examine the combination therapies of CAR T cells with other immunotherapies or targeted agents to further improve the outcome and to overcome resistance mechanisms. In the end, scientists are working to move our understanding of the long-term effects of CAR T cell therapy, such as relapse, immune reconstitution, and secondary malignancies (Wang, 2022).

15.2.1.2 TIL therapy

While TIL therapy is promising, several hurdles exist that future research wishes to overcome. At present, current TIL therapy is limited by low TIL numbers, exhaustion and immune suppression within the TME, resulting in limited toxicity with limited impact. This is followed by future directions for overcoming these hurdles. Studies to improve TIL potency address selective and expanded TIL subsets enriched for stemness, tumor recognition, and effector functions, and the genetic engineering of TIL to express CARs or other modified receptors for enhanced tumor recognition and persistence. Different approaches are being explored, including combining TIL therapy with checkpoint inhibition to restore exhausted TILs and other immunomodulatory agents to condition the TME for further infiltration and activity of TILs. Additionally, personalized TIL therapies are being developed with personalized tumor mutations identified and utilized to generate or select TILs more reactive to the patient's neoantigens (Peng et al., 2024). In addition, there is research into the manufacturing processes of TILs to further increase the yield and quality of the final product, to incorporate novel bioreactors and cell culture techniques. Finally, efforts to develop predictive biomarkers for the identification of those patients more likely to respond to TIL therapy and who are most likely responding to TIL therapy to develop more personalized and effective alternative strategies.

15.2.1.3 TCR-engineered T-cell therapy

Despite the promise of TCR-engineered T-cell therapy, it has unique hurdles that are driving the frontier of research. Whereas CAR T cells purify surface antigens, TCR-engineered T cells target MHC-presented intracellular tumor-associated antigens present in the tumors, which greatly expands the tumor landscape that can be targeted. Research in the field now includes working to solve these theoretical limitations of cross-reactivity, which can lead to off-target toxicity in healthy tissues whose peptide is similarly expressed, and their more limited persistence in most engineered T-cell products. Other means to enhance safety include TCR optimization and engineering, suicide genes, and more stringent preclinical screening. Strategies such as improving efficacy with costimulatory domain optimization, trafficking TCR therapy to solid tumor sites, and combining TCR therapy with checkpoint inhibitors or other immunomodulatory agents to overcome the immunosuppressive tumor microenvironment are being pursued to enhance efficacy. In addition, efforts are being made to expand TCR therapy to all HLA types, possibly through the generation of "universal" TCRs or HLA-independent targeting strategies. Other cellular immunotherapy or standard-of-care treatments are being combined with TCR therapy in research to exploit antitumor synergy (Neeser et al., 2023). In the final case, new tumor-associated antigens are identified by next-generation sequencing and other high-tech means, and TCR design is enhanced to be more tumor-specific and efficacious.

15.2.2 Successes and limitations of current treatments

Cellular immunotherapies like CAR T cell therapy have been very successful, and are already used to treat certain hematological malignancies—in particular, B lymphomas and some types of leukemia—by achieving high remission rates in many patients and some cures. Nevertheless, these treatments still face important limitations, including on-target, off-tumor effects that can cause toxicities such as cytokine release syndrome (CRS) and immune effector cell-associated neurotoxicity syndrome (ICANS), antigen loss that may lead to relapse, and reduced effectiveness in many solid tumors due to an immunosuppressive tumor microenvironment and limited T-cell trafficking. However, current CAR T cell therapies are still largely targeted to only a small segment of the cancer population, their extremely costly manufacturing process and limited access are suboptimal. However, even though its efficacy has been demonstrated in melanoma and some other solid tumors, TIL therapy is hindered similarly by challenges of the TME, lack of standardization of TIL quality from patient to patient, and complex and time-consuming manufacturing. In contrast to antibodies that can target only a single protein, TCR-engineered T-cell therapy increases the targetable tumor antigens, but lacks the means to find suitable targets, or faces the risk of toxicity to non-target sites.

15.3 Future advancements in cellular immunotherapy

15.3.1 Enhancing efficacy

Researchers have been experimenting with novel techniques to increase the efficiency of cellular immunotherapy. The recent advancements include Cellular Cancer Immunotherapy, especially CAR T cell therapy, which has rejuvenated the field of tumor immunology and revolutionized treatment options. There is an urgent need to enhance the efficacy of treatment. Despite successful treatment outcomes in multiple cancers, treatment failure or disease relapse still exists in some patients. Therefore approaches to enhance the efficacy of immunotherapies are likely to be the key to improving their effectiveness. Innovative technologies like CRISPR (Clustered Regularly Interspaced Short Palindromic Repeats) bring new hope for the discovery of new drugs and their manufacturing. Also, strategies to overcome the reduced T-cell persistence, developing multitargeted approach and combinational therapies are at the forefront of current research. Researchers are exploiting multiple ways to combine immunotherapy with other treatments. For example, immunotherapy was combined with checkpoint inhibitors and oncolytic viruses to increase efficacy and overcome resistance to treatment. The immune targets are widened beyond the conventional T-cell therapy to natural killer (NK) cells and gamma, delta T cells to increase the range of immune recognition and response to tackle the tumor cells adequately. With advancements in cellular therapy at a personal level, the tumor targets are identified precisely. By studying the biology of the tumor at the cellular level and immune interactions, the targets are well identified and efficiently used for tumor killing. The multifaceted action at the cellular level is essential for improving the treatment outcome and providing wider therapeutic options for cancer treatment, potentially providing a curative option for many cancers.

15.3.1.1 Improving T-cell persistence and function

Starting from the CAR T cell manufacturing process and till the infusion into the patients, there are many factors that can be manipulated to augment persistence. This includes the optimization of CAR T cell design, the conditions for expansion, enhancing T-cell stemness for adoptive T-cell transfer, modulation of CAR-T cell metabolism, altering the signaling pathways involved in T-cell differentiation, and optimizing lymphodepletion regimens prior to administration to patients, which are important contributors to persistence (López-Cantillo et al., 2022; Wittling et al., 2024). Researchers focus on enhancing cellular immunity by improving T-cell function and persistence. Gene editing techniques are used to improve the immune targeting to improve the T-cell function and identify the target location in the tumor cells. Combination therapies could potentially enhance the T-cell function and its persistence. Identifying the immune targets individually could lead to personalized treatment for patients with different cancers. Allogeneic CAR-T cells are being utilized in cancer treatment to bring treatment to a personal level as well as to increase the cost-effectiveness. By modifying the TME it is possible to increase the function of the T cells, leading to increased functionality and survival. All these approaches are useful to increase the cellular immunity and modify the immune system to correctly deliver the immune response and identify the tumor cell, leading to increased cell kill. These methods are essential to optimize the cancer treatment to bring favorable treatment outcomes.

15.3.1.2 Developing multitargeted approaches

Developing multitargeted approaches is essential to increase T cells' function, safety, and specificity. Recent research has developed novel generations of multitargeted CAR-T cells, such as Dual CAR, universal CAR, Tandem CAR, SynNotch CAR, and SUPRA CARs by modifications of the receptor that overcome immune escape due to antigen loss and target a wide range of tumor antigens, simultaneously with high efficacy and minimal toxicity. In addition to the programmable CAR structures, including Physiological CAR, split CAR, and TRUCK cells (T-redirected universal cytokine killer cells), have also emerged to improve the therapeutic function of CAR-T cells, which achieves improved antitumor response to the TME. Hence, a multitargeted approach can be employed for a longitudinal durable response rather than conventional CAR (Tahmasebi et al., 2021). Multitargeted approaches will help in efficient tumor cell kill, preventing the cancer cells from immune escape. Engineering the T cells with multiple antigens helps recognize wider tumor antigens, making the target wider and creating a comprehensive antitumor mechanism. Combining immunotherapy with other treatment modalities could be synergistic because of the different mechanisms of action of the combined agents. T cells could be more antigen-specific, whereas NK cells target the downregulated class I Major Histocompatability Complex (MHC). Combining both modalities would deliver a synergistic effect. Another option for multitargeting is the use of multitargeting tumor-specific antibodies with the use of multiple specific antibody conjugates to engage the tumor cells and the immune cells. By bridging tumor and immune cells, the tumor cells are efficiently synchronized to a pathway that leads to targeted killing.

Improving CAR T-cell technology by combining two different CAR-T cells increases treatment efficiency. Dual CART-T or CAR-T cells with two antigen-binding sites are often used to target the tumor cells. With adapter molecules attached to the CAR-T cells, the survival of the CAR-T cells can be increased by adapting to the TME, increasing the effectiveness of the therapeutic application. By developing CAR-T cells that can recognize tumor antigens and release cytokines, it helps in overcoming tumor resistance. Furthermore, by using genomics and proteomics, it is possible to identify and localize potential targets resistant to conventional treatments. Much more efficient genomic analysis and genetic screening will be required to identify potential targets and mutations that cause treatment resistance to multiple drugs. When a combination approach is selected, researchers need to be careful about the toxicity aspects of the drugs, and caution is required to prevent the toxicity overlap of the drugs. Also, associated with multitargeted techniques are complex and rigorous quality assurance procedures to assure their safety in clinical use. Clinical trial designs with animal models and preclinical studies are required for biomarker identification, target localization, screening, and monitoring during treatment. Clinicians must work in close proximity to regulatory authorities to get drug approval for newer therapies, as well as when they evolve over the course of time, their safety and efficiency are of immense importance to the development of the field of cellular immunotherapy.

15.3.1.3 Combining cellular immunotherapy with other treatment modalities

Combining immunotherapy with the standard line of treatment may result in better outcomes. Combining immunotherapy with radiotherapy has been shown to be effective as radiation delivered to the tumor site affects both tumor cells and surrounding stromal cells, which exposes tumor-specific antigens and promotes the priming and activation of cytotoxic T cells, which also facilitate the recruitment and infiltration of immune cells (Wang et al., 2018). The chemotherapy combined with immunotherapy in patients with advanced non–small-cell lung cancer resulted in an overall response rate with an increase in percentage of helper T cells and cytotoxic T cells when compared with the control group (Jiao et al., 2023).

15.3.2 Expanding target range

15.3.2.1 Identifying new tumor-specific antigens

There are a variety of antigen sources, including peptides, tumor lysate, DNA, or RNA (Huber et al., 2018). Although each antigenic source has its own advantages, employing in silico prediction for choosing peptides or peptide mixtures that cover a broad range of HLA alleles might be beneficial. Short peptides or a mixture of peptides containing multiple epitopes may generate specific effector Th1 cells but not long-lasting CD8+ cytotoxic T lymphocyte response, while whole tumor lysate has the advantage of containing a wide range of both conserved and patient-specific neoepitopes (Obleukhova et al., 2018). It also has a lower chance of tumor escape when compared to single epitope vaccines, but the possibility of poor response is also there due to factors such as molecular heterogeneity of tumor, selective tumor antigen loss, or immunosuppressive TME (Haen et al., 2020). To improve the success rate of cellular immunotherapy transcriptome analysis for cancer-specific transcripts may be helpful.

15.3.2.2 Developing therapies for solid tumors

Despite the remarkable success of CAR T-cell therapy in hematologic malignancies, adaptive cellular therapies for solid tumors face many challenges because of the heterogenous immunosuppressive TME, limited T-cell persistence, and T-cell trafficking to the tumor site that favors immune escape. Hence selecting appropriate targets to overcome the immunosuppressive TME is of utmost importance to improve the safety and efficacy in solid tumors. CAR-T cell infiltration into solid tumors can be enhanced by refining the costimulatory domains, and adapting a combination of conventional and immunotherapeutic cancer treatments may improve the therapeutic outcome of solid cancers (Comoli et al., 2019).

15.3.3 Overcoming resistance mechanisms

15.3.3.1 Addressing tumor microenvironment challenges

TME plays a crucial role in tumor progression, and response to therapeutic modalities and its composition varies between tumor type, localization and stage of disease. Alteration in the extracellular matrix causes impairment of drug delivery. Soluble factors secreted by TME favors immune escape and drug resistance (Kaemmerer et al., 2021). Cancer vaccines/ nano-vaccine, Cold atmospheric plasma, oncolytic viral therapy, bacterial therapy, antiangiogenic drugs, checkpoint inhibitors are the most effective therapies to address the immunosuppressive TME (Babar et al., 2023).

15.3.3.2 Targeting cancer stem cells

CSCs can cause multidrug resistance, radiation resistance, metastasis, and cancer relapse. CSCs can be targeted using Monoclonal antibodies (mAbs) ex.Rituximab, a CD20 mAb, for the treatment of follicular lymphoma. EpCAM is a common CSC biomarker, widely employed in clinical trials (Osta et al., 2004). The signaling pathways, such as Wnt, Notch, Hh signaling pathways, TGF-β, JAK-STAT, PI3K, and NF-κB signaling pathways, which are involved in the regulation of CSCs, can be targeted (Yang et al., 2020).

15.4 Improving accessibility and scalability

15.4.1 Developing off-the-shelf cellular therapies

Allogeneic cell therapies "off-the-shelf" therapies are derived from a donor, which has many advantages, including reduction of production costs, minimizing the manufacturing delays, the availability of therapeutic cellular products to the patient immediately at the time of need, and a standard method of preparation. The major challenge is graft-versus-host disease (GVHD) caused by allogeneic CAR-T cells, and allorejection by host immune cells causes host-versus-graft. Modification of TCR is the most frequently used approach to prevent GVHD (Moradi et al., 2023).

15.4.2 Streamlining manufacturing processes

The most critical aspect of the entire manufacturing process is scaling up of the therapeutic product while adhering to Good Manufacturing Practices and integrating them with Good Laboratory Practices for validation. Assessing the identity, stability, consistency, and potency of the therapeutic product to achieve a clinical utility is of utmost importance. Hence, quality assessment and quality assurance are critical factors of cGMP manufacturing processes. Clinical efficacy, however, cannot be predicted with a single test; hence designing a panel of complementary assays for quality control that measures phenotypic and functional attributes becomes necessary. The safety assurance of the final product is crucial for therapeutic purpose.

The guideline document for potency testing of immunotherapy products by the European Medicines Agency rightly highlights the potential problems in working with autologous cells, which range from difficulty in batch control testing and release to difficulty in minimizing the inherent variability of the cells of interest within the sourced sample that cannot be rectified during manufacturing. Hence accepting this variability is justified. However, this variability affects the validation of the potency of the cell product and warrants the establishment of an acceptance limit (European Medicines Agency, 2016). Automation of the manufacturing process improves the efficiency and the speed of the process, which enables timely treatment for the patients.

15.4.3 Reducing production costs

The production costs can be reduced by employing an automated manufacturing process, reducing raw materials, closing manufacturing facilities to overcome logistics, employing allogenic gene therapy, and in vivo gene modifications (Khang et al., 2023).

15.5 Personalization and precision medicine

15.5.1 Tailoring treatments to individual patients

Personalized cellular immunotherapy that takes into account each patient's unique characteristics is one of the most exciting developments in the field. Chemotherapy and radiation are two examples of conventional cancer treatments that frequently use blanket regimens that do not take individuals' great biological diversity into consideration. But with cellular immunotherapy, we can now tailor treatments to each patient's unique TME based on their unique genetic and cellular characteristics. By maximizing therapeutic efficacy and decreasing off-target effects, personalized treatment plans can enhance clinical outcomes (Zhou et al., 2025a). This approach involves analyzing each patient's tumor characteristics, immune profile, and genetic background, allowing clinicians to select the most effective immunotherapeutic agents. One approach is CAR T-cell therapy, which may be fine-tuned to target particular antigens shown on cancer cells (Hsiao et al., 2025). Personalized cellular therapies boost survival rates and long-term results, according to clinical evidence, especially for hematologic malignancies (Berneman et al., 2025). It is anticipated that, as research progresses, personalization will be expanded to include solid tumors in addition to hematologic cancers, thereby increasing the scope of this groundbreaking therapy modality (Sinha et al., 2025).

15.5.2 Utilizing genomic and proteomic profiling

Genomic and proteomic profiling are crucial in the personalization of cellular immunotherapy because they provide insights into a tumor's distinct molecular landscape. Genomic profiling entails sequencing the tumor genome to detect mutations, genomic rearrangements, and other changes that could be used as therapeutic targets. This method has proven critical for finding neoantigens, new antigens resulting from tumor-specific mutations that can be targeted by modified T lymphocytes (Li et al., 2025a). Proteomic profiling, on the other hand, focuses on examining the proteins expressed inside the TME, which can provide additional targets for therapy and aid in the understanding of immune evasion processes (Chang et al., 2025). Advanced approaches, including mass spectrometry-based proteomics and single-cell RNA sequencing, are increasingly used to better characterize immune cell interactions and tumor heterogeneity (He et al., 2022). Combining genomic and proteomic data allows researchers to develop highly targeted therapies that can adapt to changing tumor characteristics. TCR therapies, for example, can be customized to recognize changed peptides found on cancer cells, thereby improving immune response accuracy. Furthermore, biomarkers that predict a patient's response to immunotherapy can be identified using genomic and proteomic research, allowing for more specific patient categorization and treatment planning (Zhao et al., 2025).

15.5.3 Implementing artificial intelligence and machine learning

The application of AI and machine learning in cellular immunotherapy is revolutionizing the development and optimization of personalized treatment modalities. AI algorithms possess the capability to analyze extensive genomic, proteomic, and clinical datasets in order to discern patterns that traditional analytical methods might overlook (Zhou et al., 2025b). This proficiency is particularly advantageous for identifying complex interactions within the TME and predicting individualized responses to therapeutic interventions. Machine learning algorithms have been employed to forecast the efficacy of CAR T-cell therapy by evaluating patient-specific data, including tumor antigen expression and immune cell infiltration (Caligola et al., 2022). Moreover, AI can improve the optimization of cellular treatment production, therefore guaranteeing both scalability and consistency and preserving a high degree of therapeutic efficacy (Choudhury et al., 2025). Moreover, AI-driven methods accelerating the identification of new tumor antigens and resistance mechanisms can direct the development of next-generation drugs (Oyejide et al., 2025). Two main problems in cellular immunotherapy that might be mostly resolved by the combination of AI and precision medicine are the development of treatment resistance and patient response variability (Johnson et al., 2021). The inclusion of AI technology into clinical procedures, as it develops, should greatly enhance the success and personalization of cellular immunotherapies (Maleki Varnosfaderani & Forouzanfar, 2024).

15.6 Emerging cellular immunotherapy approaches

15.6.1 Natural killer cell therapies

Because they can identify and destroy malignant cells without prior sensitizing, NK cell-based immunotherapies have become a potent and exciting method of cancer treatment. Unlike T cells, NK cells target cancer cells by means of a balance of activating and inhibitory signals, so they are more successful against tumors with low antigen presentation capacity (Judge et al., 2025). Genetic engineering is a recent development in NK cell treatment meant to improve cytotoxicity, persistence, and tumor homing. Commonly employed in T cells, techniques like CAR engineering are currently being deployed to NK cells to produce CAR-NK therapies that can especially target tumor-associated antigens while avoiding immune evasion mechanisms (Li et al., 2025b). Furthermore, allogeneic, off-the-shelf NK cell treatments produced from induced pluripotent stem cells or donor cells solve major constraints of autologous therapies, including manufacturing complexity and expense (Zhu & Kaufman, 2019). Promising outcomes from clinical trials, especially in hematologic malignancies, have been seen; hence, continuous research seeks to expand their use to solid tumors by means of enhanced tumor infiltration and persistence in the immunosuppressive TME (Panahizadeh et al., 2025). Additionally, under study are combination treatments combining NK cells with immune checkpoint inhibitors or cytokine modification to improve antitumor activity (Chu et al., 2025).

15.6.2 Gamma delta T-cell therapies

In cancer immunotherapy, gamma delta ($\gamma\delta$) T cells are a relatively unknown yet extremely promising subpopulation of T cells. $\gamma\delta$ T cells are efficient against a variety of cancers, including those that elude typical immune responses, because they can detect antigens in a non-MHC-restricted way, unlike conventional alpha-beta ($\alpha\beta$) T cells (Kim et al., 2024). These cells bridge the gap between innate and adaptive immunity by exerting cytotoxic effects while also recruiting other immune cells to the tumor site. One of the primary focuses of current research is expanding and activating $\gamma\delta$ T cells ex vivo for therapeutic use (Li et al., 2025). The specificity, endurance, and tolerance of $\gamma\delta$ T cells to the immunosuppressive TME are enhanced through engineering (Sang et al., 2024). Combining the specificity of CAR technology with the natural targeting ability of $\gamma\delta$ receptors, CAR-modified $\gamma\delta$ T cells have shown enhanced antitumor efficacy in preclinical models (Xie et al., 2025). Combinations of $\gamma\delta$ T cells and cytokines, such as IL-2 or IL-15, are being investigated to enhance their activity and overcome difficulties, including limited in vivo expansion (Caligola et al., 2022). These treatments are currently being tested in clinical trials for both solid and hematologic malignancies, with encouraging early results (Paul et al., 2025).

15.6.3 Macrophage-based therapies

Traditionally recognized for their function in tissue remodeling and innate immunity, macrophages have drawn interest as potential therapeutic agents in the treatment of cancer. Within the TME, tumor-associated macrophages frequently display a pro-tumor M2 phenotype that suppresses the immune system and encourages angiogenesis, invasion, and immunological escape (Lendeckel et al., 2022). Recent treatment approaches, however, concentrate on reprogramming tumor-associated macrophages to take on an antitumor M1 phenotype that is proinflammatory. Numerous strategies, such as the application of small compounds, antibodies that target receptors unique to macrophages (e.g., CSF-1R inhibitors), and genetic engineering, can accomplish this reprogramming (Genard et al., 2017). CAR macrophage (CAR-M) therapy is one promising strategy in which macrophages are genetically altered to express chimeric antigen receptors, which allows them to identify and target cancer cells precisely (Lu et al., 2024). The ability to modify the TME by attracting more immune cells and eliciting long-lasting immune responses is an additional benefit of CAR-Ms over CAR T cells. Preclinical studies have demonstrated that CAR-M therapies can overcome immune suppression within solid tumors and enhance the efficacy of other immunotherapies when used in combination (Huang et al., 2024). The ability to modify the TME by attracting more immune cells and eliciting long-lasting immune responses is an additional benefit of CAR-Ms over CAR T cells (Pacheco et al., 2023).

15.7 Ethics and challenges of ethical consideration

Cellular immunotherapy, as other significant developments in cancer therapy involving targeted gene- and cell-based therapy, will have a number of attendant challenges regarding ethics. Among the ethical issues posed by the research are patient safety, the availability of an inclusive access policy, and compliance with regulations.

15.7.1 Patient safety

Cellular immunotherapy is a new antitumor treatment modality with broad indications, high therapeutic strength, and good tolerability. It is, however, important to be aware of the potential risks to patient safety. Herein, some of the most important safety factors are discussed (Lee Ventola, 2017):

- **CRS:** A clinical entity involving a sharp rise in systemic inflammation; high fever, chills, headache, and nausea are some common symptoms; however, in the worst cases, it can cause injury to several internal organs. It is a potentially life-threatening situation that occurs following CAR T-cell therapy. Proinflammatory cytokines like IL-6, TNF-α, and IFN-γ, which are released, cause this clinical syndrome. The symptoms of the patient vary from mild fever, chills, and headache to severe symptoms like hypotension and life-threatening cardiovascular compromise, leading to multiorgan failure.
- **ICANS:** It is a neurological complication that can cause confusion, seizures, and even comatose state. The pathophysiology of this syndrome is unclear, and is believed to be due to the release of inflammatory cytokines which cross the blood-brain barrier. The symptoms can vary from mild to severe neurological symptoms, which require the use of corticosteroids in severe and life-threatening cases, otherwise symptomatic and supportive care for mild symptoms.
- **Tumor Lysis syndrome:** A metabolic disorder that occurs as a consequence of killing many neoplastic cells at one time by spilling their contents into the circulation. Bleeding abnormalities, organ-related issues of the cardiac, renal, and hepatic systems, and, in some cases, seizures are all significant ramifications of this syndrome. It occurs following the infusion of the drug and within a few hours following it. There will be metabolic derangement, including hyperkalemia, hyperphosphatemia, hyperuricemia, and hypocalemia, leading to severe multiorgan dysfunction, including renal and cardiac dysfunction. Anticipation of this critical condition and prevention is the most important strategies in the management of these patients. Once it occurs, the treatment includes aggressive hydration, correction of electrolytes, allopurinol, and rasburicase to correct the serum uric acid.
- **Acute GVHD—Donor T-cell mediated:** When donor cells attack the normal body tissues of a recipient for stem cell transplantation in adoptive immunotherapy, it leads to GVHD. This results in problems such as acute fevers, skin rash, or vomiting so severe as to require hospitalization (Davis et al., 2018; Hayes, 2021). It occurs after a few days of treatment, and the symptoms can vary from mild to severe, life-threatening conditions. Mild symptoms include fever and skin rashes, whereas severe life-threatening conditions include gastrointestinal, respiratory, and infections. The pathophysiology is believed due to the interplay of T cells, antigen-presenting cells, proinflammatory cytokine-mediated immune response.

15.7.1.1 Long-term effects

Although research on cellular immunotherapy's long-term impacts is ongoing, some possible worries include:

- **Secondary malignancies:** Some forms of cellular immunotherapy include genetically modifying immune cells, which may increase the chance of secondary cancers.

- **Autoimmune disorders:** There is also a chance of getting an autoimmune disorder, which occurs when the body's own tissues are attacked by the immune system.
- **Long-term immune suppression:** Long-term immune suppression brought on by certain forms of cellular immunotherapy may raise the risk of infection.
- Long-term effects also lead to immune dysregulation, prolonged cytopenias (low blood cell counts and risk of bleeding) (Cappell & Kochenderfer, 2023; Davis et al., 2018).

15.7.2 Equitable access to treatment

CAR T-cell therapy is one example of a cellular immunotherapy that has transformed cancer treatment by utilizing the immune system. However, there are still several obstacles in the way of fair access to these life-saving treatments, given below;

- **Clinical trial eligibility:** Strict inclusion requirements may prevent many patients from participating in clinical trials, thus restricting access to those potentially life-saving therapies.
- **High cost:** The development and production of cellular immunotherapies are exceedingly costly, resulting in significant out-of-pocket expenses for patients. Access may be restricted, and financial hardship may result, especially for marginalized groups.
- **Social determinants of health:** Factors such as socioeconomic status, race, ethnicity, and geographic location can dramatically impact access to cancer therapy, especially cellular immunotherapies (Beinfeld et al., 2023).

15.7.3 Regulatory approval and process

The development and approval of cellular immunotherapies for cancer are subject to rigorous regulatory oversight to ensure their safety and efficacy. The specific requirements and processes may vary slightly between countries, but the general principles remain consistent. Here is a general overview of the regulatory process:

- **Preclinical studies:** Extensive lab and animal studies assess cellular immunotherapy's safety and effectiveness, optimal cell types, modification methods, and dosing strategies.
- **Investigational new drug application:** Sponsors submit an investigational new drug application to regulators (e.g., FDA, European Medicines Agency, and CDSCO) detailing the therapy's manufacturing, preclinical data, and clinical trial plans.
- **Clinical trials:** Phases of clinical trials evaluate the safety, tolerability, and effectiveness of therapies; 10–100 patients are tested for safety and dosage in phase I; 100–1000 patients are tested for effectiveness in phase II; and standard therapies are compared in phase III with 1000–3000 patients.
- **Biologics license application (BLA):** After Trials, the company submits a BLA to regulators with data showing the therapy's safety and efficacy.
- **Regulatory review and approval:** The agency reviews the BLA, requests more data if needed, and grants approval if the therapy is safe and effective (Center for Biologics Evaluation & Research CBER, 2024).

15.8 Conclusion

To better implement this chapter in the healthcare industry, we have broken down our conclusion on the various cellular immunotherapies for cancer treatment into the three aspects listed below;

- Recap of key future perspectives
- Potential impact on cancer treatment landscape
- Call for continued research and development

15.8.1 Recap of key future perspectives

- **CAR-T cell therapy:** According to the outcomes so far, it has the potential to be very successful therapeutic opinion. Furthermore, CAR-T cells' potential to target particular antigens on myeloma cells is being investigated in ongoing preclinical and clinical trials, which may provide patients with relapsed/refractory MM(RRMM) hope. These developments have demonstrated the potential for combination therapies or medications based on the CAR T cells to reduce adverse effects and increase treatment results (Sheykhhasan et al., 2024).
- **TIL cell therapy:** In cases of metastatic melanoma, adoptive cell treatment employing TIL has demonstrated therapeutic promise. Adult patients with metastatic melanoma can benefit from lifileucel, the first TIL treatment, which was approved by the FDA in Feb, 2024. While TIL cells therapy's studies are most clearly seen in patients with advanced melanoma, other studies are being conducted to examine its safety and effectiveness in other solid tumor indications. In order to overcome these obstacles, a number of prospective TIL cell therapy solutions are utilizing innovative technology while undergoing preclinical and early clinical development (Amaria et al., 2025).
- **TCR-engineered T-cell therapy:** The field of cancer immunotherapy has evolved as a result of T-cell engineering. One intriguing therapeutic option is T cells that express a modified T-cell receptor, or TCR T cells. Membrane proteins are not only targets; intrinsic TCR characteristics like high antigen sensitivity and near physiological signaling may enhance T-cell persistence and tumor cell detection and destruction (Baulu et al., 2023).
- Ultimately, the difficulties of allogeneic therapies can be overcome by using $\gamma\delta$ (Gamma delta) T cells and NK cells to create universal off-the-shelf cell therapy for cancer. Additionally, these cells can be used as ACT to concurrently accomplish either broad or targeted antitumor actions (Morandi et al., 2020).

15.8.2 Potential impact on cancer treatment landscape

Cellular immunotherapy can dramatically transform the treatment of cancer with the use of patients' immune cells, specially engineered to target and destroy cancer cells (Fig. 15.1). Such treatment may provide more effective, longer-lasting, and better-tailored therapies with fewer side effects than conventional therapies, particularly for cancers such as blood cancers, which have shown encouraging early result (Research, 2025).

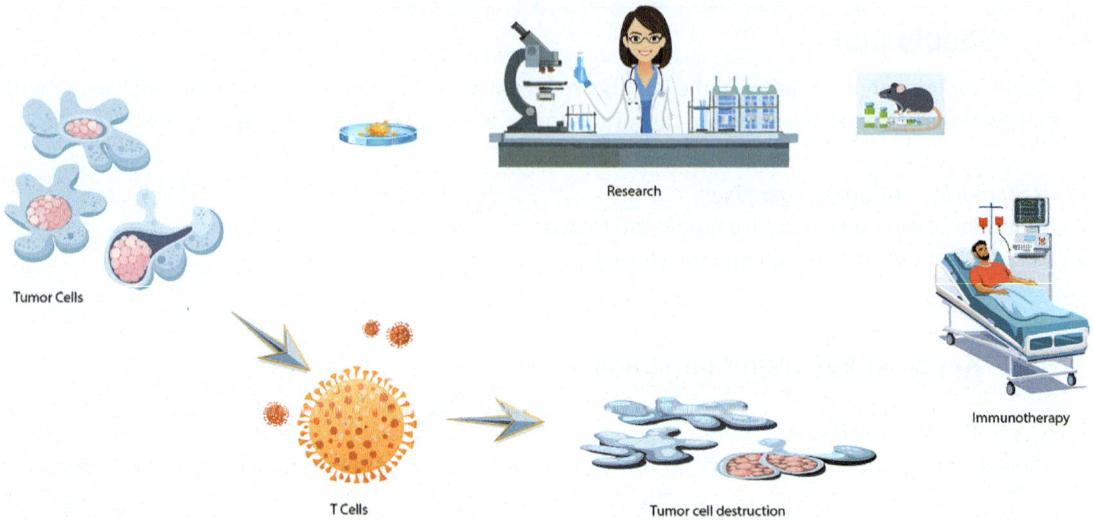

FIGURE 15.1

Future of cellular immunotherapy. Patients' T Cells are utilized and modified in the laboratory to recognize tumor cells, leading to the killing of tumor cells. The genetic modification involves making the T cells recognize the tumor antigen in the patient's body. The CAR T-cells are infused back into the patient once a sufficient quantity is obtained from the lab after multiplication. With future advancements in molecular and immunobiology, it will be possible to deliver personalized cancer treatments tailored to each patient's unique cancer biology.

15.8.3 Call for continued research and development

Increasing access, enhancing efficacy, and overcoming the current limitations in this field by researching novel target antigens, improving techniques in cell engineering, and investigating combination therapy approaches that can give durable antitumor response across a broader range of cancer should be the goals of sustained vigorous research effort, badly needed to take the development and optimization of cellular immunotherapy approaches for the cancer treatment to the next level.

References

Afrose, N., Chakraborty, R., Hazra, A., Bhowmick, P., & Bhowmick, M. (2024). AI-driven drug discovery and development. In *Future of AI in biomedicine and biotechnology* (pp. 259–277). IGI Global. https://www.igi-global.com/book/future-biomedicine-biotechnology/336539. doi:10.4018/979-8-3693-3629-8.ch013.

Amaria, R. N., Komanduri, K. V., Schoenfeld, A. J., Ramsingh, G., Burga, R. A., & Jagasia, M. H. (2025). Entering a new era of tumor-infiltrating lymphocyte cell therapy innovation. *Cytotherapy*. https://www.sciencedirect.com/science/journal/14653249. doi:10.1016/j.jcyt.2024.12.010.

Babar, Q., Saeed, A., Tabish, T. A., Sarwar, M., & Thorat, N. D. (2023). Targeting the tumor microenvironment: Potential strategy for cancer therapeutics. *Biochimica et Biophysica Acta (BBA)—Molecular Basis of Disease, 1869*(6), 166746. doi:10.1016/j.bbadis.2023.166746.

Baulu, E., Gardet, C. é, Chuvin, N., & Depil, S. té (2023). TCR-engineered T cell therapy in solid tumors: State of the art and perspectives. *Science Advances, 9*(7). doi:10.1126/sciadv.adf3700.

Beinfeld, M. T., Rucker, J. A., Jenkins, N. B., de Breed, L. A., & Chambers, J. D. (2023). Variation in Medicaid and commercial coverage of cell and gene therapies. *Health Policy OPEN, 5*. https://www.journals.elsevier.com/health-policy-open/. doi:10.1016/j.hpopen.2023.100103.

Berneman, Z. N., De Laere, M., Germonpré, P., Huizing, M. T., Willemen, Y., Lion, E., De Reu, H., Van den Bossche, J., Van den Brande, J., Specenier, P., Altintas, S., van Dam, P. A., Cools, N., Nijs, G., Stein, B., Caluwaerts, K., Snoeckx, A., Op de Beeck, B., Saevels, K., … Smits, E. L. J. M. (2025). WT1-mRNA dendritic cell vaccination of patients with glioblastoma multiforme, malignant pleural mesothelioma, metastatic breast cancer, and other solid tumors: Type 1 T-lymphocyte responses are associated with clinical outcome. *Journal of Hematology and Oncology, 18*(1). https://jhoonline.biomedcentral.com. doi:10.1186/s13045-025-01661-x.

Caligola, S., De Sanctis, F., Canè, S., & Ugel, S. (2022). Breaking the immune complexity of the tumor microenvironment using single-cell technologies. *Frontiers in Genetics, 13*. https://www.frontiersin.org/journals/genetics#. doi:10.3389/fgene.2022.867880.

Cappell, K. M., & Kochenderfer, J. N. (2023). Long-term outcomes following CAR T cell therapy: What we know so far. *Nature Reviews Clinical Oncology, 20*(6), 359–371. https://www.nature.com/nrclinonc/archive/. doi:10.1038/s41571-023-00754-1.

Center for Biologics Evaluation, & Research (CBER) (2024). *United States of Food and Drug Administration*. https://www.fda.gov/about-fda/fda-organization/center-biologics-evaluation-and-research-cber.

Chang, Z., Wang, K., Fang, Z., Tang, Y., Gao, X., & Tang, B. (2025). NanoTrackThera platform for real-time situ monitoring of tumor immunotherapy and photothermal synergistic efficacy. *Small, 21*(10), e2411705.

Choudhury, M., Deans, A. J., Candland, D. R., & Deans, T. L. (2025). Advancing cell therapies with artificial intelligence and synthetic biology. *Current Opinion in Biomedical Engineering, 34*. https://www.sciencedirect.com/science/journal/24684511. doi:10.1016/j.cobme.2025.100580.

Chu, Y., Tian, M., Saini, U., Ayala-Cuesta, J., Klose, K., Mendelowitz, A. S., Foley, K., Ozkaynak, M. F., Luo, W., Cripe, T. P., Lee, D. A., Cassady, K. A., & Cairo, M. S. (2025). Combinatorial immunotherapy with anti-ROR1 CAR NK cells and an IL-21 secreting oncolytic virus against neuroblastoma. *Molecular Therapy Oncology, 33*(1), 200927. doi:10.1016/j.omton.2024.200927.

Comoli, P., Chabannon, C., Koehl, U., Lanza, F., Urbano-Ispizua, A., Hudecek, M., Ruggeri, A., Secondino, S., Bonini, C., & Pedrazzoli, P. (2019). Development of adaptive immune effector therapies in solid tumors. *Annals of Oncology, 30*(11), 1740–1750. doi:10.1093/annonc/mdz285.

Dailah, H. G., Hommdi, A. A., Koriri, M. D., Algathlan, E. M., & Mohan, S. (2024). Potential role of immunotherapy and targeted therapy in the treatment of cancer: A contemporary nursing practice. *Heliyon, 10*(2). https://www.sciencedirect.com/science/journal/24058440. doi:10.1016/j.heliyon.2024.e24559.

Davis, T. D., Kunakom, S., Burkart, M. D., & Eustaquio, A. S. (2018). Preparation, assay, and application of chlorinase SalL for the chemoenzymatic synthesis of S-Adenosyl-L-Methionine and analogs. *Methods in Enzymology, 604*, 367–388. doi:10.1016/bs.mie.2018.02.012.

European Medicines Agency. (2016). *Guideline on potency testing of cell based immunotherapy medicinal products for the treatment of cancer (EMA/CHMP/BWP/271475/2006 rev.1). European Medicines Agency.* https://www.ema.europa.eu/en/documents/scientific-guideline/guideline-potency-testing-cell-based-immunotherapy-medicinal-products-treatment-cancer-revision-1_en.pdf.

Genard, G., Lucas, S., & Michiels, C. (2017). Reprogramming of tumor-associated macrophages with anticancer therapies: Radiotherapy versus chemo- and immunotherapies. *Frontiers in Immunology, 8*. http://journal.frontiersin.org/article/10.3389/fimmu.2017.00828/full. doi:10.3389/fimmu.2017.00828.

Golikova, E. A., Alshevskaya, A. A., Alrhmoun, S., Sivitskaya, N. A., & Sennikov, S. V. (2024). TCR-T cell therapy: Current development approaches, preclinical evaluation, and perspectives on regulatory challenges. *Journal of Translational Medicine, 22*(1). https://translational-medicine.biomedcentral.com. doi:10.1186/s12967-024-05703-9.

Haen, S. P., Löffler, M. W., Rammensee, H. G., & Brossart, P. (2020). Towards new horizons: Characterization, classification and implications of the tumour antigenic repertoire. *Nature Reviews Clinical Oncology, 17*(10), 595–610. http://www.nature.com/nrclinonc/archive/index.html. doi:10.1038/s41571-020-0387-x.

Hayes, C. (2021). Cellular immunotherapies for cancer. *Irish Journal of Medical Science (1971), 190*(1), 41–57. doi:10.1007/s11845-020-02264-w.

He, M., Gu, W., Gao, Y., Liu, Y., Liu, J., & Li, Z. (2022). Molecular subtypes and a prognostic model for hepatocellular carcinoma based on immune- and immunogenic cell death-related lncRNAs. *Frontiers in Immunology, 13*. https://www.frontiersin.org/journals/immunology#. doi:10.3389/fimmu.2022.1043827.

Hsiao, Y. J., Hsieh, M. S., Chang, G. C., Hsu, Y. C., Wang, C. Y., Chen, Y. M., Chen, Y. L., Yang, P. C., & Yu, S. L. (2025). Tp53 determines the spatial dynamics of M1/M2 tumor-associated macrophages and M1-driven tumoricidal effects. *Cell Death and Disease, 16*(1). https://www.nature.com/cddis/. doi:10.1038/s41419-025-07346-0.

Huang, T., Bei, C., Hu, Z., & Li, Y. (2024). CAR-macrophage: Breaking new ground in cellular immunotherapy. *Frontiers in Cell and Developmental Biology, 12*. doi:10.3389/fcell.2024.1464218.

Huber, A., Dammeijer, F., Aerts, J. G. J. V., & Vroman, H. (2018). Current state of dendritic cell-based immunotherapy: Opportunities for in vitro antigen loading of different DC subsets? *Frontiers in Immunology, 9*. doi:10.3389/FIMMU.2018.02804.

Jiao, J., Li, W. W., Shang, Y. H., Li, X. F., & Jiao, M. (2023). Clinical effects of chemotherapy combined with Immunotherapy in patients with advanced NSCLC and the effect on their nutritional status and immune function. *Pakistan Journal of Medical Sciences, 39*(2), 404–408. https://pjms.org.pk/index.php/pjms/article/view/6365/1670. doi:10.12669/pjms.39.2.6365.

Johnson, K. B., Wei, W. Q., Weeraratne, D., Frisse, M. E., Misulis, K., Rhee, K., Zhao, J., & Snowdon, J. L. (2021). Precision medicine, AI, and the future of personalized health care. *Clinical and Translational Science, 14*(1), 86–93. http://onlinelibrary.wiley.com/journal/10.1111/(ISSN)1752-8062. doi:10.1111/cts.12884.

Judge, S. J., Purl, M. C., Murphy, W. J., & Canter, R. J. (2025). What's in a name? Memory NK cells for cancer immunotherapy. *Journal for ImmunoTherapy of Cancer, 13*(1). http://www.immunotherapyofcancer.org/. doi:10.1136/jitc-2024-010850.

Kaemmerer, E., Loessner, D., & Avery, V. M. (2021). Addressing the tumour microenvironment in early drug discovery: A strategy to overcome drug resistance and identify novel targets for cancer therapy. *Drug Discovery Today, 26*(3), 663–676. www.elsevier.com/locate/drugdiscov. doi:10.1016/j.drudis.2020.11.030.

Kapustina, O., Burmakina, P., Gubina, N., Serov, N., & Vinogradov, V. (2024). User-friendly and industry-integrated AI for medicinal chemists and pharmaceuticals. *Artificial Intelligence Chemistry, 2*(2), 100072. doi:10.1016/j.aichem.2024.100072.

Khang, M., Suryaprakash, S., Kotrappa, M., Mulyasasmita, W., Topp, S., & Wu, J. (2023). Manufacturing innovation to drive down cell therapy costs. *Trends in Biotechnology, 41*(10), 1216–1219. doi:10.1016/j.tibtech.2023.04.006.

Kim, H. S., Kwon, S. H., Choi, O. K., & Lim, T. (2024). High-dose ascorbic acid synergizes with anti-PD1 therapy in non-small cell lung cancer in vitro and in vivo models. *Frontiers in Immunology, 15*. https://www.frontiersin.org/journals/immunology#. doi:10.3389/fimmu.2024.1512605.

Lee Ventola, C. (2017). Cancer immunotherapy, Part 2: Efficacy, safety, and other clinical considerations. *P and T, 42*(7), 452–463. https://www.ptcommunity.com/system/files/pdf/ptj4207452.pdf.

Lendeckel, U., Venz, S., & Wolke, C. (2022). Macrophages: Shapes and functions. *ChemTexts, 8*(2). doi:10.1007/s40828-022-00163-4.

Li, J., Guo, L., Feng, Y., Li, G., Sun, H., Huang, W., Tian, J., Du, Y., & An, Y. (2025a). Optical-magnetic imaging for optimizing lymphodepletion-TIL combination therapy in breast cancer. *Molecular Imaging and Biology, 27*(2), 260–273. doi:10.1007/s11307-025-01985-7.

Li, Y. R., Fang, Y., Niu, S., Zhu, Y., Chen, Y., Lyu, Z., Zhu, E., Tian, Y., Huang, J., Rezek, V., Kitchen, S., Hsiai, T., Zhou, J. J., Wang, P., Chai-Ho, W., Park, S., Seet, C. S., Oliai, C., & Yang, L. (2025b). Allogeneic CD33-directed CAR-NKT cells for the treatment of bone marrow-resident myeloid malignancies. *Nature Communications, 16*(1). https://www.nature.com/ncomms/. doi:10.1038/s41467-025-56270-6.

Li X, Dong Y, Wang T, Huang K, Guo W, Xu L, et al. Chemotherapy boosts anti-angiogenic and anti-PD-1 combination therapy through activation of cCAS-STING pathway in colon cancer. Int Immunopharmacol. 2025 Feb 3;149:114212.

Liu, B., Zhou, H., Tan, L., Siu, K. T. H., & Guan, X. Y. (2024). Exploring treatment options in cancer: Tumor treatment strategies. *Signal Transduction and Targeted Therapy, 9*(1). https://www.nature.com/sigtrans. doi:10.1038/s41392-024-01856-7.

López-Cantillo, G., Urueña, C., Camacho, B. A., & Ramírez-Segura, C. (2022). CAR-T cell performance: How to improve their persistence? *Front Immunol, 13*. doi:10.3389/FIMMU.2022.878209/BIBTEX.

Lu, J., Ma, Y., Li, Q., Xu, Y., Xue, Y., & Xu, S. (2024). CAR Macrophages: A promising novel immunotherapy for solid tumors and beyond. *Biomarker Research, 12*(1). doi:10.1186/s40364-024-00637-2.

Maleki Varnosfaderani, S., & Forouzanfar, M. (2024). The role of AI in hospitals and clinics: Transforming healthcare in the 21st century. *Bioengineering, 11*(4), 337. doi:10.3390/bioengineering11040337.

Moradi, V., Omidkhoda, A., & Ahmadbeigi, N. (2023). The paths and challenges of "off-the-shelf" CAR-T cell therapy: An overview of clinical trials. *Biomedicine & Pharmacotherapy, 169*, 115888. doi:10.1016/j.biopha.2023.115888.

Morandi, F., Yazdanifar, M., Cocco, C., Bertaina, A., & Airoldi, I. (2020). Engineering the bridge between innate and adaptive immunity for cancer immunotherapy: Focus on $\gamma\delta$ T and NK cells. *Cells, 9*(8), 1757. doi:10.3390/cells9081757.

Neeser, A., Ramasubramanian, R., Wang, C., & Ma, L. (2023). Engineering enhanced chimeric antigen receptor-T cell therapy for solid tumors. *Immuno-Oncology and Technology, 19*, 100385. doi:10.1016/j.iotech.2023.100385.

Obleukhova, I., Kiryishina, N., Falaleeva, S., Lopatnikova, J., Kurilin, V., Kozlov, V., Vitsin, A., Cherkasov, A., Kulikova, E., & Sennikov, S. (2018). Use of antigen-primed dendritic cells for inducing antitumor immune responses in vitro in patients with non-small cell lung cancer. *Oncology Letters, 15*(1), 1297–1306. http://www.spandidos-publications.com/ol/15/1/1297/download. doi:10.3892/ol.2017.7403.

Osta, W. A., Chen, Y., Mikhitarian, K., Mitas, M., Salem, M., Hannun, Y. A., Cole, D. J., & Gillanders, W. E. (2004). EpCAM is overexpressed in breast cancer and is a potential target for breast cancer gene therapy. *Cancer Research, 64*(16), 5818–5824. doi:10.1158/0008-5472.CAN-04-0754.

Oyejide, A. J., Adekunle, A. A., Abodunrin, O. D., & Atoyebi, E. O. (2025). Artificial intelligence, computational tools and robotics for drug discovery, development, and delivery. *Intelligent Pharmacy*. https://www.keaipublishing.com/en/journals/intelligent-pharmacy/. doi:10.1016/j.ipha.2025.01.001.

Pacheco, R. L., Martimbianco, A. LC., & Riera, R. (2023). What is an "ongoing" clinical trial? An analysis of different sources revealed heterogeneous definitions of when a clinical trial starts and ends: A meta-research study. *Sao Paulo Medicine Journal, 142*(1), e20220634.

Panahizadeh, R., Panahi, P., Asghariazar, V., Makaremi, S., Noorkhajavi, G., & Safarzadeh, E. (2025). A literature review of recent advances in gastric cancer treatment: Exploring the cross-talk between targeted therapies. *Cancer Cell International, 25*(1). doi:10.1186/s12935-025-03655-8.

Paul, C. D., Yankaskas, C., Thakuri, P. S., Balhouse, B., Salen, S., Bullock, A., Beam, S., Chatman, A., Djikeng, S., Yang, X. J., Wong, G., Dey, I., Holmes, S., Dockey, A., Bailey-Steinitz, L., Zheng, L., Li, W., Chandra, V., Nguyen, J., Sharp, J., Willems, E., Kennedy, M., Dallas, M. R., & Kuninger, D. (2025). Long-term maintenance of patient-specific characteristics in tumoroids from six cancer indications. *Scientific Reports, 15*(1). https://www.nature.com/srep/. doi:10.1038/s41598-025-86979-9.

Peng, L., Sferruzza, G., Yang, L., Zhou, L., & Chen, S. (2024). CAR-T and CAR-NK as cellular cancer immunotherapy for solid tumors. *Cellular and Molecular Immunology, 21*(10), 1089–1108. https://www.nature.com/cmi/. doi:10.1038/s41423-024-01207-0.

Research, C. F. B. E. A. (2025, August 15). *Center for Biologics Evaluation and Research (CBER).* U.S. Food And Drug Administration. (Retrieved: May 11, 2026). https://www.fda.gov/about-fda/fda-organization/center-biologics-evaluation-and-research-cber.

Sang, M., Ge, J., Ge, J., Tang, G., Wang, Q., Wu, J., Mao, L., Ding, X., & Zhou, X. (2024). Immune regulatory genes impact the hot/cold tumor microenvironment, affecting cancer treatment and patient outcomes. *Frontiers in Immunology, 15*. https://www.frontiersin.org/journals/immunology#. doi:10.3389/fimmu.2024.1382842.

Sheykhhasan, M., Ahmadieh-Yazdi, A., Vicidomini, R., Poondla, N., Tanzadehpanah, H., Dirbaziyan, A., Mahaki, H., Manoochehri, H., Kalhor, N., & Dama, P. (2024). CAR T therapies in multiple myeloma: Unleashing the future. *Cancer Gene Therapy, 31*(5), 667–686. doi:10.1038/s41417-024-00750-2.

Sinha, S. J., Kumar, B., Prasad, C. P., Chauhan, S. S., & Kumar, M. (2025). Emerging research and future directions on doxorubicin: A snapshot. *Asian Pacific Journal of Cancer Prevention, 26*(1), 5–15. http://journal.waocp.org/?_action=article&vol=3118&issue=3124&max_rows=100. doi:10.31557/APJCP.2025.26.1.5.

Tahmasebi, S., Elahi, R., Khosh, E., & Esmaeilzadeh, A. (2021). Programmable and multi-targeted CARs: A new breakthrough in cancer CAR-T cell therapy. *Clinical and Translational Oncology, 23*(6), 1003–1019. doi:10.1007/s12094-020-02490-9.

Wang, L. (2022). Clinical determinants of relapse following CAR-T therapy for hematologic malignancies: Coupling active strategies to overcome therapeutic limitations. *Current Research in Translational Medicine, 70*(1), 103320. doi:10.1016/j.retram.2021.103320.

Wang, Y., Deng, W., Li, N., Neri, S., Sharma, A., Jiang, W., & Lin, S. H. (2018). Combining immunotherapy and radiotherapy for cancer treatment: Current challenges and future directions. *Frontiers in Pharmacology, 9*. doi:10.3389/fphar.2018.00185.

Wittling, M. C., Cole, A. C., Brammer, B., Diatikar, K. G., Schmitt, N. C., & Paulos, C. M. (2024). Strategies for improving CAR T cell persistence in solid tumors. *Cancers, 16*(16). http://www.mdpi.com/journal/cancers/. doi:10.3390/cancers16162858.

Xie, Z., Shao, J., Shen, Z., Ye, Z., Okada, Y., Okuzaki, D., Okada, N., & Tachibana, M. (2025). HDAC1-3 inhibition triggers NEDD4-mediated CCR2 downregulation and attenuates immunosuppression in myeloid-derived suppressor cells. *Cancer Immunology, Immunotherapy, 74*(3). doi:10.1007/s00262-024-03931-y.

Yang, L., Shi, P., Zhao, G., Xu, J., Peng, W., Zhang, J., Zhang, G., Wang, X., Dong, Z., Chen, F., & Cui, H. (2020). Targeting cancer stem cell pathways for cancer therapy. *Signal Transduction and Targeted Therapy, 5*(1). doi:10.1038/s41392-020-0110-5.

Zhao, F. u, Jiang, X., Li, Y., Huang, T., Xiahou, Z., Nie, W., & Li, Q. (2025). Characterizing tumor biology and immune microenvironment in high-grade serous ovarian cancer via single-cell RNA sequencing: Insights for targeted and personalized immunotherapy strategies. *Frontiers in Immunology, 15*. doi:10.3389/fimmu.2024.1500153.

Zhao, L., & Cao, Y. J. (2019). Engineered T cell therapy for cancer in the clinic. *Frontiers in Immunology, 10*. https://www.frontiersin.org/journals/immunology#. doi:10.3389/fimmu.2019.02250.

Zhou, M., Zhou, Z., Hu, L., Chen, S., Meng, F., Chen, J., & Shen, J. (2025a). Multiplex immunohistochemistry to explore the tumor immune microenvironment in HCC patients with different GPC3 expression. *Journal of Translational Medicine, 23*(1). doi:10.1186/s12967-025-06106-0.

Zhou, Y., Wei, Y., Li, L., Yan, T., & Ye, H. (2025b). Optogenetics in medicine: Innovations and therapeutic applications. *Current Opinion in Biotechnology, 92*, 103262. doi:10.1016/j.copbio.2025.103262.

Zhu, H., & Kaufman, D. S. (2019). Engineered human pluripotent stem cell-derived natural killer cells: The next frontier for cancer immunotherapy. *Blood Science, 1*(1), 4–11. https://journals.lww.com/bls/pages/aboutthejournal.aspx. doi:10.1097/BS9.0000000000000023.

Index

Page numbers followed by "*f*" and "*t*" indicate, figures and tables respectively.

A

Acute lymphoblastic leukemia (ALL), 89
Adaptive immune responses, 217
Adaptive immunity, 5
Adipose tissue-derived mesenchymal stem cells, 121
Adoptive cell therapy, 2, 64, 193
Adoptive cell transfer, 15
Adoptive cellular immunotherapy, 187
Allogeneic chimeric antigen receptor-T cells, 79, 103
All-transretinoic acid, 300, 301
Angiogenesis, 117
Antiangiogenic therapies, 118
Anti-CD40 antibodies, 336
Antigen profiling, 20
Antitumoractivity, 336
Antitumor immunity, 188, 189
Apoptotic signaling dysregulation, 369
Armoured chimeric antigen receptor-T cells, 78, 102
Arterial thrombosis, 216
Atigen-presenting cell (APC), 177
Autophagy, 370
Axicabtagene ciloleucel, 99

B

B-cell cancer, 90
  CAR-T cell therapy, 91, 101
  tumor burden, 90
B-cell lymphomas, 72
B-cell maturation antigen (BCMA)
  CAR T-cell therapy, 72
  targeted therapy, 89
Biomarker-driven patient selection, 21
  CD19-directed CAR-T cell therapy, 21
  DNA damage repair pathways, 21
Bone marrow
  infiltration of anti-PD-1-conjugated platelets, 225
  macrophage, 320
  mesenchymal stem cells, 125

C

Cancer
  chemotherapy, 353
  hormonal therapy, 354
  immune system interactions, 188
  immunoediting, 10
  metastasis, 352
  microenvironment, 352
  molecular basis, 351
  pathophysiology, 351
  radiation therapy, 353
  surgery, 353
  tumor suppressor genes, 352
Cancer-associated fibroblasts (CAFs), 13
Cancer cell
  secreted exosomes, 254
  signaling pathways of small molecule inhibitors, 257
Cancer cell-induced platelet aggregation (CIPA), 249
Cancer immunosurveillance, 46
  adoptive cell transfer, 46
  immune checkpoint inhibitors, 35
  oncolytic viruses, 46
Cancer immunotherapy, 35, 63, 133, 174, 297, 354
  adeno-associated virus vectors, 198
  adenoviral vectors, 197
  adenoviruses, 197
  CRISPR-Cas technology, 195
  gene therapy, 195
  genome editing technologies, 195
  immuno-oncology, 197
  lentiviral vectors, 197
  lipid nanoparticles, 198
  persistence and cytotoxic capacity, 201
  polymeric gene delivery vectors, 198
  retroviral vectors, 197
  TCR-peptide-MHC complex, 65
  transcription activator-like effector nucleases, 195
  tumor-specific antigens, 65
  zinc finger nucleases, 197
Cancerous cells, 38
Cancer progression, 117
  angiogenesis, 248
  cell survival and growth, 248
  metastatic cascade, 249
  platelet-cancer cell interactions, 248
  platelet interactions, 241
  platelet support, 249
  tumor vascularization, 248
Cancer therapy, 115
  drug resistance, 354
  recurrence and metastasis, 355
  toxicity and side effects, 355
Cancer treatment, 271
  anti–PD-1 therapy, 271

cytokine therapies, 181
Cancer vaccines, 192
Carcinogenesis, 334
Cathepsin K signalling, 255
Cellular immunotherapy, 15, 199, 324, 389
activated $CD4^+$ T helper cells, 325
antigen presentation, 324
artificial intelligence, 22, 183, 397
chimeric antigen receptor-based therapies, 199
cytokine production, 325
developing multitargeted approaches, 393
enhancing efficacy, 392
IFN-γ, 325
machine learning algorithms, 397
memory T cells, 326
patient safety, 399
phagocytosis, 324
T-cell persistence and function, 392
Cellular therapy, 19
Chemoradiation therapy, 360
Chemotherapeutic drugs, 352
Chemotherapy combinations
chemotherapy/polychemotherapy, 359
hormonal therapy, 362
immunotherapy, 361
targeted therapy, 363
Chimeric antigen receptor (CAR)-natural killer cell
immunotherapies, 200
therapy, 194, 284, 397
Chimeric antigen receptor (CAR)-T cells, 89, 133
antigen-binding domains, 67
associated toxicities, 76
ex vivo expansion and quality control, 70
generations, 200
hinge, 67
immune evasion, 66
infiltration, 91
intracellular signaling, 67
nonviral gene editing, 70
on-target off-tumor toxicity, 76
products, 70
products, FDA-approved, 69
quality, 89
related encephalopathy syndrome, 101
senescence, 91
structural components, 67
T-cell collection and isolation, 69
technology, 2
viral vector–based gene transfer, 69
Chimeric antigen receptor (CAR) T-cell therapy, 16, 36, 63, 64, 71, 194, 319, 389, 390
advances and innovations, 77
antigen escape, 76
antitumor immunity, 65
B-cell malignancies, 91
efficacy, 91
immunological basis, 64
manufacturing failures, 77
multitarget and dual-target, 78
neurological side effects, 101
nonlymphoid hemopoiesis, 78
patient-related factors, 77
personalized medicine, 80
precision oncology, 80
T-cell biology, 65
toxicity reduction, 103
tumor heterogeneity, 77
Chimeric antigen receptor macrophages (CAR-M), 328
design, 328
development, 328
therapies, 200, 319
Chimeric antigen receptors (CARs), 174
hinge region/pacer region, 94
intracellular signaling domain, 94
signaling domain selection, 89
structure, 92
transmembrane domain, 94
Chromatin remodeling, 373
Chronic lymphocytic leukemia (CLL), 90
Circulating tumor DNA (ctDNA), 90
Combination therapy, 355
in cancer treatment, 359
cellular-level interactions, 357
molecular synergy, 357
pharmacodynamic interactions, 358
pharmacokinetic interactions, 358
synergistic effects, 369
therapeutic potential, 357
CRISPR-Cas9 gene editing, 2
Curcumin-based combinations, 366
Cyclin-dependent kinases (CDKs), 370
Cytokine-induced killer (CIK) cells, 16
Cytokine release syndrome (CRS), 75, 89, 100
Cytolytic T lymphocytes (CTL), 175
Cytotoxic T-lymphocyte-associated protein 4 (CTLA-4), 35, 178, 192

D

Dendritic cell-based vaccines, 17, 193
Dental pulp-derived mesenchymal stem cells (DP-MSCs), 121
Diffuse large B-cell lymphoma (DLBCL), 89
Direct receptor-receptor binding, 252
DNA vaccines, 193

Doxorubicin-loaded bone marrow-derived mesenchymal stem cells, 121

E

Early-stage B-cell cancers, 103
Engineering autologous T cells, 174
Engineering CAR-T cells, 277
Extracellular matrix (ECM) barriers, 53
Extrinsic (death receptor–mediated) apoptotic pathways, 369

F

First-generation CAR-T cells, 89
Fourth-generation chimeric antigen receptor-T cells, 102

G

Gamma delta T-cell therapies, 398
Gene-directed enzyme prodrug therapy, 124
Gene-editing–enhanced chimeric antigen receptor T cells, 80
Gene-modified immune cells, 200
Genetically modified immune cells, 202
Genomic profiling, 19, 396
Glucose-6-phosphate dehydrogenase (G6PD), 335
G protein-coupled receptors (GPCRs), 248

H

Hematological malignancies, 71
Hemostasis, 213
Humoral immune neutralization, 140
Hydrogel-based nanoscaffolds, 303

I

Immature myeloid cells (IMCs), 272
Immune
  surveillance, 8, 9
  system, 4, 181
Immune-checkpoints, 190
  blockade, 35
  inhibitors, 63, 133, 177, 179, 190
Immune defense
  $CD8^+$cytotoxic T lymphocytes, 7
  $CD4^+$helper T cells, 7
  cellular players, 6
  dendritic cells, 6
  mAb therapies, 8
  macrophages, 6
  memory T cells, 7
  natural killer cells, 6
  neutrophils, 7
  regulatory T cells, 7
  T lymphocytes, 7
  tumor-infiltrating B cells, 8
Immune effector cell-associated neurotoxicity syndrome (ICANS), 25, 75
Immune-related adverse events (IRAE), 1
Immune-related adverse reactions (IRAEs), 229
Immune tyrosine-based activation motif (ITAM), 246
Immunogenetic profiling, 1
Immunoprofiling, 3
Immunosuppressive cellular populations, 153
Immunotherapy, 1, 2, 173, 211, 389
  clinical innovation, 35
  efficacy, antioxidant systems, 51
Induced pluripotent stem cells, 18
Inhibitory receptors, 201
  lymphocyte activation gene 3, 201
  programmed cell death protein 1, 201
Innate immunity, 4, 188
  macrophages, 322
  platelets role, 216
Inside-out signaling, 251
Integrin signaling, 254
Intrinsic (mitochondrial) apoptotic pathways, 369

J

JAK/STAT pathway, 254

L

Lisocabtagene maraleucel, 99
Lysophosphatidic acid (LPA) signaling, 255

M

Macrophage, 319
  based therapies, 398
  biology, 320
  phenotypes, 322
Major histocompatibility complex (MHC), 2
Matrix metalloproteinases (MMPs), 53
Megakaryocytes, 223
Mesenchymal stem cells, 118
  based approaches, 115
  for cancer therapy, 121
  cell-based active targeting, 119
  chemotherapeutic agents, delivery of, 121
  derived exosomes, 123
  Hedgehog pathway, 118
  notch signaling pathway, 118
  utilizing oncolytic viruses, 123
  Wnt/β-catenin signaling pathways, 118
Mesenchymal stromal cells, 134
  biological characteristics and sources, 141
  encapsulation technologies, 149
  ex vivo loading, 148
  homing cascade, 143
  innate antiviral barriers, 146
  intracellular replication dynamics, 146

multiplicity of infection, 148
proinflammatory potential, 144
stress-response pathways, 147
sublytic preloading, 149
suppressive effects, 143
surface engineering, 149
as therapeutic carriers, 141
tumor-directed migration, 147
tumor tropism, 143
in vivo loading, 149
Mesenchymal stromal cells–oncolytic viruses
biodistribution, 136
cancer-associated fibroblasts, 151
immune activation cascade, 151
interactions, 156
therapy, 150
tumor targeting, 150
viral replication, 150
Microregional hypoxia, 117
MicroRNAs (miRNAs), 373
Monoclonal antibody (mAb) technology, 2
Monocytic myeloid-derived suppressor cells (M-MDSCs), 297, 299
Multiantigen CAR-T cells, 102
Multidrug resistance (MDR) transporters, 379
Multifunctional micelles, 303
Multimodal therapy, 363
Murine myeloid-derived suppressor cells, 274
Myeloid-derived suppressor cells (MDSCs), 154, 203, 271, 297, 298
expressed receptors, 298
heterogeneity and challenges, 274
immunosuppressive actions, 272
immunosuppressive function, 274, 301
mediated immunosuppression, 308
modulating vaccines, 284
strategies for targeting, 274
targeted depletion, 303

N

Nanomedicine, 298, 300
CAR-T cell therapy, 306
immune checkpoint inhibitors, 306
Nanotechnology-assisted gene and cell delivery, 189
Natural killer cell adoptive transfer, 284
Natural killer cell-based immunotherapies, 397
Natural killer cell-engaging
bispecific antibodies, 284
trispecific antibodies, 284
Neoantigen
diversity, 3
prediction, 20
Neurotoxicity, 89
Neutrophil extracellular traps (NET), 252
Nod-like receptors (NLRs), 248
Non-Hodgkin lymphoma (NHL), 89

O

"Off-the-shelf" therapies, 395
Oncolytic virotherapy, 135
engineered selectivity, 136
intrinsic or natural selectivity, 135
protein kinase R, 135
replicative cycle, 136
transcriptional targeting, 136
transductional targeting, 136
viral arming, 136
Oncolytic viruses, 133
early infection events, 145
epitope spreading/antigen spreading, 139
immunogenic cell death, 139
innate immune activation, 139
mediated antigen release, 155
receptor-mediated entry, 145
tumor-specific T cells priming, 139
Outside-in signaling, 251

P

Pancreatic cancer, 158
Patient-specific cellular immunotherapies
automated closed systems, 23
flow cytometry-based sorting, 23
genetic engineering, 23
magnetic-activated cell sorting, 23
PEGylated mesoporous silica nanoparticles (PEG-MSNs), 305
Peptide
HLA interactions, 20
vaccines, 192
Personalized cellular immunotherapy, 2, 396
advanced therapy medicinal products, 22
cell-based therapies, 22
clinical implications, 14
Personalized immunotherapy, 1
Phytochemical combinations, 367
Phytochemical epigenetic modulators, 373
Platelet, 211, 212, 241
activation, 213, 228
cancer cell interactions, 249
immune cell interactions, 217
mitochondria, 212
secretome, 212
Platelet-cell interactions
adhesion receptors, 244
receptors involvement, 243

Platelet-derived
- complement proteins, 216
- CXCL5/7 chemokines, 242
- extracellular vesicles, 213, 217
- growth factors, 251
- immune evasion via platelet interactions, 252

Platelet shielding, 212, 218
- cancer procoagulant, 220
- circulating tumor cells, 218
- platelet aggregating activity, 220
- procoagulant activity, 220
- tumor exosomes, 227
- vascular leakage, 226

P38MAPK-MMP-9 pathway, 255
Polymorphonuclear or granulocytic MDSCs (PMN-MDSCs), 297, 299
Precision medicine, 1
Programmed cell death-ligand 1 (PD-L1), 177, 224
Programmed death protein 1 (PD-1), 35, 63, 192
Proinflammatory macrophages, 337
Proregenerative M2 macrophages, 338
Protease-activated receptor (PAR) pathway, 255
Proteomics, 20, 396
P-selectins, 247, 256

Q

Quercetin, 367

R

Reactive oxygen species (ROS)
- antioxidant defense and cytoprotection, 372
- NF-κB, 372
- responsive macrophage membranes, 339

Regulatory T cells, 153, 202
Replication-competent lentivirus (RCL), 24
Resveratrol-based combinations, 367
RNA vaccines, 193

S

Single-cell RNA sequencing (scRNA-seq), 20
Switchable chimeric antigen receptors, 79
Synthetic biology, 203

T

Talimogene laherparepvec, 133
Target antigen-binding domain, 93
T-cell motility, 52
Thrombosis, 216
T-lymphocyte–based cancer immunotherapies, 175
T lymphocytes, 2, 89
Toll-like receptor-7 (TLR-7), 247
Transcriptomics, 20
Tumor
- burden, 90
- cells, 189
- heterogeneity, 3, 26
- immune evasion, 189
- immunosuppression, 189
- load, 90

Tumor-associated antigens (TAAs), 63, 89, 175
Tumor-associated carbohydrate antigens (TACAs), 21
Tumor-associated macrophages (TAMs), 26, 154, 188, 203, 332
Tumor-associated vasculature, 115, 117
Tumor-educated platelets, 243
Tumor evasion strategies, 8
Tumor immune evasion, 10
- immunosuppressive tumor microenvironment, 11
- impaired MHC class I expression, 10
- tumor antigens, 10

Tumor-induced angiogenesis, 115
Tumor-induced metabolic reprogramming, 13
Tumor-infiltrating immune cells (TIICs), 36
- antitumor immune cells, 38
- cytotoxic T lymphocytes, 38
- immune activation suppression, 45
- myeloid-derived suppressor cells, 41
- natural killer cells, 39
- in patients with colorectal cancer, 36
- regulatory T cells, 40
- research, historical development, 37*t*
- ROS-inducing treatments, 45
- therapeutic potential, 39*t*
- tumor-associated macrophages, 41
- tumor immunity, functional roles, 37, 38*t*

Tumor-infiltrating lymphocytes (TIL), 2, 52, 64, 193, 391
Tumor microenvironment (TME), 3, 115, 174
Tumor-specific antigens, 1
Tumor surveillance innate immunity, 9

U

Universal CAR T-cell systems, 78

V

Vascular endothelial growth factor-A (VEGF-A), 241
von Willebrand factor (vWF), 241

W

Warburg effect, 371
Whole-exome sequencing (WES), 19
Whole-genome sequencing (WGS), 19

Printed and bound by CPI Group (UK) Ltd, Croydon, CR0 4YY

14/09/2026

02216291-0009